ELECTRON THEORY OF METALS

ELEKTRONNAYA TEORIYA METALLOV

ЭЛЕКТРОННАЯ ТЕОРИЯ МЕТАЛЛОВ

ELECTRON THEORY
OF METALS

I. M. Lifshits, M. Ya. Azbel', and M. I. Kaganov

Translated from Russian by Albin Tybulewicz
Editor, *Soviet Physics - Semiconductors*

CONSULTANTS BUREAU • NEW YORK-LONDON • 1973

Il'ya Mikhailovich Lifshits was born in 1917. He was graduated from the Mathematics Department of Kharkov University in 1936 and from the Physico-Mechanical Faculty of the Kharkov Polytechnic Institute in 1938. From 1937 to 1968, Il'ya Lifshits worked at the Physico-Technical Institute of the Ukrainian Academy of Sciences in Kharkov; from 1941, he headed the Theoretical Division of this Institute. At this time he also held the Chair of Statistical Physics at Kharkov University. In 1968, after the death of L. D. Landau, he became Head of the Theoretical Physics Division of the Institute of Physics Problems of the Academy of Sciences of the USSR in Moscow. He is also professor in the Physics Department of Moscow University. Professor Lifshits is an Academician of the USSR and the Ukrainian Academies of Sciences. His scientific interests lie mainly in solid state theory (particularly, the electron theory of metals), quantum statistics and kinetics, the theory of disordered systems, and the physics of biopolymers. He is Chairman of the Council on Solid State Theory of the Academy of Sciences of the USSR. In 1951 Professor Lifshits was awarded the Mandelshtam Prize of the Academy of Sciences of the USSR for his work on the dynamic theory of imperfect crystals. In 1961 he received the Simon Memorial Prize of the Institute of Physics and the Physical Society (London) for his studies of the Fermi surfaces at low temperatures. For his work on the electron theory of metals and on the structure of electron energy spectra, he was awarded the Lenin Prize in 1967.

Mark Yakovlevich Azbel' was born in 1932 and was graduated in 1953 from the Physics Department of Kharkov University. In 1955 he became Candidate and in 1958 Doctor of Physico-Mathematical Sciences. Until 1964 he worked at the Physico-Technical Institute of the Ukrainian Academy of Sciences in Kharkov. In 1964 he became professor at Moscow University and in 1970 Section Head at the L. D. Landau Institute of Theoretical Physics of the Academy of Sciences of the USSR in Moscow. Professor Azbel' developed a theory of the high-frequency properties of metals in a static magnetic field (1955) on the basis of which he predicted several important physical effects: cyclotron resonance in metals (1956), which has become known as the Azbel'–Kaner resonance; high-frequency field peaks in the interior of a metal and undamped waves in metals (1960); the rapid attenuation of a static current away from the surface of a conductor, which is known as the static skin effect (1963); semimetal-semiconductor phase transitions (1965); and diamagnetic periodic structures (1967). From 1969 to 1971 Professor Azbel' turned his attention to the quasiclassical electrodynamics of superconductors and to the theory of resonances and oscillations in superconductors. At present his main interests are biophysics and the general theory of relativity.

Moisei Isaakovich Kaganov was born in 1921 and was graduated from Kharkov University in 1949. Between 1949 and 1970 he worked at the Kharkov Physico-Technical Institute. He started as a junior scientist and later became Head of the Theoretical Physics Laboratory. In 1970 he joined the Institute of Physics Problems of the Academy of Sciences of the USSR in Moscow and became professor in the Physics Department of Moscow University. Professor Kaganov's main scientific interests lie in solid state theory (the theory of metals and of magnetic phenomena) and the theory of superfluidity. Professor Kaganov's main contributions are in the theory of galvanomagnetic phenomena (together with M. Ya. Azbel' and I. M. Lifshits), in the theory of the high-frequency properties of ferromagnets and antiferromagnets (together with V. M. Tsukernik), and in the theory of the anomalous skin effect and the absorption of ultrasound in metals (with M. Ya. Azbel', A. I. Akhiezer, and G. Lyubarskii). He is a member of the Council on Solid State Theory of the Academy of Sciences of the USSR.

The Russian text, originally published by Nauka Press in Moscow in 1971, has been revised and corrected by the authors for this edition. The present translation is published under an agreement with Mezhdunarodnaya Kniga, the Soviet book export agency.

И. М. ЛИФШИЦ, М. Я. АЗБЕЛЬ, М. И. КАГАНОВ

ЭЛЕКТРОННАЯ ТЕОРИЯ МЕТАЛЛОВ

ELEKTRONNAYA TEORIYA METALLOV

Library of Congress Catalog Card Number 79-188919

© 1973 Consultants Bureau, New York
Softcover reprint of the hardcover 1st edition 1973

ISBN-13: 978-1-4615-8560-2   e-ISBN-13: 978-1-4615-8558-9
DOI: 10.1007/ 978-1-4615-8558-9

PREFACE TO THE AMERICAN EDITION

It took us a long time to write this book. In 1959, two of us (Lifshits and Kaganov) published a review of the mechanics of electrons with a complex dispersion law. About that time, geometrical terms such as extremal sections, curvatures, diameters, limiting points began to appear in papers on the electron theory of metals. They were followed by terms quite unusual in the scientific literature: monsters, pockets, arms, sheets, and so on. With their excitingly shaped figures, papers on the electron theory of metals began to resemble catalogs of exhibitions of abstract or ultramodern sculpture. The modern theory of metals was passing through its romantic period. Each newly interpreted Fermi surface and each discovery of a new structure-sensitive phenomenon was an emotional experience for the authors and readers alike. The attitude of the theoreticians was epitomized by phrases such as "This method or this phenomenon can be used to reconstruct the Fermi surface . . . ," which were found at the end of almost every paper on the electron theory of metals. The experimentalists selected convenient methods, being guided not so much by the elegance of a particular method as by its experimental capabilities. Gradually, the romantic approach gave way to a systematic activity, which resulted in the interpretation of the energy spectra of the majority of metals. There were some unavoidable disappointments. It was found that there is no single "optimal" experimental program. The different characteristics of the electron spectrum and the different types of Fermi surface demanded the employment of a range of different methods. A technique which worked excellently in the reconstruction of the Fermi surface of bismuth was found to be useless in the analysis of copper. However, the main disappointment was probably due to the most interesting experimental results, which proved that most Fermi surfaces are exceptionally complex. This complexity makes it impossible to interpret the spectra simply by applying the general formulas of the electron theory of metals to the experimental results without recourse to any additional assumptions about the nature of the Fermi surface. The Harrison model has found wide application. According to this model, the complexity of the Fermi surfaces is the consequence of the periodicity of the perturbation, i.e., of the translational symmetry of the weak lattice potential. The Harrison model has been confirmed experimentally and justified theoretically. The general formulas are still needed in the verification of the theory, its refinement, and the selection of the final numerical characteristics of the Fermi surface of a specific metal.

One further point must be made. Every investigation method should satisfy a simple but important principle: it should not alter the object being investigated. Originally, it seemed that the magnetic field would provide an excellent means of satisfying this principle: it reveals fine features of the electron energy spectrum but does not alter the energy structure of a metal right up to fantastic intensities of $\approx 10^8$ Oe. However, the complexity of the Fermi surfaces once again queered the picture: the energy barriers between different parts of the Fermi surface can be so low (even on the atomic scale) that electrons can overcome them easily in relatively weak magnetic fields. This phenomenon is known as "magnetic breakthrough" or "magnetic breakdown." A magnetic field produces new energy barriers, and these barriers are frequently investigated in relatively weak fields.

There were more exciting successes than disappointments. Gradually, everyone active in the electron theory of metals realized that the main task was not the discovery of new methods for the interpretation of the electron energy spectra but the investigation of the interesting and sometimes very unusual quantum properties of metals that had not been even suspected a few decades ago. One of the best examples is provided by the high-frequency properties of metals in strong magnetic fields, because the investigation of these properties altered drastically the traditional concept of the skin effect as a necessary attribute of the metal state.

In offering this book to American and British readers we must admit to great doubts about its viability because it will have to compete with some excellent English-language texts that have already been published. However, we are reassured by the belief that the inexhaustibility of the subject automatically ensures that there will be no "intersection": there is enough room for everybody.

I. M. Lifshits
M. Ya. Azbel'
M. I. Kaganov

PREFACE TO THE RUSSIAN EDITION

The topics which an author may include in a book on the electron theory of metals are not determined by external factors (such as size of the book and the type of reader for whom it is written) but by the picture which is associated in the author's mind with the word "metal." For us, a metal is essentially a charged liquid in an array of ionic cores which is stabilized by the liquid. This liquid consists of particles in the form of conduction electrons. When these particles are extracted from a metal, they are found to be normal electrons with a charge e, a spin $\frac{1}{2}$, and a mass m. However, a conduction electron in copper differs basically from a conduction electron in iron. The measure of their individuality is the dispersion law, which is a very complex dependence of the energy on the crystal momentum.

The dispersion law of conduction electrons affects all the properties of a metal. It determines the differences between the behaviors of various metals, particularly in the fine quantum phenomena which are currently being investigated.

This means that knowledge of the dispersion law is as necessary as knowledge of the atomic energy levels in the calculations of the properties of gases. The dispersion law of conduction electrons has already been established for many metals. A special appendix (III) to the present monograph gives a brief review of the results of published investigations together with an extensive bibliography. This appendix was kindly prepared at our request by Yu. P. Gaidukov.

Interpretation of the experimental data (which is of necessity based on more or less reliable models), i.e., the derivation of quantitative relationships between the measured quantities and the parameters of the dispersion law, is impossible without a theory of structure-sensitive phenomena such as the magnetic, the galvanomagnetic, and other properties. We have paid much attention to the theory of such phenomena, formulating it in such a way as to facilitate the interpretation of the energy spectra.

Naturally, the theory of metals is not limited to methods for the interpretation of the dispersion law and the present monograph is not a collection of such methods. The phenomena themselves are of great intrinsic interest and they form the subject matter of what are known as the electron properties of metals.

We shall not list here the subjects treated in the present monograph. They can be found by examination of the contents. However, we wish to mention the topics which have been omitted or which have been discussed only cursorily. This applies primarily to the temperature dependences of the transport coefficients such as the electrical conductivity, the thermal conductivity, etc. The brevity of treatment of these topics is in accordance with the intended style and aim of the book. We have deliberately omitted discussions of the approximate methods for the calculation of the electron energy spectra of metals in order to avoid the wasteful duplication of information in recently published specialist monographs on this subject. Finally, we have excluded the theory of collective phenomena such as ferromagnetism, antiferromagnetism, and superconductivity. By tradition, they are separate subjects.

vii

Many of the results included in the present book are based on the work of the Kharkov school of theoretical physics or on investigations which are the outgrowth of that work. Considerable progress has been made in the physics of the metal state not only as a result of successful theoretical investigations but also, to a very large degree, because of the experimental achievements. We have used the experimental results simply by way of illustration. Therefore, we are not claiming to offer a comprehensive review of all the research carried out recently on the electron properties of metals. In particular, the cited references should not be regarded as a comprehensive bibliography of the literature of the subjects discussed in this monograph.

In writing this book, we have considered carefully its potential readers. We have aimed the book at two types of reader. One of them we shall call arbitrarily an experimentalist and the other a theoretician. The experimentalist is interested in the physical essence of phenomena and, possibly, in the information on conduction electrons which can be extracted from a study of these phenomena. The theoretician is also concerned with these phenomena but he wants to know how the formulas describing them are derived. The theoretician would be expected to read all the chapters or at least those which describe the phenomena of immediate interest to him. An experimentalist would probably examine the book carefully and find that the information of interest to him is usually given at the beginning of each section.

We take this opportunity of expressing our gratitude to all our colleagues who have taken part in the development of the electron theory of metals presented in this book.

Special thanks are due to I. N. Adamenko and A. A. Shapiro, who have helped us in the preparation of the manuscript for publication.

I. M. Lifshits
M. Ya. Azbel'
M. I. Kaganov

CONTENTS

INTRODUCTION

The crucial breakthrough in the understanding of the physics of metals was the realization that metals contain free electrons. Then came the discovery that these free electrons represent a quantum gas in the ultimate state of degeneracy. Since their advent, these concepts have been modified, extended, and understood more thoroughly.

§1. Simple Model

A self-consistent theory of the metal state can be developed on the assumption that a metal contains free electrons. According to this theory, the difference between one metal and another lies in the number (density) of free electrons and in the nature of the crystal lattice. These simple ideas on electrons in metals are of purely historical importance but it is still worthwhile recapitulating some of the characteristics of a hypothetical metal based on these ideas. Such recapitulation is essential since it provides a number of important estimates and pointers which will be useful to us later.

It is usually assumed that only the valence electrons are free. This means that, in the case of metals of group I, the number of free electrons is one per atom; for metals of group II, two electrons per atom, etc. In this model, the number of electrons n per unit cell in a crystal is small, i.e., $n \approx 1/a^3$, where a is the atomic spacing. The Fermi energy ε_F of such an electron gas is of the order of 10^{11} erg or $10^5\,°K$ and this is why the electron gas is strongly degenerate at all temperatures T ($T \ll \varepsilon_F$). The distribution of electrons in the momentum space is strongly inhomogeneous: electrons fill a sphere of radius $p_F \approx \hbar/a$ and the situation is not greatly affected by the temperature. Most of the properties of metals are determined by electrons whose energies are of the order of the Fermi energy (Fermi electrons). Their velocity is $v_F \approx 10^8$ cm/sec and their de Broglie wavelength λ_B is of the order of the atomic spacing ($\lambda_B \approx a \approx 3 \cdot 10^{-8}$ cm).

The strong degeneracy of the electron gas results in quadratic temperature dependences of the thermodynamic potentials and, consequently, in a linear temperature dependence of the electronic specific heat of a metal:

$$C_{el}(1/cm^3) = \frac{\pi^2}{2}\, n\, \frac{T}{\varepsilon_F}. \qquad (1.1)$$

At sufficiently low temperatures, the specific heat of a metal is determined solely by electrons because the lattice (phonon) component of the specific heat decreases with temperature much faster than the electronic component ($C_{lat} \propto T^3$).

In the free-electron model, the interaction between the electrons and the irregularities in a crystal, specifically the crystal lattice vibrations, is limited to the scattering and can be described by introducing a mean free path l_p. The electrical conductivity σ is then given by

1

$$\sigma = \frac{ne^2 l_p}{p_F},$$

(1.2)

and the temperature dependence of the conductivity can be simply described by the temperature dependence of the mean free path $[l_p = l_p(T)]$. The subscript p indicates that the scattering mechanism is associated with the dissipation of the electron momentum.

The degenerate electron gas model explains not only the Wiedemann–Franz law

$$\frac{\varkappa}{\sigma T} = \frac{\pi^2}{3e^2}$$

(1.3)

($\varkappa$ is the thermal conductivity of the metal) but also deviations from this law. The more general expression given by the theory is

$$\frac{\varkappa}{\sigma T} = \frac{\pi^2}{3e^2} \frac{l_\varepsilon}{l_p},$$

(1.4)

where l_ε is the mean free path in the case of energy dissipation. In agreement with the experimental data, the theory predicts that $l_\varepsilon \neq l_p$ at temperatures of the order of and below the Debye value.

A model of this kind cannot describe the galvanomagnetic and thermomagnetic effects but it allows us to estimate the value of the Hall coefficient R:

$$R = 1/nec.$$

(1.5)

An electron in a magnetic field H moves along a helical path, the axis of the helix being a magnetic line of force. The radius r_H of an electron orbit in a plane perpendicular to the magnetic field is $cp_\perp/eH$, where $p_\perp$ is that component of the momentum which is perpendicular to the magnetic field. Most of the phenomena observed experimentally are dominated by those Fermi electrons for which $p_\perp = p_F$, i.e., which have zero projection of the momentum along the magnetic field. The orbit radius for these electrons is

$$r_H = cp_F/eH.$$

(1.6)

The finite nature of the motion of electrons in any plane perpendicular to the magnetic field quantizes the energy levels

$$\varepsilon = (n + {}^1/_2)\, \hbar\omega_H + \frac{p_z^2}{2m} + \sigma\mu_B H,$$

$$n = 0,\ 1,\ 2,\ \ldots;\quad \sigma = \pm 1.$$

(1.7)

Here, ω_H is the cyclotron frequency of electron revolutions in the magnetic field

$$\omega_H = eH/mc;$$

(1.8)

m is the electron mass; p_z is the projection of the momentum along the magnetic field. The last term in Eq. (1.7) is due to the electron spin, μ_B being the Bohr magneton ($\mu_B = e\hbar/2mc$).

The magnetic field intensity may be estimated in various ways. For example, we can deduce the field from the ratio of the electron orbit radius r_H to its de Broglie wavelength λ_B. Since $p_F \approx \hbar/a$, and $\lambda_B \approx a$, it follows from Eq. (1.6) that

$$\lambda_B/r_H = H/H_a;\quad H_a = c\hbar/ea^2 \approx 10^8 - 10^9 \text{ Oe}.$$

(1.9)

In practice, we always have $H \ll H_a$, and this means that $\lambda_B \ll r_H$. If we bear in mind that the condition $H \ll H_a$ implies also that the gaps between the quantized energy levels are narrow compared with the Fermi energy itself ($\hbar\omega_H \ll \varepsilon_F$), it becomes clear that in magnetic fields which can be realized in practice the motion of a conduction electron can be described by a quasiclassical approximation. This fact, which is a special characteristic of the dynamics of conduction electrons, will be used frequently in our book. We must stress that the quasiclassical nature of the motion of the Fermi electrons in external fields (and not just in a magnetic field) is related solely to the high density of electrons ($n \approx 1/a^3$) and the consequently short de Broglie wavelength ($\lambda_B \approx a$). When we speak of external (i.e., those applied from outside a crystal) fields, we mean fields which do not alter significantly the structure of the crystal. This means that we are limiting our considerations to external fields which generate forces that are weak compared with the internal atomic or crystal forces. External fields are usually fairly uniform and, therefore, the trajectories of electrons in such fields are always considerably larger than atomic distances. The foregoing considerations show that this allows us to use the quasiclassical approach. We shall consider in detail these seemingly trivial remarks since in a more complex model it would not be possible to calculate some of the observed characteristics of a metal without the use of the quasiclassical approximation.

Let us consider again an estimate of the magnetic field intensity. Equation (1.9) does not mean at all that the magnetic field can always be regarded as weak. The radius of the electron orbit r_H can quite easily become comparable with the mean free path:

$$l_p/r_H = H/H_l; \qquad H_l = H_a a/l_p. \tag{1.10}$$

At low temperatures and in sufficiently pure metals, the mean free path may be hundreds of thousands times greater than the atomic spacing and, therefore, $H_l \ll H_a$. The gaps between the quantized energy levels, which are always considerably smaller than the Fermi energy, may be of the order of or exceed the temperature[1] in relatively weak fields:

$$\hbar\omega_H/T = H/H_T; \qquad H_T = mcT/e\hbar = H_a \frac{T}{\varepsilon_F} \approx 10^4 T \ (^\circ\mathrm{K}). \tag{1.11}$$

This can be illustrated by comparing r_H and the skin depth δ. It is found that, in moderately strong magnetic fields and at reasonably high frequencies, the electron orbit radius and the skin depth are of the same order of magnitude ($\delta \sim r_H$).

It is evident from the examples given so far that the macroscopic characteristics of the electron gas (l, T, δ, etc.) are such that when they are compared with the parameters associated with the motion of an electron in a magnetic field (r_H, $\Delta\varepsilon$), we are often forced to regard the magnetic field as strong although, compared with the individual characteristics of the Fermi electrons (ε_F, $\lambda_B \approx \hbar/a$), such a field can still be weak. This situation is the cause of a great variety of interesting effects observed in metals which are subjected to sufficiently strong magnetic fields. We shall frequently use this term "strong magnetic field." We shall always understand that $H \ll H_a$, i.e., that our "strong magnetic fields" are weak compared with the atomic fields.

The nature of the motion of electrons in a magnetic field is intimately related to their energy spectrum. Therefore, investigations of the electronic properties of metals in strong magnetic fields provide one of the principal ways for making studies of the electron energy spectra. We shall give much attention to this problem. When we go over to a more realistic model of a metal, many of the estimates just given will have to be modified but the basic premise of the quasiclassical nature of the electron motion in a magnetic field will be fully retained.

---

[1] In all our formulas, the temperature will be given in energy units. The Boltzmann constant will be taken as unity. However, when talking of specific temperatures, we shall express them in degrees.

1. In developing a self-consistent theory of the metal state, we must take account of two fundamental features: the interaction between the electrons and the periodic field of the crystal lattice and the mutual interaction of the electrons.

Quantum-mechanical investigations of the motion of electrons in periodic fields have led to the establishment of a band theory whose basic conclusions are still valid. The descriptions of the nature of the intrinsic (i.e., that taking account of the mutual interaction of the electrons) energy spectrum of metals are usually based on the ideas and the terminology of the band theory. This is because an allowance for the electron–electron interaction does not affect the basic results of the band theory, which concerns the nature of the stationary states in a periodic structure. We shall now formulate briefly the principal ideas of the band theory.

The periodicity of the field U in a crystal lattice

$$U\left(\boldsymbol{r} + \boldsymbol{a}\right) = U\left(\boldsymbol{r}\right) \tag{1.12}$$

$(\boldsymbol{a} = s_1\boldsymbol{a}_1 + s_2\boldsymbol{a}_3 + s_3\boldsymbol{a}_3;\ \boldsymbol{a}_j$ are the fundamental translation vectors of the lattice; s_j are integers) results in the translational invariance of the Hamiltonian operator $\hat{\mathcal{H}}$. This allows us to classify the electron states in accordance with the behavior of a wave function during translation $(\boldsymbol{r} \rightarrow \boldsymbol{r} + \boldsymbol{a})$. We can easily show that a wave function $\psi(\boldsymbol{r})$ (which is a solution of the Schrödinger equation) should satisfy the Bloch condition:

$$\psi\left(\boldsymbol{r} + \boldsymbol{a}\right) = e^{i\boldsymbol{p}\boldsymbol{a}/\hbar}\psi\left(\boldsymbol{r}\right), \tag{1.13}$$

where $\boldsymbol{p}$ is an arbitrary real vector. This vector is known as the crystal momentum or the quasimomentum. The value of the crystal momentum $\boldsymbol{p}$ provides a convenient description of the state of an electron. However, this description is ambiguous. In fact, the general nature of the function satisfying the Bloch condition is of the form:

$$\psi\left(\boldsymbol{r}\right) = e^{i\boldsymbol{p}\boldsymbol{r}/\hbar}u\left(\boldsymbol{r}\right), \tag{1.14}$$

where $u(\boldsymbol{r})$ is a periodic function $[u(\boldsymbol{r} + \boldsymbol{a}) = u(\boldsymbol{r})]$ whose explicit form must be found by solving an equation which is derived from the Schrödinger equation by the substitution of the wave function of Eq. (1.14):

$$\hat{\mathcal{H}}_p u_p\left(\boldsymbol{r}\right) = \varepsilon\left(\boldsymbol{p}\right) u_p\left(\boldsymbol{r}\right); \quad \hat{\mathcal{H}}_p = e^{-i\boldsymbol{p}\boldsymbol{r}/\hbar}\hat{\mathcal{H}}e^{i\boldsymbol{p}\boldsymbol{r}/\hbar}. \tag{1.15}$$

For a given value of $\boldsymbol{p}$, this equation has an infinite discrete series of various solutions $u_{sp}(\boldsymbol{r})$ and, therefore,

$$\psi_{sp}\left(\boldsymbol{r}\right) = e^{i\boldsymbol{p}\boldsymbol{r}/\hbar}u_{sp}\left(\boldsymbol{r}\right), \tag{1.16}$$

and the corresponding values of the electron energy are

$$\varepsilon = \varepsilon_s\left(\boldsymbol{p}\right), \quad s = 1,\ 2,\ \ldots. \tag{1.17}$$

The subscript s, which denumerates the solutions of the Schrödinger equation (1.15), will be called the band number and the dependence of the energy on the crystal momentum for a fixed value of the band number will be referred to as the dispersion law.[2]

---

[2] An energy band is an interval between the minimum and the maximum values of the energy for a fixed value of s. Energy bands may overlap (§2).

By definition, the band number s is the number of the solution of some dispersion equation which is obtained from the original Schrödinger equation. Clearly, the different functions $\varepsilon_s(p)$ are the different branches of the same analytic function of a complex vector **p** and the formulation of the problem makes it clear which of the components of the vector **p** should be regarded as complex. This approach to the dispersion law in various energy bands makes it easier to consider the effects associated with electron transitions from one band to another.

All the functions ψ_{sp} with different values of s and **p** are orthogonal. In particular, a suitable normalization gives

$$\int u_{s'p}^{*}(r)\, u_{sp}(r)\, dv = \delta_{ss'},\qquad(1.18)$$

where the integration is carried out within one unit cell of the crystal lattice.

The functions represented by Eq. (1.16) are similar to the wave functions of a free electron $\psi = \text{const } e^{i p r / \hbar}$, which clearly demonstrate the similarity of the crystal momentum and the momentum of a free electron. This fundamental point will be discussed later and we shall consider in detail cases in which we can ignore distinction between the crystal momentum and the classical or true momentum. The definition of the crystal momentum (1.13) shows that it is many-valued. The crystal momenta, differing by a vector of the type $2\pi\hbar b$ (*b* is an arbitrary reciprocal lattice vector), are physically equivalent since they result in the identical behavior of a wave function in the operation of translation ($a \cdot b$ is an integer). This means that the functions $\psi_{sp}(\mathbf{r})$ are periodic with respect to the subscript **p**:

$$\psi_{s,\,p+2\pi\hbar b}(r) = \psi_{sp}(r),\qquad(1.19)$$

and, which is particularly important, the energy of an electron is also a periodic function of the crystal momentum:

$$\varepsilon_s(p + 2\pi\hbar b) = \varepsilon_s(p).\qquad(1.20)$$

The translation properties of the wave functions in the **r** and **p** spaces become clear when we write the expansion of a Bloch wave (1.16) in terms of plane waves (i.e., in terms of the states with fixed values of the momentum). The application of the condition (1.19) results in a very special dependence of the coefficients in the expansion on **p** and *b* :

$$\psi_{sp}(r) = \sum_b a_s(p + 2\pi\hbar b)\, e^{i(p+2\pi\hbar b)r/\hbar}.\qquad(1.21)$$

It must be stressed that the expansion of Eq. (1.21) is a direct consequence of the periodicity of the potential energy U(**r**) [see Eq. (1.12)] and it can be obtained by expanding the potential energy as a Fourier series.

The periodicity of the crystal momentum space shows that the selection of the unit cell, i.e., of the region containing one of each of the nonequivalent states, is arbitrary and governed simply by the considerations of convenience. Quite frequently (particularly in discussions of the motion of an electron in an external field), it is convenient to use an infinite periodic **p** space and then the arbitrary nature of the selection of the unit cell implies that the reciprocal lattice of a crystal can be selected in an arbitrary manner.

In some cases (especially in the analysis of the motion of an electron under the action of external forces), it is more convenient to replace the Bloch waves of Eq. (1.16) by their linear combination in the form of the Wannier functions:

$$a_s (r - R) = \frac{1}{\sqrt{N}} \sum_p e^{-ipR/h} \psi_{sp} (r). \tag{1.22}$$

Here, R is a vector whose magnitude is an integer and which is identical with the vector coordinate of the center of the unit cell; N is the number of cells in a crystal.

Summation in Eq. (1.22) is carried out over all the physically nonequivalent states. In other words, when we go over to integration with respect to p, such integration must be carried out within one reciprocal lattice cell (the edges of this cell are $2\pi\hbar b_1$, $2\pi\hbar b_2$, $2\pi\hbar b_3$, where b_j are the fundamental vectors of the reciprocal lattice). The periodicity, in respect of p, of the factor $e^{-ipR/\hbar}$ and of the function $\psi_{sp}(r)$ allows us to extend integration over all space by introducing an additional normalizing factor. We shall consider later the selection of the region in which the crystal momentum is to be determined and, in particular, the advantages which follow from the use of the whole p space.

The Wannier functions form an orthonormalized system and it is worth mentioning that the orthogonality applies to the functions associated not only with different bands ($s \neq s'$) but also with one band provided the functions are centered on different cells ($R \neq R'$).

The Wannier functions of Eq. (1.22) describe the state of an electron belonging to an s-th band and localized in the vicinity of an R-th cell in the crystal lattice. The energy of such an electron does not have a definite value. An electron in a state with definite crystal momentum is "smeared out" over the whole lattice. The probability W_{sp} of finding an electron at a point r is a periodic function. According to Eq. (1.16), $W_{sp}(r) = |u_{sp}(r)|^2$.

The one-electron approximation (we recall that we are still ignoring the interaction between electrons) provides a very simple approach and can be used, in principle, to determine the structure of the electron energy spectrum and the nature of the quantum states of electrons. The principal result of this approach is the introduction of the crystal momentum. However, direct calculations of the dispersion law and of the wave functions meet with considerable computational difficulties, which can be overcome only by making very special assumptions (tight binding, almost-free electrons, etc). The determination of the quantum states of electrons in the lattice and of their energy spectrum is basically not the final result of the electron theory of metals but only its starting point. In solving specific problems in the theory of metals, it is necessary to analyze the motion of electrons in fields external to a crystal, particularly in an external magnetic field. The exact solution of the Schrödinger equation is then practically impossible and, in most cases, it fails to give even the quantum state of an electron which is subjected to the periodic force of the lattice ions and an external aperiodic force. In these circumstances, further progress in the studies of the properties of conduction electrons can be made only because external fields are normally very weak and smoothly varying so that we can use the quasiclassical approach.

We shall now specify in detail how we can use the low intensity and the smooth variation of external fields in dealing with the motion of an electron in a crystal.

For an electron moving in the crystal lattice field $U(r)$ and an external electric field V ($-\operatorname{grad} V = eE$) the solution $\psi(r, t)$ of the nonstationary Schrödinger equation

$$-\frac{\hbar^2}{2m} \Delta\psi + U\psi + V\psi = i\hbar \frac{\partial\psi}{\partial t} \tag{1.23}$$

can be found as an expansion in terms of the Wannier functions

$$\psi (r, t) = \sum_{s, R} f_s (R, t) a_s (r - R). \tag{1.24}$$

Substituting the expansion of Eq. (1.24) into Eq. (1.23), we obtain an infinite system of equations for the coefficients $f_s(\mathbf{R}, t)$:

$$i\hbar \frac{\partial f_s(\mathbf{R}, t)}{\partial t} = \sum_{s', \mathbf{R}'} \{\delta_{ss'} \varepsilon_{s'}(\mathbf{R} - \mathbf{R}') + V_{ss'}(\mathbf{R}, \mathbf{R}')\} f_{s'}(\mathbf{R}', t), \qquad (1.25)$$

where

$$V_{ss'}(\mathbf{R}, \mathbf{R}') = \int a_s^*(\mathbf{r} - \mathbf{R}) V(\mathbf{r}) a_{s'}(\mathbf{r} - \mathbf{R}') \, d\mathbf{r} \qquad (1.26)$$

is the matrix element of the external force potential and

$$\varepsilon_s(\mathbf{R}) = \frac{1}{N} \sum_{\mathbf{p}} \varepsilon_s(\mathbf{p}) e^{-i\mathbf{p}\mathbf{R}/\hbar} \qquad (1.27)$$

is the Fourier transform of the energy.

The exact expression (1.25) can be written formally in such a way as to stress its similarity with the Schrödinger equation. This can be done by introducing a continuous function $f_s(\mathbf{r}, t)$ such that $f_s(\mathbf{r}, t)|_{\mathbf{r}=\mathbf{R}} = f_s(\mathbf{R}, t)$. Equation (1.25) then becomes

$$\left[\hat{\varepsilon}_s\left(\frac{\hbar}{i}\nabla\right) f_s(\mathbf{r}, t) - \frac{\hbar}{i}\frac{\partial f_s(\mathbf{r}, t)}{\partial t}\right]_{\mathbf{r}=\mathbf{R}} + \sum_{s'\mathbf{R}'} V_{ss'}(\mathbf{R}', \mathbf{R}) f_{s'}(\mathbf{R}', t) = 0. \qquad (1.28)$$

The operator $\hat{\varepsilon}_s\left(\frac{\hbar}{i}\nabla\right)$ may be represented by the Fourier series[3]

$$\hat{\varepsilon}_s\left(\frac{\hbar}{i}\nabla\right) = \sum_{\mathbf{R}} \varepsilon_s(\mathbf{R}) e^{-i\mathbf{R}\nabla}. \qquad (1.29)$$

Finally, we shall make some simplifying assumptions based on our knowledge that the external field $V(\mathbf{r})$ is weak and that it varies smoothly. The probability of a transition between the energy bands depends strongly on the value of the potential V. If $|V| \ll \varepsilon_a$, where ε_a is an energy of the order of the atomic value, this probability is very low and we can ignore all those matrix elements for which $s \neq s'$. If, moreover, the external field varies slowly from one point to another, we can ignore all the matrix elements except $V_{ss}(\mathbf{R}, \mathbf{R})$, which – by definition – is equal to $V(\mathbf{R}) = V(\mathbf{r})_{\mathbf{r}=\mathbf{R}}$. Thus, in the most interesting cases, we have

$$\left\{\hat{\varepsilon}_s\left(\frac{\hbar}{i}\nabla\right) + V(\mathbf{r}) - \frac{\hbar}{i}\frac{\partial}{\partial t}\right\} f_s(\mathbf{r}, t) = 0, \qquad (1.30)$$

where we assume that $\mathbf{r} = \mathbf{R}$. It is natural to extend this equation to all the points in space. The function $f_s(\mathbf{r}, t)$, which describes approximately the motion of an electron in external fields, acts as a wave function of an electron in the s-th band. The structure of this equation shows that $\varepsilon_s\left(\frac{\hbar}{i}\nabla\right)$ can be regarded as the kinetic energy operator and the $V(\mathbf{r})$ as the potential energy operator. The accuracy of the description of the motion of an electron by means of the function $f_s(\mathbf{r}, t)$ is naturally limited by the accuracy of the interpolation, i.e., by the lattice constant a. Consequently, it is permissible to use Eq. (1.30) only if the function $f_s(\mathbf{r}, t)$ varies significantly

---

[3] The transformation of Eq. (1.25) into Eq. (1.28) can be made provided certain conditions are satisfied in relation to the analytic properties of the functions $\varepsilon_s(\mathbf{p})$. We shall not elaborate this point because details can be found in, for example, [1].

over a distance considerably greater than the lattice constant. On the other hand, the de Broglie wavelength of a Fermi electron is of the order of a and this makes it possible (in considering the motion of a conduction electron in external fields) to use the quasiclassical approximation and, in most cases, simply the classical approach.[4]

In the classical treatment, we introduce the Hamiltonian

$$\mathcal{H}(p,\,r) = \varepsilon_s(p) + V(r). \tag{1.31}$$

This procedure corresponds to the inversion of the correspondence principle. In going over to the quantum treatment, the Hamiltonian is found by replacing the momentum **p** with the operator $\frac{\hbar}{i}\nabla$; here, we shall go over to the classical case by replacing $\frac{\hbar}{i}\nabla$ with the momentum **p**.

The Hamiltonian equations are written in the usual way

$$\dot{r} = \frac{\partial\mathcal{H}}{\partial p}; \quad \dot{p} = -\frac{\partial\mathcal{H}}{\partial r}. \tag{1.32}$$

The first equation is the definition of the velocity[5]

$$v = \frac{\partial\varepsilon}{\partial p}, \tag{1.33}$$

and the second equation is the analog of Newton's law

$$\dot{p} = F; \quad F = -\frac{\partial V}{\partial r}. \tag{1.34}$$

We must stress two points: 1) the velocity and the momentum are not related by the "usual" equation $v = p/m$, because the velocity is a complex periodic function of the momentum **p** given by the dispersion law (1.17); 2) Newton's equation (1.34) does not refer to the total force acting on an electron but only to the force which is external to a crystal.

If the external force is due to an electric field of intensity **E**, it follows that $\dot{p} = e\mathbf{E}$.

The influence of a magnetic field **H** can be included by replacing the right-hand side of Eq. (1.34) by the Lorentz force

$$\dot{p} = e\left\{E + \frac{1}{c}[vH]\right\}. \tag{1.35}$$

A rigorous derivation of Eq. (1.35), about which Ziman [1] remarked that "it is too good to be absolutely true," is a fairly cumbersome procedure.

However, in the classical case, Eq. (1.35) can be deduced in the standard manner using the Lagrange function (as is done, for example, in Landau and Lifshitz's "Classical Theory of Fields"). In the Lagrange function, we write $(e/c)\mathbf{A}v - e\varphi$ (rot **A** = **H**) since **A**·**p** does not give the total derivative with respect to time in the case of translation by one period of the reciprocal lattice. Moreover, the derivation is based on the constancy of the electron charge, which is

---

[4] We shall mention specially the problems which cannot be solved by the classical approach.
[5] We shall drop the subscript s, representing the band number, in all those cases in which confusion is unlikely.

a consequence of the equality of the number of quasi particles in an interacting Fermi liquid to the number of electrons (this point will be discussed later).

Thus, we obtain the usual definition of the kinematic momentum **p** in terms of the canonically conjugate momentum **P** and the vector potential **A**:

$$p = P - \frac{e}{c} A; \quad \mathrm{rot}\, A = H. \tag{1.36}$$

The Hamiltonian of a particle whose kinetic energy is $\varepsilon(\mathbf{p})$ is of the following form in a magnetic field:

$$\mathcal{H}(P, r) = \varepsilon\left(P - \frac{e}{c} A\right) + V(r); \quad V = e\varphi. \tag{1.37}$$

If we now write the Hamiltonians in terms of **P** and **r** and go over to the kinematic momentum **p**, we can easily obtain Eq. (1.35) by using the relationship between the potentials φ and **A** and between the fields **E** and **H**

$$E = -\frac{1}{c} \frac{\partial A}{\partial t} - \nabla\varphi; \quad H = \mathrm{rot}\, A,$$

as well as the definition of velocity given by Eq. (1.33). In approaching the matter this way, we wish to stress that the Lorentz force in this context is formally identical with the conventional definition of this force and is not related to the quadratic dispersion law of the electrons. We shall show later that the motion, in a magnetic field, of an electron with a complex dispersion law depends strongly on this law, that the most important manifestations of the quantum properties ("smearing out" of an electron) are very special, and that departures from the classical mechanics result from the structure of the electron energy spectrum, i.e., from the complex dependence of the energy ε on the momentum **p**.

Thus, the band theory explains the quantum states of electrons [the states with definite crystal momentum **p**, in a band of number s, as given by Eq. (1.16)], gives the principal algorithm for the calculation of the dispersion law $\varepsilon_s(\mathbf{p})$ without allowance for the interaction between electrons, and justifies the quasiclassical approach, in which the main features are the permissibility of the substitution of the true momentum for the crystal momentum, the use of the function $\varepsilon_s(\mathbf{p})$ as the kinetic energy, and the use of the classical equations of motion (1.35) after making allowance for the complex periodic dependence of the velocity on the momentum [**v** = **v(p)**]. If we further postulate that the electrons obey the Fermi–Dirac statistics, i.e., that they represent an ideal Fermi gas, we obtain a firm basis for the derivation of all the equilibrium properties of a metal (such as the thermal and magnetic properties). The introduction of collisions between the electrons and the lattice vibrations, among the electrons themselves, and between the electrons and lattice imperfections (interstitial atoms, dislocations, etc.) makes it possible to develop a theory of the transport properties of metals (the electrical and thermal conductivity, the galvanomagnetic and thermomagnetic effects, etc.).

The use of the band theory ideas can be justified by making allowance for the interaction between electrons. However, in this case, we must abandon (at least at the present stage of the development of the electron theory of metals) the possibility of the calculations of a wave function of a conduction electron (more precisely, the one-particle Green's function of electrons) and its dispersion law. The dispersion law is then assumed to be given or can be found by comparing the theoretical results with those derived from the experimental data.

2. The most difficult problem in the development of the electron theory of metals is a consistent allowance for the interaction between electrons. Various attempts to derive directly a wave function of an electron ensemble have not been very successful. A very useful approach was developed by Landau [2] for the properties of the light isotope of helium (He³) and known as the Fermi liquid method.

According to Landau, the spectrum of a Fermi liquid can be found on the assumption that gradual "intensification" of the interaction between electrons (the transition from a gas to a liquid) does not alter the level structure, i.e., the state of an electron in a crystal can still be described by specifying the crystal momentum p and the band number s.[6] The gas particles in such a liquid are replaced by elementary excitations (quasi particles). Each quasi particle has its own definite crystal momentum. Such quasi particles obey the Fermi statistics and their number is always equal to the number of particles in the liquid. A quasi particle can be regarded, in a limited sense, as a particle in a self-consistent field of the other particles. Naturally, the energy of a particle depends on the state of the surrounding particles: the energy of the whole system is not the sum of the energies of the separate constituent particles but is a functional of the distribution function. This conclusion is the basic idea underlying the Fermi liquid theory.

The criterion for the use of the Fermi liquid theory is the smallness or the deviation of the system from the ground state which is analogous to the ground state of an ideal gas: all the states below the Fermi energy are occupied and the rest are free. Excited states appear when a particle goes over from an occupied state lying below the Fermi energy to a free state. In order to be able to attribute a definite crystal momentum to each quasi particle, we must postulate that the indeterminacy of the crystal momentum is small compared not only with the momentum itself but also with the spread of the Fermi distribution Δp. We can easily see that this condition is observed if the spread is sufficiently narrow. By virtue of the Pauli principle, the probability of scattering and, therefore, the indeterminacy of the momentum are proportional to $(\Delta p)^2$. It is clear that if Δp is sufficiently small, the indeterminacy of the momentum is not large compared with the momentum itself and with Δp. The indeterminacy of the momentum may be due to electron–electron collisions as well as electron–phonon collisions. It can be shown [3] that the collisions between electrons and phonons still allow us to ascribe a definite crystal momentum to a quasi particle but, in this case, the small parameter is the quantity Θ/ε_F, where Θ is the Debye temperature of a crystal.

An excited state of a metal can be described in a natural manner by the number of electrons "above" the Fermi surface, which is equal to the number of vacant states "below" this surface. It follows from our reasoning that this number is considerably smaller than the total number of electrons per unit volume. We shall not use this description but in verbal statements about the physical phenomena we shall assume that the number of mobile electrons (those not "frozen" by the Pauli principle) is small.

The principal characteristics of a Fermi liquid are the dispersion law of the quasi particles $\varepsilon(p)$ and the correlation function $\Phi(p, p')$, introduced by the following expression which is valid for small deviations of the system from its ground state[7]:

$$E = E_0 + \int \varepsilon f_1(p)\, d\Gamma + \frac{1}{2} \int \int \Phi(p, p') f_1(p) f_1(p')\, d\Gamma\, d\Gamma',$$

$$d\Gamma = \frac{2}{(2\pi\hbar)^3}\, dp_x\, dp_y\, dp_z. \tag{1.38}$$

---

[6] Naturally, this applies if the "intensification" does not alter the nature of the states (for example, if a metal does not become semiconducting because, in this case, not only the energy level scheme is affected but a gap appears near the Fermi energy).

[7] For the sake of simplicity, we shall ignore the spin variables and omit the subscripts denoting the band number.

Here, E is the energy density in the system; E_0 is the energy density in the ground state; $f_1(p)$ is the deviation of the distribution function from the Fermi step-like form. It follows from Eq. (1.38) that $\varepsilon(p)$ is the variational derivative of the energy with respect to the distribution function and that it represents the change in the energy of the system when one particle with a momentum p is added to the system. It is this quantity $\varepsilon(p)$ that acts as the Hamiltonian of a quasi particle in the field of other particles.

We must stress that the function $\varepsilon(p)$ includes the interaction with the ionic core of the crystal lattice as well as the mutual interaction of electrons. A quasi particle characterized by a dispersion law $\varepsilon(p)$ will be called a conduction electron or, simply, an electron.

The fact that the introduction of crystal momentum is related solely to the invariance of a system under translation can be illustrated by considering the eigenstates of a Hamiltonian which exhibits translation invariance but which, because of the electron–electron interaction, may be nonlocal:

$$\hat{\mathscr{H}} = \hat{\mathscr{H}}(r, r').$$ (1.39)

Instead of the coordinate r, which can have a continuous set of values, we shall use the coordinate R, which can have only integral values representing the number of a cell in a crystal and a coordinate ρ, which varies within each cell:

$$r = R + \rho; \qquad R = s_1 a_1 + s_2 a_2 + s_3 a_3,$$ (1.40)

where a_j are the fundamental lattice vectors and s_j are any integers, which may be positive or negative.

Because of translation invariance, the Hamiltonian (irrespective of its nature) can depend only on the relative positions of the cells and not on their absolute distribution:

$$\hat{\mathscr{H}} = \hat{\mathscr{H}}(R - R', \rho, \rho').$$ (1.41)

This consequence of the translation symmetry allows us to separate the variables R and ρ. The Schrödinger equation for the wave function $\psi(R, \rho)$ has a solution of the type

$$\psi(R, \rho) = e^{ipR/\hbar}\psi_p(\rho),$$ (1.42)

where p is the crystal momentum and the function $\psi(\rho)$ satisfies the equations

$$\hat{\mathscr{H}}_p \psi_p = \varepsilon \psi_p; \qquad \hat{\mathscr{H}}_p = \sum_{R'} e^{-ipR'/\hbar}\hat{\mathscr{H}}(R', \rho, \rho').$$ (1.43)

In general, Eq. (1.43), valid within one cell, has a discrete set of solutions, each of which will be denoted by a subscript s, which is the band number:

$$\psi(R, \rho) = e^{ipR/\hbar}\psi_{sp}(\rho).$$ (1.44)

On the other hand, the wave function of an electron can be regarded as the function of a single coordinate $r = R + \rho$. Therefore, $\psi_s(\rho)$ should have the following structure: $\psi_{sp}(\rho) = e^{ip\rho/\hbar}u_{ps}(\rho)$, where $u_{ps}(\rho)$ is a periodic function. Consequently, the wave function of an electron with a specified crystal momentum is a Bloch wave [see Eq. (1.14) and the expressions that follow]. Thus, the permissibility of introducing crystal momentum is simply due to the translation symmetry of the problem and is not related to the nature of the interaction forces nor to the actual form of the Hamiltonian.

The successes achieved in the modern theory of metals follow from the fact that the energy levels of a conduction electron $\varepsilon_s(\mathbf{p})$ can be treated, to a high degree of precision, as the kinetic energy of a free particle and the crystal momentum $\mathbf{p}$ can often be regarded as the classical or true momentum. We shall consider this point again and shall proceed in such a manner as to leave no doubt that the approximations used are not related to any particular form of the Hamiltonian (or with the one-electron approximation).

In order to consider the motion of an electron in external fields, it is necessary to introduce the following coordinate operator in the p–s representation

$$\hat{\mathbf{r}} = \hat{\mathbf{R}} + \hat{\boldsymbol{\rho}}. \tag{1.45}$$

It follows from the wave functions given by Eq. (1.42) that

$$\hat{\mathbf{R}} = -\frac{\hbar}{i}\frac{\partial}{\partial \mathbf{p}}. \tag{1.46}$$

This means that the crystal momentum $\mathbf{p}$ and the discrete coordinate $\mathbf{R}$ are canonically conjugate. The crystal momentum $\mathbf{p}$ is regarded, in accordance with the terminology used in mechanics, as the momentum related to the generalized discrete coordinate $\mathbf{R}$.

The operator $\hat{\boldsymbol{\rho}}$ is diagonal with respect to the crystal momentum $\mathbf{p}$ and is determined by its matrix elements

$$\rho_{ss'}(\mathbf{p}) = (\psi_{ps}^{*}\boldsymbol{\rho}\psi_{ps'}). \tag{1.47}$$

Since, in the p-s representation, the Hamiltonian of Eq. (1.41) is specified by the matrix elements of $\varepsilon_s(\mathbf{p})$, it follows from Eqs. (1.45)-(1.47) that the electron velocity operator has the following matrix elements:

$$\mathbf{v}_{ss'} = \frac{\partial \varepsilon_s}{\partial \mathbf{p}}\delta_{ss'} + \frac{\varepsilon_s - \varepsilon_{s'}}{\hbar}\boldsymbol{\rho}_{ss'}. \tag{1.48}$$

Hence, the average value of the velocity is $\partial \varepsilon_s/\partial \mathbf{p}$, in accordance with Eq. (1.33).

We shall now consider the motion of a conduction electron under the action of an external force $\mathbf{F} = -\partial V/\partial \mathbf{r}$ and this should be regarded as an additional justification of the procedure for using the classical approach [see Eqs. (1.32)-(1.35)]. The Hamiltonian of a conduction electron consists of the kinetic energy operator $\varepsilon_s(\mathbf{p})$ and the potential energy operator $\hat{V}(\mathbf{r})$:

$$\hat{\mathscr{H}} = \varepsilon_s(\mathbf{p}) + \hat{V}(\mathbf{r}). \tag{1.49}$$

Commuting $\mathbf{p}$ with the Hamiltonian of Eq. (1.49) and noting that the commutation relationships between the coordinate and the crystal momentum $\mathbf{p}$ are the same as those between the coordinate and the true momentum, we obtain a relationship which is analogous to the Ehrenfest theorem:

$$\hat{\dot{\mathbf{p}}} = -\frac{\partial \hat{V}}{\partial \mathbf{r}}. \tag{1.50}$$

However, we must stress once again that Eq. (1.50), like Eq. (1.34), includes only forces external to the crystal. We shall now use the fact that these forces are generally weak. In our case, this means that $V(\mathbf{r}) \equiv V(\mathbf{R} + \boldsymbol{\rho})$ is a fairly smooth function of its own argument and that it can be expanded in powers of $\boldsymbol{\rho}$, neglecting the terms containing $\boldsymbol{\rho}$ because the matrix elements $\rho_{ss'}$

are of the order of the atomic spacing [see Eq. (1.47)]. Thus, in a relatively "rough" description of the motion of an electron, the equations of motion contain only the coordinate $\mathbf{R}$ whose discrete nature may be ignored. Therefore, the crystal momentum can be regarded as the true momentum. For the sake of consistency, we should also omit the nondiagonal elements from the velocity operator of Eq. (1.48). However, the nondiagonal elements of the velocity are of the same order as the diagonal elements. The nondiagonal elements can be neglected because external fields vary smoothly and, therefore, transitions from one band to another occur very rarely, i.e., in all cases the motion of a conduction electron in an external field obeys the criteria of validity of the quasiclassical approach.

We shall conclude our introduction by formulating certain assumptions which will be used later.

A. The spectrum of conduction electrons in a metal is of the Fermi type. This means that the charge is transported by quasi particles (conduction electrons) which obey the Fermi statistics, i.e., the valence electrons interacting with one another and with the crystal lattice field can be replaced (in respect of their statisticothermodynamic and transport properties) by a Fermi liquid whose properties are described by the dispersion law $\varepsilon(\mathbf{p})$ and the correlation function $\Phi(\mathbf{p}, \mathbf{p}')$ [see Eq. (1.38)]. The charge of each quasi particle is equal to the electron charge and the number of quasi particles is equal to the total number of electrons. In the absence of an external magnetic field, each electron state is doubly degenerate, provided the crystal is not magnetically ordered. Therefore, we may attribute a spin of $\frac{1}{2}$ to a conduction electron but, because of the spin–orbit coupling, the splitting of the energy levels in a magnetic field $\delta\varepsilon_H$ is not equal to $2\mu_B H$. This circumstance is described by the introduction of a g factor which is different from the corresponding factor of a free electron.

B. In addition to the fermion branches of the energy spectrum, a metal always has boson branches (for example, phonons) but all the available experimental data support the hypothesis that the boson part of the spectrum is not involved in charge transport in nonsuperconducting crystals.

C. In view of the symmetry of a crystal, the energy of a quasi particle $\varepsilon(\mathbf{p})$ should be a periodic function of the crystal momentum and the period should be equal to that of the reciprocal lattice [see Eq. (1.20)]. In general, the formula just quoted is many-valued because it represents different electron energy bands. This aspect of the band theory is not related to the one–electron Bloch model but is of much more general validity.

D. The motion of an electron in external fields is usually quasiclassical and this allows us frequently to ignore the difference between the crystal momentum and the true momentum and to regard $\varepsilon(\mathbf{p})$ as the kinetic energy of conduction electrons.

3. In developing the electron theory of metals, it is frequently stressed that an electron in a metal behaves similarly to a free electron and it is quite usual to ignore those aspects of the behavior of a conduction electron which are associated with the complex nature of its dispersion law. Investigations carried out in the last few decades show that many properties of metals can be explained only if we assume that the electron energy spectrum of a metal is complex and basically different from the energy spectrum of free electrons. The complexity of the dispersion law is manifested primarily in the nature of the motion of an electron in external fields. Therefore, the electron theory of metals should start with a study of the mechanics of an electron characterized by a complex dispersion law. The mechanics has several special features which can be explained very generally without going into the details of the wave functions and the dispersion law. On the other hand, the special nature of the motion of conduction electrons is reflected in a great variety of macroscopic properties of metals.

In view of this, our aim will be to consider the various consequences which follow from the Fermi nature of the spectrum of conduction electrons in a metal, leaving aside the origin of this spectrum and avoiding any specific model for the quasi particles. It is found that a large proportion of the transport and statisticothermodynamic properties of a metal (including its magnetic properties) can be expressed in terms of parameters representing quasi particles (the number of such particles, the dispersion law, etc.). Moreover, we can formulate and solve the converse problem, which is the determination of the energy spectrum of the quasi particles (conduction electrons) from the experimental data obtained for various macroscopic properties of a metal. The most sensitive means for the determination of the energy spectrum are provided by the various properties of a metal in a magnetic field, when the radius of curvature of the electron trajectory is considerably shorter than the mean free path, so that between successive collisions an electron can manifest its dynamic properties (low-temperature oscillations of the susceptibility, galvanomagnetic effects in strong fields, high-frequency properties, resonance effects). An allowance for the interaction of conduction electrons with the lattice field and with other electrons does not (in most metals) alter significantly the estimated Fermi energy of electrons: ε_F is of the order of 10^4-10^5 °K. This means that the temperature can always be regarded as low and the inequality $T \ll \varepsilon_F$ is practically always satisfied. Moreover, we have mentioned already that, in all these effects, the important particles are those whose energies are close to the Fermi value ($\varepsilon \approx \varepsilon_F$) and, consequently, only those aspects of the dispersion law are manifested which are related to the behavior of the function $\varepsilon(\mathbf{p})$ near the constant-energy surface $\varepsilon(\mathbf{p}) = \varepsilon_F$ in the momentum space. Such a surface separates the vacant and the occupied regions of the $\mathbf{p}$ space at $T = 0$ and is known as the Fermi surface. The characteristics of the Fermi surface determine many of the properties of metals. In addition to the shape of the Fermi surface, the energy spectrum of electrons is characterized by the quantity $\partial\varepsilon/\partial\mathbf{p} = \mathbf{v}$ which has the meaning of the electron velocity. Therefore, when we speak of the determination of the energy spectrum of conduction electrons, we shall always regard the Fermi surface and the electron velocity as the two most important characteristics.

The correlation function $\Phi(\mathbf{p}, \mathbf{p}')$ will be shown later to have no effect on most of the static properties of metals and, therefore, we can use the terminology of the theory of gases in investigations of the static and quasistatic properties of metals. However, we must remember that the dispersion law $\varepsilon(\mathbf{p})$ includes the interaction between electrons in the ground state.

It is clear from the foregoing discussion that the quantum state of an electron can be described by attributing to it some crystal momentum. This follows from the translation symmetry (periodicity) of the crystal lattice.

The quantum state corresponding to a specified crystal momentum can be described by a modulated plane wave. This provides a very clear illustration of the statistical nature of elementary excitations and of the transport phenomena associated with such excitations. Thus, for example, the concept of collisions between particles and of the mean free path is introduced in relation to a change in the crystal momentum. The whole existing terminology is based on these concepts. It follows that all we have said so far applies basically to crystals.

Nevertheless, the transport phenomena in amorphous materials (disordered solid solutions, liquid metals, etc.) have often the same nature as those in crystals. This means that many of the properties of solids must be related to aspects other than the periodic structure. Much work has been done in recent years on the energy spectra of such disordered systems. However, this problem is outside the scope of the present monograph.

4. The plan of the book is as follows. An analysis of the structure of the electron energy spectrum is followed, in the first chapter, by the presentation of the classical and the quantum mechanics of conduction electrons. The second chapter deals with the statistical thermodynamics of the electron gas, specifically with the thermal and magnetic properties of metals under equilibrium conditions. The third chapter provides a discussion of the transport properties with the main stress laid on the galvanomagnetic effects. The high-frequency and, particularly, the resonance phenomena are considered in the fourth chapter.

The material is selected in such a way as to cover all those properties which are intimately connected with the nature of the electron energy spectrum and can be used as methods for the determination of this spectrum.

Many publications (including reviews) have recently appeared on the subject of the energy spectrum of conduction electrons. Some are based on the almost-free electron approximation, which has been found to be very effective in deducing the Fermi surfaces of polyvalent metals (Harrison's model). The present monograph does not deal with calculations of the Fermi surfaces of specific metals or with comparisons between the calculated and the experimental data. Readers interested in this aspect of the theory of metals are directed to the interesting monograph by Harrison [4].

CHAPTER I

CLASSICAL AND QUANTUM MECHANICS
OF CONDUCTION ELECTRONS

A charge carrier in a metal (a conduction electron) is a quasi particle with a complex periodic dependence of its energy on the crystal momentum. It differs from a free electron in its behavior in fields external to the crystal. For example, in a magnetic field, some conduction electrons do not rotate about the field in a plane perpendicular to the field but travel to infinity.

The characteristic features of the classical and the quasiclassical motion of an electron can be formulated in terms of its dispersion law. A quantum-mechanical treatment of a conduction electron can be provided on the basis of the energy band theory, in which the crystal momentum is a numerical parameter. This makes it possible to solve, in principle, any problem concerned with the motion of a conduction electron in external fields.

We shall present the mechanics of a conduction electron in such a way as to stress the differences in its behavior from that of a free electron.

§2. Geometry of Constant-Energy Surfaces

The dispersion law for conduction electrons differs basically from the corresponding law for free electrons. In analytic studies of the structure of the electron energy spectrum or, more precisely, in the actual derivation of the functions $\varepsilon_s(\mathbf{p})$, which is possible only if some specific assumptions are made, the results of the calculations are usually presented in the form of the dependence of the energy on the crystal momentum of an electron moving along a fixed direction. However, it would be much more convenient to consider not the dependence of the energy on the crystal momentum but the constant-energy surfaces $\varepsilon_s(\mathbf{p}) = \varepsilon$, which also provide comprehensive information on the dispersion law. Knowing the dependence of the energy on the crystal momentum, we can plot a constant-energy surface but this is usually difficult and requires special consideration because, in most cases, the dependence $\varepsilon_s(\mathbf{p})$ is known for a few very special directions of the crystal momentum. If, moreover, we take into account the fact that analytic expressions are derived from not very reliable models, it becomes obvious that it would be much more natural to consider the structure of the electron energy spectrum on the basis of very general topological considerations. This approach will be used in the present section and in much of the book.

A very important circumstance which helps one to analyze the structure of the constant-energy surfaces is the fact that the periodicity of the distribution of the atoms in a crystal lattice results in a periodic dependence of the energy on the crystal momentum [see Eq. (1.20) in §1]. The symmetry elements of a crystal usually affect the symmetry of the functions $\varepsilon_s(\mathbf{p})$. Moreover, we must bear in mind that the dispersion law can also have its own symmetry elements. For example, invariance of the quantum-mechanical equations under time inversion demands that $\varepsilon_s(-\mathbf{p}) = \varepsilon_{s'}(\mathbf{p})$. If the energy bands do not overlap, this condition is satisfied for s = s' and this means that constant-energy surfaces have a center of inversion. The energy bands do frequently overlap (i.e., min $\varepsilon_{s'}$ < max ε_s < max $\varepsilon_{s'}$), but the individual nature of each band is retained since each has its own dispersion law. Formally, the overlap of the energy bands represents degeneracy: different states have the same energy. However, in general, this does not result in singularities in the spectrum because different crystal momenta correspond to the same energies. In geometrical language, which is convenient in discussions of the structure of the energy spectrum of electrons, this means that, in the general case of overlap, the corresponding constant-energy surfaces [$\varepsilon_s(\mathbf{p}) = \varepsilon$ and $\varepsilon_{s'}(\mathbf{p}) = \varepsilon$] are located in different parts of the p space. An important and very interesting case of degeneracy is that of the intersection of constant-energy surfaces, i.e., a situation when at some points in the momentum space the equation $\varepsilon_s(\mathbf{p}) = \varepsilon$ has a solution for several different band numbers s. We shall refer only to this case as degenerate and shall analyze it in detail later.

Let us now consider the structure of the constant-energy surfaces within an energy band. At certain values of the crystal momentum, the energy reaches its minimum and maximum values. Near a minimum or maximum, the energy can be expanded in a series of the degrees of deviation from the value of the momentum at which the energy has its extremal value. If the degeneracy is ignored, the expansion is of the form

$$\varepsilon(\mathbf{p}) = \varepsilon(\mathbf{p}_0) + \frac{1}{2}\left(\frac{\partial^2 \varepsilon}{\partial p_i \, \partial p_k}\right)_{\mathbf{p}=\mathbf{p}_0} (p_i - p_{i0})(p_k - p_{k0}). \tag{2.1}$$

Here, $\left(\frac{\partial^2 \varepsilon}{\partial p_i \, \partial p_k}\right)_{\mathbf{p}=\mathbf{p}_0}$ is a symmetrical tensor of the second rank; if $\mathbf{p} = \mathbf{p}_0$ corresponds to an energy minimum, the principal values of this tensor are positive but if this condition corresponds to a maximum the corresponding values are negative. The components of the tensor have the dimensions of the reciprocal of mass and the tensor itself is called the tensor of reciprocal effective masses, denoted by m_{ik}^{-1}.

Thus,

$$\varepsilon(\mathbf{p}) = \varepsilon(\mathbf{p}_0) + \frac{1}{2}\, m_{ik}^{-1}(p_i - p_{i0})(p_k - p_{k0}). \tag{2.1a}$$

Near such extremal points, the constant-energy surfaces in the crystal momentum space are closed and in the immediate vicinity of these points the surfaces are ellipsoids, as indicated by Eq. (2.1) or (2.1a).

We must bear in mind that any closed constant-energy surface near a minimum encloses a region of the momentum space in which the value of the energy is less than that on the surface, whereas near a maximum a closed surface surrounds a region in which the value of the energy is greater than that on its surface. This means that near a minimum the velocity vector $\mathbf{v} = \partial \varepsilon / \partial \mathbf{p}$ is directed along the outward normal to the constant-energy surface, whereas near a maximum it is directed along the inward normal.

Away from a minimum or a maximum, a constant-energy surface is somewhat deformed but remains closed as long as $|\varepsilon - \varepsilon_{\min}|$ or $|\varepsilon_{\max} - \varepsilon|$ are small. We must remember that because of the periodicity of the function $\varepsilon(\mathbf{p})$ these constant-energy surfaces are repeated throughout the reciprocal lattice.

Obviously, these topologically simple surfaces must be interspersed by more complex surfaces of the self-intersecting and open-type (the latter extending throughout the whole of the reciprocal lattice) because, otherwise, it would be impossible to proceed continuously from surfaces surrounding minima to those surrounding maxima.

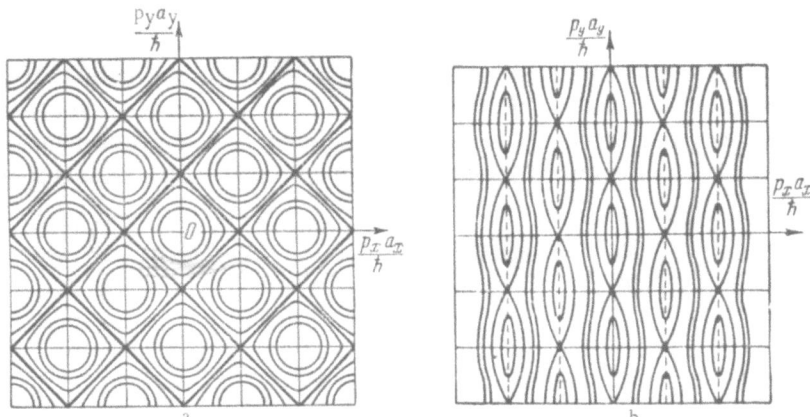

Fig. 1. Constant-energy curves.

As an example, Fig. 1 shows the constant-energy "surfaces" (two-dimensional curves) for the dispersion law

$$\varepsilon(p) = A_1 \cos \frac{p_x a_x}{\hbar} + A_2 \cos \frac{p_y a_y}{\hbar}.$$

Clearly, if $A_1 = A_2$ we have one "open surface" (the system of straight thick lines for Fig. 1a). However, if $A_1 \neq A_2$, there is a layer of such surfaces (Fig. 1b).

In the three-dimensional case, the open surfaces are encountered very frequently. For example, a "simple" dispersion law such as

$$\varepsilon = \varepsilon_0 \left\{ 1 - \frac{1}{3} \left(\cos \frac{p_x a_x}{\hbar} + \cos \frac{p_y a_y}{\hbar} + \cos \frac{p_z a_z}{\hbar} \right) \right\} \qquad (2.2)$$

gives rise to layers of open surfaces, which fill about one-third of the total volume of the reciprocal lattice (Fig. 2).

The dispersion law (2.2) can be regarded as representing the first terms in the expansion of a more general dispersion law $\varepsilon = \varepsilon(p)$ in terms of a Fourier series. In the theory of

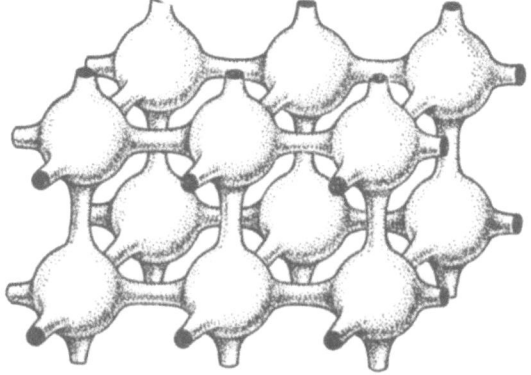

Fig. 2. Typical open constant-energy surface (three-dimensional network). The dependence of the energy on the crystal momentum is given by Eq. (2.2).

tightly-bound Bloch electrons, this law corresponds to the inclusion of the interaction between electrons of a given atom with the nearest neighbors in a simple cubic lattice [1]. It is shown in [5, 6] that the retention of additional terms in the expansion, which would correspond to the inclusion of the interaction with the second-nearest and further neighbors, complicates considerably the constant-energy surfaces. In particular, one very frequently encounters constant-energy surfaces connected in a complex manner. Figure 3 shows such constant-energy surfaces for the dispersion law

$$\varepsilon = A_0 + A_1\left(\cos\frac{p_x a}{\hbar} + \cos\frac{p_y a}{\hbar} + \cos\frac{p_z a}{\hbar}\right) + A_2\left(\cos\frac{p_x+p_y}{\hbar}a + \cos\frac{p_x+p_z}{\hbar}a + \cos\frac{p_y+p_z}{\hbar}a\right) +$$

$$+ A_3\left(\cos\frac{p_x+p_y-p_z}{\hbar}a + \cos\frac{p_x-p_y+p_z}{\hbar}a + \cos\frac{p_x-p_y-p_z}{\hbar}a\right) + A_4\left(\cos\frac{2p_x}{\hbar}a + \cos\frac{2p_y}{\hbar}a + \cos\frac{2p_z}{\hbar}a\right)$$

for different values of the parameters A_j and the energy ε. Figure 3c shows a special case, in which one part of the constant-energy surface is contained within another; part of the outer surface is cut open to show this more clearly.

Open surfaces can have a great variety of shapes: for example, they can be singly or multiply connected. Some examples of open surfaces are shown in Figs. 2 and 4.

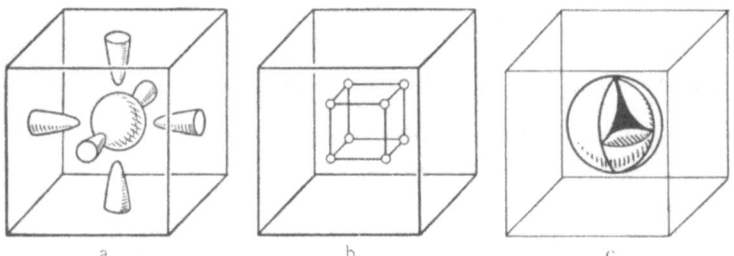

Fig. 3. Constant-energy surfaces corresponding to different values of the parameters A_j and different energies ε.

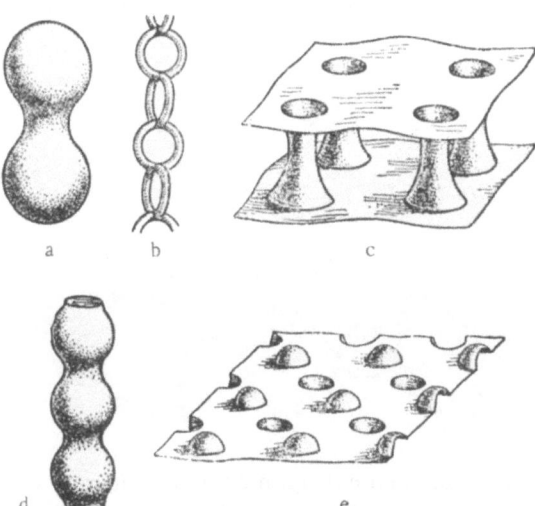

Fig. 4. Various types of constant-energy surface.

The topology of constant-energy surfaces or, more precisely, the nature of the curves formed by planar sections of these surfaces determines the dynamics of an electron in a magnetic field. Therefore, it is convenient to classify constant-energy surfaces not only as open or closed but also in accordance with the nature of the two-dimensional curves formed by planar sections of these surfaces. One such classification is given below.

1. Closed Nonintersecting Surfaces

These surfaces can be of two types:

a) closed surfaces without self-intersecting two-dimensional curves (a simple example of this type is a sphere);

b) closed surfaces which include self-intersecting two-dimensional curves (Fig. 4a).

2. Closed Self-Intersecting Surfaces

We must point out that, in general, self-intersection is possible only at one point.

3. Open Surfaces

Such surfaces can be of several types:

a) open surfaces without open sections (for example, Fig. 4b)[1];

b) surfaces with open curves only for fixed directions of the normal to the secant plane (Fig. 4c);

c) a one-dimensional set (dihedral angle) of directions of normals to the secant plane resulting in open curves (Fig. 4d, which shows a "corrugated" cylinder with a dihedral angle of 2π);

d) a two-dimensional set (a solid angle) of directions resulting in open curves (Fig. 4e, which shows a "corrugated" plane with an angle of 4π, or Fig. 2 which shows a "three-dimensional network" whose solid angle is less than 4π).

In the case of complex constant-energy surfaces the directions of the normals to the planes which produce open sections can be represented conveniently by a stereographic projection (Fig. 5).[2]

The points in the p space at which the energy has its minimum ε_{min} and maximum ε_{max} values are the simplest singular points. In an energy band, there are always some singular values of the energy. We shall call them the critical energies and denote them by ε_c. The topology of the constant-energy surface changes at $\varepsilon = \varepsilon_c$. We shall show later that at all critical points, particularly at $\varepsilon = \varepsilon_{min}$ and $\varepsilon = \varepsilon_{max}$, there is a singularity in the density of states of electrons. The topology may change in the following manner: at some point in the p space at $\varepsilon = \varepsilon_c$ either a new sheet of the constant-energy surface appears or disappears (Fig. 6a), or the surface changes its connectivity. In particular, an open surface can split into

---

[1] By analogy with open surfaces, the term "open sections" is used for sections which extend over the whole reciprocal lattice.

[2] A stereographic projection is the image of a hemisphere of unit radius projected on a circle of unit radius on which the position of a point is specified by two polar coordinates ρ and χ. The correspondence with points on the hemisphere is given by the following relationships: $\rho = 2\theta/\pi$, $\chi = \varphi$, where θ and φ are the azimuthal and polar angles ($0 \le \theta \le \pi/2$, $0 \le \varphi \le 2\pi$).

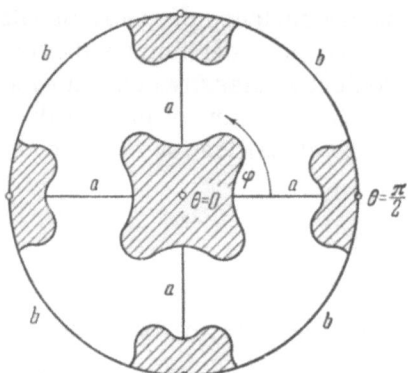

Fig. 5. Stereographic projection of the directions of the normals to secant planes for which open sections are obtained in the constant-energy surface shown in Fig. 2. The directions shown correspond to the shaded regions, lines a, and curves b.

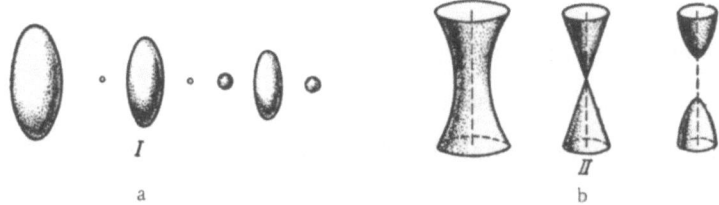

Fig. 6. Appearance of a new sheet (a) and the rupture of a neck (b) in a constant-energy surface. The surfaces denoted by I and II correspond to the critical energy ε_c. Surface II has a conical point.

a series of closed surfaces, a toroidal one can split into a system of ellipsoids (Fig. 6b), etc. In those cases when the connectivity changes, the critical constant-energy surface $\varepsilon(\mathbf{p}) = \varepsilon_c$ always includes one or more conical points $\mathbf{p} = \mathbf{p}_c$ near which the dispersion law can be approximated satisfactorily by the equation of a two-sheet hyperboloid (after the rupture of its "neck") and by the equation of a one-sheet hyperboloid (before breaking). The appearance or the disappearance of a new sheet of the constant-energy surface always occurs at the point $\mathbf{p} = \mathbf{p}_c$, where the energy (regarded as a function of the crystal momentum) has its minimum or maximum value. The constant-energy surfaces passing close to this point can be described satisfactorily by the equation of an ellipsoid. In this case, the principal directions of the effective masses are positive if the energy has its minimum value and negative if the energy has its maximum value. Figure 7 shows the appearance and the disappearance of a sheet of a constant-energy surface shown in two dimensions. In this case, the constant-energy surface is a curve.

We shall now analyze the structure of constant-energy surfaces under degenerate conditions. Let us assume that at some point in the $\mathbf{p}$ space (we shall denote it by $\mathbf{p}_0$) we have

$$\varepsilon_1(\mathbf{p}_0) = \varepsilon, \qquad \varepsilon_2(\mathbf{p}_0) = \varepsilon. \tag{2.3}$$

The numbers of the degenerate bands will be arbitrarily denoted by "1" and "2." We shall show that, in general, only two constant-energy surfaces can intersect at the same time. Multiple

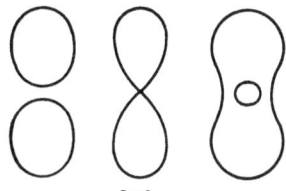

Fig. 7. Two-dimensional model of the appearance (disappearance) of a sheet in a constant-energy surface.

$\varepsilon = \varepsilon_C$

intersection is totally forbidden. In general, surfaces of this kind intersect at isolated points. They approach each other in such a way that, particularly in investigations of electron dynamics, they can conveniently be considered as one complex self-intersecting surface.

According to Eq. (1.43), the Schrödinger equation in the p representation is of the form

$$\hat{\mathcal{H}}_p \psi_p(\rho) = \varepsilon \psi_p(\rho), \tag{2.4}$$

and the assumption about degeneracy means that the Hamiltonian operator at the point $\mathbf{p} = \mathbf{p}_0$ has two identical eigenvalues:

$$\mathcal{H}^{11}_{p_0} = \mathcal{H}^{22}_{p_0} = \varepsilon(\mathbf{p}_0). \tag{2.5}$$

We shall consider the values of p close to $\mathbf{p}_0$:

$$\mathcal{H}^{ss'}_p \approx \varepsilon(\mathbf{p}_0)\delta_{ss'} + V^{ss'}. \tag{2.6}$$

The matrix $V^{ss'}$ is a function of the difference $\mathbf{p} - \mathbf{p}_0$ and $V^{ss'} \equiv 0$ when $\mathbf{p} - \mathbf{p}_0 = 0$.

To determine the correction to the energy $\Delta\varepsilon$, we can use an equation found by equating to zero the determinant of Eq. (2.4):

$$\begin{vmatrix} \Delta\varepsilon - V^{11} & V^{12} \\ V^{21} & \Delta\varepsilon - V^{22} \end{vmatrix} = 0. \tag{2.7}$$

Hence,

$$\Delta\varepsilon = \frac{V^{11} + V^{22}}{2} \pm \sqrt{\frac{(V^{11} + V^{22})^2}{4} + |V^{12}|^2}. \tag{2.8}$$

We shall use the Hermitian properties of the Hamiltonian $V^{12} = (V^{21})^*$. We shall introduce

$$\varepsilon(\mathbf{p}_0) + \frac{V^{11} + V^{22}}{2} = A(\mathbf{p}); \quad \frac{V^{11} + V^{22}}{2} = B(\mathbf{p}); \quad |V^{12}|^2 = C^2(\mathbf{p}).$$

Equation (2.8) then assumes the form

$$\varepsilon = A(\mathbf{p}) \pm \sqrt{B^2(\mathbf{p}) + C^2(\mathbf{p})}. \tag{2.9}$$

Hence, we can see that, at some definite value of the energy ε, the point of degeneracy $\mathbf{p}_0$ (the point of intersection of the constant-energy surfaces) is given by the three equations

$$A(\mathbf{p}) = \varepsilon; \quad B(\mathbf{p}) = C(\mathbf{p}) = 0. \tag{2.10}$$

In general (if no special assumptions are made), the points of degeneracy are located along some line in the $\mathbf{p}$ space (each value of the energy ε corresponds to one value of the crystal momentum $\mathbf{p}_0$) and these points are the singular conical points of a self-intersecting surface given by Eq. (2.9). If the expansion of the matrix $V^{ss'}$ in powers of $\mathbf{p} - \mathbf{p}_0$ does not begin from the first term (this is usually because of some symmetry requirement), the self-intersection is replaced by contact. The intersection of two constant-energy surfaces along a curve or the intersection of more than two surfaces is the result of the special symmetry properties of a given crystal (for example, if B or C in the above expressions is identically equal to zero).

Figure 8 shows the vicinity of a point of intersection, and for each of the branches of Eq. (2.9), i.e., for each of the bands, it is shown whether the energy exceeds or is less than ε.

Figure 9 shows schematically a self-intersecting constant-energy surface. The limiting cases of the family in Fig. 9 are shown separately (a and b). The energies corresponding to these surfaces represent the maximum or the minimum values of the energy in a band. It may seem surprising that the surfaces associated with different energies can intersect. However, we can easily show that the intersection occurs between surfaces corresponding to different energy bands. Thus, an electron whose crystal momentum is, for example, $\mathbf{p}'$ (Fig. 9) can have an energy ε or ε', depending on the band to which it belongs, i.e., its energy can be $\varepsilon_s(\mathbf{p}')$ or $\varepsilon_{s'}(\mathbf{p}')$. Usually (in the absence of degeneracy), the constant-energy surfaces of each band (cor-

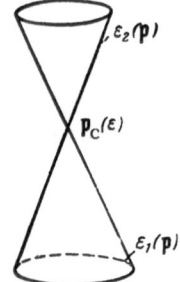

Fig. 8. Constant-energy surfaces $\varepsilon = \varepsilon_{1,2}(\mathbf{p})$ intersecting at a point $\mathbf{p} = \mathbf{p}_c(\varepsilon)$. Outside the cone defined by $\varepsilon_1(\mathbf{p}) = \varepsilon$, the energy is $\varepsilon_1(\mathbf{p}) > \varepsilon$; within the cone defined by $\varepsilon_2(\mathbf{p}) = \varepsilon$, the energy is $\varepsilon_2(\mathbf{p}) > \varepsilon$.

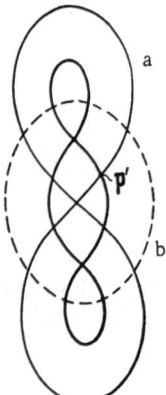

Fig. 9. Constant-energy surface containing two points of degeneracy. The thin curve (a) and the dashed curve (b) represent the critical surfaces of a family with two points of degeneracy.

responding to a particular value of s) are plotted in the band's own **p** space. This is inconvenient under degenerate conditions. The two bands ("1" and "2") are given the same **p** space.

Degeneracy is very frequently the result of a particular symmetry of a crystal and is usually observed along certain selected lines in the **p** space. Moreover, an analysis of the symmetry of particular points in the **p** space of a crystal can be used to find all the points at which degeneracy must occur and to determine the dependence of the elements of the matrix $V^{ss'}$ on the components of the difference $\mathbf{p} - \mathbf{p}_0$. This allows us to establish the structure of a constant-energy surface near a point of degeneracy without invoking any specific model [7].

It frequently happens that the inclusion of relatively small terms in the Hamiltonian of a conduction electron (for example, the spin–orbit coupling) can alter significantly the symmetry of the Hamiltonian and can lift the degeneracy. In such cases, it is naturally found that a constant-energy surface consists of two sheets lying close to each other.

The values of the energy corresponding to self-intersecting constant-energy surfaces are not singular: usually, self-intersecting surfaces form a layer. We must remember that each self-intersecting surface has a singular (conical) point. However, we shall show later that an important consideration in the classification of the singular points is whether the velocity vanishes. Critical surfaces (those corresponding to $\varepsilon = \varepsilon_c$) always include a point at which $v = 0$. At a point of self-intersection, $v \neq 0$. Therefore, the special cases considered in the preceding paragraphs (the appearance or disappearance of a new sheet at a point; the rupture of a neck or the merging to form a neck; the disappearance or appearance of self-intersecting surfaces) represent all the possible types of singular constant-energy surface and all the types of singular point in the **p** space.

In some cases, singular points in the **p** space can occupy a complete curve, known as a loop of extrema. For example, in the case of wurtzite-type crystals, a loop of extrema is a circle with its center on the symmetry axis [8]. A similar situation occurs in other crystals (CdS, InSb, etc.). Figure 10 shows the electron energy structure of wurtzite-type crystals whose dispersion law for electrons with energies close to the extrema is of the form [8]

$$\varepsilon^{\pm} = a\left(p_{\perp} \pm p_0\right)^2 + b p_z^2, \tag{2.11}$$

where a, b, p_0 are positive constants; $p_{\perp} = \sqrt{p_x^2 + p_y^2}$; and the axis p_z coincides with the axis of the crystal.

At energies below $a p_0^2$ (but exceeding zero), there is only one branch of the energy spectrum corresponding to the minus sign in Eq. (2.11). In this case, the constant-energy surfaces

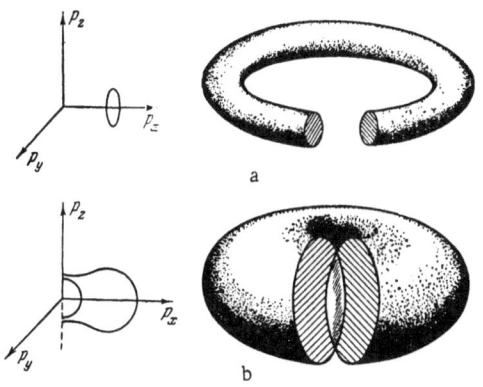

a

b

Fig. 10. Constant-energy surfaces in wurtzite-type crystals are obtained in two ways: a) by rotating an ellipse about the p_z axis in the case when $0 < \varepsilon < a p_0^2$ [see Eq. (2.11)], or b) by rotating a more complex figure about the same axis in the case when $\varepsilon > a p_0^2$.

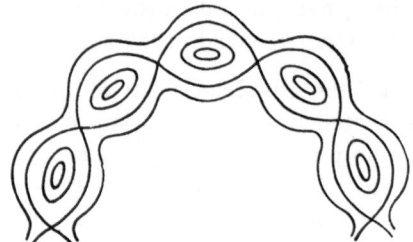

Fig. 11. Transition from ellipsoidal constant-energy surfaces to "corrugated" tori (section in the plane $p_z = 0$).

are tori with elliptical cross sections in planes passing through the p_z axis. The line along which the energy has its minimum value is a loop of radius p_0 (we note that the value of p_0 is related to the spin–orbit interaction parameter β by $p_0 = \beta/2a$ [8]). The critical value of the energy is taken to be zero ($\varepsilon_c = \varepsilon_c^{(-)} = 0$). The dispersion law of Eq. (2.11) gives rise also to an additional critical energy $\varepsilon_c = \varepsilon_c^{(+)} = ap_0^2 = \beta^2/4a$. At this energy, the topology of the surfaces $\varepsilon^{(-)}(\mathbf{p}) = \varepsilon$ changes: the torus "fills up." At the same time, an ovaloid sheet appears [this sheet corresponds to the plus sign in Eq. (2.11)]. It follows that the origin ($p = p_z = 0$) is also a singular point but we shall show later that the singularity at this point is weaker than at, for example, that point in the $\mathbf{p}$ space where a neck ruptures. A more detailed analysis, in which we include the small terms rejected in Eq. (2.11), would result in the appearance of a new band starting from a system of ellipsoids which would deform and merge into a toroidal surface, as shown in Fig. 11. [9].

Much of what we have said is based on the assumption that the $\mathbf{p}$ space is of infinite extent. However, we note that physically nonequivalent states occupy only one cell in this space. In some cases, the question arises of the selection of the appropriate unit cell. Usually, it is convenient to select a unit cell in such a way as not to break the closed surfaces. Therefore, it may be found that different bands (and even different energy intervals within one band) correspond to different unit cells. In the actual plotting of the energy spectrum, it is convenient to use the Brillouin zone (more precisely, the first Brillouin zone) which is that part of the $\mathbf{p}$ space which includes only nonequivalent vectors $\mathbf{p}$ selected on the following basis. Out of all the equivalent vectors, we select the shortest one. In such selection and comparison of the crystal momenta, one must bear in mind all the symmetry elements of a crystal of a given class. For most crystals, the first Brillouin zone is not a parallelepiped but a more complex figure which demonstrates clearly the symmetry elements of that crystal. Naturally, the volume of the first Brillouin zone is equal to the volume of a unit cell in the reciprocal lattice multiplied by $(2\pi\hbar)^3$. We should also note that the whole $\mathbf{p}$ space can be filled with the first Brillouin zones.[3]

The shape of the Brillouin zone is often very important in the determination of the nature of the symmetry of the constant-energy surfaces. Frequently one can use the periodicity of the energy $\varepsilon(\mathbf{p})$ to limit the analysis to the first Brillouin zone. In these circumstances, it is very difficult to consider the motion of electrons in external fields when the crystal momentum is not conserved. The question immediately arises as to what happens to an electron when it reaches the Brillouin zone boundary. This difficulty can be bypassed but, in most cases, it is

---

[3] The Brillouin zones have been introduced in connection with investigations of the energy spectra of almost-free electrons. Consideration of the second, third, etc., zones would be pointless because they are too closely connected with the properties of the almost-free electrons.

more convenient to regard the crystal momentum of electrons as defined throughout the reciprocal lattice space and to assume that the energy and other characteristics of the electron (for example, its velocity) are periodic functions of $\mathbf{p}$.

§ 3. Density of States per Unit Energy Interval

One of the most important characteristics of the energy spectrum, which is essential in the statistical thermodynamics of the electron gas, is the density of states per unit energy interval $\nu(\varepsilon)$. This quantity is closely related to the dispersion law $\varepsilon_s(\mathbf{p}) = \varepsilon$.

In a crystal of volume V, the number of electron states in the s-th band, taken per element of volume in the momentum space, is

$$dn_{\mathbf{p}}^{(s)} = \frac{2V}{(2\pi\hbar)^3}\, dp_x dp_y dp_z. \tag{3.1}$$

It is frequently more convenient to dispense with the Cartesian system of coordinates in the momentum space and to use instead a system in which the coordinate and the constant-energy surfaces coincide. In such a system, the state of an electron is given by its position on a constant-energy surface $\varepsilon_s(\mathbf{p}) = \varepsilon$ and an element of volume $dp_x dp_y dp_z$ is equal to $dSd\varepsilon/v_s$, where dS is an element of area on the constant-energy surface and v_s is the modulus of the electron velocity.[1]

Integrating Eq. (3.1) over a constant-energy surface, we find that the number of electron states $dn_{\varepsilon}^{(s)}$ per energy interval $d\varepsilon$ is

$$dn_{\varepsilon}^{(s)} = \frac{2V}{(2\pi\hbar)^3} \oint\limits_{\varepsilon_s(\mathbf{p})=\varepsilon} \frac{dS}{v_s}\, d\varepsilon. \tag{3.2}$$

The density of electron states in the s-th band is then

$$\nu_s(\varepsilon) = \frac{2V}{(2\pi\hbar)^3} \oint\limits_{\varepsilon_s(\mathbf{p})=\varepsilon} \frac{dS}{v_s}. \tag{3.3}$$

Irrespective of whether the constant-energy surface is closed or open, the integration indicated in Eq. (3.3) is carried out within the limits of one unit cell in the reciprocal lattice (one Brillouin zone). In those cases when the surface $\varepsilon_s(\mathbf{p}) = \varepsilon$ consists of several sheets, $\nu_s(\varepsilon)$ is naturally the sum of the corresponding integrals.

Equation (3.3) can be rewritten in the form

$$\nu_s(\varepsilon) = \frac{2V}{(2\pi\hbar)^3} \frac{d}{d\varepsilon} \Lambda_s(\varepsilon), \tag{3.4}$$

where

$$\Lambda_s(\varepsilon) = \int E\left[\varepsilon_s(\mathbf{p}) - \varepsilon\right] d\tau_p. \tag{3.5}$$

The function E(x) is defined by

$$E(x) = \begin{cases} 1 & x < 0, \\ 0 & x > 0. \end{cases}$$

---

[1] $|\nabla_{\mathbf{p}}\varepsilon_s(\mathbf{p})|^{-1} \equiv v_s^{-1}$ is the Jacobian for the transformation of the variables p_x, p_y, and p_z to the variables ε, ξ, and η, where ξ and η represent any orthogonal system of coordinates on the surface $\varepsilon_s(\mathbf{p}) = \varepsilon$ (dS = $d\xi\, d\eta$).

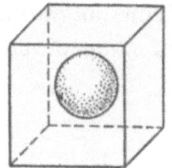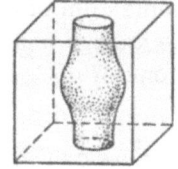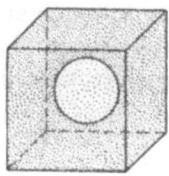

Fig. 12. Volume of that part of the Brillouin zone where the electron energy exceeds ε is shown dotted.

Irrespective of the topology of the constant-energy surface $\varepsilon_s(\mathbf{p}) = \varepsilon$, the quantity $\Delta_s(\varepsilon)$ is the volume of that part of the Brillouin zone where the electron energy is less than ε (Fig. 12).

The quantity $[2V/(2\pi\hbar)^3]\,\Delta_s(\varepsilon)$ gives the number of those electron states in the s-th band for which the energy is less than ε. We shall denote it by $N_s(\varepsilon)$. Then,

$$\nu_s(\varepsilon) = \frac{dN_s(\varepsilon)}{d\varepsilon}. \tag{3.6}$$

The density of electron states $\nu_s(\varepsilon)$ is a complex function of its own argument, which is the electron energy. Its explicit form can be obtained if we know the dispersion law of conduction electrons.

Even if we make no specific assumptions about the dispersion law, we can still draw some conclusions about the singularities of the function $\nu_s(\varepsilon)$ specified in the interval $(\varepsilon_0, \varepsilon_1)$, where $\varepsilon_0 = \varepsilon_{\min}^{(s)}$, and $\varepsilon_1 = \varepsilon_{\max}^{(s)}$. We have mentioned in §2 that the constant-energy surfaces near a minimum $(\varepsilon \gtrsim \varepsilon_0)$ or a maximum $(\varepsilon \lesssim \varepsilon_1)$ are ellipsoids. Therefore,

$$\nu(\varepsilon) \approx \begin{cases} \dfrac{\sqrt{2}\,V}{\pi^2\hbar^3}(m_1 m_2 m_3)^{1/2}(\varepsilon - \varepsilon_0)^{1/2}, \\[2ex] \dfrac{\sqrt{2}\,V}{\pi^2\hbar^3}(m_1' m_2' m_3')^{1/2}(\varepsilon_1 - \varepsilon)^{1/2}. \end{cases} \tag{3.7}$$

The square-root singularity,[2] which is exhibited by $\nu(\varepsilon)$ when $\varepsilon \to \varepsilon_0$ and $\varepsilon \to \varepsilon_1$, is typical of all those values of the energy at which there is a change in the topology of the constant-energy surfaces. On this basis, a minimum can be regarded as a point at which a new sheet of the constant-energy surface appears and a maximum is a point at which a sheet disappears. We have seen that the appearance or the disappearance of a sheet may also occur in the middle of a band at certain critical values of the energy $\varepsilon = \varepsilon_c$. At energies close to ε_c, the density of electron states $\nu(\varepsilon)$ can be represented by

$$\nu(\varepsilon) = \nu_0(\varepsilon) + \delta\nu(\varepsilon), \tag{3.8}$$

where $\nu_0(\varepsilon)$ is a smooth function of the energy and $\delta\nu(\varepsilon)$ does not vanish on that side of the critical value for which the number of sheets is greater by unity than the number of sheets on the other side. To be specific, we shall assume that the number of sheets increases with increasing energy. Then,

$$\delta\nu(\varepsilon) = \begin{cases} 0 & (\varepsilon < \varepsilon_c), \\[2ex] \dfrac{\sqrt{2}\,V}{\pi^2\hbar^3}(m_1 m_2 m_3)^{1/2}(\varepsilon - \varepsilon_c)^{1/2} & (\varepsilon \geqslant \varepsilon_c). \end{cases} \tag{3.9}$$

---

[2] More precisely, the derivative of $\nu(\varepsilon)$ becomes infinite because it is proportional to $1/x^{1/2}$ when $x \to 0$.

We have shown in the preceding section that an increase in the number of sheets of a constant-energy surface may result not only from the creation of a new sheet (Fig. 6a) but also from the rupture of a neck (Fig. 6b).[3] We shall consider this change in topology in more detail. A constant energy surface $\varepsilon(\mathbf{p}) = \varepsilon_c$ includes a conical singular point $\mathbf{p} = \mathbf{p}_c$. At energies below ε_c, the part of the surface near the point $\mathbf{p} = \mathbf{p}_c$ forms a two-sheet hyperboloid and in the range $\varepsilon > \varepsilon_c$ forms a one-sheet hyperboloid. If the coordinate axes are chosen in a suitable manner, the equation of that part of the constant-energy surface which is located close to $\mathbf{p} = \mathbf{p}_c$ ($\varepsilon \approx \varepsilon_c$) can be described by

$$\varepsilon = \varepsilon_c + \frac{p_1^2}{2m_1} + \frac{p_2^2}{2m_2} - \frac{p_3^2}{2m_3} \qquad (m_1,\ m_2,\ m_3 > 0). \tag{3.10}$$

A calculation of the volume bounded by the surface described by Eq. (3.10) and by the plane $p_3 = $ const (the p_3 axis lies along the axis of the neck) gives

$$\Delta(\varepsilon, p_3) = \begin{cases} \frac{2\pi p_3^3}{3}\frac{(m_1 m_2)^{1/2}}{m_3} + 4\pi (m_1 m_2)^{1/2}(\varepsilon - \varepsilon_c)p_3 & (\varepsilon > \varepsilon_c), \\ \frac{2\pi p_3^3}{3}\frac{(m_1 m_2)^{1/2}}{m_3} + 4\pi (m_1 m_2)^{1/2}(\varepsilon - \varepsilon_c)p_3 + \frac{8\pi}{3}(2m_1 m_2 m_3)^{1/2}(\varepsilon_c - \varepsilon)^{3/2} & (\varepsilon < \varepsilon_c). \end{cases} \tag{3.11}$$

In this case, the density of electron states $\nu(\varepsilon) = [2V/(2\pi\hbar)^3][d\Delta(\varepsilon)/d\varepsilon]$ can be represented as the sum of two terms: a smoothly varying $\nu_0(\varepsilon)$ and an irregular correction $\delta\nu(\varepsilon)$ which has a square-root singularity and does not vanish at the energies at which the number of sheets increases (see also the comments made in the preceding paragraphs). Equation (3.9) is still valid [the meaning of m_1, m_2, and m_3 follows from Eq. (3.10)].

In the preceding section, we considered the self-intersection of constant-energy surfaces. At the energies at which such intersection occurs, the density of states $\nu(\varepsilon)$ has no singularities. In this case, the singular or the critical energies are those at which there is a change in the topology of the constant-energy surface. For example, the surfaces denoted by a and b in Fig. 9 are singular in the sense just defined. We can easily show that in such cases $\delta\nu \sim |\varepsilon - \varepsilon_c|^{1/2}$ and that it vanishes only on one side of $\varepsilon = \varepsilon_c$. However, if the singular point $\mathbf{p} = \mathbf{p}_c$ is a degenerate point, the equation for a constant-energy surface in the vicinity of this point is more complex than we have assumed earlier:

$$\varepsilon = \varepsilon_c + (\mathbf{p} - \mathbf{p}_c)^2\varphi(\mathbf{n}),$$

where $\varphi(\mathbf{n})$ is some function of a unit vector $\mathbf{n} = (\mathbf{p} - \mathbf{p}_c)/|\mathbf{p} - \mathbf{p}_c|$, corresponding to the symmetry of the point $\mathbf{p} = \mathbf{p}_c$. Thus, for example, the energy spectrum near the valence band maxima of Ge and Si crystals is of the form

$$\varepsilon_1 - \varepsilon = Ap^2 \pm \sqrt{B^2 p^4 + C^2(p_x^2 p_y^2 + p_x^2 p_z^2 + p_y^2 p_z^2)},$$

where A, B, C are constants and the crystal momentum $\mathbf{p}$ is measured from the point at which the energy reaches its maximum value $\varepsilon(\mathbf{p}=0) = \varepsilon_1$.

The special cases (the appearance of a new sheet and the rupture of a neck) represent all the possible singular points in the $\mathbf{p}$ states. However, we have seen that, under certain fully

---

[3] A change in the topology of a constant-energy surface may take place without a change in the number of sheets. For example, it may take place by the rupture of a neck in a toroidal surface. In such cases, the connectivity of the surface changes at a point $\varepsilon = \varepsilon_c$ and $\delta\nu(\varepsilon) \neq 0$ in that range of energies where the connectivity decreases.

justified assumptions, the singular points in the **p** space can be regarded not as isolated points but as points located on a curve known as a loop of extrema. Let us consider the density of electron states in such cases, restricting our analysis to energies close to their extremal values. To be specific, we shall assume that the dispersion law is given by Eq. (2.11). We recall that there are two critical values of the energy ($\varepsilon_c = 0$ and $\varepsilon_c = a p_0^2$). When a torus is formed ($\varepsilon_c = 0$), the density of states changes discontinuously:

$$\nu^-(\varepsilon) = \begin{cases} 0 & \varepsilon < 0, \\ \nu_0 = \dfrac{\beta}{4\pi\hbar^3 a^{3/2} b^{1/2}} & \varepsilon > 0. \end{cases} \qquad (3.12)$$

When a torus "fills up" ($\varepsilon_c = a p_0^2$), the density of states of the branch considered has a weak singularity

$$\delta\nu^-(\varepsilon) \approx \frac{\nu_0}{3\pi}\left(\frac{\varepsilon - \varepsilon_c}{\varepsilon_c}\right)^{3/2} \qquad (\varepsilon \gtrsim \varepsilon_c).$$

We note that a new sheet (an ovaloid) appears at the same energy. Its appearance is accompanied by a similar weak singularity:

$$\delta\nu^+(\varepsilon) \approx \frac{\nu_0}{3\pi}\left(\frac{\varepsilon - \varepsilon_c}{\varepsilon_c}\right)^{3/2} \qquad (\varepsilon \gtrsim \varepsilon_c).$$

Hence,

$$\delta\nu(\varepsilon) \approx \frac{2\nu_0}{3\pi}\left(\frac{\varepsilon - \varepsilon_c}{\varepsilon_c}\right)^{3/2} \qquad (\varepsilon \gtrsim \varepsilon_c). \qquad (3.13)$$

If an analysis of this kind is made more rigorously we find that the appearance of a band corresponding to the minus sign in Eq. (2.11) starts with ellipsoids. At the appropriate critical point, the density of states has the standard square-root singularity. In a very narrow range of energies, the density of states reaches a value corresponding to a torus and when the ellipsoids merge, $\nu(\varepsilon)$ exhibits a second square-root singularity (Fig. 13).

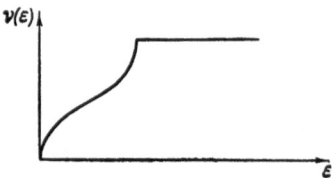

Fig. 13. Behavior of the density of states on transition from
ellipsoidal to toroidal constant-energy surfaces.

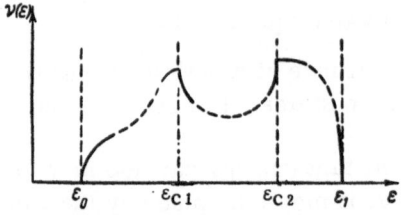

Fig. 14. Simplest dependence of the density of states
on the electron energy.

Summarizing our discussion, we can state that the density of states $\nu(\varepsilon)$ has at least two square-root singularities within each band [10]. The simplest form of $\nu(\varepsilon)$ is shown in Fig. 14. We find that between the layers of closed surfaces there must be at least one layer of open surfaces. When we go over from closed to open surfaces ($\varepsilon = \varepsilon_{c1}$), the derivative $d\nu/d\varepsilon$ assumes a negative infinite value and on transition from open to closed surfaces ($\varepsilon = \varepsilon_{c2}$) the derivative passes through a positive infinity. If we bear in mind that $\nu(\varepsilon_0) = \nu(\varepsilon_1) = 0$ and $\nu(\varepsilon) \geq 0$, we find that the density of states in the intervals (ε_0, ε_{c1}) and (ε_{c2}, ε_1) has maxima or reaches its highest values and between these intervals it has at least one minimum or some other singularity.

The total density of electron states is naturally the sum of all the states in the s bands:

$$\nu(\varepsilon) = \sum_s \nu_s(\varepsilon), \tag{3.14}$$

and the "inclusion" of a new band (in the case of band overlap) is indistinguishable in respect of the density-of-states singularities from the appearance of a new sheet of a constant-energy surface within a band of the rupture of a neck of such surface.

By analogy with Eq. (3.14), we can introduce the total number of electrons whose energy is less than a certain specified value:

$$N(\varepsilon) = \sum_s N_s(\varepsilon). \tag{3.15}$$

Concluding this section, we shall give the expressions for the density of states in a free-electron gas

$$\nu(\varepsilon) = \frac{\sqrt{2}\,V}{\pi^2 \hbar^3}\, m^{3/2} \varepsilon^{1/2} \tag{3.16}$$

and for the number of electrons whose energies do not exceed the specified value:

$$N(\varepsilon) = \frac{2\sqrt{2}\,V}{3\pi^2 \hbar^3}\, m^{3/2} \varepsilon^{3/2}. \tag{3.17}$$

§4. Classical Mechanics of a Particle with an Arbitrary Dispersion Law

We discussed in §1, in detail, the possibility of applying classical mechanics to the motion of a conduction electron in external fields. The results of a procedure analogous to the inversion of the correspondence principle are that $\varepsilon(\mathbf{p})$ should be regarded as the kinetic energy of an electron and that the equations of motion in electric and magnetic fields can be written in the standard form:

$$\dot{\mathbf{p}} = \mathbf{F}, \quad \mathbf{F} = e\left\{ \mathbf{E} + \frac{1}{c}\,[\mathbf{v}\mathbf{H}] \right\}, \quad \dot{\mathbf{r}} = \mathbf{v}, \quad \mathbf{v} = \frac{\partial \varepsilon}{\partial \mathbf{p}}. \tag{4.1}$$

These equations are the foundations of the classical mechanics of an electron with a complex dispersion law.

In order to avoid misunderstanding, we must stress again that a conduction electron is undoubtedly a quantum-mechanical particle, i.e., its laws of motion can be derived only on the basis of quantum mechanics. In particular, only quantum mechanics can explain the "freedom" of a conduction electron, i.e., its ability to move in a crystal lattice over macroscopic distances, and makes it possible to introduce such concepts as quasi particle, crystal momentum (quasi-momentum), dispersion law, etc. However, since such concepts have already been formulated and we can speak of an electron in a metal as a particle with a complex dispersion law, we find that it is possible to solve some problems in a purely classical manner. We shall not consider the conditions which must be satisfied before this can be done (they were discussed in detail in §1). We must mention, however, that the classical approach to the motion of an electron with a specified dispersion law and subjected to external fields can be used only in those cases when the dimensions of an electron trajectory exceed considerably the atomic spacings. In other words, external fields should be weak compared with the internal and atomic fields.

We shall now consider several cases of electron motion in external fields. Although we cannot use the Newtonian equations, they are still very useful in clarifying the motion of a particle with a complex dispersion law. Differentiating the second of the equations in the system (4.1) with respect to time, we obtain

$$\ddot{x}_i = \frac{\partial^2 \varepsilon}{\partial p_i \partial p_k} F_k. \tag{4.2}$$

This system of equations is inconvenient because the tensor $\partial^2\varepsilon/\partial p_i\partial p_k$, known as the reciprocal effective-mass tensor, depends generally on the momentum and, consequently, on time. This makes it extremely difficult to integrate a system of equations of the type given by Eq. (4.2). If the conditions of the problem allow us to ignore the terms which are nonlinear functions of the applied forces, the components of the reciprocal effective-mass tensor may be regarded as constants. It is worth mentioning that at an arbitrary point in the **p** space the tensor in question has three principal values which differ in magnitude and sign. Near these values of the momentum at which the energy reaches its minimum or maximum, the components of the reciprocal effective-mass tensor are constants and the principal values of the tensor have the same sign (§2). In this case, we have an additional justification for using Eq. (4.2) with a constant value of the reciprocal effective-mass tensor.

Even in the simplest case (when the terms are linear in respect of the forces), we may encounter a very interesting situation: the same force acting in one direction may accelerate an electron but acting in another direction may decelerate it.

Equation (4.2) demonstrates clearly the point that the kinematic properties of an electron with a complex dispersion law cannot be represented by a single mass. The velocity of a conduction electron is not proportional to its momentum: this velocity **v** is a complex periodic function of the momentum and the coefficient of proportionality between the force and the acceleration is a second-rank tensor, which depends in a complex manner on the momentum. In this section, we have introduced the concept of an effective mass, which is convenient in discussing the motion of a particle in a magnetic field. In some very special cases, the "normal" heavy mass of an electron can be used in the formulas. This occurs when the effect considered is governed by the true relativistic momentum of an electron $p_r = \mathscr{E}v/c^2$, where c is the velocity of light and $\mathscr{E}$ is the total energy of the electron, which includes its rest mass. Since the energies of the interactions between electrons and the lattice and between electrons and other particles are considerably smaller than the energy corresponding to the rest mass m_0c^2, it follows that $p_r = m_0(v)$ (§24). We shall now consider the motion in a constant and uniform electric field. In this case, the equations of motion simplify considerably to the following expressions:

$$\dot{p} = eE, \quad \dot{r} = v. \tag{4.3}$$

It follows that if **E** = const, an electron moves at a constant velocity in the **p** space:

$$p = p_0 + eEt. \tag{4.4}$$

It follows from the law of conservation of energy

$$\varepsilon(p) - e\mathbf{E}\mathbf{r} = \text{const} \tag{4.5}$$

that the motion of a conduction electron in a constant and uniform electric field is finite along this field. If the direction of the field coincides with one of the crystallographic axes, the electron oscillates at a frequency $\omega_E = 2eE/\Delta p_E$, where Δp_E is the period of the function $\varepsilon(p)$ along the electric field. Since $\Delta p_E \approx 2\pi\hbar/a$ (a is the lattice parameter), it follows that $\omega_E \approx eEa/\hbar$. The amplitude of the oscillations is $\Delta\varepsilon/eE$ ($\Delta\varepsilon$ is the forbidden band width). In general, when the direction of the field does not coincide with one of the crystallographic axes, the motion of the electron is nearly periodic. For reasonable values of the field, the oscillation amplitude is extremely high: it can be millions of times greater than the mean free path ($\Delta\varepsilon/eE \approx 10^6$ cm for $E = 10^{-8}$ cgs esu,[1] and $\Delta\varepsilon \approx 10^{-12}$ erg). Therefore, in calculating the resistivity of a metal and in similar cases, the periodic nature of the motion of electrons can be ignored. Over short distances, the motion of an electron can be regarded as translation.

In dielectrics, the finite nature of the motion of an electron in an electric field may be manifested by an increase in the probability of the band-band breakdown.

The anisotropic nature of the dispersion law means that the electron motion is not free in a plane perpendicular to the applied field: the velocity in this plane varies with the projection of the momentum along the electric field. This is one of the basic causes of the anisotropy of the electrical resistivity, thermal conductivity, and other transport coefficients.

We shall now consider the motion in a constant and uniform magnetic field ($\mathbf{E} = 0$, $\mathbf{H} \neq 0$).

In this case, the Lorentz equation is of the form:

$$\frac{d\mathbf{p}}{dt} = \frac{e}{c}[\mathbf{v}\mathbf{H}]. \tag{4.6}$$

Multiplying this equation scalarly once by $\mathbf{v}$ and once by $\mathbf{H}$, we find that — as in the case of a free electron — the energy of a conduction electron ε and the projection of its momentum $\mathbf{p}$ along the magnetic field are conserved during the motion in such a field. If the z axis is taken along the magnetic field, we find that

$$\varepsilon = \text{const}, \quad p_z = \text{const}. \tag{4.7}$$

The equalities of Eq. (4.7) describe the trajectory of an electron in the momentum space. Depending on the topology of the constant-energy surface, the trajectory can be either closed (i.e., it can be split into closed curves, each of which is located within one unit cell in the reciprocal lattice[2]) or open (i.e., it can pass continuously through the whole reciprocal lattice). The classification of the various curves is given in §2.

The position of an electron on a trajectory [Eq. (4.7)] in the momentum space can be determined conveniently by measuring the time of revolution from some arbitrary point on the trajectory. Projecting Eq. (4.6) onto the (x, y) plane, we obtain

$$\frac{dp_\perp}{dt} = -\frac{eH}{c}v_\perp. \tag{4.8}$$

---

[1] $E = 10^{-8}$ cgs esu corresponds to a current density j $= 10^2$ A/cm^2 in a metal whose resistivity is $\rho = 10^{-8}$ $\Omega \cdot$cm.

[2] We note that in those cases when the (x, y) plane does not coincide with any of the crystallographic planes, the curves located in different cells of the reciprocal lattice are not identical. This is not in conflict with the periodicity of the function $\varepsilon = \varepsilon(p)$.

Here, $dp_\perp$ is an element of the arc of the curve described by Eq. (4.7) and $v_\perp = \sqrt{v_x^2 + v_y^2}$. We recall that the z axis coincides with the direction of the magnetic field. The minus sign is selected because the electron charge is negative ($-e > 0$).

The integration of Eq. (4.8) gives

$$t = -\frac{c}{eH} \int \frac{dp_\perp}{v_\perp}. \tag{4.9}$$

If the curve (4.7) is closed, the electron moves periodically in a magnetic field and the period is

$$T_H = -\frac{c}{eH} \oint \frac{dp_\perp}{v_\perp}. \tag{4.10}$$

This expression can be transformed using the circumstance that $\mathbf{v}_\perp$ (which is the vector whose components are v_x and v_y) is normal to the planar curve of Eq. (4.7). We shall use $S(\varepsilon, p_z)$ to denote the area bounded by the curve (4.7) in the case when this curve is closed and nonintersecting.

Since

$$S(\varepsilon, p_z) = \int\int dp_x\, dp_y = \int d\varepsilon \oint \frac{dp_\perp}{v_\perp},$$

it follows that

$$\oint \frac{dp_\perp}{v_\perp} = \frac{\partial S}{\partial \varepsilon},$$

and hence

$$T_H = -\frac{c}{eH} \frac{\partial S}{\partial \varepsilon}. \tag{4.11}$$

The quantity $(1/2\pi)\, \partial S/\partial \varepsilon = m^*$ can be called the effective mass of an electron in a magnetic field. Here, m^* is a function of ε and p_z, which remain constant in the magnetic field, and this makes m^* much more convenient than the tensor of the reciprocal effective masses, which generally depends on all the components of the vector $\mathbf{p}$.

The period and the frequency of the Larmor or cyclotron revolution of an electron are given by expressions similar to those for the period and the frequency of revolution of a free electron

$$T_H = \frac{2\pi m^* c}{|e| H}; \qquad \omega_H = \frac{|e| H}{m^* c}. \tag{4.12}$$

The effective mass of a free electron is $m^* = m$. If $\varepsilon = p^2/2m$, it follows that Eq. (4.7) represents a circle whose area is $S = \pi(2m\varepsilon - p_z^2)$. Hence, $\partial S/\partial \varepsilon = 2\pi m$.

Since the effective mass m^* depends on ε and p_z, this mass is naturally different for different conduction electrons. Therefore, in contrast to the free-electron gas, a gas of conduction electrons does not rotate in a magnetic field with a velocity common to all the electrons. Different electrons perform different periodic motions. Those electrons that move along open trajectories in planes perpendicular to the magnetic field (in the momentum space) may go to infinity.

We have ignored the direction of motion of electrons which is obviously determined by the relative directions of the vectors $\mathbf{v}_\perp$ and $\mathbf{p}_\perp$. Using Eq. (4.6), we can easily show that the direction of motion is related to the sign of the effective mass: if $m^* > 0$ the electron moves along a left-handed helix (like a free electron) but if $m^* < 0$ it moves along a right-handed helix in the same manner as a positively charged particle. The effective mass has the same sign as the derivative $\partial S/\partial \varepsilon$ and this sign obviously depends on whether the energy within the surface $\varepsilon(p) = \varepsilon$ is smaller or larger than ε. In the former case, the derivative and the effective mass m^* are positive, whereas in the latter case they are negative.

We must stress that the concept of an effective mass in a magnetic field associated with the period of revolution along an orbit cannot be introduced for open trajectories. However, we can establish a rule for the determination of the direction of motion of an electron which will apply to the closed and open trajectories: the electron moves in such a way that the region to the right of its direction of motion has a lower energy and this applies to every point on the trajectory.

We have considered so far the motion of an electron in the momentum space. It follows from the Lorentz equation that the trajectory of an electron in the momentum space is closely related to the projection of the electron trajectory in the coordinate space onto a plane perpendicular to the magnetic field. The expressions in Eq. (4.6) show that the velocity of an electron at each moment is perpendicular to the velocity represented in the momentum (p) space. This means that the projection of the electron trajectory in the coordinate space onto a plane perpendicular to the magnetic field can be found from its trajectory in the momentum space by rotating it through an angle of $\pi/2$ and altering the scale, which has to be multiplied by c/eH. In particular, this means that the period and the frequency of motion of an electron in the coordinate space are T_H and ω_H.

The motion of a conduction electron in a magnetic field has certain interesting features in those cases when the constant-energy surface is open. For example, in this case, the trajectory of an electron in the momentum space can be open. This means that in the coordinate space the same electron escapes to infinity in a plane perpendicular to the magnetic field.

The basic difference between a free electron and an electron in a crystal becomes manifest also in those few cases when the electron trajectory in a magnetic field is closed but passes through a saddle point of the constant-energy surface. At such a point, the projection of the velocity onto a plane perpendicular to the normal is zero. Therefore, when an electron moves along a trajectory (4.7) which passes formally through a saddle point, we find that the electron can only approach this point asymptotically. An electron actually located at the saddle point cannot move at all. It is therefore clear that the electrons which move along trajectories passing near a saddle point spend quite a long time near this point, which means that the period is determined mainly by the nature of the motion of such an electron near the saddle singularity. The equation of the trajectory of an electron in a magnetic field (in the momentum space) near a saddle point can be expressed in the following way if the axes are chosen suitably:

$$\varepsilon - \varepsilon_0(p_z) = \frac{1}{2}\left(\frac{p_1^2}{m_1} - \frac{p_2^2}{m_2}\right). \tag{4.13}$$

Here, $\varepsilon_0(p_z)$ is the value of the energy at the saddle point; the positions of axes 1 and 2 are shown in Fig. 15; the p_z axis is, as usual, directed along the magnetic field; $1/m_1$ and $1/m_2$ are the values of the corresponding second derivatives of ε with respect to $\mathbf{p}$ at the saddle point (m_1, $m_2 > 0$). Since, over much of the trajectory, the motion proceeds at a finite velocity and near the point denoted by A the velocity is infinitesimally small, it follows that

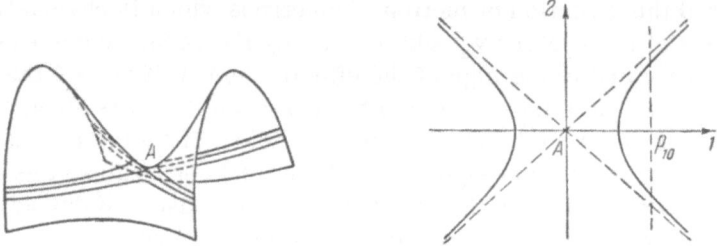

Fig. 15. Trajectory of an electron near a saddle point
A, where its velocity is zero.

$$T_H \cong \frac{c}{|e|H} \int_{-p_0}^{p_0} \sqrt{\frac{1+\left(\frac{dp_1}{dp_2}\right)^2}{v_1^2 + v_2^2}}\, dp_2 = \frac{2c}{|e|H} \sqrt{\frac{m_1+m_2}{m_2}} \int_0^{p_0} \frac{dp_2}{\sqrt{\frac{p_1^2}{m_1^2} + \frac{p_2^2}{m_2^2}}} \qquad (4.14)$$

(p_0 is the value of p_2 at which the trajectory moves sufficiently far away from the saddle point; see Fig. 15). Expressing p_1 in terms of $\Delta\varepsilon = \varepsilon - \varepsilon_0(p_z)$ and integrating, we obtain:

$$T_H \cong \frac{2c}{|e|H} (m_1 m_2)^{1/2} \left| \ln \frac{\Delta\varepsilon}{\varepsilon_0} \right|. \qquad (4.15)$$

Thus, the period increases logarithmically when the trajectory approaches a saddle point. It is clear from the derivation of Eq. (4.15) that the approach to such a singularity may result from a change in the energy ε or a change in the projection of the momentum p_z (which are the cases considered here), as well as from a change in the direction of the magnetic field.

The logarithmic increase in the period of revolution can be described by introducing an effective mass, which can become infinite:

$$m^* = \frac{1}{\pi} (m_1 m_2)^{1/2} \ln \left| \frac{\varepsilon_0}{\Delta\varepsilon} \right|. \qquad (4.15a)$$

The motion along a trajectory passing through a saddle point is similar to infinite motion: a particle takes an infinite time to traverse such a trajectory. The trajectories considered can be considered as the boundaries between trajectories of different types. Their "boundary" role is particularly important in the motion of an electron in a slowly varying magnetic field (this will be considered in the next section).

Open trajectories or open sections are possible only for electrons on open constant-energy surfaces. However, even in the case of open surfaces, the sections may be open or closed, depending on the direction of the magnetic field and the value of p_z. We shall now determine the law which predicts an infinite period in the case when the parameters of motion are altered in such a way that a closed trajectory becomes open. For clarity, we shall consider a "corrugated" cylinder surface, which is a good approximation to any open surface along an open direction. Moreover, to be specific, we shall assume that the direction of the magnetic field may vary. When the $\varphi = \pi/2 - \theta$ (θ is the angle between the cylinder axis and the magnetic field) is reduced, the period for a smooth (not corrugated) cylinder increases in inverse proportion to $\sin \varphi$ or, near an open trajectory, in proportion to φ. However, the presence of corrugations basically alters the situation. Now, when the angle θ approaches 90°, the trajectory must pass through saddle points (Fig. 16). On approaching a saddle point, the period increases logarithmically to infinity. The angular dependence of the period now becomes quite complex:

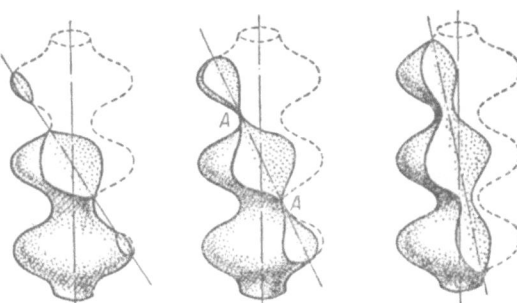

Fig. 16. Changes in the nature of the trajectory of an electron with the angle between the direction of the magnetic field and the axis of a corrugated cylinder. The saddle points are denoted by A.

it is shown schematically in Fig. 17. The points at which the period becomes infinite correspond to those angles (for fixed values of ε and p_z) at which the trajectory passes through a saddle point.

It is worth mentioning an interesting feature of the motion of an electron in a magnetic field near a conical point on a self-intersecting constant-energy surface. Using Eq. (2.9), we can show [11] that the period and the effective mass, tend to zero on approaching a conical point. The magnetic field should not deviate too much from the axis of the cone. Several effects should follow from the vanishing of the effective mass, i.e., from the infinite value of the Larmor frequency.

It is known that slow variation of the conditions of motion does not alter the adiabatic invariants. Therefore, in the case of electrons with closed orbits, the motion in a plane perpendicular to a magnetic field is periodic and the adiabatic invariant is the integral $I = \frac{1}{2\pi} \oint P_\perp \, dr$, calculated for the first period of motion, where $P_\perp$ is the projection of the generalized momentum onto the plane of motion [Eq. (1.36) in §1].

A closed expression for the adiabatic invariant can be obtained for any dispersion law. We shall assume that the vector potential is

$$A_x = -Hy; \quad A_y = A_z = 0. \tag{4.16}$$

Hence

$$P_x = p_x - \frac{eH}{c} y, \quad P_y = p_y; \quad P_z = p_z. \tag{4.17}$$

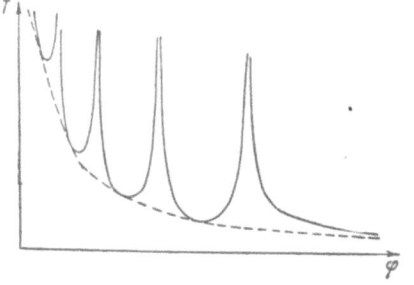

Fig. 17. Dependence of the period of revolution of an electron on the angle between the direction of the magnetic field and the axis of the cylinder shown in Fig. 16. The dashed curve represents the corresponding dependence for a smooth cylinder.

When the vector potential is selected in the form of Eq. (4.16), the coordinate x is cyclic and, therefore, P_x = const and dy = (c/eH) dp_x. Using these relationships, we obtain easily

$$I = \frac{1}{2\pi} \frac{c}{eH} \oint p_y \, dp_x.$$

The integral in the above equation defines an area $S(\varepsilon, p_z)$ of a section of a constant-energy surface by a plane perpendicular to the magnetic field. Therefore,

$$I = \frac{cS(\varepsilon, p_z)}{2\pi eH}. \tag{4.18}$$

For a square isotropic dispersion law, this area is $S(\varepsilon, p_z) = \pi p_\perp^2$ and

$$I = \frac{cp_\perp^2}{2eH}. \tag{4.19}$$

The adiabatic invariant is identical, to within a constant factor, with the projection of the magnetic moment resulting from the orbital motion onto the magnetic field (this projection is averaged out over the trajectory).[3] If we use the bar to denote the averaging and employ the similarity of the trajectories in the momentum and the coordinate spaces, we find that

$$M_z = \frac{e}{2c} \overline{[\boldsymbol{r}, \, \boldsymbol{v}]}_z = \frac{e}{2c} \overline{(xv_y - yv_x)} = -\frac{1}{2H} \overline{(v_x p_x + v_y p_y)}.$$

On the other hand, we can easily show (for details see §27) that

$$\overline{p_x v_x} = \overline{p_y v_y} = \frac{S(\varepsilon, \, p_z)}{2\pi m^*}. \tag{4.20}$$

Finally, we obtain

$$M_z = \frac{e}{2m^* c} I. \tag{4.21}$$

It is worth stressing that the magnetic moment is defined in such a way that the coordinate is measured from the "center" of an orbit.

It follows from our discussion that $S(\varepsilon, p_z)$, which is the area of a section of a constant-energy surface by a plane p_z = const, is an important quantity in the dynamics of the motion of a particle in a magnetic field. We shall need later several formulas which include the derivatives of $S(\varepsilon, p_z)$.

We shall now calculate the derivative of this area with respect to the energy: $(\partial S/\partial p_z)_\varepsilon$. By definition,

$$\left(\frac{\partial S}{\partial p_z}\right)_\varepsilon = \oint \left(\frac{\partial p_y}{\partial p_z}\right)_{p_x', \, \varepsilon} dp_x'.$$

On the other hand, by equating the differential $d\varepsilon$ to zero (i.e., by assuming that ε = const), we find that $(\partial p_y/\partial p_z)_{p_x, \varepsilon} = -v_z/v_y$. Thus,

$$\left(\frac{\partial S}{\partial p_z}\right)_\varepsilon = -\oint \frac{v_z}{v_y} dp_x. \tag{4.22}$$

---

[3] Equations (4.21)-(4.25) will be derived using the results reported in [12].

By definition, the area $S(\varepsilon, p_z)$ depends on the direction of the magnetic field $\xi = \mathbf{H}/H$. We can show that the change in the cross-sectional area δS resulting from a change in the direction of the field by an amount $\delta \xi$ is given by the formula:

$$\delta S = \oint \frac{v_z p \, d\xi}{v_y} \, dp_x. \tag{4.23}$$

In fact,

$$\left(\frac{\partial S}{\partial \xi_i} \right)_{\varepsilon, \, p_z} = \oint \left(\frac{\partial p_y}{\partial \xi_i} \right)_{p'_x, \, \varepsilon, \, p_z} dp'_x. \tag{4.24}$$

We shall now use the circumstance that the dependence $\varepsilon = \varepsilon(p_x, p_y, p_z, \xi)$ on ξ is related only to the special selection of our coordinate system: the z axis is always directed along the magnetic field. Therefore, a change to a new direction $\xi + d\xi$ corresponds to a rotation of the p space by an "angle" $\delta\varphi$;

$$\delta p = [p, \ \delta \varphi],$$

and $\delta\varphi$ is related to $d\xi$ by a similar expression

$$d\xi = [\xi, \ \delta \varphi].$$

From the constancy of the energy

$$d\varepsilon = v_x \, dp_x + v_y \, dp_y + v_z \, dp_z + \frac{\partial \varepsilon}{\partial \xi_i} \delta \xi_i$$

and from the fact that $\frac{\partial \varepsilon}{\partial \xi_i} = \frac{\partial \varepsilon}{\partial p_i} \delta p_i$, we find that

$$\left(\frac{\partial p_y}{\partial \xi_y} \right)_{p_x, \, \varepsilon, \, p_z} = - p_z + \frac{v_z p_y}{v_y},$$

and this gives us Eq. (4.23).

We shall now give, without further proof, another useful relationship:

$$\bar{v}_z = \left(\frac{\partial \varepsilon}{\partial p_z} \right)_S. \tag{4.25}$$

Let us now consider briefly the motion of a conduction electron in crossed electric and magnetic fields, which we shall assume to be mutually perpendicular. It is known that a free electron subjected to crossed fields drifts along a direction perpendicular to both fields. The drift velocity, i.e., the average velocity of the electron, is then given by

$$v_{EH} = \frac{c}{H^2}[\mathbf{E}, \ \mathbf{H}]. \tag{4.26}$$

Strictly speaking, Eq. (4.26) gives the value of the velocity only for the direction perpendicular to the vectors $\mathbf{E}$ and $\mathbf{H}$. Moreover, this free electron can move along the magnetic field since the projection of the momentum (and of the velocity) along the magnetic field is conserved. Therefore, the free electron drifts at an angle to the magnetic field in a plane perpendicular to the electric field. Let us now consider the motion of an electron subject to an arbitrary dispersion law. Using Eq. (4.26), we can rewrite Eq. (4.1) in the form

$$\frac{d\boldsymbol{p}}{dt} = \frac{e}{c}\,[v - \boldsymbol{v}_{EH},\ \boldsymbol{H}].\tag{4.27}$$

Averaging this equation with respect to time, we find that in those cases when the trajectory in the momentum space is closed dp/dt = 0, the average velocity in a plane perpendicular to the magnetic field is equal to v_{EH}, as in the case of a free electron, i.e.,

$$\bar{\boldsymbol{v}} = \boldsymbol{v}_{EH} + \boldsymbol{v}_{\parallel};\quad \boldsymbol{v}_{\parallel} = \frac{\boldsymbol{H}\,(v\boldsymbol{H})}{H^2}\,.$$

However, in those cases when the trajectory is open, $\overline{d\boldsymbol{p}}/dt \neq 0$, and $v_{\perp} \neq v_{EH}$. We must stress that we are not speaking of a trajectory in the magnetic field but of one in crossed fields.

It is evident from Eq. (4.27) that the motion of a particle whose dispersion law is $\varepsilon = \varepsilon(\boldsymbol{p})$ and which is subjected to crossed electric and magnetic fields can be treated in the same way as the motion of a particle subjected solely to a magnetic field but which has a different dispersion law:

$$\varepsilon^{*}(\boldsymbol{p}) = \varepsilon(\boldsymbol{p}) - \boldsymbol{v}_{EH}\boldsymbol{p}.\tag{4.28}$$

In this case, the equations of the trajectory in the momentum space are

$$\varepsilon(\boldsymbol{p}) - \boldsymbol{v}_{EH}\boldsymbol{p} = \text{const};\quad p_z = \text{const}.\tag{4.29}$$

We can easily see that, even when the trajectory in a magnetic field is closed, the trajectory of Eq. (4.29) in crossed fields may be open.[4] Knowing the trajectory in the momentum space, we can easily plot the trajectory of a conduction electron in the coordinate space if we note that one of the consequences of the law of conservation of energy

$$\frac{d\varepsilon}{dt} = v\,\frac{d\boldsymbol{p}}{dt} = ev\boldsymbol{E}$$

is that the velocity $\mathbf{v}$ is always perpendicular to the vector dp*/dt, where $\mathbf{p^{*}} = \mathbf{p} - e\mathbf{E}t$. In those cases when the electron trajectory in the momentum space is closed, we can easily determine the period of revolution of an electron T_{EH}. It follows from Eq. (4.28) that, as in the absence of an electric field,

$$T_{EH} = -\frac{c}{eH}\,\frac{\partial S^{*}}{\partial \varepsilon^{*}}\,.$$

Here, S* is an area bounded by a curve defined by the two expressions in Eq. (4.29). In general, the value of $\partial S^{*}/\partial \varepsilon^{*}$ depends on the electric field. It is interesting to note that this dependence vanishes if the dispersion law is quadratic. Since the electric field is practically always weak compared with the magnetic field,[5] we shall be interested (and then only in the case of semiconductors and Bi-type semimetals) in the change of the period under the action of a weak electric field. This change should be calculated retaining terms quadratic in the electric field (the linear terms vanish in the calculation of specific effects because of the averaging over all the electrons).

---

[4] Only the presence of open trajectories can explain the fact that in crossed fields the total current of a completely filled energy band is zero.

[5] We note that the condition E/H ≪ 1 is an essential condition for the use of the classical non-relativistic mechanics (see, for example, [13]).

We must stress that the static and the high-frequency electrical conductivities of a metal in a magnetic field can be calculated regarding an electric field as a perturbation. Therefore, the macroscopic properties of a metal in crossed fields are determined by the dynamics of conduction in the magnetic field.

§5. Motion of Conduction Electrons in Nonuniform Fields

In the preceding section, we analyzed the motion of quasi particles with a complex dispersion law in constant and uniform electric and magnetic fields. We found some features which are typical only of particles with a complex dependence of their energy on the crystal momentum. An even greater variety of behavior is observed when particles move in nonuniform fields [14-16].[1]

We shall consider the motion of quasi particles in electric and magnetic fields which vary slowly in time and space. These fields satisfy the conditions

$$T_H \ll T_0; \quad r_H \ll L_0; \quad E \ll \frac{v}{c} H, \tag{5.1}$$

where T_0 and L_0 are, respectively, the characteristic time and length of variation of the fields **E** and **H**. In practice, these conditions are satisfied right up to very high gradients and frequencies (for example, in the case of $H \approx 10^3$ Oe, we can have gradients up to $|\nabla H| \approx 10^5$ Oe/cm and frequencies up to $\omega_0 = 2\pi/T_0 \approx 10^8$ sec^{-1}).

We must stress that, in accordance with Eq. (5.1), the electric field is assumed to be weak. Thus, in the present section, we shall assume that the unperturbed motion is the motion in a constant and uniform magnetic field.

The conditions of Eq. (5.1) allow us to separate the actual motion of a quasi particle into the sum of two motions: one is the fast motion in a moderately strong magnetic field, and the other is the slow variation of the characteristics of the fast motion. The fast motion can best be illustrated in the case of closed trajectories in the momentum space. In this case, the slow motion is the drift, rotation, and deformation of the orbit followed by the fast-moving particle.

If a constant-energy surface is not everywhere convex, it follows that a certain effect, which can be interpreted as the "scattering of a quasi particle by the saddle points of the constant-energy surface," may be observed in an electromagnetic field which varies slowly in space and time. This scattering is not related to the presence of any force center in the **r** space. It is due to the fact that a saddle point is a stationary point in the motion of a quasi particle in a constant and uniform magnetic field. In an electromagnetic field satisfying the conditions of Eq. (5.1), this point separates several regions of the **p** space in which the conditions of motion are radically different. When the path of a particle passes through such a singularity, the conditions of motion change sharply and the initial conditions determine whether a particle then enters a given part of the **p** space. In this case, the physical interest lies in the determination of the probabilities of a given particle entering different regions of the momentum space. Let us consider first the case when a fast-moving particle follows a closed trajectory in the momentum space. The motion of this quasi particle can be examined conveniently by replacing the Cartesian components of the momentum with the quantities p_ξ, ε, and τ, where p_ξ is the projection of the momentum along a unit vector of the magnetic field $\boldsymbol{\xi} = \mathbf{H}/H$, ε is the energy, and τ is the angular variable which determines the position of the particle on the fast-trajectory motion defined by $p_\xi = $ const, $\varepsilon = $ const (see §4).

---

[1] No experimental effects have yet been discovered which manifest these properties. Therefore, on the first reading of this book, the present section may be omitted.

We shall write the functions $\mathbf{r}(t)$, $p_\xi(t)$, and $\varepsilon(t)$ in a form which separates explicitly the fast-motion component:

$$\mathbf{r}(t) = \bar{\mathbf{r}}(t) + \mathbf{\rho}(t); \quad p_\xi(t) = \bar{p}_\xi(t) + \tilde{p}_\xi(t); \quad \varepsilon(t) = \bar{\varepsilon}(t) + \tilde{\varepsilon}(t), \tag{5.2}$$

where a bar over a symbol denotes averaging over a period of fast motion $\bar{x} = \dfrac{1}{T_H} \int\limits_{t}^{t+T_H} x(t')\, dt'$.

The "average" quantities $\mathbf{R}(t) = \bar{\mathbf{r}}(t)$, $P_\xi(t) = \bar{p}_\xi(t)$, and $\mathcal{E}(t) = \bar{\varepsilon}(t)$ define the slow (smooth) motion of the particle; here, $\mathbf{R}$ is the coordinate of the center of the orbit for which the unit vector is $\xi = \mathbf{H}(\mathbf{R}, t)/H(\mathbf{R}, t)$, whereas the quantities $\mathcal{E}(t)$ and $P_\xi(t)$ give the position of the orbit in the $\mathbf{p}$ space. The period $T_H = 2\pi m^* c/eH$ is determined by the instantaneous values of $\mathcal{E}$, P_ξ, and ξ, whereas the functions $\mathbf{\rho}$, $\tilde{p}_\xi$, and $\tilde{\varepsilon}$ are fluctuations whose average values are zero. On the strength of Eq. (5.1) we find that $|\tilde{p}_\xi|$ and $|\tilde{\varepsilon}|$ are considerably smaller than P_ξ and $\mathcal{E}$, whereas $|\mathbf{\rho}|$ is a correction of the order of linear dimensions of the Larmor orbit ($|\mathbf{\rho}| \approx r_H$). The exact equations of motions have their usual form (4.1). Let us now rewrite them in terms of our new variables. For this purpose we multiply

$$\dot{\mathbf{p}} = e\mathbf{E} + \frac{e}{c}[\mathbf{vH}] \tag{5.3}$$

by the unit vector $\xi(\mathbf{r}, t)$ along the magnetic field

$$\xi\dot{\mathbf{p}} = eE_\xi.$$

By definition,

$$\dot{p}_\xi = \frac{d}{dt}(\xi\mathbf{p}) = \dot{\xi}\mathbf{p} + \xi\dot{\mathbf{p}},$$

and

$$\frac{d\xi}{dt} = \frac{\partial \xi}{\partial t} + (\mathbf{v}\nabla)\,\xi.$$

Therefore,

$$\dot{p}_\xi = eE_\xi + \mathbf{p}\left(\frac{\partial \xi}{\partial t} + (\mathbf{v}\nabla)\,\xi\right).$$

Since $\xi(d\xi/dt) = 0$, it follows that

$$\dot{p}_\xi = \mathbf{p}_\perp(\mathbf{v}\nabla)\,\xi + \mathbf{p}_\perp \frac{\partial \xi}{\partial t} + eE_\xi, \tag{5.4}$$

where $\mathbf{p}_\perp = \mathbf{p} - (\mathbf{p}\xi)\,\xi$ is the projection of the vector $\mathbf{p}$ onto a plane perpendicular to the magnetic field $\mathbf{H}(\mathbf{r}, t)$. Equation (5.4) must be supplemented by the law which gives the time dependence of the energy ε:

$$\dot{\varepsilon} = e\mathbf{Ev} \tag{5.5}$$

and by an equation which gives the time dependence of the coordinates of the particle:

$$\dot{\mathbf{r}} = \mathbf{v}_\perp + v_\xi\xi, \quad \mathbf{v}_\perp = \mathbf{v} - (\mathbf{v}\xi)\,\xi. \tag{5.6}$$

Equations (5.4)-(5.6) represent the complete system of the equations of motion of a particle in terms of the new variables.

The quantities which describe the position of a particle ($\mathbf{r}$ and $\mathbf{p}$) satisfy a relationship of the type $\bar{\dot{x}} \cong \dot{\bar{x}}$. To the same degree of accuracy, we can carry out averaging by replacing integration with respect to real time by integration with respect to τ along the contour $\varepsilon = \mathscr{E}$, $p_\xi = P_\xi$ (we can easily show that $dt = d\tau$ to within quantities $\sim cE/vH$; see §27). When we average over τ, we must know the dependence $\mathbf{p} = \mathbf{p}(\tau)$. It can be found from the equations of motion in the average field:

$$\frac{d\mathbf{p}}{d\tau} = \frac{e}{c} [v\mathbf{H}], \qquad \mathbf{H} = \mathbf{H}(\mathbf{R}),$$

where the magnetic field $\mathbf{H}$ is assumed to be constant. Hence

$$v_{\substack{x \\ (y)}} = (\mp) \frac{c}{eH} \frac{\partial p_{y\,(x)}}{\partial \tau}.$$

The x and y axes are taken in a plane perpendicular to the magnetic field. Consequently, $\overline{v}_\perp = 0$. Moreover, $\overline{v_x p_y} = \overline{v_y p_x} = 0$; on the other hand, $\overline{v_x p_x} = \overline{v_y p_y}$ and are equal to $S/2\pi m^*$, where S and $m^* = (1/2\pi)(\partial S/\partial \varepsilon)$ are the functions of $\mathscr{E}$ and P_ξ (see §4).

We shall now perform the averaging in Eqs. (5.4)-(5.6). In Eq. (5.4), difficulty may be encountered only in the averaging of the first term on the right. We shall consider the averaging procedure of this term in detail. We shall select the x, y, and z axes in such a way that $\xi_z = 1$, and $\xi_x = \xi_y = 0$. Then, $\nabla \xi_z = 0$ and

$$\overline{\mathbf{p}_\perp (v\nabla)\,\boldsymbol{\xi}} = \frac{S}{2\pi m^*} \left(\frac{\partial \xi_x}{\partial x} + \frac{\partial \xi_y}{\partial y} \right) + \overline{v_z \left(\mathbf{p}_\perp \frac{\partial \boldsymbol{\xi}}{\partial z} \right)}.$$

We shall use the relationship $\overline{v_x p_x} = \overline{v_y p_y} = S/2\pi m^*$ mentioned in §4. The last term can be transformed by going over from integration with respect to τ to integration with respect to p_x:

$$\overline{v_z \left(\mathbf{p}_\perp \frac{\partial \boldsymbol{\xi}}{\partial z} \right)} = \frac{1}{2\pi m^*} \oint \frac{v_z \left(\mathbf{p}_\perp \frac{\partial \boldsymbol{\xi}}{\partial z} \right)}{v_y}\, dp_x.$$

Using Eq. (4.24), we obtain

$$\overline{v_z \left(\mathbf{p}_\perp \frac{\partial \boldsymbol{\xi}}{\partial z} \right)} = \frac{1}{2\pi m^*} (\boldsymbol{\xi}\nabla)\, S.$$

On the other hand, since div $\mathbf{H} = 0$, we can show that

$$\frac{\partial \xi_x}{\partial x} + \frac{\partial \xi_y}{\partial y} = H (\boldsymbol{\xi}\nabla) \frac{1}{H}.$$

Thus,

$$\overline{\mathbf{p}_\perp (v\nabla)\,\boldsymbol{\xi}} = \frac{H}{2\pi m^*} (\boldsymbol{\xi}\nabla) \frac{S}{H} \tag{5.7}$$

and the averaged form of Eq. (5.4) becomes

$$\dot{P}_\xi = \frac{H}{2\pi m^*} (\boldsymbol{\xi}\nabla) \frac{S}{H} + \overline{\mathbf{p}}_\perp \frac{\partial \boldsymbol{\xi}}{\partial t} + eE_\xi. \tag{5.8}$$

Let us now consider Eq. (5.5). The electric field $\mathbf{E}(\mathbf{r})$ is approximately equal to $\mathbf{E}(\mathbf{R})$ + $(\rho\nabla)\mathbf{E}$. Bearing in mind that $\bar{v}_x = \bar{v}_y = 0$, we find that

$$\overline{e\mathbf{E}(\mathbf{r})\,\mathbf{v}} = eE_\xi \bar{v}_\xi + e\overline{\tilde{v}_i \rho_k}\,\frac{\partial E_i}{\partial x_k}.$$

Since $\tilde{v}_i = \dot{\rho}_i$, we can integrate by parts and show easily that $\overline{\tilde{v}_i\rho_k}$ is an antisymmetric tensor. Therefore,

$$\overline{\tilde{v}_i\rho_k}\,\frac{\partial E_i}{\partial x_k} = \frac{1}{2}\,\overline{\tilde{v}_i\rho_k}\left(\frac{\partial E_i}{\partial x_k} - \frac{\partial E_k}{\partial x_l}\right) = \frac{1}{2}\,\overline{[\rho\tilde{v}]}_i\,(\operatorname{rot}\mathbf{E})_i.$$

However, according to Maxwell's equations

$$\operatorname{rot}\mathbf{E} = -\frac{1}{c}\,\frac{\partial\mathbf{H}}{\partial t},$$

and $(e/2c)\overline{[\rho,\tilde{\mathbf{v}}]} = \mathbf{M}$ is, as already mentioned, the average magnetic moment of the particle path relative to the "orbit center" $\mathbf{R}$. Therefore

$$\dot{\mathscr{E}} = eE_\xi\bar{v}_\xi - \mathbf{M}\,\frac{\partial\mathbf{H}}{\partial t}. \tag{5.9}$$

Equation (5.6) is averaged automatically:

$$\dot{\mathbf{R}} = \bar{v}_\xi\boldsymbol{\xi}. \tag{5.10}$$

Equations (5.8)-(5.10) represent the complete system of the averaged equations of motion of an electron. The right-hand sides of these equations consist of functions which depend only on P_ξ, $\mathscr{E}$, and $\mathbf{R}$. To find the explicit forms of these functions, we can sometimes use the formulas deduced in §4.

The system of equations (5.8)-(5.10), supplemented by the relationships

$$\bar{v}_\xi = -\frac{1}{2\pi m^*}\,\frac{\partial S}{\partial P_\xi}; \qquad \frac{\partial S}{\partial\xi} = \oint\frac{p_\perp v_\xi}{v_\perp}\,dp_l,$$

where integration is carried out along the average trajectory, can be used to demonstrate easily that the ratio $S(P_\xi, \uparrow, \xi)/H(\mathbf{R}, t)$ is an integral of motion. This shows, in particular, that the nature of the adiabatic invariant in a nonuniform magnetic field is the same as in a constant field [see Eq. (4.18)].

Another important property of motion in an electromagnetic field which varies slowly in space and time is the fact that the velocity of the center of the orbit is parallel to the magnetic field direction, as shown by Eq. (5.10).

The knowledge of the averaged equations allows us to investigate completely the nature of the motion and the deformation of the particle path in slowly varying fields. The nature of such motion is particularly clear in the simplest case when a static magnetic field is a slow function of the coordinates. An electron then moves on a constant-energy surface in such a way that the ratio S/H remains constant. The center of the particle path moves along a line of force. In particular, we may observe interesting oscillations which are analogous to the oscillations of particles in a magnetic trap. However, in a

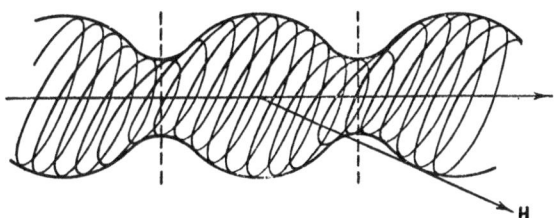

Fig. 18. Trajectory of an electron in a nonuniform magnetic field.

metal, periodic motion need not be due to a special magnetic field distribution but to the actual nature of the dispersion law, because of which S may become a periodic function [14] (Fig. 18).

If the equations p_ξ = const, ε = const describe an open trajectory, we must distinguish the following two cases.

1. In this case, the trajectory is a periodic curve with the "open" direction parallel to some vector in the reciprocal lattice. The principal difference from the case of closed sections is the fact that the velocity is not directed along a magnetic line of force, so that S/H is not an adiabatic invariant. However, the general nature of the motion resembles very closely the motion along a closed trajectory. Thus, the period is now the time in which an electron crosses one unit cell in the reciprocal lattice.

2. In the second case, the trajectory is not a periodic curve: the "open" direction is not parallel to any of the reciprocal lattice vectors. Averaging along an aperiodic trajectory involves integration along an open trajectory which passes through an infinite number of unit cells[2] and which can be replaced by the sum of the integrals over the equivalent parts within one cell (Fig. 19). In the case of an aperiodic trajectory, these parts fill completely a unit cell. It follows that the average quantities characterizing slow motion cannot depend on p_H, the projection of the momentum along the magnetic field. Naturally, the dependences of the average quantities on p_H can be neglected also in the case of periodic trajectories if the period exceeds considerably $\hbar/a$ (a is the average distance between atoms). This circumstance (the independence of the average values of p_H) alters significantly the nature of the slow motion (for a detailed consideration of this point see [15]).

The most interesting situation arises when a transition occurs from one set of conditions to another during motion in a field which varies slowly in space and time. As mentioned earlier, we encounter, in this case, a phenomenon which resembles very closely the scattering by a force center [15, 16].

The regions in the **p** space characterized by different conditions of motion are separated by parts of a self-intersecting trajectory which is formed when the surface ε = const is cut by

---

[2]In fact, the derivation of the equations describing the slow motion (such as drift) is related to the averaging over a time interval T satisfying the inequality $T_0 \ll T \ll \tau$, where T_0 is the transit time of a particle across a unit cell and τ is the time taken to travel one mean free path. However, in averaging finite quantities, the integration procedure can be extended to the whole trajectory because difference

$$\lim_{T' \to \infty} (2T')^{-1} \int_{-T'}^{T'} f(t')\, dt' - \frac{1}{T} \int_{t}^{t+T} f(t')\, dt'$$

is a rapidly oscillating function of the order of $f \cdot T_0/T$.

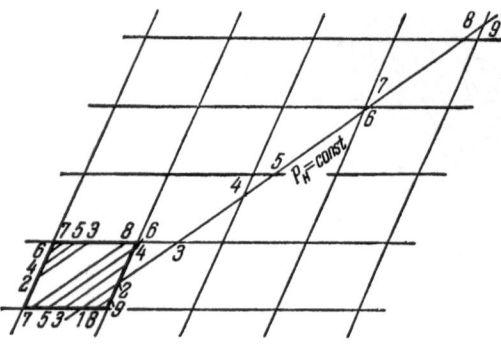

Fig. 19. One of the crystallographic planes in the reciprocal lattice cut by the plane p_H = const (equivalent points are denoted by the same numbers).

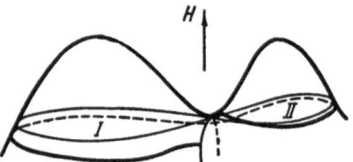

Fig. 20. Motion of an electron in a nonuniform magnetic field near a saddle point of a constant-energy surface (splitting of a "current lobe").

a plane tangent to the hyperbolic points on the constant-energy surface. These points are the stationary points which occur in the motion in a uniform and constant magnetic field. The period of motion of a particle diverges logarithmically in the limit $p_H, \rightarrow p_H^k$, where p_H^k is the vector in the **p** space drawn through a singularity [see Eqs. (4.15) and (4.15a)].

The essence of the scattering phenomenon can be illustrated best by considering the motion of a particle in a weakly nonuniform but constant (independent of time) magnetic field which has a straight line of force. Let F be a saddle point on the constant-energy surface $\varepsilon = \varepsilon_0$ and let the normal to the surface at this point be parallel to **H**. The cross section $\varepsilon = \varepsilon_0$, $p_H = p_H^k$ is a "figure of eight," whose point of intersection coincides with F. If a "current lobe" comes in contact with the constant-energy surface at the saddle point, it splits into two lobes, which are denoted by I and II in Fig. 20 and are separated by the singularity. The conditions of motion in regions I and II are quite different. Depending on the exact "microscopic" initial conditions, a particle may enter either lobe I or lobe II. These initial conditions alternate in such a way that in each macroscopic element of a constant-energy surface, defined by the average values of the "coordinates," there are points from which a particle may enter the first or the second region. Therefore, we can regard the entry of a particle into either of these regions as a random process and we can speak of the scattering of particles near a singularity: the probabilities of scattering into regions I and II (W_I and W_{II}) have definite values.

In order to determine these probabilities, we shall consider a classical ensemble of particles distributed in accordance with some impact parameter, which we shall define later. In the last revolution before reaching region I or II, each particle intersects the principal curvature line passing through the point F at some value $p_H(0)$. After a complete revolution along one of the loops of the "figure of eight," a particle is again in the vicinity of the point of intersection. Depending on the sign of the difference $p_H(t) - p_H^k$ at this moment, the particle enters region I or II and the value of $p_H(t) - p_H^k$ is defined uniquely by the value of $p_H(0)$ at the beginning of the revolution. It follows that $p_H(0)$ can be used as the initial impact parameter. Regions I and II correspond to the intervals δ_I and δ_{II} of the values of $p_H(0)$ which define the scattering of particles into these regions. The scattering probabilities W_I and W_{II}, i.e., the relative numbers of particles reaching regions I and II, are proportional to the respective fluxes of the particles corresponding to δ_I and δ_{II}. For a sufficiently smooth distribution function, these fluxes are proportional to the magnitudes of the intervals themselves (this is justifiable as a first approximation).

The intervals δ_I and δ_{II} are defined by

$$p_H(t) = p_H(0) + \int_0^t \dot{p}_H \, dt'.$$

Hence, we find that

$$\delta_I = \int_0^{T_1} \dot{p}_H \, dt', \qquad \delta_{II} = \int_{T_1}^{T_2} \dot{p}_H \, dt' \tag{5.11}$$

(here, T_1 is the period of motion around the first loop of the "figure of eight" and T_2 is the period of motion around the whole "figure of eight").

The main contribution to p_H during one revolution along the "figure of eight" contour is made by those parts of the trajectory which are far from the singularity. This is because, in the case considered, $\dot{p}_H = 0$ at the singularity.

Using the equation of motion for a "current lobe," we can show that the expressions in Eq. (5.11) yield [15]:

$$W_I/W_{II} = \delta_I/\delta_{II} = S_I/S_{II}, \tag{5.12}$$

where S_I and S_{II} are the areas of the loops of the "figure of eight." It follows from Eq. (5.12) that

$$W_I = \frac{S_I}{S_I + S_{II}}; \qquad W_{II} = \frac{S_{II}}{S_I + S_{II}}.$$

In general, there are several types of transition from one set of conditions to another, including transitions from a closed to an open trajectory (periodic or aperiodic), and conversely. In the case of motion in an arbitrarily weak nonuniform electromagnetic field which varies slowly with time, the probabilities of a given set of transitions are somewhat more complex than those in the case just discussed. Lifshits et al. [15] used a general formula to deduce compact expressions for the probabilities in the most interesting special cases. Thus, for example, in the case of motion in a weak electric field, which is parallel to a constant and uniform magnetic field, we obtain

$$\frac{W_I}{W_{II}} = \left(\frac{dS_I/dp_H}{dS_{II}/dp_H} \right)_{p_H = p_H^k} \tag{5.13}$$

§6. Collisions of Quasiparticles. Scattering Processes

In the preceding sections, we considered the classical motion of a conduction electron in electric and magnetic fields. We assumed that the fields are such that the characteristic dimensions of a quasiparticle trajectory are much larger than the atomic spacings: this made it possible to speak of a definite trajectory of a conduction electron. However, a conduction electron can move in other ways. Interactions with various inhomogeneities whose dimensions are of the order of the unit cell size a may alter radically the state of an electron (its momentum and energy) over distances of the order of a. In such cases, we speak of the scattering resulting from the collisions of an electron with local irregularities in the periodicity of a crystal. Since the de Broglie wavelength of a conduction electron is of the order of the atomic spacings, the problem of the scattering of an electron by a local inhomogeneity in a crystal is, essentially, of a quantum-mechanical nature. However, some important conclusions on the scattering of an electron can be drawn simply from the laws of conservation. The problem of the effective scattering cross section cannot be solved in its general form: the cross section depends strongly on the structure of the local inhomogeneity, i.e., on the nature of the energy of interaction between an electron and an inhomogeneity.

Even in classical mechanics, the effective scattering cross section of a particle can be calculated completely only in the simplest case of a central field. The anisotropy of a crystal makes this problem much more difficult.[1]

Usually, the local inhomogeneities in a crystal are impurity atoms (sometimes isotropic impurities) or defects such as vacancies or dislocations. Their mobility is much lower than the mobility of electrons because these inhomogeneities are heavier. Therefore, the scattering by such inhomogeneities may be regarded as the scattering by a force center. If the collision in question is elastic (we shall restrict our treatment to elastic collisions), the crystal momentum of an electron is altered but its energy is conserved. The complex dispersion law results in a characteristic effect: the scattering angle is determined by the direction of the velocity of an electron (but not its momentum) after the collision. If the constant-energy surface is convex, each value of the velocity corresponds to one value of the momentum (Fig. 21 shows how to find this momentum). However, if the geometry of the constant-energy surface is more complex, there must be directions along which the scattering has a property not encountered in the free-particle case: the same scattering angle corresponds to several values of the momentum of the scattered particle (Fig. 22). The actual momentum of the scattered electron can then be determined only by an analysis of the scattering event itself, i.e., by an analysis of the nature

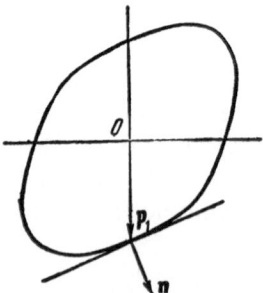

Fig. 21. Determination of the momentum
of an electron scattered along the direction
n. The reference plane is in contact with
the constant-energy surface $\varepsilon = \varepsilon(p)$ at
the point denoted by p_1.

---

[1] The quantum-mechanical problem of the scattering of particles with an arbitrary dispersion law is considered in §9.

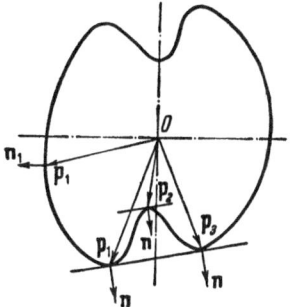

Fig. 22. Determination of the momentum of electrons scattered along the directions **n** and $\mathbf{n}_1$. We can see that the more complex geometry of the constant-energy surface results in the appearance, for some directions, of several values of **p**, all of which correspond to the same scattering direction.

of the motion of an electron in the field of an impurity. We must stress that such an analysis is complicated additionally by the periodic nature of the dispersion law because the result of the elastic scattering of an electron may be the transfer of the electron to a different sheet of the constant-energy surface, located in one of the neighboring cells in the momentum space.[2] In the quasiclassical approach, we can determine exactly the final state of an electron by specifying precisely its initial conditions. In most cases, this is unnecessary and we shall calculate only the effective scattering cross section. However, in some physical problems (ionic transport, electromechanical effects, etc.), we must consider the problem of the change in the crystal momentum resulting from a single scattering event [17].

If a particle is scattered through a small angle, we can find an analytic expression for the change in the momentum due to the scattering event without making any special assumptions about the nature of the dispersion law [17]. In such a case, calculations can be carried out quite satisfactorily using the classical equations of motion since the scattering through small angles takes place at relatively large values of the impact parameter ρ ($\rho \gg a$) when the quasiclassical criteria are satisfied by conduction electrons (§1). We shall make one simplifying assumption by postulating that the force acting on an electron is central. Then, the change in the crystal momentum in a scattering event is

$$\boldsymbol{p}' - \boldsymbol{p} = \boldsymbol{n}\delta_{\parallel} + \boldsymbol{v}\delta_{\perp}, \tag{6.1}$$

where $\delta_{\parallel}$ and $\delta_{\perp}$ are the changes in the momentum parallel to the velocity $\mathbf{n} = \mathbf{v}/v$ and at right-angles to this direction $\nu = \rho/\rho$. Since the energy of an electron suffering an elastic collision is conserved, it follows that

$$\varepsilon\left(\boldsymbol{p} + \boldsymbol{n}\delta_{\parallel} + \boldsymbol{v}\delta_{\perp}\right) = \varepsilon\left(\boldsymbol{p}\right). \tag{6.2}$$

Expanding ε as a series in $\delta_{\parallel}$ and $\delta_{\perp}$, we obtain the following expression, which is accurate up to second-order terms:

$$0 = v\left(\boldsymbol{n}\delta_{\parallel} + \boldsymbol{v}\delta_{\perp}\right) + \frac{1}{2}\left(m^{-1}\right)_{ik}\left(n_i\delta_{\parallel} + v_i\delta_{\perp}\right)\left(n_k\delta_{\parallel} + v_k\delta_{\perp}\right), \tag{6.3}$$

where $(m^{-1})_{ik} = d_2\varepsilon/dp_i dp_k$ is the tensor of the reciprocal effective masses (§4). The scalar product of the vectors **v** and ν vanishes and $\mathbf{v} \cdot \mathbf{n} = v$; the terms containing $\delta_{\parallel}^2$ and $\delta_{\parallel}\delta_{\perp}$ can be

---

[2] In quantum-mechanical language, this means that the scattering is accompanied by an umklapp process (this point will be discussed later).

dropped because they are small compared with the term linear in δ. Therefore,

$$v\delta_\parallel = -\frac{1}{2}(m^{-1})_{ik}\nu_i\nu_k\delta_\perp^2. \tag{6.4}$$

When one considers an electron flux, it is usually necessary to know the average change in the momentum. Averaging Eq. (6.4) over the azimuthal angle and noting that $\overline{\nu_i\nu_k} = 1/2\delta_{ik}$ (i, k = x, y; the z axis is directed along the electron trajectory), we obtain

$$\delta_\parallel = -\frac{1}{4v}(m_{xx}^{-1} + m_{yy}^{-1})\delta_\perp^2 (\rho). \tag{6.5}$$

We note that $\delta_\parallel$ is the change in the average momentum of unit electron flux with the impact parameter ρ, taken per unit time. The quantity $\delta_\perp$ can be found directly from the equations of motion

$$\delta_\perp = \int_{-\infty}^{\infty} F_\rho\, dt = \frac{1}{v}\int_{-\infty}^{\infty} F_\rho\, dz \tag{6.6}$$

or, with the same degree of accuracy, from

$$\delta_\perp = -\frac{\rho}{v}\int_{-\infty}^{\infty}\frac{\partial V}{\partial z}\frac{dz}{r}; \quad r = \sqrt{z^2 + \rho^2}. \tag{6.7}$$

The other limiting case which can be analyzed quite simply is the head-on collision of an electron with an impurity. Even in this very simple case, an electron may be scattered from one sheet of a constant-energy surface to another. If the dispersion law of an electron is

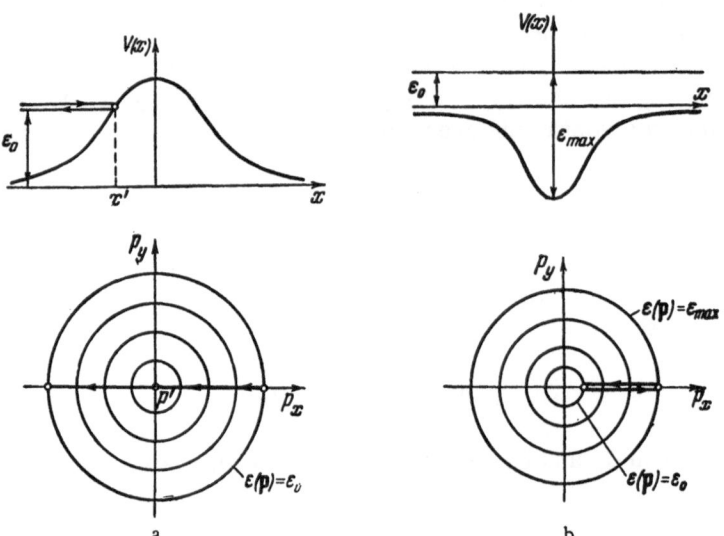

Fig. 23. Scattering of an electron with a quadratic dispersion law by: a) a repulsive potential; b) an attractive potential. Here, ε_0 is the kinetic energy of an electron at infinity; x' is the turning point in the **x** space (it corresponds to a point p' in the momentum space). In the first case (a), the electron is transferred to the opposite side of the constant-energy surface, whereas in the second case (b) it returns to the same point in the **p** space.

$\varepsilon = p^2/2m$ (this applies to an electron in vacuum), a head-on collision may have one of the following two consequences: if the electron has a turning point in the coordinate space, it is transferred to the opposite point on the constant-energy sphere (Fig. 23a); if the electron has no turning point, it returns to the same point on the constant-energy sphere at which the scattering process started (Fig. 23b). In the case of an electron whose dispersion law is periodic, the situation is more complex because such an electron may be reflected not only at a point at which the energy has its minimum value but also at a point at which it has its maximum value, because the velocity vanishes at both points. Figure 24 shows several cases of motion of an electron in the field of an impurity. For simplicity, a very elementary dispersion law has been selected. When we analyze the motion in the momentum space, we must bear in mind that, in those cases when there is a turning point at which an electron is instantaneously at rest (Fig. 23a), the electron is still acted on by a force of the same sign (there is no point of instantaneous rest in the momentum space); and in those cases when there is no turning point (Fig. 23b), the force changes its sign and, therefore, $\dot{p}$ vanishes, i.e., there is a turning point in the $\mathbf{p}$ space. It is clear from Fig. 24 that, very frequently, a particle is scattered into the neighboring cell in the reciprocal lattice.

In addition to collisions with local inhomogeneities, conduction electrons in a metal may be scattered by other quasi particles such as electrons, phonons, spin waves, etc. In these collisions, the laws of conservation of the crystal momentum and energy should be satisfied:

$$p_i = p_e + 2\pi\hbar b, \qquad \varepsilon_i = \varepsilon_e,$$

where $\mathbf{b}$ is an arbitrary vector of the reciprocal lattice; p_i and p_e are, respectively, the sums

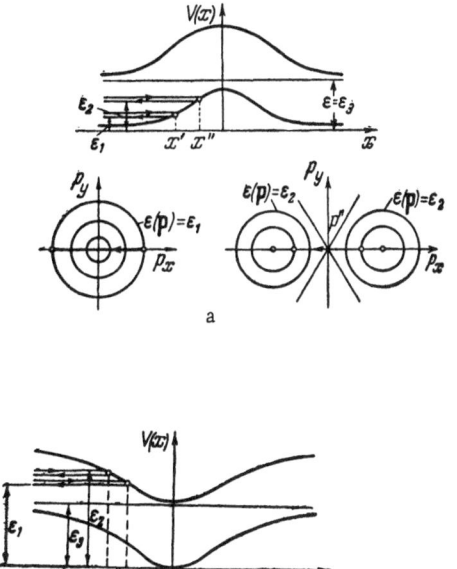

a

b

Fig. 24. Scattering of an electron with a dispersion law $\varepsilon(p) = \varepsilon$ by: a) a repulsive potential; b) an attractive potential (the energies are measured from the top of a band).

a) If the electron energy is close to its minimum value $\varepsilon = \varepsilon_1$ (i.e., far from the impurity causing the scattering), the electron remains within the same cell in the reciprocal lattice. If the electron energy is close to its maximum value ($\varepsilon = \varepsilon_2$), the electron is scattered into the neighboring cell in the reciprocal lattice. If the energy is such that there is no turning point ($\varepsilon = \varepsilon_3$), the electron always returns to that point in the $\mathbf{p}$ space at which the scattering by the repulsive potential begins.

b) If the electron energy is close to its minimum value ($\varepsilon = \varepsilon_1$), the electron is scattered into another cell; if the energy is close to its maximum value ($\varepsilon = \varepsilon_2$), the electron remains in its own cell. At sufficiently low energies, there is no turning point in the $\mathbf{x}$ space (compare with Fig. 23b).

of the crystal momenta of the particles before and after the collision (for example, in the case of the collision of two electrons $p_i = p_{1i} + p_{2i}$ and ε_i, ε_e are the energies of the particles before and after the collision).

All the possible collision processes are usually divided into normal or N (b = 0) and umklapp or U (b ≠ 0) processes. Since a unit cell in the reciprocal lattice can be selected in an arbitrary manner (as discussed in §§1 and 2), such a division is also arbitrary. When we are talking of a single collision event, we can always select the cell in such a manner as to suppress the umklapp nature of the process. However, in practice, we must consider a great variety of different collisions (for example, collisions of a given electron with all phonons, etc.). Naturally, in this case, all the collision processes must be referred to a fixed reference frame in the reciprocal lattice. It is usual to select a reciprocal lattice cell in such a manner so as to ensure the maximum probability for the normal processes. The periodicity of the wave functions in the p space ensures that all the quantities encountered in such calculations (electrical conductivity, thermal conductivity, etc.) are single-valued.

The number of bosons before and after a "collision" may be different: the creation and annihilation of phonons, spin waves, and other quasi particles may occur. The number of electrons in such interactions is conserved if the metal in question contains no fermions of the opposite sign ("holes") which can act as the antiparticles of electrons. If a metal contains holes or if they can be created, it is found that electrons and holes are annihilated and created in pairs so that the difference between the number of electrons and holes remains constant.

The concept of a "hole" in the electron theory of metals may be ambiguous and, therefore, we shall try to refine it. We have demonstrated in §4 that electrons with energies close to the maximum for a given band have a negative effective mass. In many cases, these electrons act as particles with a positive charge (for example, in a constant magnetic field they rotate in a direction opposite to that expected for negative particles, they are slowed down by an electric field which accelerates free electrons, etc.). Usually, the final formulas for the electrical conductivity, the Hall coefficient, etc., include not the number of electrons with a negative effective mass but the number of free states with a negative effective mass. It is usual to call the latter the number of holes. However, the full discussion can be carried out ignoring the existence of holes and simply taking account of the nature of the dispersion law near maximum values of the energy.[3] If, as a result of a collision, an electron is transferred from a state with a positive effective mass to one with a negative effective mass, this can be regarded as the annihilation of an electron and a hole but one can also avoid the introduction of new concepts (this will be more convenient in our approach) and simply allow for the fact that, in the final state, the electron has a negative effective mass.

The concept of a "hole" is very frequently used in a different sense. One can develop a theory in such a way that it starts from the ground state of a metal. It is known that at absolute zero all the states below a certain limit are occupied whereas all the states above this limit are vacant. This condition can be regarded as "vacuum" in relation to elementary excitations. Then, any excitation of the electron system is related to the simultaneous appearance of filled states at energies above the limit and of empty states at energies below the limit. Naturally, such excitations can appear only in pairs. The particles whose energy exceeds the limit are usually called electrons and the vacant states are referred to as holes. However, in this treatment, we cannot avoid considering electron–hole annihilation processes. Although this approach has certain advantages (in particular, there is no need to consider specially the law of conservation of particles because it is satisfied automatically) and is widely used, it will not

---

[3] See §27 where the theory of galvanomagnetic effects in metals is developed in this way for a complex dispersion law of conduction electrons.

be employed in the present book. Thus, the word "hole" in our discussions will simply mean a vacant state with a negative effective mass.

§7. Quasiclassical Energy Levels

Investigations of the motion of electrons with a complex dispersion law show that in some cases electrons move along finite paths. Under these circumstances, some of the electron energy associated with such motion becomes quantized. When the criteria for the quasiclassical approximation are satisfied (this is the most important case, as shown in §1), the quantized energy levels can be found without a consistent quantum-mechanical treatment of conduction electrons.

The most important quantization of the energy of an electron in a metal is the one encountered in the motion of an electron in a magnetic field. The criteria for the quasiclassical treatment of this case were discussed in §1 and it was shown that the restriction imposed on the magnetic field is really of little importance: the magnetic field H should be less than the atomic magnetic field $H_a \approx 10^8\text{-}10^9$ Oe [see Eq. (1.9)]. However, we must bear in mind that this estimate is based on the assumption that the electron energy has the Fermi value and this value is of the order of $10^{-12}\text{-}10^{-11}$ erg. In some metals (for example, in Bi), the number of electrons is much lower than in other metals (the electron concentration in Bi is of the order of 10^{-5} per unit cell in the crystal lattice) and the Fermi energy is low so that the quasiclassical criteria may be violated in fields which are easily reached in the laboratory ($H \approx 10^4\text{-}10^5$ Oe). Therefore, the formulas given in the present section should be treated with caution, particularly when they are applied to metals such as Bi.

The quasiclassical energy levels can be determined using Bohr's quantization rule

$$\frac{1}{2\pi} \oint \boldsymbol{P}_\perp \, d\boldsymbol{r} = n\hbar, \tag{7.1}$$

where n is an integer.

Since the adiabatic invariant $\oint \boldsymbol{P}_\perp \, d\boldsymbol{r}$ has already been calculated in §4, we shall apply Eq. (4.18):

$$S(\varepsilon, p_z) = \frac{2\pi\hbar e H}{c} \, n. \tag{7.2}$$

The expression (7.2) should be regarded as applying to the electron energy ε. The solution of this equation permits us to find the quantized energy levels $\varepsilon_n(p_z)$. Sometimes, the energy ε is specified by the conditions of the problem. Then, Eq. (7.2) represents the quantization of the momentum $p_{zn}(\varepsilon)$. The actual energy levels can be found only in those cases when we know the dispersion law $\varepsilon = \varepsilon(\boldsymbol{p})$ by means of which we can calculate the dependence of the cross-sectional area $S(\varepsilon, p_z)$ on the energy and on the projection of the momentum along the magnetic field. However, in many cases (see, for example, §15), it is sufficient to know Eq. (7.2) in order to calculate the thermodynamic or the transport properties of a metal.

Equation (7.2) is valid for $n \gg 1$ and many problems in the electron theory of metals can be solved ignoring the higher terms in the expression for the quasiclassical energy levels. However, in some cases, the first correction to such levels is of importance. This correction can be calculated within the framework of the quasiclassical approximation. Usually, the inclusion of the higher terms in the expansion requires the replacement of n in Eq. (7.1) with $n + \frac{1}{2}$. An analysis shows that, in the most common cases, the substitution is justified also for an electron with an arbitrary dispersion law. Hence

$$S(\varepsilon, p_z) = \frac{2\pi\hbar eH}{c}(n + 1/2). \tag{7.3}$$

However, in this approximation it is necessary to include the quantization of the energy due to the electron spin. If there is no spin–orbit coupling, an allowance for the spin splitting is very simple: each level is split into two and the splitting is $(e\hbar/m_0 c)\,H$, where m_0 is the mass of a free electron.[1]

Thus,

$$\varepsilon_{n\sigma}(p_z) = \varepsilon_n(p_z) + \sigma\frac{e\hbar}{2m_0 c}H \qquad (\sigma = \pm 1), \tag{7.4}$$

where $\varepsilon_n(p_z)$ are the roots of Eq. (7.3).

If $\varepsilon = p^2/2m_0$, it follows that $S(\varepsilon, p_z) = \pi(2m_0\varepsilon - p_z^2)$ and we arrive at the well-known formula obtained by Landau in 1930:

$$\varepsilon_{n\sigma}(p_z) = \left(n + \frac{1}{2}\right)\hbar\omega_H + \sigma\mu_0 H + \frac{p_z^2}{2m_0}, \tag{7.5}$$

where $\omega_H = eH/m_0 c$; $\mu_0 = e\hbar/2m_0 c$ is the Bohr magneton.

If an effective mass is introduced into the dispersion law, i.e., if we assume that $\varepsilon = p^2/2m^*$, we find that m^* occurs in the first and third terms of Eq. (7.5). We must stress that, if $m^* \neq m_0$, we do not observe the well-known double degeneracy resulting from the coincidence of the energies of the states with $\sigma = 1$ and $\sigma = -1$, which differ by unity in respect of the quantum number n.

If the dispersion law is quadratic but anisotropic $(\varepsilon = p_1^2/2m_1 + p_2^2/2m_2 + p_3^2/2m_2)$, the formula for the quantized energy levels becomes somewhat more complicated because of the difference between the longitudinal and the transverse effective masses

$$\varepsilon_{n\sigma}(p_z) = \left(n + \frac{1}{2}\right)\hbar\omega_H + \sigma\mu_0 H + \frac{p_z^2}{2m_\parallel}, \quad \hbar\omega_H = \frac{eH}{m_\perp c}, \tag{7.6}$$

and $m_\perp$, $m_\parallel$ depend in the following manner on the magnetic field:

$$\left.\begin{array}{l}
m_\perp = \left(\dfrac{\alpha_1^2}{m_2 m_3} + \dfrac{\alpha_2^2}{m_1 m_3} + \dfrac{\alpha_3^2}{m_1 m_2}\right)^{-1/2}, \\[2mm]
m_\parallel = m_1\alpha_1^2 + m_2\alpha_2^2 + m_3\alpha_3^2,
\end{array}\right\} \tag{7.7}$$

where α_j are the direction cosines of the magnetic field relative to the principal axes of the effective mass tensor $\left(\sum\limits_{j=1}^{3}\alpha_j^2 = 1\right)$.

---

[1] In the presence of the spin–orbit coupling, the splitting of the levels depends strongly on ε and p_z, and the quantization condition can be formulated as follows:

$$\frac{c}{e\hbar H}S(\varepsilon, p_z) \pm \frac{\pi}{2}g(\varepsilon, p_z) = 2\pi\left(n + \frac{1}{2}\right),$$

where the function $g(\varepsilon, p_z)$ is the equivalent of the g factor and is equal to 2 for a free electron. If $S = \pi(2m_0\varepsilon - p_z^2)$, the function $g(\varepsilon, p_z)$ can be calculated in some specific cases (see [18] where this calculation is carried out for Bi).

Using Eq. (7.3), we can calculate the separations $\Delta\varepsilon$ between the energy levels. Since n ≫ 1, it follows that $\Delta\varepsilon \ll \varepsilon$ (the quasiclassical criterion) and Eq. (7.3) yields

$$\Delta\varepsilon = \frac{2\pi\hbar e H}{c \dfrac{\partial S}{\partial \varepsilon}} = \frac{\hbar e H}{m^{*}\,(\varepsilon,\,p_z)\,c},$$

because, by definition, m* = $(1/2\pi)\partial S/\partial\varepsilon$ (§4). In other words, the separation between the levels is equal to the classical frequency of revolution ω_H multiplied by the Planck constant $\hbar$:

$$\Delta\varepsilon = \hbar\omega_H. \tag{7.8}$$

This formula does not mean that the energy levels are equidistant. They are only quasi-equidistant or locally equidistant. The effective mass depends on the energy ε and on p_z, which is the projection of the momentum along the magnetic field. This means that the energy depends in a complex manner on the quantum number n. This dependence is particularly strong near those values of the energy or momentum at which the nature of the section changes significantly. By way of example, we shall determine the dependence of the energy levels on the quantum number n close to a self-intersecting trajectory. We demonstrated earlier that the time of revolution for such a trajectory approaches infinity logarithmically since the point of intersection is a turning point. To be specific, we shall assume that the energy is known and we shall seek the dependence on n of the quantum values of the projection of the momentum along the magnetic field. The value $p_z = p_{zk}$ corresponds to the self-intersecting trajectory (thick line in Fig. 25).

The quasiclassical energy levels can be obtained quite accurately from the usual condition for quantization of the area. However, in calculating the energy levels, we must bear in mind that at the point $p_z = p_{zk}$ the area S has a singularity of the type $\Delta p_z \ln \Delta p_z$, where $\Delta p_z = p_z - p_{zk}$. On one side of the plane $p_z = p_{zk}$ we have two systems of quasiequidistant levels given by the equations

$$S_1(\varepsilon,\,p_z) = \frac{2\pi\hbar e H}{c}\,n_1, \qquad S_2(\varepsilon,\,p_z) = \frac{2\pi\hbar e H}{c}\,n_2, \tag{7.9}$$

and on the other side we have one system given by the equation

$$S(\varepsilon,\,p_z) = \frac{2\pi\hbar e H}{c}\,n. \tag{7.10}$$

"Restructuring" of the energy levels occurs near $p_z = p_{zk}$ and is described by the system of equations:

$$\left.\begin{aligned}
S_{1k} + p_0\,\Delta p_z \ln\left|\frac{\Delta p_z}{p_0}\right| &= \frac{2\pi\hbar e H}{c}\,n_1, \\
S_{2k} + p_0\,\Delta p_z \ln\left|\frac{\Delta p_z}{p_0}\right| &= \frac{2\pi\hbar e H}{c}\,n_2, \\
S_{k} + p_0\,\Delta p_z \ln\left|\frac{\Delta p_z}{p_0}\right| &= \frac{2\pi\hbar e H}{c}\,n,
\end{aligned}\right\} \tag{7.11}$$

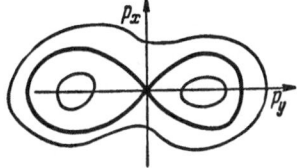

Fig. 25. Trajectories of an electron in a magnetic field near a point of self-intersection.

where $p_0 = 2v_z(p_{zk})\sqrt{m_1 m_2}$. Using

$$\frac{cS_{1k}}{2\pi\hbar eH} = n_{1k}, \qquad \frac{cS_{2k}}{2\pi\hbar eH} = n_{2k}, \qquad \frac{cS_k}{2\pi\hbar eH} = n_k,$$

we obtain from the system (7.11)

$$\left.\begin{aligned}
\Delta p_z &= \frac{2\pi\hbar eH}{cp_0} \frac{n_1 - n_{1k}}{\ln \frac{2\pi\hbar eH}{cp_0}(n_1 - n_{1k})}, \\[2mm]
\Delta p_z &= \frac{2\pi\hbar eH}{cp_0} \frac{n_2 - n_{2k}}{\ln \frac{2\pi\hbar eH}{cp_0}(n_2 - n_{2k})}, \\[2mm]
\Delta p_z &= \frac{2\pi\hbar eH}{cp_0} \frac{n - n_k}{\ln \frac{2\pi\hbar eH}{cp_0}(n - n_k)}.
\end{aligned}\right\} \tag{7.12}$$

It is interesting to note that the separation between the energy levels near p_{zk} is markedly not equidistant and that the change in the topology of the section results in a nonanalytic dependence of the energy (or p_z, if the energy is fixed) on the quantum number n. This result follows from the quasiclassical quantization conditions even when terms of the order of $1/n$ are ignored compared with unity (the anomalies which are due to such a change in the topology of the section appear in terms of the order of $\ln n/n \gg 1/n$). The system (7.11) is applicable to n_1, n_2, $n \gg 1$ whereas the solutions given by the system (7.12) are valid only for n_1, n_2, $n \gg |n - n_{1k}|$, $|n - n_{2k}|$, $|n - n_k| \gg 1$. A rigorous quantum-mechanical analysis of the motion of electrons which move along "figure of eight" trajectories in the quasiclassical approximation ([19] and, particularly, [20], where a full treatment is given) yields results similar to those obtained in the quasiclassical approach. Refinements of the energies of the quasiclassical levels (such as the inclusion of $\frac{1}{2}$ in the Bohr quantization rule) demonstrate an interesting oscillatory dependence of the separation between the levels on the magnetic field [20]. This dependence is essential in the calculation of the oscillatory components of various thermodynamic quantities (§15).

The formulas given in the preceding paragraphs demonstrate a significant difference between the structure of the energy spectrum of a free electron (dispersion law $p^2/2m$) and an electron with a complex dispersion law $\varepsilon = \varepsilon(\mathbf{p})$ when both of them are subjected to a magnetic field. This difference is found to be particularly great when we consider the existence of open constant-energy surfaces and open sections: the quantization then applies only to those electrons in a metal which follow finite paths in a plane perpendicular to the magnetic field. The presence or absence of discrete energy levels (corresponding to a fixed value of p_z) is then determined not only by the shape of the constant-energy surface but also by the direction of the magnetic field and − if the field is fixed − by the value of p_z. The transition from the closed to the open trajectories is discussed in detail in §4. The discussion given in that section shows clearly that the structure of the energy spectrum becomes very complex when the usual closed sections are replaced by the open ones. The logarithmic singularities which appear on approaching the saddle points (Fig. 17) are then superimposed on the general tendency, which is a reduction in the separation between the levels due to an increase in the period of revolution.

The formulas for the energy of an electron in a magnetic field thus show that the infinite degeneracy which is observed in the case of free electrons (the energy levels are independent of the conserved component of the momentum P_x) is still retained for an electron with an arbitrary dispersion law if the quasiclassical approximation is employed. However, we must bear in mind that the energy levels of a free electron calculated in the quasiclassical approximation are exact. This property, which is the result of a strictly equidistant distribution of the energy levels of an oscillator, is lacking in the case of electrons with a complex dispersion law. Therefore, in particular, the inclusion of higher quantum corrections [21] lifts the degeneracy in respect of P_x and this results in the broadening of the energy levels of Eq. (7.3). Usually, this broadening is very small $\Delta\varepsilon/\varepsilon \approx (H/H_a)^2$ [see Eq. (1.9)]. The dependence of ε on P_x and, consequently, the broadening of the energy levels are related to the fact that P_x determines the position of the electron trajectory in space [in the case of a free electron cP_x/eH is the center of the orbit in the (x, y) plane along which the electron is traveling]. In free space (in the case of free electrons), all the points in space are equivalent (we shall assume that the magnetic field is uniform) but this is not true in a periodic structure such as a crystal.

In those cases when the trajectory of an electron in a magnetic field is open in the momentum space, the quasiclassical approximation does not result in the quantization of the electron energy. However, a more rigorous analysis [21] shows that the periodic dependence of the energy on the crystal momentum gives rise to characteristic discontinuities in the continuous spectrum of the electron. These discontinuities occur at points given by an equation which resembles the quantization rule of Eq. (7.2):

$$S\left(\varepsilon,\ p_z\right) = \frac{2\pi\hbar eH}{c} n \quad \text{(n is an integer)}. \tag{7.13}$$

However, in this case, S is that part of the area described by the electron which is contained in one cell in the momentum space. The relative width of the discontinuity for moderately narrow "necks" is exponentially small $\left(\Delta\varepsilon/\varepsilon \approx e^{-\beta H_a/H};\ \beta \sim 1\right)$.

The energy levels near the orbits passing through the whole crystal lattice are of a rather special nature. In this case, we obtain a system of split levels, sublevels, etc.

In addition to the mechanism of the energy level broadening which has been referred to above (the dependence of ε on P_x), there is another mechanism which is due to the transitions of the electrons from one classical orbit to another. This phenomenon is known as the magnetic breakthrough or breakdown and it will be dealt with in a special section (§10).[2] We shall point out only that, usually, the broadening of the levels due to such transitions is exponentially small $\Delta\varepsilon/\varepsilon \sim e^{-\beta' H_a/H};\ \beta' \sim 1$ and can be ignored. However, in the case of open constant-energy surfaces, the broadening effect associated with transitions from closed to open sections is very important: the broadening may be so strong that the discrete spectrum becomes continuous.

It was shown in §4 that the motion in crossed electric and magnetic fields can be regarded as the motion of a particle with a dispersion law

$$\varepsilon^*\left(p\right) = \varepsilon\left(p\right) - V_{EH}p \tag{7.14}$$

in the magnetic field alone. If the trajectory in the momentum space is closed, the particle moves along a finite trajectory parallel to the y axis (the z axis is, as usual, taken along the magnetic field and the x axis along the electric field). This allows us to apply the results obtained in the preceding sections to the motion of a particle in crossed fields. Following the previous procedure, we obtain the quantization condition in the form

$$S^*\left(\varepsilon^*,\ p_z\right) = \frac{2\pi\left|e\right|\hbar H}{c} n, \tag{7.15}$$

---

[2] The transitions between different bands, induced by the application of a magnetic field, will be considered in that section.

where S* is the cross-sectional area of the $\varepsilon*(\mathbf{p})$ = const surface cut by a plane p_z = const. The separation between the levels $\Delta\varepsilon$ can easily be shown to be equal to the frequency ω_{EH} multiplied by the Planck constant $\hbar$, i.e.,

$$\Delta\varepsilon = \hbar\omega_{EH} = \frac{2\pi|e|\hbar H}{c\dfrac{\partial S^*}{\partial\varepsilon^*}}. \qquad (7.16)$$

We recall that the frequency ω_{EH} for an arbitrary dispersion law depends on the applied electric field and that this dependence vanishes for the quadratic dispersion law. Consequently, the separation between the discrete energy levels of an electron in crossed fields should depend on the electric field (this is an analog of the Stark effect).

If we ignore the transitions between the energy bands, we find that a conduction electron subjected to a uniform electric field undergoes finite motion (§4). This finite motion should be associated with quantum energy levels whose separation can be found quite easily from the correspondence principle

$$\Delta\varepsilon = \hbar\omega_E.$$

Obviously, the effects which would result from such quantization cannot be observed. Even if we assume that transitions between the energy bands are extremely unlikely, an electron would be unable to complete a revolution because of its collisions with lattice irregularities.

The crystal lattice in which an electron is moving is always limited by the boundaries of a sample. However, usually the dimensions are so large in relation to the atomic spacings that a sample may be regarded as infinite. Nevertheless, we may encounter cases when it is necessary to take account of the finite dimensions of a metal specimen, for example, in considering the properties of thin films. The surface of a metal is always slightly deformed, i.e., some departures from the periodicity of the structure must always occur near the surface. An investigation of the behavior of an electron near the surface is a complex problem and one that has been clearly formulated only relatively recently.

Frequently, the boundary of a sample is regarded as an infinitely high potential barrier, without any rigorous justification. In other words, an electron in a metal is sometimes regarded as a particle in a potential box.

In recent years, there have been several experiments whose results are regarded as the proof that electrons are reflected specularly from an inner boundary in a metal [22]. Specular reflection evidently occurs when the de Broglie wavelength of an electron is large compared with the atomic spacings (this happens in Bi and other semimetals), or when an electron impinges the surface at a low angle, i.e., specular reflection may be observed for almost glancing incidence. For these electrons, the projection of the momentum, which determines the characteristic de Broglie wavelength, is also small (this point will be considered later). Moreover, there may be an additional reason why electrons are reflected specularly in Bi-type metals [23]: this may be due to band bending over distances of the order of the Debye–Hückel radius, which is anomalously large for such metals (it is much larger than the lattice constant).

A free electron reflected from an infinite barrier also suffers specular reflection and the normal projection of its momentum is reversed. For an electron with a complex dispersion law, the nature of its reflection at an infinite barrier is a much more complex event. We must bear in mind that the energy and $p_\perp$ (the projection of the momentum on the surface of the metal) are conserved in reflections. If we consider the reflection of an electron as a process in which the kinetic energy is first reduced and then increased to its initial value, we can find the change in the normal projection of the momentum due to reflection.

The projection of the momentum along the z axis (this axis is taken to be perpendicular to the boundary) before and after collision satisfies the relationship

$$\varepsilon\left(\boldsymbol{p}_\perp,\, p_z\right)=\varepsilon, \tag{7.17}$$

where the initial value $p_z^{(i)}$ is defined by the condition $v_z^{(i)} > 0$ and the final value $p_z^{(f)}$ by the condition $v_z^{(f)} < 0$.

Equation (7.17) may have more than two roots. In such a case, all the roots can be divided into pairs so that the kinetic energy is always less than the initial energy when we go over from $p_z^{(i)}$ to $p_z^{(f)}$ (this should be compared with the scattering of an electron in the field of an impurity, which is considered in §6). For a spherical Fermi surface, we have $\Delta p_z = p_z^{(i)} - p_z^{(f)} = 2\sqrt{2m\varepsilon - p_\perp^2}$.

The quasiclassical quantization procedure imposed by the finite dimensions of a sample can be followed in the standard manner by equating the adiabatic invariant I to the Planck constant multiplied by an integer. In this case,

$$I = \frac{1}{2\pi} \oint p_z\, dz = \frac{1}{2\pi}\left| p_z^{(i)} - p_z^{(f)} \right| L = \frac{L\,|\Delta p_z|}{2\pi},$$

where L represents the transverse dimension of the sample (the thickness of a film). Hence,

$$\Delta p_z = \left| p_z^{(i)} - p_z^{(f)} \right| = \frac{2\pi\hbar n}{L} \qquad (n = 0,\ 1,\ 2,\ \ldots). \tag{7.18}$$

This equation should be regarded as an expression which applies to the energy at a fixed value of $\boldsymbol{p}_\perp$. The solution of this equation gives

$$\varepsilon = \varepsilon_n\left(\boldsymbol{p}_\perp\right), \tag{7.19}$$

which is a system of quantized energy levels. If Eq. (7.17) has several pairs of roots, we obtain several systems of energy levels [a similar situation occurs in the case of quasiclassical quantization in a magnetic field when the constant-energy surface has a saddle point: see Eq. (7.12)].

We can obtain compact equations for the determination of the energy levels also in those cases when an electron is subjected to a constant and uniform magnetic field as well as a force perpendicular to this field. If we assume that the magnetic field is directed along the z axis, the existence of an additional force can be allowed for by introducing a potential field U(y). The vector potential can be taken in the form $A_x = -Hy$ and $A_y = A_z = 0$ [see Eq. (4.16)]. In this case, x is a cyclic coordinate and, therefore, the momentum P_x is conserved. Using the Bohr quantization condition, we can now obtain the following equation for the levels of the energy $\mathscr{E}$:

$$S(\mathscr{E},\, p_z,\, P_x) = \frac{2\pi\hbar\,|e|\,H}{c}\, n. \tag{7.20}$$

Here, $S(\mathscr{E},\, p_z,\, P_x)$ is the area bounded by the electron trajectory in the kinematic momentum space. This trajectory is given by the integrals of motion

$$\varepsilon(p_x,\, p_y,\, p_z) + U\left[\frac{c}{eH}(p_x - P_x)\right] = \mathscr{E},\quad p_z = \text{const},\ P_x = \text{const}. \tag{7.21}$$

Equations (7.20) and (7.21) give the dependences of the energy levels on the quantum number n, on the components of the momentum p_z and P_x, and on the magnetic field. It is assumed that the trajectory of Eq. (7.21) is a closed curve.

There is no degeneracy in respect of P_x: the trajectories corresponding to the same values of $\mathscr{E}$ and p_z are not equivalent because of the presence of the field U(y). Equations (7.20) and (7.21) can be used to determine the dependences of the electron energy levels in a magnetic field on the dimensions of a plate parallel to the field (as usual, we shall assume that the reflection from the boundaries of the plate is specular). We shall postulate that the field is defined by

$$U(y) = \begin{cases} 0 & |y| < d, \\ \infty & |y| \geqslant d. \end{cases} \tag{7.22}$$

In other words, we shall represent the metal by an infinitely deep potential well. The expressions in Eq. (7.22) are equivalent to the assumption that $|y|$ cannot have values exceeding d. Since $y = (c/eH)(p_x - P_x)$, the trajectory of an electron in the momentum space is now given by

$$\varepsilon(p_x, p_y, p_z) = \mathscr{E}, \quad |p_x - P_x| < \frac{|e|Hd}{c}, \\ P_x = \text{const}, \quad p_z = \text{const.} \tag{7.23}$$

We shall use $p_x^{(\min)}(\mathscr{E}, p_z)$ and $p_x^{(\max)}(\mathscr{E}, p_z)$ to denote the extreme left- and right-hand p_x coordinates of the curve $\varepsilon(p_x, p_y, p_z) = \mathscr{E}$, $p_z = \text{const}$. If

$$p_x^{(\max)}(\mathscr{E}, p_z) - \frac{|e|Hd}{c} < P_x < p_x^{(\min)}(\mathscr{E}, p_z) + \frac{|e|Hd}{c}, \tag{7.24}$$

we find that the trajectory drawn in the coordinate space lies entirely within the plate and the electron is not affected by the presence of the boundaries. In this case, the energy levels are identical with those of an electron in an infinite sample. If the trajectory does not lie within the plate, the area in Eq. (7.20) is bounded by the curve $\varepsilon(p_x, p_y, p_z) = \mathscr{E}$, $p_z = \text{const}$ and truncated by one or more lines of the type $p_x = P_x \mp (|e|Hd/c)$. If the magnetic field is sufficiently strong, the majority of the closed trajectories lie completely within the plate: the electrons moving along closed trajectories are practically unaffected by the boundaries of the plate. The situation is different in the case of those electrons which move along open trajectories: they always collide with one of the walls, irrespective of the value of the field. This is the reason why the energies of such electrons are quantized.

Recent studies have established that surface electron levels appear in metals subjected to high-frequency magnetic fields. We shall leave the discussion of the high-frequency properties to Chap. IV (§46) and restrict ourselves to the derivation of an expression for the surface energy levels on the assumption that the magnetic field is applied parallel to the surface.

In a study of the surface states, the nature of the scattering of electrons by the surface of a conductor is of basic importance. The nature of the scattering process is determined by the ratio of the de Broglie wavelength λ_B of the electrons incident on the surface to the characteristic dimensions of the surface irregularities. For good metals, the value of λ_B is of the order of the atomic spacings and, therefore, the reflection from the surface can be regarded as diffuse if the angle of incidence is not too small. This means that the natural width of the levels is of the same order as the separations between them (because it is determined by the same frequency) and the spectrum of such electrons is not quantized. As the angle of incidence φ is gradually decreased, the de Broglie wavelength measured along the normal to the surface increases proportionally to φ^{-1} (Fig. 26). This means that at sufficiently small angles of incidence the reflection of electrons from the surface can be nearly specular.

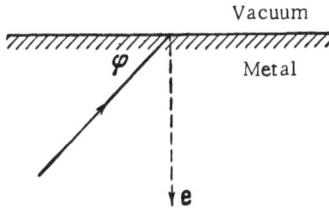

Fig. 26. Incidence of an electron on a metal-vacuum boundary.

Fig. 27. Quantization occurs in
the shaded area between dashed
lines.

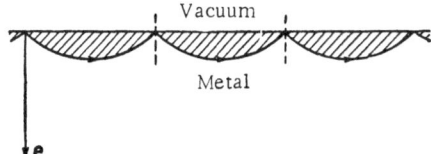

We shall be interested only in those electrons which are incident on the surface at a very small angle (these electrons dominate the high-frequency properties of metals).

In the quasiclassical approach, the quantization of the levels of those electrons which collide with the surface in a magnetic field which is parallel to the surface are found in the same way as the levels in an infinitely large sample, i.e., the result is the same as before:

$$S(\varepsilon, p_z, P_x) = \frac{2\pi\hbar\,|\,e\,|\,H}{c}\,n,\tag{7.25}$$

where S is the area bounded by the orbit of an electron colliding with the surface of a sample (we are referring here to the momentum space). This area naturally depends on the conserved value of the momentum P_x which determines the position of the orbit $y = (c/eH)(p_x - P_x)$, $e < 0$. The area S is shown shaded in Fig. 27, where it is bounded by

$$\varepsilon(p_x, p_y, p_z) = \varepsilon, \quad p_z = \text{const}, \quad p_x \leqslant P_x.\tag{7.26}$$

The area of an orbit in the momentum space differs from the corresponding area in the coordinate space by the factor $(eH/c)^2$. It follows from the equations of motion that $y = cp_x/eH + y_0$, $x = -cp_y/eH + x_0$. For small values of φ, we have

$$y = y_{max} - \frac{x^2}{R}, \qquad y_{max} = x_{max}\varphi\tag{7.27}$$

(where R is the radius of curvature at the point $y = y_{max}$), so that

$$y_{max} = R\varphi^2, \qquad x_{max} = R\varphi.$$

The quantization of the orbit area gives rise to the quantization of the angle φ:

$$\frac{2}{3}R^2\varphi^3 = \frac{nch}{eH}, \qquad \varphi_n = \left(\frac{3}{2}\frac{nch}{eHR^2}\right)^{1/3}.\tag{7.28}$$

Since R is of the order of the Larmor radius and the energy can be taken as equal to the Fermi value ($\varepsilon = \varepsilon_F$, $p \approx p_F$), it follows that

$$\varphi_n \propto \left(\frac{eHh}{cp_F^2} n \right)^{1/3} \propto \left(\frac{\hbar\omega_H}{\varepsilon_F} \right)^{1/3} n^{1/3}. \tag{7.29}$$

Since $x = v_x^0 t = v_F t$, the period of revolution is $T_\varphi = 2\Delta x_{max}/v_x^0 = 2R\varphi_n/v_x^0$ and the frequency of revolution ω_n along such orbits is

$$\omega_n(\varepsilon_F, p_z) = \frac{\pi v_x^0}{R\varphi_n} \propto \omega_H \left(\frac{\varepsilon_F}{\hbar\omega_H} \right)^{1/3} \frac{1}{n^{1/3}}. \tag{7.30}$$

The separation between the corresponding levels is

$$\Delta\varepsilon_n = \hbar\omega_n. \tag{7.31}$$

The natural width of the levels $\delta\varepsilon$ is determined by the "bulk" mean free time, $\delta\varepsilon_T \propto \hbar/\tau$, and by the nonspecular reflection. If the probability of diffuse scattering in each collision event is q, where $q \ll 1$, it follows that when the number of collisions is such that the final result is almost completely diffuse motion, we find that $(1 - q)^n \propto \frac{1}{2}$ and hence $n \propto 1/q$ so that $\delta\varepsilon_q \propto \hbar\omega_n/q$. Thus, the necessary conditions for the existence of quasistationary levels (7.31) becomes

$$q(\varphi_n) \ll 1; \qquad \omega\tau \left(\frac{\varepsilon_F}{\hbar\omega} \right)^{1/3} \frac{1}{n^{1/3}} \gg 1. \tag{7.32}$$

This may be satisfied in weak fields H provided $q(\omega\tau)^{-1/2} \ll 1$.

All the quantization cases considered so far are the special cases of a very general quasi-classical approach, in which we find the adiabatic invariants I_j and equate them to $n_j\hbar$, where n_j is an integer. Then, the energy levels (which can be expressed in terms of I_j) are given by

$$\varepsilon_{n_1, n_2, n_3} = \varepsilon(n_1\hbar; n_2\hbar; n_3\hbar). \tag{7.33}$$

This expression is valid if the motion is finite in all three degrees of freedom. If such motion is infinite in one or two degrees of freedom, the energy depends only on, respectively, two or one discrete quantum numbers.

§8. Quantum Mechanics of an Electron with an Arbitrary Dispersion Law

In spite of the fact that the quasiclassical approach is usually sufficient in tackling specific problems in the electron theory of metals, the development of a consistent quantum mechanics of conduction electrons is of fundamental interest.

If we start from the assumption that a complex dispersion law is the result of the motion of an electron in a periodic field, it would be natural to approach this problem by introducing external fields into the Schrödinger equation and solving the problems arising out of this approach. However, a conduction electron is not a simple particle. Its dispersion law makes allowance for the interaction of all the electrons with one another. Therefore, in developing the quantum mechanics of a conduction electron, we must first specify the framework within which we shall be working. The quantum mechanics cannot be formulated simply in terms of the dispersion law, although the p representation is extremely useful. The Schrödinger equation is usually written in the coordinate representation since the kinetic energy operator can be expressed very simply in terms of the differentiation operator. This makes the Schrödinger equation a linear partial differential equation and the theory of such equations has been fully developed. In the case of an electron in a metal, the kinetic energy is $\varepsilon(\mathbf{p})$, which is a complex periodic function of the crystal momentum. On the other hand, the problem in a metal is simplified by the fact that external forces are, in most cases, relatively simple functions of the coordinates. This encourages the formulation of the quantum mechanics of a conduction electron in terms of the Hamiltonian written in the $\mathbf{p}$ representation:

$$\hat{\mathcal{H}} = \varepsilon(\mathbf{p}) + \hat{\mathcal{U}}, \tag{8.1}$$

where $\hat{\mathcal{U}}$ is the potential energy operator related to the external forces. The use of this operator is inconsistent with the quantum-mechanical approach because the transition from the quasiclassical to the quantum approach demands that we not only take account of the fact that the potential energy is an operator, i.e., that we replace the coordinate $\mathbf{r}$ ($\mathcal{U}$ is a function of this coordinate) with the operator $-(\hbar/i)(\partial/\partial\mathbf{p})$, but also that we allow for the difference between the crystal momentum $\mathbf{p}$ and the true momentum in free space. According to Eqs. (1.45) and (1.46), this means that the coordinate operator should be of the form

$$\hat{\mathbf{r}} = -\frac{\hbar}{i}\frac{\partial}{\partial\mathbf{p}} + \hat{\rho}, \tag{8.2}$$

and the operator $\hat{\rho}$, which is diagonal in terms of the crystal momentum $\mathbf{p}$, is specified in terms of matrix elements associated with band–band transitions [see Eq. (1.47)]. Therefore, the final form of the Hamiltonian should be the sum of the operators representing the "kinetic" energy $\hat{\mathcal{K}}$ and the potential energy $\hat{\mathcal{U}}$:

$$\mathcal{H} = \hat{\mathcal{K}} + \hat{\mathcal{U}}, \tag{8.3}$$

where the operator $\hat{\mathcal{K}}$ is specified by its matrix elements in the $\mathbf{p}$ representation:

$$\mathcal{K}_{ss'}^{(\mathbf{p})} = \varepsilon_s(\mathbf{p})\,\delta_{ss'}, \tag{8.4}$$

and the operator $\hat{\mathcal{U}}$ is defined by the dependence on the coordinate operator $\mathbf{r}$ [see Eq. (8.2)]. The Hamiltonian of a conduction electron in an external field allows us to solve, in principle, any problem in the quantum mechanics of an electron with a complex dispersion law. Naturally, the results will be expressed not only in terms of the dispersion law, i.e., in terms of the functions $\varepsilon_s(\mathbf{p})$ and their derivatives, but also in terms of the matrix elements $\rho_{ss'}(\mathbf{p})$. This approach, based on the dispersion law $\varepsilon_s(\mathbf{p})$ and the matrix elements $\rho_{ss'}(\mathbf{p})$ is completely consistent but we must specify not $\rho_{ss'}(\mathbf{p})$ but a set of eigenfunctions of the electron Hamiltonian $\psi_{s\mathbf{p}}(\mathbf{r})$ which is unperturbed by external fields. The matrix elements $\rho_{ss'}(\mathbf{p})$ can be expressed in terms of these eigenfunctions. Using Eq. (1.47) and the explicit form of the wave function given by Eq. (1.16), we obtain:

$$\rho_{ss'}(\mathbf{p}) = i\hbar \int u_{s\mathbf{p}}^*(\mathbf{r})\frac{\partial}{\partial\mathbf{p}}u_{s'\mathbf{p}}(\mathbf{r})\,dv. \tag{8.5}$$

This expression can be given in a different form if we use the Bloch-wave expansion (1.16) in the form of a Fourier series (1.21):

$$\rho_{ss'}(\mathbf{p}) = i\hbar \sum_b a_s^*(\mathbf{p}+2\pi\hbar\mathbf{b})\frac{\partial}{\partial\mathbf{p}}a_{s'}(\mathbf{p}+2\pi\hbar\mathbf{b}). \tag{8.6}$$

This form is convenient in the solution of specific problems based on the assumption of weak interaction. Here $\psi_{s\mathbf{p}}(\mathbf{r})$ are the wave functions of a conduction electron in the coordinate representation. The coefficients in the expansion of the function $\psi(\mathbf{r},\,t)$ (representing an arbitrary state) in terms of the wave functions $\psi_{s\mathbf{p}}(\mathbf{r})$ are themselves wave functions in the p representation. We shall denote them by g(s, $\mathbf{p}$, t). Thus, the wave function of an electron in a state with a specified value of the crystal momentum ($\mathbf{p} = \mathbf{p}_0$) in a particular band (s = s_0) is

$$g\,(s,\,\mathbf{p},\,t) = e^{-i\varepsilon_{s_0}\,(\mathbf{p}_0)\,t/\hbar}g\,(s,\,\mathbf{p}), \left.\begin{array}{c} \\ \\ \end{array}\right\}$$
$$g\,(s,\,\mathbf{p}) = A\delta_{ss_0}\delta(\mathbf{p} - \mathbf{p}_0), \qquad (8.7)$$

where A is a normalization constant.

Using the definition of the Hamiltonian given by Eq. (8.3), we can easily calculate the velocity operator in the coordinate space $\hat{\dot{\mathbf{r}}} = \hat{\mathbf{v}}$ and in the crystal-momentum space $\hat{\dot{\mathbf{p}}}$. The velocity operators are discussed in §1. We shall not give their values again but refer the reader to Eqs. (1.48) and (1.50). We note only that these equations allow us, at least in principle, to determine the time dependence of the state of a conduction electron subjected to external fields. A problem of this type, i.e., a determination of the change in the state with time, can be solved also by means of wave functions, i.e., simply by finding the time dependence of a wave function. Let us consider a simple example. We shall find the time dependence of the wave function of an electron subjected to a constant and uniform force F = eE, where E is the applied electric field. According to Eqs. (8.3), (8.4), and (8.2), the system of equations for the functions g(s, $\mathbf{p}$, t) now becomes

$$-\frac{\hbar}{i}\frac{\partial g\,(s,\,\mathbf{p},\,t)}{\partial t} = \varepsilon_s g_s + \frac{e\hbar}{i}E\frac{\partial g\,(s,\,\mathbf{p},\,t)}{\partial p} - eE\sum_{s'}\rho_{ss'}(p)\,g\,(s',\,\mathbf{p},\,t). \qquad (8.8)$$

In some cases (and practically always for metals), this equation can be solved by the method of successive approximations by considering the last sum on the right-hand side as a small perturbation. This is justified, in the final analysis, by the weakness of the applied field compared with the atomic fields. In estimating a dimensionless small parameter, we must remember that the matrix elements are $\rho_{ss'}^{x,y,z} \approx a$ (see §1). However, before we can use the method of successive approximations, we must transform Eq. (8.8) by replacing the time t with a new variable $\xi = p_x - eEt$ (the x axis is selected to lie along the electric field and p_x is denoted simply by p). Then,

$$-\frac{\hbar}{i}\,eE\left(\frac{\partial g_s}{\partial p}\right)_\xi = \varepsilon_s g_s - eE\sum_{s'}\rho_{ss'}^x g_{s'} \qquad (8.9)$$

(the discrete argument s is written in the form of an index). The projections of the crystal momentum p_y and p_z are present in the above equation as parameters. The functions $g_s(p, \xi)$ will be found in the form

$$g_s\,(p,\,\xi) = e^{-\frac{i}{eE\hbar}\int\varepsilon_s\,(p')\,dp'}\varphi_s\,(p,\,\xi). \qquad (8.10)$$

The system of equations for the functions $\varphi_s(p, \xi)$ is extremely simple:

$$\left(\frac{\partial\varphi_s}{\partial p}\right)_\xi = \sum_{s'}q_{ss'}\varphi_{s'}\,(p,\,\xi), \qquad (8.11)$$

where

$$q_{ss'} = \frac{i}{\hbar}\,e^{\frac{i}{eE\hbar}\int\varepsilon_s\,dp'}\rho_{ss'}^x e^{-\frac{i}{eE\hbar}\int\varepsilon_{s'}\,dp'}. \qquad (8.12)$$

We shall now apply the method of successive approximations. In the zeroth approximation, which corresponds to the case when we neglect the transitions between the energy bands, we obtain

$$\left(\frac{\partial \varphi_s^{(0)}}{\partial p}\right)_\xi = 0,$$
(8.13)

i.e.,

$$\varphi_s^{(0)} = \varphi_s^{(0)}(\xi) \equiv \varphi_s^{(0)}(p - eEt).$$
(8.14)

We shall now satisfy the initial conditions of the problem. If, at t = 0, the wave function of the electron in question is $g_s(p, 0) = g_s(p)$, it follows from Eqs. (8.14) and (8.10) that

$$g_s^{(0)}(p, t) = g_s(p - eEt)\, e^{-\frac{i}{eE\hbar} \int \varepsilon_s(p')\, dp'}.$$
(8.15)

The quantity $|g_s(p, t)|^2$ determines the probability that an electron belonging to the s-th band has, at a moment t, a crystal momentum in the range (p, p + dp).

The last equation shows that, in the zeroth approximation (when the transitions between the energy bands are neglected), the probability distribution moves as a whole in the crystal momentum space at a constant velocity $\dot{p} = eE$. In particular, this applies to a δ-like distribution of the probabilities: if an electron initially has a definite value of the crystal momentum, it follows that at any moment t this electron has a different definite value of the momentum, which can be found from the classical equation of motion:

$$\frac{dp}{dt} = eE.$$
(8.16)

This conclusion provides additional justification for the usually employed classical and quasiclassical approaches.

Substituting the zeroth approximation on the right-hand side of the system (8.11), we can determine the functions $g_s(p, t)$ in the first approximation with respect to the electric field and calculate the probability of a band—band electron transition [24]. However, we shall not consider this problem in detail.

We must make some comments on the equation of motion

$$\frac{dp}{dt} = -\frac{\partial \hat{U}(r)}{\partial r},$$

which is a more general form of Eq. (8.16). The use of this equation of motion and its derivation presupposes that the p space is infinite: the possibility of the crystal momentum being "restored" to a specific unit cell in the reciprocal lattice is excluded. The physical equivalence of the states in different cells, which is ensured by the periodicity of the wave function describing a state with a specified crystal momentum, makes the results single-valued.

On the other hand, in calculations of the probability of various processes, we should be using the full (but not "overfull") set of wave functions, i.e., having selected a given cell in the reciprocal lattice, we should employ only the wave functions associated with that cell.[1] We can then find that umklapp processes are allowed, i.e., they have a nonzero probability.

---

[1]If necessary, different cells should be taken for the initial and the final states.

The description of the same process by means of an equation of motion and a probability scheme are complementary. The actual description selected depends on the problem.

We have not yet considered the quantum properties of an electron in a magnetic field. The most interesting problems associated with the transition of an electron from one classical trajectory to another ("magnetic breakthrough") will be considered in §10. Here, we shall make only some general comments.

The components of the kinematic momentum operator[2]

$$p = P - \frac{e}{c} A \tag{8.17}$$

do not commute with one another because the vector potential depends on the coordinate $\mathbf{r}$ (rot $\mathbf{A} = \mathbf{H}$). Using the commutation rules between the coordinate $\mathbf{r}$ and the canonically conjugate momentum $\mathbf{P}$, we can determine the commutation relationships between the components of the kinematic momentum ("kinematic crystal momentum"):

$$\{p_l, p_k\} = \frac{e\hbar}{ic} \, \varepsilon_{ikl} H_l, \tag{8.18}$$

where ε_{ikl} is a unit antisymmetric tensor of the third rank and the braces represent commutation. We note that the magnetic field H and Planck constant $\hbar$ occur in this expression as a product. The stronger the magnetic field, the stronger are the quantum properties. We have drawn a similar conclusion from a comparison of the radius of the electron orbit with its de Broglie wavelength. It is evident from the commutation rules (8.18) that this conclusion is independent of the nature of the trajectory of an electron in a magnetic field. We must draw attention to the fact that the components of the kinematic momentum are canonically conjugate (at least to within a constant factor).

If we start from the dispersion law $\varepsilon_s(\mathbf{p})$ and derive the Hamiltonian of a conduction electron in a magnetic field, we encounter the problem of the meaning of $\varepsilon_s(\mathbf{p})$ as a function of the noncommuting variables $\hat{p}_x, \hat{p}_y, \hat{p}_z$. The situation is complicated by the fact that the coordinate $\mathbf{r}$ is not conserved. Formally, this can be regarded as the absence of commutation between the kinematic momentum and the "band number operator" (the band number operator $\hat{s}$ is the operator whose eigenvalues are the quantum numbers which label an energy band). If the vector potential is expressed in the form $\mathbf{A} = \frac{1}{2}[\mathbf{Hr}]$, we can show that

$$\{\hat{s}, \widehat{p}\}_{ss'} = -\frac{e}{2c}(s - s')[H, \rho_{ss'}(p)]. \tag{8.19}$$

The general form of the Hamiltonian of an electron in a magnetic field cannot be derived simply from the law of dispersion. The structure of the Hamiltonian is related in a fundamental manner to the origin of the Fermi branch of the energy spectrum. However, if we ignore the transitions between the energy bands[3] [if we ignore ρ compared with $\mathbf{R}$ in Eq. (1.45)], the nature of the symmetrization of the function $\varepsilon(\mathbf{p})$ can be found on the basis of fairly general considerations. We shall start from the expansion of $\varepsilon(\mathbf{p})$ in terms of a Fourier series:

---

[2] Strictly speaking, this relationship between $\mathbf{P}$ and $\mathbf{p}$ is valid only for the momentum in free space (the true momentum) but we can show that it applies also to the crystal momentum, at least for an electron in the periodic field of the crystal lattice [21] (see also §1).

[3] Rigorous criteria for this approximation cannot be established. Obviously, the bands should not overlap and the gap between them should be sufficiently large.

$$\varepsilon(\boldsymbol{p}) = \sum_a A_a e^{i a p/h}, \tag{8.20}$$

where the summation is carried out over all the periods $\boldsymbol{a}$ of the crystal lattice. In going over to the Hamiltonian, i.e., in replacing the components of the kinematic momentum by the operators obeying the commutation rules (8.18), it is natural to use the expansion (8.20), regarding each term as an exponential function of one operator $\boldsymbol{a} \cdot \mathbf{p}$. In other words, if we ignore the band–band transitions, it follows that

$$\varepsilon(\widehat{\boldsymbol{p}}) = \sum_a A_a e^{i a \widehat{p}/h}. \tag{8.21}$$

The expression (8.21) represents the "total symmetrization" of the Hamiltonian in respect of the operators of the kinematic momentum components. The need for the "total symmetrization" is supported by the following considerations. The component of the kinematic momentum along the magnetic field is conserved and it commutes with $\hat{p}_x$ and $\hat{p}_y$. In a plane perpendicular to the magnetic field, the axes are not fixed and the canonically conjugate variables can be any linear combinations of p_x and p_y. The vector $\mathbf{a}$ can always be represented as $\boldsymbol{a} = \boldsymbol{a}_\perp + \boldsymbol{a}_\parallel$, where $\boldsymbol{a}_\parallel = (\boldsymbol{a}\mathbf{H})\mathbf{H}/\mathrm{H}^2$. If the x axis is now directed along $\mathbf{a}_\perp$, we find that the argument of the exponential function contains only one operator. These considerations are supported by the results of the investigation reported in [21].

The commutation rules (8.18), which we shall regard as applying to the components of the true (free-space) kinematic momentum (and not the crystal momentum), show that the quantum nature of the problem becomes more pronounced as the intensity of the magnetic field increases. Strictly speaking, making no additional assumptions, the structure of the quantum states of an electron in a magnetic field can be formulated only in two cases.

1. Weak Magnetic Fields. This is the quasiclassical situation: an electron moves along a classical orbit and has a definite dispersion law in the absence of a magnetic field; the quantization isolates certain sections of the constant-energy surfaces [see Eqs. (7.2) and (7.3)].

2. Magnetic Fields Sufficiently Strong to Ignore the Effects of the Internal Crystal Fields. In this case, an electron no longer "sees" the crystal lattice field. The energy levels of the electron are given by the Landau formula [see Eq. (7.5)]. This case cannot be realized in practice because it requires fields of the order of 10^8-10^9 Oe. However, the special nature of the energy spectra of some metals is such that the Fermi electrons behave as "free" particles (i.e., particles which are not affected by the internal crystal fields) in moderately strong magnetic fields (this point is discussed in § 10).

We shall conclude this section by giving a quantum-mechanical solution of the problem of surface levels in a magnetic field, which we have solved by the quasiclassical approach in the preceding section. The need for such a solution is due to the fact that the first levels are of importance in high-frequency properties (§46) and the calculation of these levels by means of Eqs. (7.25)-(7.31) is far too approximate.

In this case, the Schrödinger equation, obtained by replacing the classical quantities with the operators, is relatively simple since a displacement along a normal to the surface of a sample is small (glancing angles) and, therefore, we can use the expansion $\varepsilon(p_x, p_y, p_z)$ near the point

$$\varepsilon(p_x^0, p_y^0, p_z) = \varepsilon_0, \quad v_y(p_x^0, p_y^0, p_z) = 0 \quad (\dot{v}_y < 0):$$

$$\left\{ -\alpha \left(\frac{c\hbar}{eH}\right)^2 \frac{\partial}{\partial y} + (y - y_{\max}) \right\} \psi = 0, \quad y_{\max} = y(0), \tag{8.22}$$

where

$$\alpha \equiv \alpha(\varepsilon, p_z) = -\left.\frac{\partial^2 y}{\partial x^2}\right|_{x=0} = -\frac{eH}{c}\left.\frac{\partial^2 p_x(\varepsilon, p_z, p_y)}{\partial p_y^2}\right|_{\substack{p_y=p_y^0(\varepsilon, p_z)\\ v_y=0,\ \dot v_y<0}}. \tag{8.23}$$

In the specular reflection of electrons, we obtain

$$\psi\,|_{y=0}=0, \quad \psi\,|_{y=\infty}=0. \tag{8.24}$$

Equation (8.22) can be altered by the substitution

$$\psi=\psi_1\left(\frac{eH}{c\hbar}(y-y_{\max})\right) \tag{8.25}$$

so that it becomes an equation in terms of the Airy functions and the boundary conditions (8.24) determine $y_{\max}$. However, $y_{\max}=\frac{p_x^{\max}(\varepsilon, p_z)-P_x}{eH/c}$, so that we can find $\varepsilon=\varepsilon_n(p_z, p_x)$. The explicit form of ε_n can be found using the expansion $\varepsilon(p_x, p_y)$. Consequently, we obtain

$$\varepsilon_n=\hbar v_x^0(p_z)\left(\frac{eH}{\hbar}\right)^{2/3}(2R(p_z))^{-1/3}a_n, \quad Ai(-a_n)=0, \tag{8.26}$$

where $Ai(x)$ is the Airy function.

Equation (8.26) solves the problem of finding the surface energy levels of electrons in a magnetic field directed parallel to the surface of a sample.

§9. Quantum Theory of the Scattering of Electrons with an Arbitrary Dispersion Law

The development of the scattering theory requires the calculation of the effective cross sections of various processes, and investigations of the wave surface of the scattered particle at large distances from the scattering center. The first part of this problem (the calculation of the effective cross sections) cannot be solved without knowledge of the structure of the inhomogeneity acting as a scatterer. However, the second part (the investigation of the form of the wave surface at large distances) can be formulated and solved making very general assumptions about the nature of the inhomogeneity, and the asymptotic behavior of the wave function of an electron is found to be determined by the shape of its constant-energy surface.

We shall consider the scattering of an electron in the state (s_0, p_0), i.e., an electron whose wave function is $\psi_{s_0 p_0}^{(0)}(r)=e^{ip_0 r/\hbar}u_{s_0 p_0}(r)$, and we shall postulate that this electron is scattered by a local inhomogeneity. If we assume that the scattering inhomogeneity (an impurity atom, a vacancy, etc.) is infinitely heavy, we find that the perturbed wave function of the electron is:

$$\psi_{s_0 p_0}(r)=\psi_{s_0 p_0}^{(0)}(r)-\sum_s\int\frac{C_{s_0 p_0}^{sp}e^{ipr/\hbar}d\tau_p}{\varepsilon_s(p)-\varepsilon_{s_0}(p_0)}. \tag{9.1}$$

The quantities $C_{s_0 p_0}^{sp}$ are determined by the nature of the scattering center, which is assumed to be located at the point $r=0$. If the inhomogeneity of a crystal is described by the perturbation operator $\hat\Lambda$, we find that in those cases when the perturbation is δ-like [25], these quantities are

$$C_{s_0 p_0}^{sp}=\frac{\Lambda_{s_0 p_0}^{sp}u_{sp}(r)}{1+\sum_s\int\frac{\Lambda_{s_0 p_0}^{sp}}{\varepsilon_s(p)-\varepsilon_{s_0}(p_0)}d\tau_p}.$$

Here, $\Lambda^{sp}_{s_0 p_0}$ is the matrix element of the operator $\hat{\Lambda}$.

If we can use the perturbation theory, we find that

$$C^{sp}_{s_0 p_0} = \Lambda^{sp}_{s_0 p_0} u_{sp}(r).$$

We shall use the notation

$$I_s(r) = \int \frac{C^{sp}_{s_0 p_0} e^{ipr/\hbar}}{\varepsilon_s(p) - \varepsilon}\, d\tau_p; \quad \varepsilon \equiv \varepsilon_{s_0}(p_0). \tag{9.2}$$

The specific structure of the integrals $I_s(r)$ allows us to investigate the scattered wave at large distances from the scatterer.

The integration indicated in Eq. (9.2) may be carried out over energy and over the constant-energy surfaces. It is found that

$$I_s(r) = \int \frac{I_s(r, \varepsilon')\, d\varepsilon'}{\varepsilon' - \varepsilon}, \tag{9.3}$$

where

$$I_s(r, \varepsilon') = \int\limits_{\varepsilon_s(p)=\varepsilon'} \frac{C^{sp}_{s_0 p_0} e^{ipr/\hbar}\, dS_{\varepsilon'}}{|\nabla \varepsilon_s(p)|}. \tag{9.4}$$

Here, $dS_{\varepsilon'}$ is an element of area of the constant-energy surface $\varepsilon_s(p) = \varepsilon'$, and the integration in Eq. (9.3) is carried out between the limits permissible for the s-th energy band (from $\varepsilon_{s\,\text{min}}$ to $\varepsilon_{s\,\text{max}}$).

In order to calculate the asymptotic value (at large values of r) of the integral in Eq. (9.4), it is convenient to describe the constant-energy surface $\varepsilon_s(p) = \varepsilon'$ by a system of coordinates which is linked to the lines of intersection of this surface by the planes $n \cdot p = \text{const}$ ($n = r/r$). We shall use $p \cdot n = u$ and integrate first over a strip lying between u and $u + du$ and then over

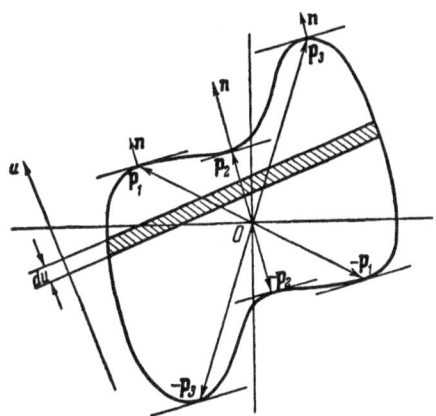

Fig. 28. Section of the surface $\varepsilon_s(p) = \varepsilon'$ cut by a plane $n \cdot p = \text{const}$. Here p_ν and $-p_\nu$ ($\nu = 1, 2, 3$) are the points of contact of a reference plane perpendicular to the direction of propagation.

the variable u (Fig. 28). This gives

$$I_s(r, \varepsilon') = \int f_s(u)\, e^{iur/h}\, du,$$

$$f_s(u) = \int\limits_{\substack{\varepsilon_s(p)\,=\,\varepsilon' \\ u\,<\,pn\,<\,u+du}} \frac{C_{s_0p_0}^{sp}}{|\nabla \varepsilon_s(p)|}\, dS_{\varepsilon'}. \qquad (9.5)$$

It is clear from this expression that the terms of the order of $1/r$, which correspond to the divergent (scattered) waves, appear as a result of integration near discontinuities of the function $f_s(u)$. We can easily show that the only points of discontinuity of the function $f_s(u)$ are the points of contact between the constant-energy surface $\varepsilon_s(p) = \varepsilon'$ and a reference plane perpendicular to the direction of propagation n (Fig. 28). Denoting these points of contact by p_ν, we obtain

$$I_s(r, \varepsilon') \approx \sum_\nu \frac{2\pi C_{s_0p_0}^{sp} e^{ip_\nu r/h}}{|\nabla \varepsilon_s(p_\nu)|\sqrt{K_\nu}} \frac{1}{r}, \qquad (9.6)$$

where K_ν is the Gaussian curvature of the surface $\varepsilon_s(p) = \varepsilon'$ at the point $p = p_\nu$.

The points of contact p_ν are defined by the equations

$$\varepsilon_s(p_\nu) = \varepsilon', \quad [n, v_s(p_\nu)] = 0. \qquad (9.7)$$

Substituting Eq. (9.6) into Eq. (9.3), we can demonstrate that the integrals over energy remain finite in the limit $r \to \infty$ only if the values $\varepsilon \equiv \varepsilon_{s0}(p_0)$ are located within the range of energies permissible in the s-th band. In this case, the integrals (9.3) or (9.8) should be regarded as the limiting values of the integrals

$$I_s(r) = \int \frac{I_s(r, \varepsilon')\, d\varepsilon'}{\varepsilon' - \varepsilon + i\gamma} \qquad (9.8)$$

in the case of vanishingly weak damping ($\gamma \to +0$).

The main contribution to the integrals of Eqs. (9.3) or (9.8) is made by the region near a pole $\varepsilon' = \varepsilon$.

Bearing this in mind, we can finally obtain an asymptotic expression for the perturbed wave function:

$$\psi_{s_0p_0}(r) \approx \psi_{s_0p_0}^{(0)} - \frac{(2\pi)^{3/2} i}{r^{1/2}} \sum_{s,\,\nu} \frac{C_{s_0p_0}^{sp} e^{ip_\nu r/h}}{|\nabla \varepsilon_s(p_\nu)|\sqrt{K_\nu}}, \qquad (9.9)$$

where the values of p_ν are given by Eq. (9.7) if ε' is replaced with $\varepsilon \equiv \varepsilon_0(p)$.

The results obtained here are in agreement with the conclusions drawn in §6 on the basis of the classical (corpuscular) approach. The procedure for finding p_ν of Eq. (9.7) is as follows: we have to find those points on the constant-energy surface at which the electron velocity $v_s = \nabla \varepsilon_s(p)$ coincides in direction with a specified scattering direction n.

The expression obtained for a perturbed wave function allows us to draw some conclusions on the nature of the scattered waves.

In most cases, the wave function of an electron scattered by a single inhomogeneity consists of several superimposed waves whose number is equal to the number of possible solutions of Eq. (9.7) for $\varepsilon' = \varepsilon_{s0}(p_0)$. Each of these waves is of specific form and travels at its own velocity (Fig. 29). Even in those cases when the energy of the scattered electron lies within the

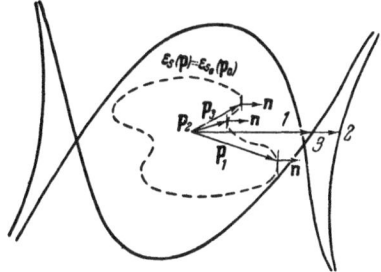

Fig. 29. Section of the wave surface of a scattered electron (continuous curves). The dashed curve represents a section of the constant-energy surface.

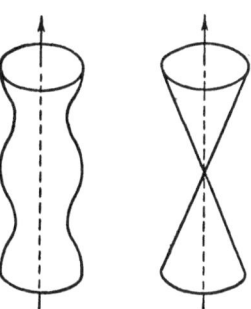

Fig. 30. Range of forbidden angles for a constant-energy surface of the corrugated cylinder type.

original energy band, the number of scattered waves exceeds unity because the constant-energy surfaces are usually not convex but complex. The amplitude of each of the scattered waves is determined not only by the dispersion law but also by the characteristics of the scattering center (via the quantities $C_{s_0 p_0}^{sp}$, which are related to the perturbation engendered by the scattering center). The phase of the wave, i.e., the shape of the wave surface is determined only by the dispersion law and the surface of each wave is polar with respect to the constant-energy surface of the electron $\varepsilon_s(\mathbf{p}) = \varepsilon$.

Knowing the wave function of the scattered electron we can calculate the effective scattering cross section along the direction $\mathbf{n}$ within a solid angle do:

$$d\sigma(\mathbf{n}) = \frac{16\pi^4}{|\nabla \varepsilon_{s_0}(\mathbf{p}_0)|} \sum_{s,\,\nu} \frac{\left| B_{s_0 p_0}^{sp_\nu} \right|^2}{|K_\nu||\nabla \varepsilon_s(\mathbf{p}_\nu)|}\, do, \tag{9.10}$$

where the quantities $B_{s_0 p_0}^{sp_\nu}$ are related in a simple manner to the coefficients $C_{s_0 p_0}^{sp_\nu}$.

Each of the terms in Eq. (9.10) determines the cross section of the process in which the scattered electron has a definite value of the crystal momentum and belongs to a definite energy band.

We must draw attention to the following point. In the case of open constant-energy surfaces, there may be directions along which a particle cannot move at all. Figure 30 shows the range of angles which are not allowed for a surface of the corrugated cylinder type. In this case, an electron cannot be scattered along a forbidden direction. We can state this more precisely as follows: a scattering event along a forbidden direction occurs only if the particle is transferred to a different sheet of the constant-energy surface.

§10. Magnetic Breakthrough

One of the most interesting features of the quantum motion of an electron in a magnetic field is undoubtedly the possibility of a transition from one classical orbit to another. This transition is known as the magnetic breakthrough or as the magnetic breakdown. The cause of this effect is the same as that of the tunnel effect: the wave nature of the electron. However, in the effect we are considering, the motion in a magnetic field is of special importance. In particular, the quantum nature of this motion becomes more pronounced as the intensity of the magnetic field increases (§7). Therefore, the probability of magnetic breakthrough increases with the magnetic field intensity.

Before presenting a rigorous theory of magnetic breakthrough, we must consider the most important features of the phenomenon. The transition from one classical orbit to another has a low but finite probability W only if these trajectories are located sufficiently close to each other and the potential barrier between them is not too high. These closely spaced trajectories may be of different origin. Both may belong to the same (Fig. 31a) or different (Fig. 31b) bands. The existence of closely spaced trajectories belonging to the same band is self-explanatory. All that is necessary is that a constant-energy surface should have a saddle point. Then, at values of p_z ($H_z = H$, $H_x = H_y = 0$) close to the critical value $p_z = p_{zk}$ (the trajectory for $p_z = p_{zk}$ is a "figure of eight"), the trajectories are close to one another (Fig. 31a). If the two trajectories in question belong to different bands, their closeness is always due to their proximity to a point of degeneracy. Sometimes, such degeneracy (if it is not essential) can be lifted. Different cases of closely spaced trajectories are shown in Fig. 31. It is worth noting the direction of motion of an electron in Figs. 31a and 31b. The transition of an electron from one classical trajectory to another is always accompanied by a change in the direction of motion if both trajectories belong to the same band. If the transition takes place from one band to another, the direction of motion does not change, and the electron seems to continue its motion along a classical trajectory. This can be seen particularly clearly in the weak interaction approximation. If the interaction between an electron and the lattice is ignored completely, the constant-energy surfaces are spherical (Fig. 32). An allowance for the periodic field of the lattice makes it necessary to replace the true momentum with the crystal momentum and this makes the points in the p space which differ by $2\pi\hbar b$ physically equivalent. This means that the spheres may be drawn not only around the origin of the coordinates but around any point $p = 2\pi\hbar b$ (Fig. 32). If the value of the energy is such that the neighboring spheres do not intersect, the interaction with the lattice is of little importance. However, if the spheres do intersect, there is some degeneracy which is lifted by an allowance for the periodic potential. Figure 32 shows the case of double degeneracy. When the degeneracy is lifted, we obtain two constant-energy surfaces which refer to two different bands. Far from the points of degeneracy, the surface can be described quite accurately by the original sphere. Figure 32 shows the direction of motion of an electron in a magnetic field perpendicular to the plane of the figure. In a sufficiently strong magnetic field, the electron does not "see" the potential barrier and moves along a circle. In weaker fields, the trajectory is far from circular. In particular, the trajectory may be open.

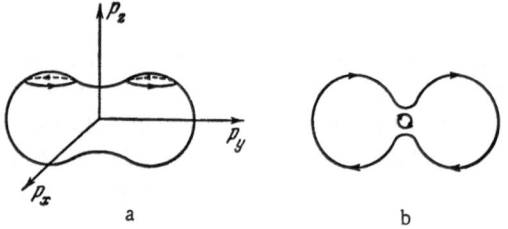

a   b

Fig. 31. Different nature of closely spaced trajectories: a) constant-energy surface with a "neck"; b) one surface located within another (here, the closely spaced trajectories belong to different bands). The arrows indicate the directions of motion.

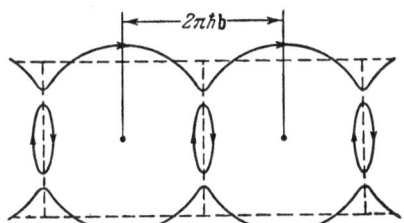

Fig. 32.   Appearance of closely
spaced trajectories in the model
of almost-free electrons.

The following points should be noted when comparing the interband (band—band) magnetic breakthrough with the intraband mechanism. In the intraband case, the breakthrough probability W tends to $\frac{1}{2}$ when the magnetic field is increased. The quantum limit corresponds to the "splitting of an electron" at the point of closest approach. In the interband transition, $W \to 1$ when the magnetic field tends to infinity (this is a more rigorous formulation of the statement: "the electron does not 'see' the potential barrier"). The most important point is that the critical breakthrough fields H_b are quite different in the interband and intraband cases. In the intraband breakthrough effect, we have

$$H_b = \Delta\varepsilon/\mu, \qquad (10.1)$$

where $\Delta\varepsilon$ is the potential barrier and μ is the Bohr magneton, whereas in the interband case, we have

$$H_b = (\Delta\varepsilon)^2/\varepsilon\mu \ll \Delta\varepsilon/\mu. \qquad (10.2)$$

In other words, the interband magnetic breakthrough occurs in much weaker fields than the intraband effect. The interpretation of Eq. (10.1) is quite simple. It means that when the magnetic field reaches $H \approx H_b$ for the intraband case the broadening of the trajectory due to the indeterminacy relationship is of the order of the distance between the trajectories. It follows from Eq. (8.18) that $\Delta p_x \Delta p_y \sim e\hbar H/c$. The indeterminacy of one component of the crystal momentum is governed by the dimensions of the trajectory $\Delta p_x \sim p$, whereas $\Delta p_y \sim \Delta\varepsilon/v$, where v is the electron velocity; since $p/v \sim m$ and $\mu \approx e\hbar/mc$, the intraband breakthrough field (10.1) can be estimated from the indeterminacy relationship. The fact that the interband breakthrough occurs in much weaker fields is a consequence of a rigorous theory of such breakthrough, which we shall now present.[1]

We shall start by considering in greater detail the structure of an energy spectrum with a small gap between the bands whose dispersion laws are $\varepsilon_1(p)$ and $\varepsilon_2(p)$. Let us assume that

$$\Delta(p) = |\varepsilon_1(p) - \varepsilon_2(p)|. \qquad (10.3)$$

We shall be interested in that region of the p space where $\Delta \ll \varepsilon$ ($\varepsilon \approx \varepsilon_{1,2}$). Considerations similar to those given in §2 can be applied to the structure of the spectrum near a point of degeneracy and they show that at a point p close to p', so that $\Delta(p') \ll \varepsilon(p')$, the dispersion law can be written in the form

---

[1] We shall follow the treatment given in [26] although the first calculation of the probability of the interband breakthrough was given by Blount [27]. A detailed presentation of the theory of magnetic breakthrough is given in a recent review [28].

$$\varepsilon_{1,2}(p) = \frac{1}{2}(\varepsilon_1 + \varepsilon_2) + A\,\delta p \pm \sqrt{\frac{(\Delta + B\,\delta p)^2}{4} + |\,C\,\delta p\,|^2},$$
(10.4)

which shows that the dispersion laws of different bands are simply different branches of the same analytic function. In Eq. (10.4), we have

$$A = \frac{1}{2}(v_1 + v_2), \quad B = v_1 - v_2, \quad C = v_{12} = v_{21}^*, \quad \delta p = p - p'.$$

All the quantities ε_1, ε_2, A, B, etc., are taken at the point p'. The presence of a small parameter Δ in the radicand of Eq. (10.4) shows that near p' the dispersion laws $\varepsilon_{1,2}(p)$ are "sharp" functions of their argument. A characteristic quantity in this respect is of the order of $\Delta 2\pi\hbar b/\varepsilon_0$, where ε_0 is of the order of the band width. The space groups of the crystal lattices of all metals have a center of inversion. We can show[2] that this ensures that the vector C is real at all points in the p space [26]. Therefore, we shall assume that $C = 0$. We must now consider two topologically different cases.

1. **The Vectors B and C Are Parallel.** Near the point p' there is no degeneracy because $\Delta(p) \neq 0$ and min $\Delta(p)$ is reached in a plane which is perpendicular to the vector $B(p')$ (and C) and passes close to the point p', so that

$$B\,\delta p = -\frac{B^2}{B^2 + 4C^2}\,\Delta.$$
(10.5)

In other words, the points where $\Delta \ll \varepsilon_0$ are concentrated near a surface of the closest approach between the bands, which we shall call the M surface. On the M surface and near it, we have $\Delta(p) \neq 0$. It follows from Eq. (10.4) that the spectrum near the M surface is given by the formula

$$\varepsilon_{1,2}(p) = \frac{1}{2}[\varepsilon_1(p_M) + \varepsilon_2(p_M)] + \frac{1}{2}[v_1^n(p_M) + v_2^n(p_M)]\,\delta p_n \pm \sqrt{\frac{\Delta^2(p_M)}{4} + (\delta p_n v_{12}^n)^2},$$
(10.6)

where $\delta p_n = n(p - p_M)$, n is a unit vector along the normal to the M surface at the point p_M lying on this surface, and $v_{12}^n = n v_{12}(p_M)$ (to within terms of the order of Δ_M, where Δ_M is a characteristic value of the parameter Δ on the M surface). It follows from Eq. (10.6) that the shortest distance between the classical trajectories in a magnetic field $\varepsilon_1(p) = \varepsilon$, $p_z = p_{z0}$ and $\varepsilon_2(p) = \varepsilon$, $p_z = p_{z0}$ is reached at the points of intersection of these trajectories with the M surface. This minimum distance between the trajectories, $p_{min} \simeq \Delta_M 2\pi\hbar b/\varepsilon_0$, is a smooth function of the energy ε and of the projection of the crystal momentum onto the magnetic field p_{z0}.

In the model of almost-free electrons, the M surface coincides with the boundaries of the Brillouin zone and the smallness of the parameter Δ is ensured by the low value of the energy of interaction between a conduction electron and the crystal lattice.

2. **The Vectors B and C Are Not Parallel.** In this case, the difference $\Delta(p)$ vanishes on a line passing near the point p' and formed by the intersection of the planes $\Delta(p') + B(p')\delta p = 0$ and $C(p')\delta p = 0$. Thus, a spectrum of this type is characterized by the existence, in the p space, of a line of p_0 points (a p_0 line) along which $\Delta(p_0) = 0$, i.e., we are dealing with double degeneracy $\varepsilon_1(p_0) = \varepsilon_2(p_0) = \varepsilon(p_0)$.

---

[2] The proof does not apply to magnetically ordered metals.

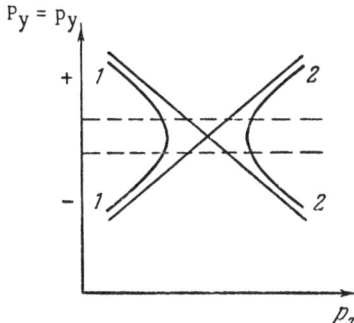

Fig. 33. Trajectories of an electron near the point of closest approach.

We can show (see §2 and [26] that the constant-energy surface $\varepsilon_s(\mathbf{p}) = \mathscr{E}$, which passes through the point $\mathbf{p}_0(\mathscr{E})$, is an elliptic cone whose axis coincides with the $\mathbf{p}_0$ line (Fig. 8). The two sheets (parts) of this cone separated by the point $\mathbf{p}_0(\mathscr{E})$ represent the constant-energy surfaces in the first and second bands. We have already mentioned that the situation described here may result because of the essential degeneracy associated with a given type of symmetry of the crystal lattice of a metal. Such degeneracy should be observed in all metals with the hcp structure. There are many such metals (Zn, Cd, Mg, etc.).

An experimental observation of a spectrum of this type depends on whether the Fermi surface (§1) passes through the $\mathbf{p}_0$ line. In the case of graphite, the Fermi surface does indeed pass through the $\mathbf{p}_0$ line. For some metals, the Fermi surface is close to the line of essential degeneracy and may pass through it.

We note that in both cases the classical trajectories in the region of closest approach are different branches of the same hyperbola (Fig. 33).

Our calculations will be based on the quasiclassical approach to the motion of an electron and we shall ignore the interband (band–band) transitions over a large part of the trajectory. Far from the region of closest approach of the trajectories, we can use the stationary Schrödinger equation with the Hamiltonian derived as described in §8, i.e., by replacing the components of the kinematic momentum by the operators obeying the commutation rules (8.18) and following this by the symmetrization of the expression obtained. In the representation in which $\hat{p}_y$ is diagonal, the Schrödinger equation can be written in the form[3]:

$$\varepsilon_s\left(P_x + i\sigma\,\frac{\partial}{\partial p_y}\,;\; p_y;\; p_z\right) g_s(\mathbf{p}) = \mathscr{E} g_s(\mathbf{p}). \tag{10.7}$$

The parameter $\sigma = e\hbar H/c$, the root of which ($\sqrt{\sigma}$) is always small compared with the characteristic change in the crystal momentum, can be used as the quasiclassical parameter [21].

The quasiclassical solution of Eq. (10.7), corresponding to the classical motion along trajectories defined by $\varepsilon_s(\mathbf{p}) = \mathscr{E}$, $P_z = p_{z0}$, can be represented by

$$\left.\begin{aligned}
g_s^{\pm}(\mathbf{p}) &= \sum_{\nu} f_{s\nu}^{(\pm)}(p_y)\,\delta(p_x - P_{x0})\,\delta(p_z - p_{z0}), \\
f_{s\nu}^{(\pm)}(p_y) &= \frac{C_{\nu,\,s}^{(\pm)}}{\sqrt{\left|\dfrac{\partial \varepsilon_s}{\partial P_x}\right|}}\exp\left\{\frac{i}{\sigma}\left(P_{x0}p_y + \int^{p_y} p_x^{(\nu,\,s)}(p_y')\,dp_y'\right)\right\}.
\end{aligned}\right\} \tag{10.8}$$

_____

[3] We shall employ the usual vector potential [see Eq. (4.16)].

The index ν denotes one of the solutions of the equation

$$\varepsilon_s(p_x(p_y),\, p_y,\, p_z) = \mathscr{E}, \qquad (10.9)$$

and the sign $\pm$ refers to the two quasiclassical ranges of the values of p_y separated by a small interval within which Eq. (10.7) is invalid because the band–band transitions are important in this interval. As is usual in the quasiclassical approximation, the constant coefficients $C_{s\nu}^{(\pm)}$ should be related to one another by an approximate solution of the Schrödinger equation in that region where the motion differs considerably from the classical case.

This region is shown bounded by dashed lines in Fig. 33. The relationship between the amplitudes $C_{s\nu}^{(\pm)}$ can be described by introducing special "matching" matrices $\tau_{ss'}^{(\nu)}$:

$$C_{s\nu}^{(+)} = \sum_{s'=1,\,2} \tau_{ss'}^{(\nu)} C_{s'\nu}^{(-)}. \qquad (10.10)$$

The "matching" must be carried out independently for each pair of trajectories (the different values of ν are not related).

Thus, an investigation of the dynamics of a conduction electron under magnetic breakthrough conditions reduces to finding the matrix $\hat\tau$ whose elements are functions of the magnetic field H and of two quantities which are conserved: the energy $\mathscr{E}$ and the projection of the momentum p_{z0} along the magnetic field.

C a s e I. Since we are interested in local properties, we may assume – without a loss of generality – that the origin of the coordinates lies on the M surface which is a plane. The p_x axis is parallel to the M plane and the p_z axis (the magnetic field is, as usual, directed along the z axis) makes an angle θ with this plane.

We can show [26] that, near the line of intersection of the plane $p_z = p_{z0}$ with the M plane, the system of equations for the wave functions $g_s(\mathbf{p})$ can be written as follows:

$$\left\{\varepsilon_s\left(P_x + i\sigma\frac{\partial}{\partial p_y},\, \bar p_y,\, p_z\right) - \mathscr{E}\right\} g_s(\boldsymbol{p}) + \delta p_y \sum_{s'=1,\,2} v_{ss'}^{(y)}\left(P_x + i\sigma\frac{\partial}{\partial p_y},\, \bar p_y,\, p_z\right) g_{s'}(\boldsymbol{p}) = 0, \qquad (10.11)$$

where

$$\delta p_y = p_y - \bar p_y, \quad \delta p_y \ll \bar p_y = p_{z0}\tan\theta,$$

which is derived bearing in mind the quasiclassical nature of the motion, i.e., on the assumption that the parameter σ is small. However, the terms associated with magnetic breakthrough (containing $v_{ss'}$ for $s \neq s'$) are not rejected and full use is made of the structure of the energy spectrum near the point of closest approach. Solving this system in the specified approximation and matching the solution with the functions (10.8), we find – after making fairly complex calculations – the following expressions for the elements of the matrix $\tau_{ss'}^{(\nu)}$:

$$\hat\tau^{(\nu)} = \begin{pmatrix} \tau_\nu e^{i\omega_\nu} & -\rho_\nu \\ \rho_\nu & \tau_\nu e^{-i\omega_\nu} \end{pmatrix}, \quad \tau_\nu^2 + \rho_\nu^2 = 1, \qquad (10.12)$$

where

$$\rho_\nu = e^{-\pi\gamma_\nu/2}; \quad \gamma_\nu = \frac{c\bar\Delta_\nu^2(\mathscr{E},\, p_{z0})}{e\hbar H\left|\bar v_x^\nu \bar v_{12}^{n\nu}\right|\cos\theta}, \qquad (10.13)$$

and

$$\omega_\gamma = \left[\frac{\pi}{4} + \frac{\gamma_\nu}{2} - \frac{\gamma_\nu}{2}\ln\frac{\gamma_\nu}{2} + \arg\Gamma\left(\frac{i\gamma_\nu}{2}\right)\right]\text{sign }\bar{v}_x^\nu. \tag{10.14}$$

The bar above Δ_ν and v means that they are taken at a point lying on the M plane, half-way between the trajectories $\varepsilon_{1,2}(\mathbf{p}) = \mathscr{E}$, $p_z = p_{z0}$; $\Gamma(z)$ is the Riemann gamma function. The unitarity of the matrix $\hat{\tau}$ ensures that the law of conservation of particles (the continuity of the density of the probability flux) is obeyed. The quantity ω_ν represents a phase discontinuity of the wave function occurring during magnetic breakthrough, whose probability W is determined by the interband breakthrough parameter γ_ν. When $\gamma_\nu \to 0$ (i.e., when the probability of break-through tends to unity) the phase discontinuity tends to zero. By definition,

$$W_\nu = |\tau_{12}^\nu|^2 = |\tau_{21}^\nu|^2 = \rho_\nu^2 = e^{-H_{bM}^\nu/H}, \tag{10.15}$$

where

$$H_{bM}^\nu = H_{bM}^\nu(\mathscr{E}, p_{z0}) = \frac{c\pi\overline{\Delta}_\nu^2}{e\hbar\,|\bar{v}_x^\nu\bar{v}_{12}^{n,\,\nu}|\cos\theta}. \tag{10.16}$$

Equation (10.16) confirms our earlier estimate represented by Eq. (10.2). The formulas obtained are valid for any value of the magnetic field. According to Eqs. (10.15) and (10.16), the probability W_ν decreases when the angle θ between the magnetic field H and the M plane is increased. As $\theta \to \pi/2$, we find that $H_{bM}^\nu \to \infty$, and $W_\nu \to 0$, but when the magnetic field is almost perpendicular to the M plane the classical trajectories approach in a region which is not small and the method given above is inapplicable. An analysis of our conclusions shows that Eqs. (10.12)-(10.16) can be used if $\Delta\theta \gg \mu H/\varepsilon^0$ where $\Delta\theta = \pi/2 - \theta$, which does not impose very severe restrictions on these formulas, particularly if we bear in mind that the magnetic fields are weak compared with the internal atomic fields.

Case II. This case is concerned with the values of p_y for which the transitions located near the point, $p_0^{(\gamma)}(p_{z0})$.are important. An approximate system of equations, valid for small values of $\delta p_y = p_y - p_0^{(\gamma)}(p_{z0})$, can be solved in the quasiclassical approximation and matched to give the matrix $\hat{\tau}$, which is still given by Eq. (10.12) but now the "frequency" ω and the break-through parameter γ are of the form

$$\omega = \frac{\pi}{4} + \frac{\gamma}{2} - \frac{\gamma}{2}\ln\frac{\gamma}{2} + \arg\Gamma\left(\frac{i\gamma}{2}\right), \quad \gamma = B\,\delta p_z^2/\sigma. \tag{10.17}$$

Here, $\delta p_z = p_0^z(\mathscr{E}) - p_{z0}$, and B is a dimensionless parameter which depends on the direction of the magnetic field and on the shape of the elliptic cone near the point of degeneracy. The formulas (10.17) are valid only if the direction of the magnetic field is such that the classical trajectories near the point of degeneracy have two branches and, consequently, magnetic break-through is possible. In this range of directions of the magnetic field, the dimensionless parameter is $B \sim 1$, but on approach to the boundary of this region the parameter B tends to infinity ($W \to 0$). The second of the formulas in (10.17) shows that the probability of the interband breakthrough is close to unity for $\delta p_z \sim \sqrt{\sigma}$. This means that electrons moving along trajectories lying in a narrow layer near the point of degeneracy participate in the interband breakthrough. This is reflected in macroscopic effects. In the same manner as Eqs. (10.12)-(10.16), the expressions in Eq. (10.17) are valid for any value of the magnetic field. However, we must remember that full magnetic breakthrough can occur only when the probability of a jump from one trajectory to another due to collisions is low compared with W. In other words, magnetic breakthrough may by observed only at low temperatures in perfect samples of a metal.

The structure of the electron energy spectrum in a magnetic field (§7) is closely related to the nature of the classical trajectories of electrons. In the quasiclassical approximation, the closed trajectories result in discrete energy levels which are infinitely degenerate (the multiplicity of degeneracy is infinite) and the open trajectories are not quantized. The problem of the energy spectrum of electrons under magnetic breakthrough conditions has not yet been solved completely. The difficulty is that, if the magnetic field direction is arbitrary, aperiodic trajectories appear even if the breakthrough probability is close to unity (this comment applies to the case when the breakthrough converts a closed to an open trajectory; see, for example, Fig. 32). However, the problem can be solved along certain directions of the magnetic field. We shall assume that the magnetic field is parallel to one of the simple crystallographic directions and that the trajectories are open in the first band and closed in the second (Fig. 32). The open direction (the p_y axis) is parallel to one of the reciprocal lattice vectors. The period along this direction is $2\pi\hbar b$ and the trajectories are symmetrical with respect to the p_y axis. In this case (case I according to our classification), the energy levels can be obtained from the condition of periodicity of a wave function along the p_y axis:

$$g_s(p_y) = g_s(p_y + 2\pi\hbar b).\tag{10.18}$$

We shall again omit the intermediate steps (which can be found in [26]) and give the final result, which is an equation for the determination of the energy levels of an electron:

$$\frac{S_2}{2\sigma} - \omega = \pi n - \arccos\frac{\rho^2 \sin\Phi_1}{\sqrt{[2\tau\cos\varphi - (1+\tau^2)\cos\Phi_1]^2 + \rho^4\sin^2\Phi_1}},\tag{10.19}$$

where

$$\Phi_1 = \frac{S_1}{2\sigma} + \omega, \qquad \varphi = \frac{2\pi\hbar b P_x}{\sigma}.$$

Here, n is an arbitrary integer; $S_2(\mathscr{E}, p_z)$ is the area inside a closed trajectory $\varepsilon_2(\mathbf{p}) = \mathscr{E}$, $p_z = $ const; $S_1(\mathscr{E}, p_z)$ is the area of a figure formed by two open trajectories in the first band and by the boundaries of the cell; the quantities ω, $\rho^2 = W$, $\tau^2 = 1 - W$ are defined by Eqs. (10.13) and (10.14). For fixed values of n and p_z, Eq. (10.19) gives a single-valued function $\varepsilon = \varepsilon_n(p_z, \varphi)$, which is periodic in φ. A new band structure appears and the width of the "magnetic" band is of the order of $\hbar\omega_H\sqrt{1 - W^2}$, whereas the distance between the bands is of the order of $W\hbar\omega_H$ (ω_H is the cyclotron frequency). The structure of the spectrum can be made clear by considering the limiting cases. It is convenient to start with the case of strong magnetic fields. When the magnetic field reaches a sufficiently high value, the breakthrough parameter becomes very small and W tends to unity. Since the "frequency" $\omega \to 0$ and $\gamma \to 0$, Eq. (10.19) becomes

$$S_1(\varepsilon, p_z) + S_2(\varepsilon, p_z) = 2\pi\sigma(n + 1/2), \quad \sigma = e\hbar H/c.\tag{10.20}$$

This equation describes discrete energy levels (for a fixed value of p_z) corresponding to the motion along the classical orbit consisting of parts of trajectories in the first and the second bands. The energy levels gradually broaden as the magnetic field is reduced. In the opposite limiting case ($\gamma \to \infty$, $W \to 0$), Eq. (10.19) splits into two: one part describes infinite motion along an open trajectory in the first band, and the other describes discrete energy levels corresponding to the classical motion along a closed trajectory in the second band.

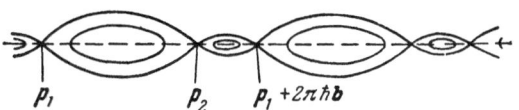

p_1 p_2 $p_1 + 2\pi\hbar b$

Fig. 34. Constant-energy surfaces with one self-intersecting trajectory per surface (it is assumed that p_z is constant). The points of self-intersection are distributed periodically.

The states described by Eq. (10.19) involve the transport of charge. The average velocity along the x axis is not zero and, to within the accuracy specified earlier, it is equal to $\partial \varepsilon_n / \partial P_x \approx v_0 \sqrt{1 - W^2}$, where $v_0 = \varepsilon_0 / 2\hbar b$ is a characteristic velocity in a band for H = 0.

Case II can be considered also for special directions of the magnetic field. Figure 34 shows a configuration for which the complete electron energy spectrum can be calculated: in this case, the magnetic field is perpendicular to the p_0 line, on which a set of points of degeneracy is distributed in a periodic manner. For a fixed value of p_z, each constant-energy surface includes one self-intersecting open trajectory. The other trajectories (corresponding to other values of p_z) split into a system of closed curves. An allowance for magnetic breakthrough spreads the discrete levels that correspond to the motion along closed trajectories into a band whose state is determined by the value of P_x.

This means that the average velocity in a plane perpendicular to the magnetic field (along the x axis) is not equal to zero and is of the order of Wv_0. In contrast to case I, the velocity depends strongly on the value of p_z and it decreases exponentially away from the critical trajectory.

We have considered essentially the two simplest cases of the change in the electron energy spectrum due to magnetic breakthrough, in the form of the disappearance and appearance of open trajectories. In a real crystal, we may encounter a much greater variety of situations. When the magnetic field is increased, we should observe the gradual rupture of increasingly wider "necks" and this should result in a significant departure from the simple quasiclassical description given in §7. We shall stress the role of magnetic breakthrough in subsequent discussions of the behavior of metals in strong magnetic fields.

CHAPTER II

STATISTICAL MECHANICS OF CONDUCTION ELECTRONS

In this chapter, we shall consider the equilibrium properties of metals at low temperatures. In keeping with the rest of the book, we shall concentrate on those phenomena and properties which are sensitive to the dispersion law of conduction electrons.

We shall present our treatment on the basis of the "gas model," i.e., we shall always assume (unless stated otherwise) that conduction electrons form an ideal gas of charged particles. This approach has the advantage of simplicity and is justified by the fact that, in the most interesting cases (low temperatures and strong magnetic fields), the results derived from the Fermi liquid model (§1) are identical with those deduced from the gas approximation. All cases of discrepancy between these two models will be mentioned specially. However, we must remember that, strictly speaking, the concept of a conduction electron as an elementary excitation with a definite crystal momentum has a meaning only for excitations whose energies are of the order of the Fermi energy (§1). In the course of derivation, we shall frequently use the gas-kinetic terminology for states far from the Fermi surface but most of the final results obtained will apply to electrons whose energies are of the order of the Fermi energy. Some of the formulas will include the volume of the Fermi surface. According to the Landau–Luttinger theorem [1], this volume is not affected by interactions. Therefore, such concepts as the "number of electrons" or the "number of holes" have a physical meaning.

§11. Criterion Separating Metals from Dielectrics. Fermi Energy. Fermi Surface. Number of Electrons

We described in Chap. I the general structure of the electron energy spectrum and considered the behavior of a single electron in external fields. We shall now discuss a statistical ensemble of electrons. The properties of such an ensemble depend strongly on whether a crystal is a dielectric or a metal. The electron band structure, which is the basis of our discussion, allows us to formulate the criterion by means of which we can distinguish a metal (conductor) from a dielectric (insulator). The presence of bands which are partly filled with electrons at absolute zero is typical of metals. Dielectrics have either completely filled or completely empty energy bands. The complete filling of an energy band represents uniform occupancy of all the crystal momentum (**p**) states. Since one cell of the phase space cannot contain more than two electrons (with opposite directions of the spins), it is clear that each band cannot contain more than $2\aleph$ electrons, where $\aleph$ is the number of unit cells in a crystal. If there were no overlap between the energy bands, all the crystals with an even number of electrons per unit cell would be dielectrics and all those with an odd number would be metals. The band overlap makes this classification quite unreliable and explains why, in the crystalline state, most of the elements are metals.

As a rule, metals have several partly filled bands. These are known as the conduction bands. They are responsible for the metallic properties, particularly the electrical conductivity. The electrons fill uniformly the **p** space but those located in the deep bands take practically no part in the thermal motion,[1] because energies of the order of several electron-volts are needed to excite them, i.e., to transfer them to the conduction band. Therefore, one of the most important characteristics of a metal is the number of electrons in the partly filled bands (N), which – as indicated by the preceding discussion – should depend very weakly on the temperature. The total number of electrons in the partly filled bands should, naturally, be equal to an integer multiplied by the number of atoms in the crystal.

The electron system has a single chemical potential ζ, which determines the level of electron occupancy of the energy bands at absolute zero. The chemical potential is frequently measured from the bottom of an appropriate band. In this case, each band has its own chemical potential $\zeta_i = \zeta - \varepsilon_{0i}$, where ε_{0i} is the energy representing the bottom of the band. At T = 0, all the electron states below a certain limit are occupied and this limit is known as the Fermi energy ε_F and the corresponding constant-energy surface is known as the Fermi surface.

The Fermi energy of a metal is located within one of the energy bands, The Fermi surface is complex and periodic and – for most metals – passes continuously through the whole reciprocal lattice (this is an open surface in the terminology of §2). A closed Fermi surface is repeated periodically in each cell of the **p** space.

At T = 0, electrons occupy only those parts of the **p** space where $\varepsilon < \varepsilon_F$ and, therefore, the number of conduction electrons per unit cell is

$$n = 2\frac{\Delta_F}{\Delta}; \quad \Delta_F = \int_{(\Delta)} E(\varepsilon - \varepsilon_F)\, d\tau_p. \tag{11.1}$$

Here, the integration is carried out over the volume Δ of one unit cell in the **p** space and the function E(x) is unity for x > 0 and zero for x ≤ 0 (§3).

The electron density N/V (N is the number of free electrons in a crystal of volume V) is

$$N/V = 2\Delta_F/(2\pi\hbar)^3, \tag{11.2}$$

or, if we use the density-of-states definition (§3), we can write

$$N = \int_0^{\varepsilon_F} \nu(\varepsilon)\, d\varepsilon. \tag{11.2a}$$

For a closed Fermi surface, Δ_F denotes the volume which is enclosed by the Fermi surface. This surface can consist of several sheets located within the same cell.

In those cases when a metal has several partly filled bands, the Fermi surface splits into several subsurfaces (the number of these subsurfaces is the same as the number of bands) and the nature of intersection of these subsurfaces (if they intersect at all) is rigorously defined by the principle of the nonintersection of terms (§2). The quantity Δ_F^s for each of these surfaces now represents the number of electrons per unit cell in the relevant (s-th) band:

$$n_s = 2\Delta_F^s/\Delta. \tag{11.3}$$

Naturally,

$$\sum_s n_s = n. \tag{11.4}$$

---

[1] More exactly, they move together with the lattice ions. The division of the whole electron-ion system into two independent subsystems – electrons and ions (lattice) – is justified by the considerable difference between the velocities of electrons and ions, which is due to the difference between their masses. This division is known as the adiabatic approximation.

It is at present impossible to calculate theoretically the shape of the Fermi surface for all metals. Therefore, it is natural to determine the Fermi surfaces on the basis of the available experimental data. However, even this approach requires the use of some trial models. In developing these models, one usually employs either the almost-free electron approximation (due to Brillouin) or the tight-binding approximation (due to Bloch) [2]. The most important aspect of these two methods is the careful allowance for the symmetry of a crystal, which makes it possible to predict the general outline of the surface and to find the points of essential degeneracy, which makes it possible to determine the structure of the split-off parts of the Fermi surface (see, for example, [3]).

The idea behind the first method (using the concept of almost-free electrons) is presented briefly in §10. The principal consequence of this approach is that the anisotropy of the Fermi surface is entirely the result of the periodicity of a crystal. In the zeroth approximation, the Fermi surface is a set of spheres of radius $2\pi\hbar (3N/8\pi V)^{1/3}$ whose centers are located at points in the $\mathbf{p}$ space which are equivalent to $p = 0$, i.e., at the points $p = 2\pi\hbar\mathbf{b}$, where $\mathbf{b}$ is an arbitrary reciprocal lattice vector. An allowance for the interaction reduces to the lifting of the degeneracy (if the diameter of a sphere is greater than the dimensions of a cell) and gives rise to considerable changes in the Fermi surface, such as the formation of open surfaces. The use of the method of almost-free electrons in the construction of the Fermi surfaces [3] is additionally justified by the fact that, in the case of polyvalent metals, for which the results of the construction are in reasonable agreement with the experimental data, the Fermi energy is very high and the lattice field is largely compensated by the conduction electrons. Figure 35 shows the Fermi surfaces of several polyvalent metals. It is worth noting the great variety of shapes and the fact that the surfaces consist of segments of a sphere.

The other approximation (the tight-binding approach) is based on the expansion of the electron energy as a Fourier series and only the terms satisfying all the symmetry elements of a crystal are retained. The Fermi surfaces of several metals (Ag, Au, etc.) are considered in [4] and [5]. These Fermi surfaces match the experimental results [6] and, in some cases, resemble very closely the surfaces deduced from the almost-free electron model. The results obtained by these two different methods are similar because all the symmetry elements of crystals are included correctly in both cases.

Sometimes, the shape of the Fermi surface or some part of it can be investigated without the use of a specific model or approximation. This is true when the Fermi surface is located near singular points in the $\mathbf{p}$ space (§§2, 3). In the simplest case, the Fermi surface is close to the bottom or the top of a band. In other words, the Fermi surface lies close to that point in the $\mathbf{p}$ space where the energy in a given band reaches its minimum or maximum value. If we are speaking of the whole Fermi surface, we find that this applies only to the case of a very slight overlap of the energy bands. This is observed for group V metals (Sb, As, Bi). The crystal lattices of these metals are of the same type. They all belong to the rhombohedral system with two atoms per unit cell and the special feature of these lattices is that they can be derived from simple cubic lattices with one atom per cell by a slight displacement of the atoms. Crystals with an odd number of electrons per cell (we recall that Sb, As, and Bi have an odd number of electrons per atom because they belong to group V) should be good metals, i.e., the Fermi energy should be located somewhere in the middle of an allowed band because the band contains only two electrons for each unit cell. However, the change in the translation symmetry which results in the doubling of the cell volume reduces greatly the number of electrons in the conduction band and it can, in principle, even convert a metal into a dielectric. We recall that doubling the cell volume in the coordinate space halves its volume in the $\mathbf{p}$ space.

Figure 36 shows, by way of example, the effects of doubling the period in the one- and two-dimensional cases. Group-periodic analysis methods [7] demonstrate that few very gen-

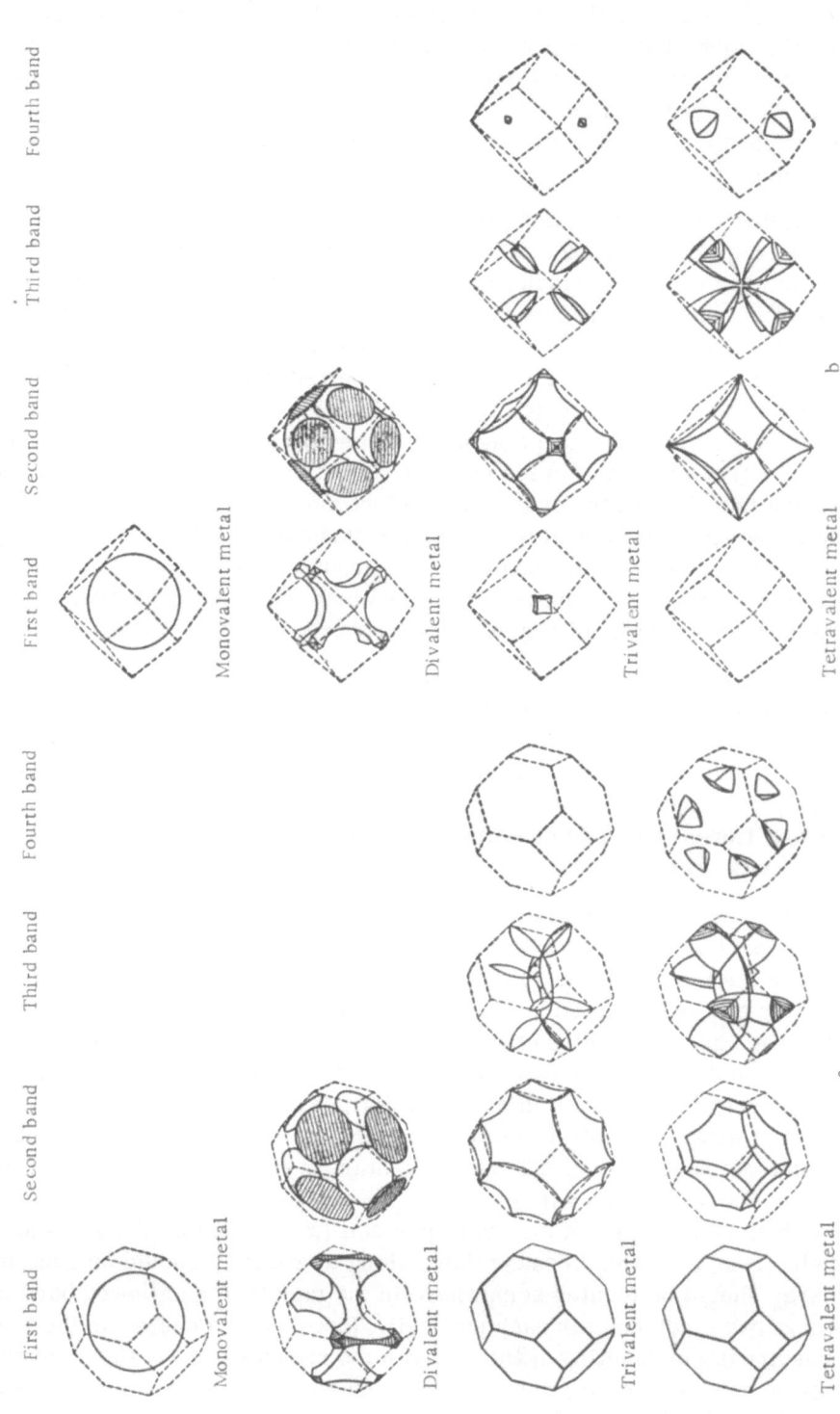

Fig. 35. Fermi surfaces given by Harrison [3]: a) fcc crystals; b) bcc crystals.

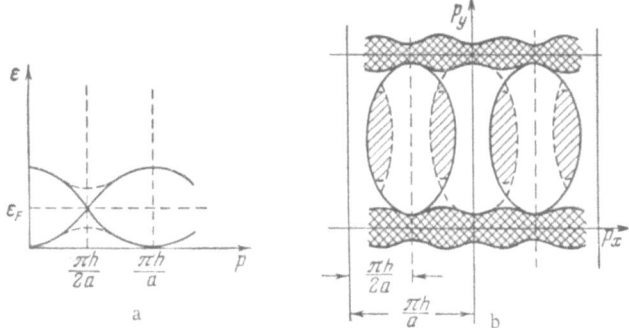

Fig. 36. Changes in the electron energy spectrum
resulting from doubling of the lattice period. a)
One-dimensional case: doubling of the period converts a "metal" with a half-filled band into a dielectric. b) Two-dimensional case: doubling of
the period along the x axis reduces greatly the
number of conduction electrons in a "metal" with
an initially half-filled band; small electron bands
(single hatching) and small hole bands (double
hatching) appear in the structure

eral assumptions can be used to show that the energy considerations favor the appearance of
very small (but finite) internal deformations in the cubic lattice. These deformations reduce
the symmetry of the lattice and double the volume of a unit cell (see also [7a]). One of the conditions for such a conversion is that the Fermi surface of the original cubic metal should be
located in the $\mathbf{p}$ space quite close to the points of essential degeneracy. Obviously, there is no
need to seek any special physical justification for the explanation of the fact that the Fermi
surface passes close to these points (located on the boundary of the Brillouin zone) because the
surface of a sphere whose volume is in accordance with the number of electrons per unit cell
must pass very close (in the numerical sense) to the boundaries of the Brillouin zone. Since
the almost-free model provides a good approximation to the experimentally determined shapes
of the Fermi surfaces of polyvalent metals, the arguments based on the almost-free electron
approximation are quite convincing.[2]

The spectra of group V metals given in [7] are very similar but this similarity is due to
degeneracy (§§2, 10). If the degeneracy is ignored, it is found that the Fermi surface located
close to that value of $\mathbf{p} = \mathbf{p}_0$ at which the energy in a given band is minimal, $\varepsilon(\mathbf{p}_0) = \varepsilon_0$, is an ellipsoid whose semiaxes are $\sqrt{2m_1(\varepsilon_F - \varepsilon_0)}$, $\sqrt{2m_2(\varepsilon_F - \varepsilon_0)}$, $\sqrt{2m_3(\varepsilon_F - \varepsilon_0)}$, where m_1, m_2, m_3 are
the principal values of the effective mass tensor $(\partial^2 \varepsilon / \partial p_i \, \partial p_k)_{\mathbf{p}=\mathbf{p}_0}$. It follows easily from Eq.
(11.3) that

$$\varepsilon_F - \varepsilon_0 = \left(\frac{3N}{8\pi V} \right)^{2/3} \frac{(2\pi\hbar)^2}{2 \, (m_1 m_2 m_3)^{1/3}}. \tag{11.5}$$

If the Fermi surface is located close to $\mathbf{p} = \mathbf{p}_1$, where $\varepsilon(\mathbf{p})$ reaches its maximum $\varepsilon(\mathbf{p}_1) = \varepsilon_1$, the
surface is again an ellipsoid. Its semiaxes are now $\sqrt{2m_1'(\varepsilon_1 - \varepsilon_F)}$, $\sqrt{2m_2'(\varepsilon_1 - \varepsilon_F)}$, $\sqrt{2m_3'(\varepsilon_1 - \varepsilon_F)}$;

---

[2] Detailed presentation of the theory of the electron energy structure of Bi and a comparison
with the experimental results are given in [7b].

where $m_1^!$, $m_2^!$, $m_3^!$ are the principal values of the effective mass tensor taken with the sign reversed. The quantity $\varepsilon_1 - \varepsilon_F$ is frequently associated with the number of free electron states in a given band (N'):

$$\varepsilon_1 - \varepsilon_F = \left(\frac{3N'}{8\pi V}\right)^{2/3} \frac{(2\pi\hbar)^2}{2\left(m_1' m_2' m_3'\right)^{1/3}}. \tag{11.6}$$

An additional comment should also be made about the limited band overlap [8]. As already mentioned, the electrons located close to a maximum in a band have negative effective masses and, therefore, their behavior resembles that of positive particles. In particular, they are attracted by electrons with positive effective masses. This Coulomb attraction may result in the formation of a quasi atom consisting of two electrons (having positive and negative effective masses) and the elimination of the band overlap with the consequent appearance of a forbidden band, i.e., of an energy gap between the allowed bands. The condition ensuring that such a gap does not appear is

$$n^{1/3}/m^* > 2e^2/(3\pi^2)^{2/3}\hbar^2$$

and can be obtained as the condition that the Fermi (kinetic) energy is higher than the energy of the Coulomb attraction. Thus, if there is some limited band overlap (small values of n), it follows that the electrons or the holes should have very small effective masses. This condition is indeed satisfied by charge carriers in bismuth and in other group V metals.

We have considered those cases in which the whole Fermi surface is an ellipsoid. This ellipsoid may only be a small part of a complex Fermi surface. It is clear from our discussion that this happens when the Fermi energy ε_F is close to one of the critical values of the energy ε_c at which a new sheet of the surface splits off. A part of the energy surface located near $p = p_c$, i.e., the point at which the new sheet appears, is described satisfactorily by the equation for an ellipsoid. We note that the number of electrons (if $\varepsilon_F > \varepsilon_c$) or the number of vacant states (if $\varepsilon_F < \varepsilon_c$) within an ellipsoid is related to the quantity $\varepsilon_F - \varepsilon_c$ by Eq. (11.5) in the first case and by Eq. (11.6) in the second. If a conical point ε_c is located near the Fermi surface, i.e., if the Fermi energy is close to one of its critical values ε_c at which the topology of the constant-energy surfaces changes, the shape of the whole Fermi surface cannot be deduced from general considerations. However, that part of the surface which is located close to p_c can be approximated satisfactorily by the equation of a two-sheet hyperboloid. We shall show later that such a part of the Fermi surface plays an important role because the velocity vanishes at $p = p_c$ (§§2, 13).

We have already considered (§§2, 10) the structure of the energy spectrum near the points of degeneracy. The very special nature of the intersection of constant-energy surfaces forces us to regard the surfaces passing through a point of degeneracy as single self-intersecting surfaces. The identity of the Fermi energy ε_F with one of the isolated critical values of the energy ε_c at which there is a change in the topology of the constant-energy surfaces is accidental (see also §13). However, in the case of band overlap, it is very likely that the Fermi surface is self-intersecting because the degeneracy usually occurs at values of the energy which fill a certain finite interval. Moreover, we must remember that essential degeneracy occurs in some crystals along certain lines and on certain planes in the p space. Therefore, if the Fermi surface has a point of degeneracy, it is natural to assume that both parts of the Fermi surface (those lying in the two bands) are located in the same p space and that part of this p space is doubly filled with electrons. Those electrons which fill uniformly the whole of the p space, i.e., the electrons in completely filled bands, are ignored. Figure 37 shows the Fermi surface of graphite [9]. All the p space outside the "hole" sheet is filled with electrons and the electron sheets are doubly filled. We note also that Eq. (11.3), which relates the volume enclosed by the Fermi

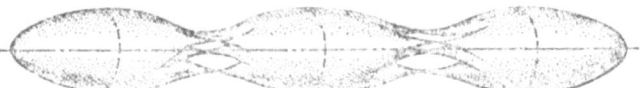

Fig. 37. Fermi surface of graphite.

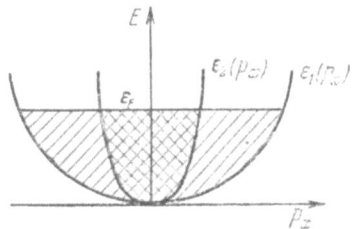

Fig. 38. Occupancy of energy bands
with a common minimum. In spite of
the same occupancy level ($\varepsilon_{1,2} < \varepsilon_F$),
the number of electrons in the two
bands are different.

surface Δ_F^s to the number of electrons in a band n_s, needs no correction or generalization in the degenerate case.

The special nature of the values of the crystal momentum at which degeneracy is observed means that, at these values, the energy usually has its minima or maxima. If two bands have a common minimum, the numbers of electrons in these bands are not equal although they have the same extremal energies (Fig. 38).

We must now say a few words about the Fermi energy of a dielectric. If we assume that a dielectric can be derived from a metal by the gradual reduction in the number of electrons or "holes"(for example, by the introduction of acceptor or donor impurities), it naturally follows that the Fermi surface should contract to a point when the number of carriers is reduced and that, in the limit, the Fermi energy should coincide with the band edge. However, it is usual to define the Fermi energy of a dielectric[3] as the limiting value of the chemical potential of electrons which is reached as the temperature tends to zero. According to this definition, the Fermi level lies in the forbidden band, i.e., between the last filled and the first empty bands (if there are only two bands, the Fermi energy is located exactly in the middle of the forbidden band).

§12. Thermodynamics of Conduction Electrons

Knowing the dispersion law and, consequently, the density of the electron states (§3) it is possible to develop the thermodynamics of conduction electrons.

We have already mentioned that the structure of the electron energy spectrum of a metal in its nonsuperconducting state is such that the charged elementary excitations which we call conduction electrons form a gas of quasi particles which obeys the Fermi-Dirac statistics. This means that the equilibrium distribution function of electrons $n_F(\varepsilon)$ is the Fermi function

---

[3] When the electron properties of a dielectric are considered, it is usually called an intrinsic semiconductor.

$$n_F(\varepsilon) = \frac{1}{e^{\frac{\varepsilon - \zeta}{T}} + 1}, \tag{12.1}$$

where T is the temperature measured in ergs and ζ is the chemical potential defined by the normalization condition

$$N = \int \frac{\nu(\varepsilon)\,d\varepsilon}{e^{\frac{\varepsilon - \zeta}{T}} + 1}. \tag{12.2}$$

The integration should be extended to all energies since the band structure is represented by the function $\nu(\varepsilon)$: in the forbidden regions $\nu(\varepsilon) \equiv 0$.

A thermodynamic theory can be stated conveniently by defining the thermodynamic potential [10]

$$\Omega = -T \int \nu(\varepsilon) \ln\left[1 + e^{\frac{\zeta - \varepsilon}{T}}\right] d\varepsilon, \tag{12.3}$$

whose derivative with respect to the chemical potential is equal to the total number of particles $(-\partial\Omega/\partial\zeta = N)$.

After integration by parts, we obtain

$$\Omega = -\int \frac{N(\varepsilon)\,d\varepsilon}{e^{\frac{\varepsilon - \zeta}{T}} + 1}. \tag{12.4}$$

On the other hand, the total energy of the electron gas is

$$E = \int \frac{\varepsilon \nu(\varepsilon)\,d\varepsilon}{e^{\frac{\varepsilon - \zeta}{T}} + 1}. \tag{12.5}$$

A comparison of Eqs. (12.4) and (12.5) shows that the well-known relationship $PV = \frac{2}{3}E$, where P is the pressure (we recall that $\Omega = -PV$) applies only to free electrons which satisfy

$$\nu(\varepsilon) = dN/d\varepsilon \sim \varepsilon^{1/2} \quad \text{[see formulas (3.16) and (3.17)].}$$

Since $\Omega = -PV$, Eqs. (12.2) and (12.4) can be regarded as parametric forms of the equation of state of a gas of conduction electrons. The parameter is the chemical potential ζ. It is evident from these formulas that the equation of state depends strongly on the density of states $\nu(\varepsilon)$, i.e., it depends on the dispersion law.

We shall now consider the most important case, which is that of low temperatures, i.e., we shall assume that the electron gas is strongly degenerate.

The condition of degeneracy of the free-electron gas is very simple: $T \ll \varepsilon_F$. In the case of conduction electrons, which have a complex dispersion law, the condition of degeneracy implies that

$$T \ll \min|\varepsilon_F - \varepsilon_c|, \tag{12.6}$$

where ε_c represents special (critical or extremal) values of the energy.

If the condition (12.6) is satisfied, the integrals containing the Fermi function $n_F(\varepsilon)$ can be calculated using the well-known expansion as a power series in temperature:

$$\int \varphi(\varepsilon)\, n_F(\varepsilon)\, d\varepsilon \approx \int_0^\zeta \varphi(\varepsilon)\, d\varepsilon + \frac{\pi^2}{6}\, T^2 \left(\frac{\partial \varphi}{\partial \varepsilon}\right)_{\varepsilon = \zeta} + \cdots. \tag{12.7}$$

It follows from Eqs. (12.4), (12.7), and (12.6) that

$$\Omega = \Omega_0 - \frac{\pi^2}{6} \nu(\zeta) T^2 + \dots \tag{12.8}$$

Here, Ω_0 is the thermodynamic potential Ω at T = 0:

$$\Omega_0 = -\int_0^{\zeta} N(\varepsilon)\, d\varepsilon. \tag{12.8a}$$

The second term in Eq. (12.8) can be regarded as a small correction to Ω_0. If the chemical potential ζ is expressed in terms of the density in the zeroth approximation, i.e., if ζ is replaced by the Fermi energy ε_F, which is related to the electron density by Eq. (11.2), the free energy can be written in the form

$$F = F_0 - \frac{\pi^2}{6} \nu(\varepsilon_F) T^2 + \dots, \qquad F_0 = \int_0^{\varepsilon_F} \nu(\varepsilon)\, \varepsilon\, d\varepsilon. \tag{12.9}$$

Hence, we can find the entropy S_e and the specific heat C_e of the conduction-electron gas:[1]

$$S_e = \frac{\pi^2}{3} \nu(\varepsilon_F) T, \qquad C_e = \frac{\pi^2}{3} \nu(s_F) T. \tag{12.10}$$

Thus, at low temperatures, the electron specific heat C_e depends linearly on the temperature [Eq. (12.6)]. This conclusion is not related to the dispersion law of conduction electrons. Moreover, Eq. (12.10) remains valid even when an allowance is made for the interaction between electrons in the same sense as in Landau's theory of Fermi liquids [11].

Equations (12.8)-(12.10) can be used to find approximate expressions for the energy E in the thermodynamic potential Φ of the conduction-electron gas:

$$E \approx E_0 + \frac{\pi^2}{6} \nu(\varepsilon_F) T^2, \qquad E_0 = \int_0^{\varepsilon_F} \nu(\varepsilon)\, \varepsilon\, d\varepsilon; \tag{12.11}$$

$$\Phi \approx \Phi_0 - \frac{\pi^2}{6} \nu(\varepsilon_F) T^2, \qquad \Phi_0 = N\varepsilon_F. \tag{12.12}$$

The last equation allows us to determine the thermal expansion coefficient α, which is $-(1/V)(\partial V/\partial T)_P$:

$$\alpha \approx \frac{\pi^2}{3} V^{-1} \frac{\partial}{\partial P} [\nu(\varepsilon_F)] T. \tag{12.13}$$

We note that Eq. (12.13) is no less general than Eq. (12.10). In particular, Eq. (12.13) applies if the interaction between electrons is taken into account. In fact, $\alpha \sim \partial^2\Phi/\partial T\partial P = \partial^2\Phi/\partial P\partial T = \partial S/\partial P$, and the entropy of electrons, which is a quantity with combinatorial meaning, is determined by the system of states which is not affected by the transition from a Fermi gas to a Fermi liquid.

For most metals, the condition of degeneracy is satisfied at practically all temperatures. Therefore, conduction electrons make a very small contribution to the thermodynamic parameters of a metal (specific heat, internal energy, etc.). However, at low temperatures, when the internal energy of the vibrational degrees of freedom tends rapidly ($\sim T^4$) to zero, the role

---

[1]Since the temperature is measured in ergs, the specific heat is a dimensionless quantity.

played by conduction electrons increases significantly. At low temperatures,[2] defined by $T \ll \Theta\sqrt{\Theta/\varepsilon_F}$, where Θ is the Debye temperature, conduction electrons actually determine the thermodynamic properties of metals. This means, in particular, that the specific heat and the thermal expansion coefficient of a metal depend linearly on the temperature if the temperature is sufficiently low. This is in accordance with well-known experimental and theoretical investigations.

At low temperatures, the constant γ in the Grüneisen "law" (i.e., the temperature-independent ratio of the thermal expansion coefficient α to the specific heat C) for metals is given by the pressure dependence of the density of states of electrons.[3]

$$\gamma \approx \frac{\partial}{\partial P} \ln \nu(\varepsilon_F). \tag{12.14}$$

It is evident from Eq. (12.10) that measurements of the specific heat of metals at low temperatures should make it possible to determine the very important characteristic of the electron energy spectrum, which is the density of electron states at the Fermi energy, whereas measurements of the constant γ should yield information on the pressure dependence of the density of electron states [Eq. (12.14)].

The de Haas–van Alphen and similar effects (§§15-17) indicate that most metals have bands with anomalously low occupancies or completely filled bands. In other words, the Fermi surface of most metals is located near the extremal points of one or more bands in the **p** space.

The band in which the number of electrons or the number of vacant states N_a is considerably less than the number of unit cells in a crystal $\Re$ ($N_a \ll \Re$) will be called anomalous. The existence of anomalous bands leads to special temperature dependences of the thermodynamic properties of crystals, which are due to the strong temperature dependence of the number of electrons in such bands.

By definition,

$$N_a(T) = \int \frac{\nu_a(\varepsilon)\, d\varepsilon}{e^{\frac{\varepsilon - \zeta(T)}{T}} + 1}, \tag{12.15}$$

and the temperature dependence of the chemical potential common to all bands is very weak:

$$\zeta(T) \approx \varepsilon_F - \frac{\pi^2}{6} \frac{\nu'(\varepsilon_F)}{\nu(\varepsilon_F)} T^2. \tag{12.16}$$

Here, $\nu(\varepsilon)$ is the total density of electron states in all the bands and $\nu_a(\varepsilon)$ is the density of electron states in an anomalous band. The anomaly appears because $|\varepsilon_{extr}^{(a)} - \varepsilon_F| \ll \nu(\varepsilon_F)/\nu'(\varepsilon_F)$, where $\varepsilon_{extr}^{(a)}$ is the minimum value of the energy in an almost empty band or the maximum value in an almost full band, and $\nu(\varepsilon_F)/\nu'(\varepsilon_F)$ represents the "gap" between the Fermi energy ε_F and the critical values of the energy ε_c in the principal (normal) bands.

---

[2] We shall assume that $\nu(\varepsilon_F) \approx N/\varepsilon_F$ and $N \approx \Re$.

[3] The Grüneisen constant is determined as the ratio of the total expansion coefficient to the total specific heat and, therefore, at temperatures $\Theta\sqrt{\Theta/\varepsilon_F} \ll T \ll \Theta$ the Grüneisen constant is naturally governed by the pressure dependence of the Debye temperature.

It follows from Eqs. (12.16), (12.15), and (12.7) that

$$
N_a(T) \approx
\begin{cases}
N_a(0) + \dfrac{\pi^2}{6}\, \nu_a(\varepsilon_F)\, \dfrac{d}{d\varepsilon_F}\left[\ln \dfrac{\nu_a(\varepsilon_F)}{\nu(\varepsilon_F)}\right] T^2, \quad T \ll |\varepsilon_F - \varepsilon_{\text{extr}}^{(a)}|, \\[2ex]
N_a(0)\, \dfrac{3(\sqrt{2}-1)}{4\sqrt{2}}\, \sqrt{\pi}\, \zeta\!\left(\dfrac{3}{2}\right)\left\{\dfrac{T}{|\varepsilon_F - \varepsilon_{\text{extr}}^{(a)}|}\right\}^{3/2}, \\[2ex]
\dfrac{\nu(\varepsilon_F)}{\nu'(\varepsilon_F)} \gg T \gg |\varepsilon_F - \varepsilon_{\text{extr}}^{(a)}|.
\end{cases}
\tag{12.17}
$$

The above formulas are derived on the assumption that, according to Eq. (3.7) $\nu_a(\varepsilon) \sim |\varepsilon - \varepsilon_{\text{extr}}^{(a)}|^{1/2}$.

The expansion of the specific heat of a metal in powers of the temperature begins with a linear term. In the presence of anomalous bands, the next term in the expansion should not contain T^3 (which is the normal case) but $T^{3/2}$ [12]. The anomalous temperature dependence $N_a(T)$ may have an even greater influence on other properties of a metal. There are some cases in which the properties of a metal are entirely or principally determined by the electrons located in anomalous bands. The total temperature dependence of a given property is then simply a consequence of the temperature dependence of the number of electrons in an anomalous band.

We have so far ignored the fact that every sample is finite. This approach is fully justified in considering the bulk effects in those cases when the dimensions of a sample are far larger than the lattice constant.

The physical properties of small metal samples (particularly thin films and wires) have been investigated in detail under a great variety of conditions. Even a sketchy description of these investigations is outside the scope of the present monograph. However, there are several problems which are basically similar to those just considered. We shall deal with them briefly in the present and following sections (detailed treatment will be found in the references cited).

The energies of electrons in a film are quantized (§7) and we can show [12] that each surface (boundary) of a metal film contributes the following term to the density of states:

$$
\nu_\sigma(\varepsilon) = -\frac{2\sigma}{(2\pi\hbar)^2} \oint_{C(\varepsilon,\, e)} \frac{dl}{v},
\tag{12.18}
$$

where σ is the area of the surface of the film and the integration is carried out over a contour $C(\varepsilon, e)$ which is the projection onto a plane with a normal e (the unit vector e is perpendicular to the surface) of a curve defined by[4]

$$
\varepsilon(p) = \varepsilon; \quad ev = 0.
\tag{12.19}
$$

For a sample of arbitrary shape whose radius of curvature R is considerably greater than the de Broglie wavelength (i.e., $R \gg \lambda_B$), Eq. (12.18) can be generalized to

$$
\nu_\sigma(\varepsilon) = -\frac{2}{(2\pi\hbar)^2} \int d\sigma \oint_{C(\varepsilon,\, e)} \frac{dl}{v},
\tag{12.20}
$$

---

[4]For a convex surface, the curve defined by Eq. (12.19) is the boundary between light and dark on the surface $\varepsilon(p) = \varepsilon$ illuminated by a parallel beam of light along the direction e.

where the integration is taken over the whole surface of the metal (**e** is a function of the coordinates of this surface).

An allowance for the surface density of states makes it possible to calculate corrections to all the thermodynamic properties of a metal (specific heat, thermal expansion coefficient, etc.). These corrections are naturally small because the ratio a/L is small (L is the smallest dimension of a sample). It is very interesting to calculate the electron component of the surface tension of a metal $\varkappa_e = -\partial F_e/\partial\sigma$ which is given by [12]

$$\varkappa_e = \frac{1}{8\pi^2\hbar^2} \int\limits_0^{\varepsilon_F} S(\varepsilon, \, e) \, d\varepsilon, \tag{12.21}$$

where

$$S(\varepsilon, \, e) = \int E[\varepsilon - \varepsilon_0(p_\perp)] \, dp_\perp, \qquad p_\perp = p - (pe)\, e,$$

and the definition of $\varepsilon_0(p_\perp)$ is given by Eqs. (7.19) and (7.18).

The temperature-dependent terms omitted from Eq. (12.21) usually begin with a term proportional to T^2 $[\varkappa_e(T) \approx \varkappa_e(0)[1 + \beta(T/\varepsilon_F)^2], \, \beta \approx 1]$. We must stress that the electron component of the surface tension of a metal can be expressed, like other thermodynamic quantities, in terms of the dispersion law $\varepsilon = \varepsilon(p)$.

§13. Anomalous Electron Properties of Metals at High Pressures

A change in the topology of the constant-energy surfaces gives rise to a singularity in the density of states $\nu(\varepsilon)$, as shown in §3. Generally speaking, the values ε_c are located sufficiently far from the Fermi energy ε_F and the presence of singularities of critical energies ε_c can be established only from x-ray spectra. However, if there is a continuously variable parameter which would make $\varepsilon_F - \varepsilon_c$ pass through zero, i.e., which would make it possible to alter the topology of the Fermi surface, the singularities in the density of states $\nu(\varepsilon)$ and in the dynamics of electrons near the critical surface would lead to characteristic anomalies in the thermodynamic and transport properties of the electron gas in a metal [14].

One such parameter is the deformation of the lattice, particularly that caused by strong hydrostatic compression. It is known that high pressures reduce the anisotropy of most of the properties and, therefore, we may expect that the "corrugated cylinder" Fermi surfaces, typical of layered structures, would be gradually deformed into closed surfaces even if the total number of electrons in the conduction band were to remain the same (Fig. 39). Clearly, other changes in the topology of the Fermi surface could also occur. It ought to be stressed that these changes have nothing in common with the changes in the symmetry of the lattice and, therefore, they do not correspond to phase transitions of the second kind.

On the other hand, the original lattice may cease to be thermodynamically stable even before the critical Fermi surface is approached so that a phase transition of the first kind may occur before the appearance of the anomalies associated with changes in the Fermi surface topology.[1] However, since the time needed to change the electron structure during deformation is short in relation to the time needed to change the lattice in a phase transition, the anomalies

---

[1] This case evidently occurs in "isomorphous" transitions in which the lattice parameters vary discontinuously in spite of the fact that the crystal structure does not change.

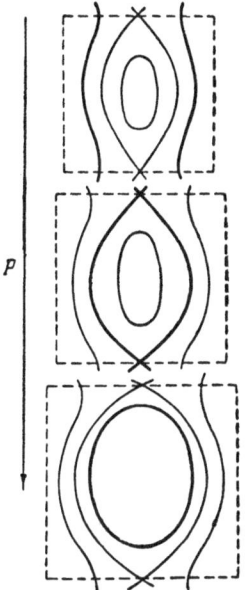

Fig. 39. Gradual transition from an open Fermi surface (shown by a thick line) to a closed surface. It is worth pointing out that an increase in the pressure P increases somewhat the dimensions of a unit cell in the reciprocal lattice.

in question may be observed briefly under metastable conditions. Therefore, we shall not consider the problem of stability or metastability of a state near the point of change in the topology of the Fermi surface.

The concentration of impurities or components in an alloy is another parameter which would seem to be capable of causing large changes in the chemical potential ζ and the Fermi surface. However, in the case of a disordered lattice, the Fermi surface concept itself loses its characteristic meaning in view of the absence of the translation symmetry and, therefore, a singularity in $\nu(\varepsilon)$ may broaden and actually disappear.[2]

Let us now consider the properties of a metal near an electron transition caused by a change in the topology of the Fermi surface.

If the chemical potential ζ is close to ε_c it follows from Eqs. (12.4) and (3.8) that the thermodynamic potential Ω is of the following form:

$$\Omega(\zeta, T) = \Omega_0(\zeta, T) + \delta\Omega, \tag{13.1}$$

where

$$\delta\Omega = -\int_0^\infty \frac{\delta N(\varepsilon)\, d\varepsilon}{e^{\frac{\varepsilon - \zeta}{T}} + 1}, \tag{13.2}$$

and

$$\delta N = \begin{cases} 0 & (\varepsilon < \varepsilon_c), \\ \frac{2}{3}\alpha(\varepsilon - \varepsilon_\kappa)^{3/2} & (\varepsilon > \varepsilon_c). \end{cases} \tag{13.3}$$

---

[2] This does not apply to Bi alloys, in which considerable changes in the Fermi surface are observed after the introduction of acceptor and donor impurities [15].

The meaning of the parameter α is defined by Eqs. (3.9) and (3.6). We recall that, to be specific, we are assuming that a new sheet of the Fermi surface appears if $\varepsilon > \varepsilon_c$. Assuming that $\zeta - \varepsilon_c = z$, we find that at low temperatures ($T \ll z$) near the point $z = 0$ the change in the thermodynamic potential Ω can be represented in the form

$$\delta\Omega = \begin{cases} -\dfrac{V\sqrt{\pi}}{2}\,\alpha T^{5/2} e^{-\frac{|z|}{T}}, & \text{(I)} \\[2mm] -\dfrac{4}{15}\,\alpha\,|z|^{5/2} - \dfrac{\pi^2}{6}\,\alpha T^2 |z|^{1/2}. & \text{(II)} \end{cases} \tag{13.4}$$

The transition from region I to region II corresponds to the appearance of a new sheet of the $\varepsilon(\mathbf{p}) = \zeta$ surface or to a reduction in its connectivity.

The two formulas in Eq. (13.4) are valid for $T \ll |z|$. At absolute zero, we have

$$\delta\Omega = \begin{cases} 0 & \text{(I)}, \\[2mm] -\dfrac{4}{15}\,\alpha\,|z|^{5/2} & \text{(II)}. \end{cases} \tag{13.5}$$

This means that the second derivatives of Ω at the point $z = 0$, which is the point of the electron anomaly, have a vertical kink, and the third derivatives approach infinity as $z^{-1/2}$.

Since the number of electrons in a conduction band is constant, it is more convenient to study the thermodynamic anomalies by means of the free energy $F(T, V, N)$ and not the potential Ω. The volume V is then a parameter which is related directly to the applied pressure. The critical energy ε_c is a direct function of the volume $\varepsilon_c = \varepsilon_c(V)$ and the chemical potential ζ is also a function of V because the number of particles is constant[3]:

$$N(\zeta, V) = N. \tag{13.6}$$

We shall use V_c to denote the volume at which there is a change in the Fermi surface topology, i.e.,

$$\zeta(V_c) = \varepsilon_c(V_c). \tag{13.7}$$

It follows from Eq. (13.6) and (13.7) that the modulus of z, which occurs in Eqs. (13.4) and (13.5), can be expressed in terms of $|V - V_c|$:

$$z = \gamma\,|V - V_c|; \quad \gamma = \left| \nu^{-1}(\varepsilon_c)\left\{ \frac{\partial N}{\partial V} + \nu(\varepsilon_c)\frac{\partial \varepsilon_c}{\partial V} \right\} \right|, \tag{13.8}$$

where we have used $\partial N/\partial\zeta = \nu(\zeta)$.

We shall write the free energy F in the form

$$F = F_0 + \delta F,$$

where F_0 is a smooth component of the free energy calculated from the density of states $\nu_0(\varepsilon)$. We can easily show that δF is quantitatively equal to an irregular correction $\delta\Omega$, which can be expressed in terms of V and T, i.e., δF can be found from Eqs. (13.4) and (13.5), in which $|z|$

---

[3] In the case of several overlapping bands, $N(\zeta, V)$ is the total number of particles in all the bands.

should be replaced in accordance with Eq. (13.8):

$$\delta F = \begin{cases} -\frac{\sqrt{\pi}}{2} \, \alpha T^{1/2} e^{-\frac{|z|}{T}}, & \text{(I)} \\[2mm] & (T \ll |z|, \ |z| = \gamma |V - V_c|) \\[2mm] -\frac{4}{15} \alpha |z|^{5/2} - \frac{\pi^2}{6} \alpha T^2 |z|^{1/2}. & \text{(II)} \end{cases}$$

Hence, when the temperature tends to absolute zero, we obtain

$$\delta \frac{C_e}{T} = -\delta \frac{\partial^2 F}{\partial T^2} = \begin{cases} 0, & \text{(I)} \\[2mm] \frac{\pi^2}{3} \alpha |z|^{1/2}, & \text{(II)} \end{cases} \tag{13.9}$$

$$\delta \frac{\partial P}{\partial V} = -\delta \frac{\partial^2 F}{\partial V^2} = \begin{cases} 0, & \text{(I)} \\[2mm] \alpha \gamma^2 |z|^{1/2}, & \text{(II)} \end{cases} \tag{13.10}$$

$$\delta \left(\frac{\partial P}{\partial T} \Big/ T \right) = -\frac{1}{T} \delta \frac{\partial^2 F}{\partial T \, \partial V} = \begin{cases} 0, \\[2mm] \frac{\pi^2}{3} \alpha \gamma |z|^{-1/2}. \end{cases} \tag{13.11}$$

The expressions in Eq. (13.9) describe the anomaly in the electron specific heat and the expressions in Eq. (13.10) describe the anomaly of the electron component of the compressibility.

The total pressure in a metal is the sum of the electron pressure and the partly compensated pressure of the lattice "core." However, since the lattice component of the compressibility $\partial P_L/\partial V$ is generally continuous at the point $z = 0$, a singularity in the total compressibility can be found from the expression for the electron component of the compressibility.[4]

It is evident from Eq. (13.11) that the strongest singularity is exhibited by the thermal pressure coefficient $\partial P/\partial T$. Bearing in mind the fact that the electron component of this coefficient predominates at low temperatures, we find that the nonsingular part of this coefficient is given by

$$\left(\frac{\partial P}{\partial T} \right)_V = AT, \qquad A = \frac{\pi^2}{3} \frac{\partial \nu}{\partial V}.$$

Thus, near a singularity we have

$$\left(\frac{\partial P}{\partial T} \right)_V = \frac{\pi^2}{3} T \left\{ \frac{\partial \nu}{\partial V} \pm \frac{1}{2} \alpha \gamma |z|^{-1/2} \right\}, \tag{13.12}$$

where the plus sign applies if $V - V_c > 0$ in region II, and the minus sign applies if $V - V_c < 0$ in region II. A similar singularity is exhibited also by the thermal expansion coefficient

$$\left(\frac{\partial V}{\partial T} \right)_P = -\left(\frac{\partial P}{\partial T} \right)_V \Big/ \left(\frac{\partial P}{\partial V} \right)_T = \frac{\pi^2 T}{3 \varkappa_0} \left\{ \frac{\partial \nu}{\partial V} \pm \frac{1}{2} \alpha \gamma |z|^{-1/2} \right\}, \tag{13.13}$$

where

$$\varkappa_0 = -\left(\frac{\partial P}{\partial V} \right)_{T=0}.$$

It is worth noting that, when an increase in the pressure causes the appearance of a new sheet at $P = P_c$, the thermal expansion coefficient near $P = P_c$ ($V = V_c$) is negative on the high-pressure side and that its absolute value increases to infinity when $P \to P_c$ (Fig. 40).

---

[4]The "lattice" component of the cohesive energy due to the conduction electrons may have a singularity similar in nature to that of the thermodynamic potential of electrons. This would give rise to a small correction to $\partial P/\partial V$ which would not alter the results to any significant extent.

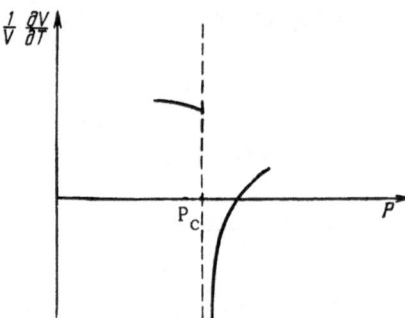

Fig. 40. Behavior of the thermal expansion coefficient (at $T = 0$) near a critical pressure P_c if the number of sheets of the Fermi surface increases at $P = P_c$.

In order to consider all the anomalies in Eqs. (13.9)-(13.11) along the pressure scale, it is sufficient to use $P - P_c = -\varkappa_0(V - V_c)$ and then

$$|z| = \gamma |V - V_c| = \frac{\gamma}{\varkappa_0}|P - P_c|. \tag{13.14}$$

It is convenient to estimate the coefficients γ, $\gamma/\varkappa_0$, and the critical pressure P_c by expressing them in terms of the initial energy difference $z_0 = (\zeta - \varepsilon_c)_{P=0}$ (at zero pressure) and in terms of the critical deformation $(V - V_c)/V_0$, at which the transition occurs:

$$\left.\begin{aligned} |z| &= \frac{|z_0||P - P_c|}{P_c}, \\ P_c &= \varkappa_J|V - V_c|, \\ \gamma &= \frac{z_0}{|V_0 - V_c|}. \end{aligned}\right\} \tag{13.15}$$

The expressions in Eq. (13.15) are simply estimates because the relationship between P and $V - V_0$ is not linear over a wide range of deformations. If we assume that critical deformations correspond to $(V_0 - V_c)/V_0 \sim 0.05$-$0.1$, it follows that $P_c \sim 5 \times 10^4$-10^5 kgf/cm^2.[5]

Strictly speaking, Eqs. (13.9)-(13.11) are valid only at absolute zero. At all other temperatures, the singularities of all the thermodynamic quantities are less sharp. The thermal broadening of the anomalies is $\Delta z \propto T$; on the pressure scale this gives

$$\frac{\Delta P}{P_c} \sim \frac{T}{|z_0|} = \frac{T}{|\zeta - e_c|_{P=0}}. \tag{13.16}$$

It follows from Eq. (13.10) that in the region of an anomaly the negative quantity $(\partial P/\partial V)_0 = -\varkappa_0$ has a positive correction $\sim z^{1/2}$. If, as a result of this correction, $P(V)$ ceases to be a monotonic function and the derivative $(\partial P/\partial V)$ becomes positive, we observe an isomorphous phase transition of the first kind which is accompanied by a discontinuity in the volume. At low positive values of z and at temperatures close to zero, the derivative $(\partial P/\partial V)$ can be expanded in the form

---

[5] Less symmetrical deformations (such as uniaxial compression or stretching) might be expected to alter the geometry of the Fermi surface much more strongly and thus significant changes may occur at lower stresses. In particular, a departure from the original surface of a crystal may result, even at very low deformations, in the splitting of the Fermi surface at the points of essential self-intersection.

$$\frac{\partial P}{\partial V} = -\varkappa_0 + \alpha\gamma^2 z^{1/2} - \varkappa_1 z,$$

which shows that a labile region $\partial P/\partial V > 0$ may be observed if $\alpha\gamma^2 > 2\sqrt{\varkappa_0\varkappa_1}$ and that this region lies on the right of the point $z = 0$ in the interval

$$\alpha\gamma^2 - \sqrt{(\alpha\gamma^2)^2 - 4\varkappa_0\varkappa_1} < 2\varkappa_0 z < \alpha\gamma^2 + \sqrt{(\alpha\gamma^2)^2 - 4\varkappa_0\varkappa_1}.$$

Under these conditions, the electron transition point $z = 0$ can lie either in the metastable or the stable region.

Since a singularity at the point $z = 0$ becomes broader with rising temperature, one should not call an electron transition at this point a phase transition. This is why we are speaking of "anomalies" at the point $z = 0$ although, in accordance with Ehrenfest's terminology, we could call these anomalies transitions of the type $2\frac{1}{2}$ at $T = 0$ because the second derivatives of the thermodynamic potentials have a singularity proportional to $z^{1/2}$ and the third derivatives a singularity proportional to $z^{-1/2}$.

We demonstrated in §2 the possibility of a discontinuity in the density of electron states at some energy $\varepsilon = \varepsilon_c$ [see also Eq. (3.12)]. If hydrostatic pressure or some other deformation of a crystal causes the Fermi energy to become equal to this critical energy, an anomaly should be manifested by a discontinuity in the electron specific heat and in other second derivatives of the thermodynamic potential. In this special case, the electron anomaly does indeed resemble very closely a phase transition of the second kind.

The critical pressure cited in this section may be a gross overestimate. Recent investigations have clearly demonstrated that the Fermi surfaces of most metals have narrow "necks," very small split-off pockets, etc. (this is the fine structure of the electron energy spectrum). A change in the topology associated with the fine structure of the Fermi surface may be observed at relatively low pressures. This point of view is supported by some experimental investigations [15a]. Electron anomalies resulting from changes in the Fermi surface topology should appear not only in those cases which are described by Eqs. (13.9)-(13.11) but also in the magnetic properties (this point will be considered in the next section). They should also appear very clearly in the pressure dependences of the galvanomagnetic characteristics (§27). Moreover, the superconducting transition temperature has unexpectedly been found to be sensitive to changes in the surface topology [16].

A change in the Fermi surface topology clearly makes the dependence of the chemical potential on the external parameter (in this case, the pressure) a nonanalytic function. A sudden change in the chemical potential may be observed in its dependence on the magnetic field (this would also apply to other thermodynamic quantities), and the fact that anomalies in the dependence on the magnetic field can occur in fields attainable in the laboratory is related to the existence of a fine structure in the electron energy structure (§19).

The surface density of states governed by the reflection of electrons from a boundary of a sample is even more sensitive to changes in the Fermi surface geometry than is the bulk density of states. Equation (12.18) or (12.20) can be used to analyze the dependence of the density of states $\nu_\sigma(\varepsilon)$ near the critical values of the energy ε_c. It is then found that the singularities of $\nu_\sigma(\varepsilon)$ are sharper than the singularities of the bulk density of states. Thus, the appearance of a new sheet at $\varepsilon = \varepsilon_c$ is accompanied by a discontinuity in the surface density of states, and if a "neck" is broken at $\varepsilon = \varepsilon_c$ [this means that there is a conical point on the surface $\varepsilon(p) = \varepsilon_c$], the surface density of states $\nu_\sigma(\varepsilon)$ has a logarithmic discontinuity ($\delta\nu_\sigma(\varepsilon) \sim \ln |\varepsilon - \varepsilon_c|$). Naturally, this "sharpening" of the density-of-states singularities gives rise to a corresponding "sharpening" of the singularities in the thermodynamic properties at high pressures causing changes in the Fermi surface topology. For example, the coefficient in the expression for the surface component of the specific heat of a metal C_σ/T should have a discontinuity if a Fermi surface sheet disappears or appears at $P = P_c$. If a "neck" is broken at $P = P_c$, C_σ/T should have a logarithmic singularity [12].

§14. Paramagnetism and Diamagnetism in Weak Magnetic Fields

If we subject a metal to a magnetic field, many of its properties are significantly affected. Thus, for example, a magnetic field of the order of tens of kilo-oersteds may increase the low-temperature resistivity of a pure metal tens and sometimes even thousands of times. As a rule, the transport coefficients of metals (their electrical resistivity, thermal conductivity, etc.) are affected most. However, the thermodynamic equilibrium properties of a metal are also altered by the application of a magnetic field. In particular, all metals exhibit a magnetic moment in a magnetic field, and at low temperatures this moment depends in a complex manner on the magnitude and the direction of the magnetic field.

It is known [17] that the dependences of the thermodynamic properties on the magnetic field are due to the quantization of the energy of electrons and nuclei. In the case of electrons this applies also to electrons with a complex dispersion law, i.e., conduction electrons. In the classical approach an allowance for the magnetic field is made by the use of the kinematic momentum $\mathbf{p} = \mathbf{P} - (e/c)\mathbf{A}$ (see §§1, 4). In calculations of the thermodynamic potentials, the integration is carried out over the whole $\mathbf{P}$ space. By changing the variables (going over from $\mathbf{P}$ to $\mathbf{p}$), we can show that the thermodynamic potentials are independent of the magnetic field.

In the quantum approach, the dependence on the magnetic field arises from two causes. First, the electrons and nuclei have intrinsic magnetic moments (spins); secondly, the orbital (spatial) motion of charged particles becomes quantized in a magnetic field. The presence of intrinsic magnetic moments is responsible for paramagnetism and similar phenomena, and the quantization of the orbital motion gives rise to diamagnetism.

The quantity that will be discussed mainly in the present section is the static magnetic moment of a metal. This magnetic moment includes contributions not only from the conduction electrons (the electrons in partly filled energy bands) but also from the nuclei (their contribution is small because the nuclear magnetic moments are small), as well as from the electrons in filled bands which participate in the establishment of the diamagnetic moment.[1]

The diamagnetic susceptibility of conduction electrons can be expressed, using the gas model, in terms of the dispersion law [18]. However, the expression obtained is extremely complex and it depends on the properties of all the electrons, not only on those with $\varepsilon \sim \varepsilon_F$. On the other hand, the actual concept of the dispersion law is applicable only near the Fermi surface (§1). Therefore, the expression obtained in [18] is not of basic significance. In the case of the quadratic and isotropic dispersion law the diamagnetic susceptibility given in [18] becomes naturally identical with the value of the susceptibility calculated by Landau [19]:

$$\varkappa_d = \frac{1}{16 \cdot 3^{2/3} \pi^{4/3}} \frac{N^{1/3}}{m^* c^2} \tag{14.1}$$

where m* is the effective mass.

---

[1] We can distinguish the paramagnetic and the diamagnetic moments only in weak fields in which the moment is a linear function of the field.

The spins of the electrons in the filled bands are compensated.[2] The absence of vacant states in the filled bands results in an exponential temperature dependence of the paramagnetic moments of these bands. Therefore, the conduction electrons are those which dominate the paramagnetism of a metal (the Pauli magnetism).

The paramagnetic susceptibility of a metal $\varkappa_p$ depends weakly on the temperature [20] and at low temperatures $T \ll \varepsilon_F$. It is determined by the density of electrons per unit energy interval at the Fermi energy:

$$\varkappa_p = \mu^2 \nu\,(\varepsilon_F). \tag{14.2}$$

This formula is derived on the basis of the "gas model" of a metal, i.e., ignoring the interaction between electrons. An allowance for the electron–electron interaction alters Eq. (14.2) because the coefficient of $\nu(\varepsilon_F)$ is no longer μ^2.

The problem of the paramagnetic susceptibility is complicated still further by the need to allow for the spin–orbit coupling, which may alter significantly the value of the effective magnetic moment (this situation is observed in Bi [21]). In fact, under these conditions, the diamagnetic and the paramagnetic susceptibilities can no longer be calculated separately. We are speaking here of a consistent calculation of a magnetic moment of a metal. Basically, the division of the magnetic susceptibility into its diamagnetic and paramagnetic components is meaningless since these quantities are of the same order in a metal.[3] For example, the diamagnetic susceptibility of the free-electron gas is only one-third of the paramagnetic susceptibility. The values of the diamagnetic and the paramagnetic susceptibilities are comparable as a direct consequence of the electron gas degeneracy ($T \ll \varepsilon_F$).

In any case, some part of the magnetic susceptibility of a metal (in a very simple theory, this is the paramagnetic component) is proportional to the density of electron states per unit energy interval (at the energy $\varepsilon = \varepsilon_F$). This means that the spin paramagnetism has an anomaly when the Fermi surface topology is altered by pressure (see §13 as well as [14]). Let us consider the nature of this anomaly.

For simplicity, we shall limit our discussion to the case of an ideal Fermi gas. Then,

$$M_p = \tfrac{1}{2}\,\mu\,\{N\,(\zeta + \mu H) - N\,(\zeta - \mu H)\},$$

where $N(\varepsilon)$ is the number of electrons whose energies are less than ε (§3). Therefore, the paramagnetic susceptibility $\varkappa_p$ near an anomaly is

$$\varkappa_p = \varkappa_p^{(0)} + \tfrac{1}{2}\,\mu^2\,\{\delta\nu\,(\zeta + \mu H) + \delta\nu\,(\zeta - \mu H)\},$$

where

$$\delta\nu\,(z) = \begin{cases} 0 & (z < 0), \\ a z^{1/2} & (z > 0). \end{cases}$$

---

[2] The filled-band structure is typical of diamagnets. In the case of paramagnets, the filled bands contain electrons with uncompensated spins (this is analogous to the case of the inner-shell electrons with uncompensated spins in paramagnetic atoms). The exchange interaction between spins usually results in their ordering, i.e., in the appearance of ferromagnetism or antiferromagnetism, if the temperature is sufficiently low.

[3] Once again, bismuth is an exception: in the case of bismuth and its alloys, the diamagnetic susceptibility is anomalously high because of the small gaps between the bands [18].

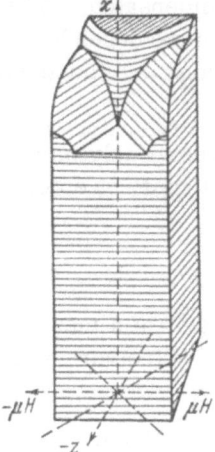

Fig. 41. Anomaly of the paramagnetic susceptibility near the point $P = P_c$.

The meaning of the parameter α is clear from Eq. (3.9). Thus, the irregular part of the paramagnetic susceptibility $\delta \varkappa_p$ is:

$$\delta \varkappa_p = \begin{cases} 0, & z \pm \mu H < 0, \\ \dfrac{1}{2} \alpha \mu^2 (z + \mu H)^{1/2}, & z + \mu H > 0, \ z - \mu H < 0, \\ \dfrac{1}{2} \alpha \mu^2 (z - \mu H)^{1/2}, & z + \mu H < 0, \ z - \mu H > 0, \\ \dfrac{1}{2} \alpha \mu^2 \left\{ (z + \mu H)^{1/2} + (z - \mu H)^{1/2} \right\} & z \pm \mu H > 0. \end{cases}$$

Although the anomaly of the magnetic susceptibility of a metal is complicated somewhat by the superposition of diamagnetism, the basic nature of the anomaly is not affected by the diamagnetic component. In particular, it is worth noting the highly unusual dependence of the magnetic susceptibility on the magnetic field near the critical pressure. This anomaly should be observed in a very narrow range of pressures $\Delta P \sim (\mu H / \varepsilon_F) P_c$.

The dependences of $\delta \varkappa_p$ on z and on the magnetic field are shown in Fig. 41.

The presence of an anomaly in the magnetic susceptibility at high pressures is not the only manifestation of the special nature of the behavior of the electron spectrum of metals in weak fields. For example, the temperature dependence of the magnetic susceptibility of some metals can be understood [13] only on the assumption that the electron spectra of these metals have a fine structure (bands with an anomalously low occupancy, small gaps between bands, etc.). Similar conclusions can be drawn by investigating the magnetic susceptibility of alloys as a function of their composition [22, 15].

§15. de Haas–van Alphen Effect in Strong Magnetic Fields

Investigations of the diamagnetism and the paramagnetism of metals (§14) have demonstrated that the development of a rigorous theory of magnetism in weak fields is a complex problem which cannot be solved without postulating some specific models: the magnetic susceptibility depends strongly on the interaction of electrons with one another and the final result cannot, in principle, be expressed simply in terms of the dispersion law. Moreover, the situation is complicated by the fact that the magnetic susceptibility of the conduction electrons is, generally speaking, of the same order of magnitude as the susceptibility of the electrons in the filled bands. Therefore, it is also difficult to determine experimentally the role of conduction electrons.

In strong fields ($\mu H \gtrsim T$), the situation is quite different. All metals exhibit an oscillatory dependence of the magnetic moment and other characteristics on the magnetic field, which is not observed for dielectrics or semiconductors. We shall show later that the principal parameters of the oscillations are so closely related to the Fermi surface that we can find the shape of the Fermi surface and the distribution of velocities on the surface from the experimental data. Most of the experimental work has been carried out on the oscillatory dependence of the magnetic moment (de Haas–van Alphen effect). However, this is not the only oscillatory effect. The specific heat, the electrical resistivity (Shubnikov–de Haas effect), and other quantities oscillate periodically with the magnetic field. The nature of these oscillations is the same as that of the magnetic moment. Therefore, we shall call all these phenomena the de Haas–van Alphen effect. Several oscillatory phenomena of a basically different nature (cyclotron resonance, geometrical resonance in the absorption of sound, etc.) have also been observed in magnetic fields. At the end of the present section, we shall formulate the criteria by means of which the de Haas–van Alphen effect can be distinguished from the other oscillatory effects.

Let us now define more precisely the conditions necessary for the observation of the de Haas–van Alphen effect. It is observed only in sufficiently pure single crystals. The criterion of the required purity is the ratio between the mean free path l and the radius of the orbit r_H in a magnetic field. We must satisfy the condition $l \ll r_H$ (the single-crystal requirement will be discussed later). Moreover, the magnetic field must be sufficiently strong ($\mu H \gtrsim T$) but not too strong. It should satisfy the following conditions:

$$kT \lesssim \mu H \ll \varepsilon_F, \qquad (15.1)$$

and the second of the above inequalities allows us to regard $\mu H / \varepsilon_F$ (§1) as the small parameter of the problem. We note that the oscillatory dependence is not a manifestation of some special state of a metal and there are, in fact, no limits to the conditions under which the de Haas–van Alphen effect exists. The conditions just specified give the range in which this effect is observed most easily; outside this range, the oscillation amplitude is exponentially small.

Before we present a consistent theory of the de Haas–van Alphen effect, we must consider the basic nature of the phenomenon. Like all other magnetic phenomena, the de Haas–van Alphen effect is, in the final analysis, of quantum origin. It is due to the quantization of the orbital motion of conduction electrons traveling in closed orbits in a magnetic field. This quantization of the energy means that the number of populated energy levels with a fixed value of the projection of the momentum along the magnetic field p_z changes by unity when the magnetic field (more precisely, the reciprocal of the magnetic field) is increased by an amount $\Delta(1/H)$ equal to $2\pi|e|\hbar/cS(\varepsilon_F, p_z)$ [see the quantization condition given by Eq. (7.3)]. The dependence of $\Delta(1/H)$ on p_z weakens somewhat the strong dependence on the magnetic field. However, the oscillatory effects are dominated by the electrons whose values of p_z are such that $S(\varepsilon_F, p_z)$ varies slowly with p_z, i.e., the values for which $S(\varepsilon_F, p_z)$ has an extremum. Thus, the most important characteristic of the de Haas–van Alphen effect, which is its oscillation period, is determined by the extremal sections of the Fermi surface (Fig. 42):

$$\Delta \frac{1}{H} = \frac{2\pi|e|\hbar}{cS_{\text{extr}}(\varepsilon_F)}. \qquad (15.2)$$

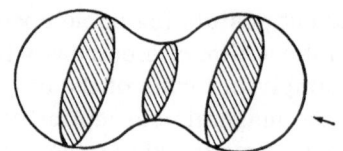

Fig. 42. Example of extremal
cross sections of the Fermi
surface. The direction of the
magnetic field is shown by the
arrow.

The involved shape of the Fermi surface (the presence of many different extremal sections) of most metals makes the magnetic moment an extremely complex function of the magnetic field, which cannot be analyzed in any simple manner.

We shall now present a consistent theory of the de Haas−van Alphen effect. We must calculate the thermodynamic potential Ω which will be used to find the magnetic moment, the specific heat, and other properties of interest to us. We shall use the "gas" approximation and we shall demonstrate in the next section (§16) that the inclusion of the Fermi-liquid interaction does not alter the nature of the phenomenon. According to general formulas

$$\Omega = -T \sum_{(s)} \ln\left\{ 1 + \exp\frac{\zeta - \varepsilon(s)}{T} \right\}. \tag{15.3}$$

The summation is carried out over all the states of an electron[1]: over the quantum number n [see Eq. (7.3)], over the two projections of the spin σ ($\sigma = \pm 1$), over all values of the z projection of the crystal momentum p_z, and over the number s of the partly filled bands. Since p_z varies continuously, the summation over p_z can be replaced by integration.

The number of states dn_{p_z} in the interval $(p_z, p_z + dp_z)$ can be easily calculated, keeping the other parameters (s, n, σ) fixed, by noting that

$$dn_{p_z} = \frac{V}{(2\pi\hbar)^3} \left\{ \iint dp_x\, dp_y \right\} dp_z, \tag{15.4}$$

where the integration must be carried out over all the values of p_x and p_y located between two neighboring classical trajectories of an electron in the momentum space, i.e., between the trajectories whose values of n differ by unity. According to Eq. (7.3), the required area is $2\pi|e|\hbar H/c$.

Therefore,

$$dn_{p_z} = \frac{|e|HV}{(2\pi\hbar)^2 c}\, dp_z. \tag{15.5}$$

The use of quasiclassical quantization formula does not imply that Eq. (15.5) is valid only in the quasiclassical approximation. According to Eq. (15.4), the density of states dn_{p_z}/dp_z is determined by the "cell area" in the p_x-p_y space $\left(\iint dp_x\, dp_y = \Delta p_x\, \Delta p_y \right)$, which depends only on the conditions of commutation of the operators $\hat{p}_x$ and $\hat{p}_y$. The application of the commutation rules of Eq. (8.18) gives again Eq. (15.5).

We must bear in mind that Eq. (15.5) is derived on the assumption that the electron energy spectrum is not affected by the magnetic field. This sets an upper limit to the magnetic

---

[1] We shall not consider the integration over the coordinate space because this reduces to the multiplication of the expressions obtained by the volume of a sample V.

field. However, significant changes in the electron energy spectra of most metals occur in fields $H \gtrsim 10^7$-10^8 Oe, which are not normally attainable in the laboratory. The only exceptions to this rule are metals of the Bi type (for details, see §19).

The quasiclassical quantization formulas (7.3) and the expression for the density of states (15.5) can be used to calculate the density of states $\nu_H(\varepsilon)$ per energy interval $d\varepsilon$ in a quantizing magnetic field

$$\nu_H(\varepsilon) = \frac{|e| HV}{(2\pi\hbar)^2 c} \frac{dp_z}{d\varepsilon}.$$

In order to calculate $dp_z/d\varepsilon$, we must sum over all the Landau subbands. Each subband is defined by the magnetic quantum number n and its dispersion law is the solution of Eq. (7.3) in the form $\varepsilon = \varepsilon_n(p_z)$. It is evident from Eq. (7.3) that for those values of $p_z^{(n)}$ for which $S(\varepsilon_n, p_z^{(n)}) = (2\pi\hbar e H)/c (n+\frac{1}{2})$ and $\partial S(\varepsilon_n, p_z^{(n)})/\partial p_z = 0$, the function $\varepsilon_n(p_z)$ has an extremum. If $S(\varepsilon_n, p_z)$ has its maximum value at $p_z = p_z^{(n)}$, the dispersion law $\varepsilon_n(p_z)$ has a minimum, whereas the minimum value of $S(\varepsilon_n, p_z)$ corresponds to a maximum of the dispersion law. It is clear that the singularities of the density of states should occur in each Landau subband at extremal values of the energy. We can easily show (§3) that the density $\nu_n(\varepsilon)$ approaches infinity at $\varepsilon = \varepsilon_n$ by the square-root law: $\nu(\varepsilon_n) \propto |\varepsilon - \varepsilon_n|^{-1/2}$. The presence of density-of-states singularities depending on the magnetic field gives rise to a periodic dependence of the thermodynamic properties on the reciprocal of the magnetic field [Eq. (15.2)].

Using the value of dn_{p_z}/dp_z, we can rewrite Eq. (15.3) in the following manner:

$$\Omega = \sum_{n=0}^{\infty} \varphi(n), \tag{15.6}$$

where

$$\varphi(n) = \frac{V |e| HT}{(2\pi\hbar)^2 c} \sum_{(\sigma, s)} \int_{-\infty}^{\infty} \ln\left\{1 + \exp\frac{\zeta - \varepsilon_{s\sigma}(p_z, n)}{T}\right\} dp_z. \tag{15.7}$$

It will now be convenient to use the Poisson formula [23][2]

$$\sum_{n_0}^{\infty} \chi(n) = \int_a^{\infty} \chi(n)\, dn + 2\operatorname{Re} \sum_{k=1}^{\infty} \int_a^{\infty} \chi(n)\, e^{2\pi i k n}\, dn. \tag{15.8}$$

Here, $\chi(n)$ is an arbitrary function and a is a number which lies between $n_0 - 1$ and n_0. Under our conditions, $n_0 = 0$ and it is convenient to assume that $a = -\frac{1}{2}$.

It then follows from Eqs. (15.6) and (15.8) that

$$\Omega = \int_{-1/2}^{\infty} \varphi(n)\, dn + 2\operatorname{Re} \sum_{k=1}^{\infty} \int_{-1/2}^{\infty} \varphi(n)\, e^{2\pi i k n}\, dn. \tag{15.9}$$

---

[2] The Poisson formula can be easily obtained from

$$\sum_{(n)} \delta(x - n) = \sum_{(k)} e^{2\pi i k x}.$$

The summation is carried out over positive and negative integral values of n and k. The equality just given is the expansion of the periodic function $\sum_{n=-\infty}^{\infty} \delta(x-n)$ as a Fourier series.

Further calculations can be carried out only if we know the dependence of the electron energy on the quantum numbers. We shall assume that the energy levels are given by Eqs. (7.4) and (7.3). We must point out that in the case of spin–orbit coupling we must make allowance for the difference between the g factor of electrons and 2.

The quasiclassical energy levels can be used because, in moderate magnetic fields, the oscillatory component of the thermodynamic potential is dominated by the contribution of those electrons which have high quantum numbers n, i.e., such values of ε and p_z that the following condition is satisfied:

$$cS/2\pi |e| \hbar H \gg 1. \tag{15.10}$$

The first term in Eq. (15.9) describes the monotonic dependence of the thermodynamic potential on the magnetic field. The oscillatory dependence is included in the remaining terms, which we shall now consider. We shall, therefore, drop all the monotonic terms. The oscillatory component of the potential Ω will be denoted by $\tilde{\Omega}$.

According to Eqs. (15.8) and (15.9), we have

$$\tilde{\Omega} = \sum_{\sigma} \tilde{\Omega}_{\sigma}, \quad \tilde{\Omega}_{\sigma} = 2 \operatorname{Re} \sum_{k=1}^{\infty} I_k, \tag{15.11}$$

where

$$I_k = \frac{V|e|HT}{(2\pi\hbar)^2 c} \int\limits_{-1/2}^{\infty} dn \int\limits_{-\infty}^{\infty} dp_z \ln\left\{ 1 + \exp\frac{\zeta_{\sigma} - \varepsilon_n (p_z)}{T} \right\} e^{2\pi i k n}, \tag{15.12}$$

and

$$\zeta_{\sigma} = \zeta - \sigma\mu H, \quad \mu = \frac{e\hbar}{4m_0 c} g. \tag{15.13}$$

The expression (15.12) can be transformed if Eq. (7.3) is used to go over to integration over energy (instead of the integration over n), and then the integration over p_z is replaced by the integration over n, again using Eq. (7.3). In this way, we obtain

$$I_k = \frac{2\pi V}{(2\pi\hbar)^3} \int\limits_0^{\infty} \frac{d\varepsilon}{\exp\left(\frac{\varepsilon - \zeta_{\sigma}}{T}\right) + 1} \int\limits_0^{\varepsilon} d\varepsilon' \sum \int\limits_{n_{\min}}^{n_{\max}} \frac{m^*(\varepsilon', n)}{\left|\frac{\partial n}{\partial p_z}\right|} e^{2\pi i k n} \, dn. \tag{15.14}$$

The sign of the sum indicates summation over the intervals in which the function $n(\varepsilon_n, p_z)$ varies monotonically when ε is fixed.

The oscillatory dependence can be isolated because all the quantities which occur in the integral of Eq. (15.14), with the exception of $[(\varepsilon - \zeta_{\sigma})/T + 1]^{-1}$ and $e^{2\pi i k n}$, vary slowly with ε and n. When $\varepsilon \sim \zeta_{\sigma}$, the values of $n_{\min}$ and $n_{\max}$ are usually much larger than unity,[3] i.e.,

---

[3] For each value of the energy, there is a value $n_{\min} = 0$ (it corresponds to that value of p_z at which the secant plane p_z = const just touches the constant-energy surface). These points will not be of interest to us because integration near these points makes no contribution to the oscillatory part of the potential [24]. In the case of the quadratic and isotropic dispersion law, we have

$$\frac{cS_{\text{extr}}}{2\pi |e| \hbar H} = \frac{\varepsilon_F}{2\pi\mu H_k}.$$

$cS_{\text{extr}}/2|e|H\hbar \gg 1$. This corresponds to the fact that an electron whose energy is close to the Fermi value describes a path enclosing an area which is much larger than the square of the atomic spacing. This condition is in agreement with the criterion for the validity of the quasi-classical approximation.

Since $n \gg 1$, we can use the method of steepest descents and show that the main contribution to the inner integral is made by the integration in the vicinity of the extremal points $n_m(\varepsilon')$. Our calculations give

$$I_k \approx \frac{VeHe^{-i\pi/2}}{2\pi c\,(2\pi\hbar)^2\,k^{5/2}} \sum_m e^{\pm\,i\pi/4} \int_0^\infty \frac{\exp\left(2\pi i k n_m(\varepsilon)\right)}{\exp\left(\dfrac{\varepsilon-\zeta_\sigma}{T}\right)+1} \left|\frac{\partial^2 n}{\partial p_z^2}\right|_{p_{zm}}^{-1/2} d\varepsilon. \qquad (15.15)$$

The summation is carried out over all the extremal points of the function $n(\varepsilon, p_z) = cS(\varepsilon, p_z)/2\pi|e|\hbar H$ for a fixed value of ε. The points at which $n_m = 0$ are ignored. The plus sign in the expression $e^{\pm i\pi/4}$ is used when the extremal point is a minimum and the minus sign when it is a maximum.

We shall now use the fact that the main contribution to the oscillatory component I_k is made by the integration near the point where the Fermi function is strongly nonstationary, i.e., near $\varepsilon = \zeta_\sigma$. This makes it possible to write the oscillatory component in the form

$$I_k \approx \frac{V|e|He^{-i\pi/2}}{2\pi c\,(2\pi\hbar)^2\,k^{5/2}} \sum_m \exp\left(-2\pi i k n_m(\zeta_\sigma) \pm \frac{i\pi}{4}\right) \left|\frac{\partial^2 n}{\partial p_z^2}\right|_{\substack{\varepsilon=\zeta_\sigma \\ p_z=p_z^m}}^{-1/2} \int_0^\infty \frac{\exp\left\{2\pi i k\left(\dfrac{\partial n_m}{\partial \varepsilon}\right)_{\varepsilon=\zeta_\sigma}(\varepsilon-\zeta_\sigma)\right\}}{\exp\left(\dfrac{\varepsilon-\zeta_\sigma}{T}\right)+1} \,d\varepsilon.$$

Introducing a new integration variable $x = (\varepsilon - \zeta_0)/T$ and using the fact that $T \ll \zeta_\sigma$, we shall extend the integration from $-\infty$ to $+\infty$. After integration, we shall substitute the expression obtained into Eq. (15.11) and this gives

$$\tilde{\Omega} \approx \frac{V}{2\pi^3 V \sqrt{2\pi}\,\hbar^3} \left(\frac{e\hbar H}{c}\right)^{5/2} \sum_m \left|\frac{\partial^2 S}{\partial p_z^2}\right|_{\substack{\varepsilon=\zeta \\ p_z=p_z^m}}^{-1/2} \frac{1}{m^*(\zeta, p_z^m)} \sum_{k=1}^\infty \frac{\psi(k\lambda)}{k^{5/2}} \cos\left\{k\left(\frac{cS_m}{e\hbar H}-\pi \pm \frac{\pi}{4}\right)\right\}\cos\left(\pi k\,\frac{m^*}{m_0}\right), \quad (15.16)$$

where

$$\psi(z) = \frac{z}{\operatorname{sh} z}, \qquad \lambda = 2\pi^2\,\frac{T}{\hbar\omega_H}, \qquad \omega_H = \frac{eH}{m^*c}.$$

In the derivation of Eq. (15.16), we used the fact that, in the summation over two directions of the spin [see Eq. (15.11)], we can everywhere, except in the argument of the cosine, replace ζ_σ with ζ, because $\mu H \ll \zeta$. In the argument of the cosine, we can expand $S_m(\zeta_\sigma)$ in powers of μH and retain only the term linear in μH. An allowance for a small change in S_m is essential because, in accordance with the condition $cS/2\pi e\hbar H \gg 1$, even a small change in S_m can result in a large change in the argument.

Since $\psi(z)$ is a monotonically decreasing function of the order of or less than unity [$\psi(z) = 1$ for $z = 0$ and $\psi(z) = 2\pi e^{-z}$ for $z \gg 1$] and $\partial^2 S/\partial p_z^2 \sim 1$, the order of magnitude of Ω at moderately high temperatures is determined by the factor $(V/\hbar^3)(|e|\hbar H/c)^{5/2}\,1/m^*$. The monotonic part of the thermodynamic potential in moderately strong fields ($\mu H \ll \varepsilon_F$) is of the same order as its classical value, i.e., it is of the order of $\int_0^\zeta N(\varepsilon)\,d\varepsilon$ [see Eq. (12.8a)]. If the Fermi surface is not extremely anisotropic, the last expression is $\sim VS^{5/2}/\hbar^3 m^*$. Therefore, the oscillatory part of the potential represents a small correction $\sim (e\hbar H/cS)^{5/2}$ to its regular (basically the classical) value. It follows that the oscillatory part of the free energy $\widetilde{F}$ is numerically equal to $\widetilde{\Omega}$ if the chemical potential ζ in Eq. (15.16) is expressed in terms of the total number of electrons N and the volume V in the zeroth approximation with respect to the magnetic field and the temperature. This means that ζ should simply be replaced by the Fermi energy ε_F.

The chemical potential is affected by the magnetic field even if this field is not very strong [see condition (15.1)]. This dependence on the magnetic field should be included in the calculation of such quantities as the contact potential difference, the emission current in a magnetic field, etc. (§22).

Thus, Eq. (15.16) can be used directly to calculate the quantities which are of interest to us. The oscillatory component of the z projection of the magnetic moment[4] is $\widetilde{M}^z = -\partial\widetilde{\Omega}/\partial H$ and the corresponding component of the specific heat is $\widetilde{C} = -T\partial^2\widetilde{\Omega}/\partial T^2$.

Equation (15.16) is valid for any relationship between $\hbar\omega_H = |e|\hbar H/m^*c$ and the temperature T. In deriving it, we assume only that T, $\hbar\omega_H \ll \varepsilon_F$.

At absolute zero (T $\ll \hbar\omega_H$), we have

$$\widetilde{\Omega} \approx \frac{V}{2\pi^3\sqrt{2\pi}\,\hbar^3}\left(\frac{e\hbar H}{c}\right)^{5/2}\sum_m \frac{1}{m^*}\left|\frac{\partial^2 S}{\partial p_z^2}\right|_m^{-1}\sum_{k=1}^{\infty}k^{-5/2}\cos\left\{k\left(\frac{cS_m}{e\hbar H}-\pi\right)\pm\frac{\pi}{4}\right\}\cos\left(\pi k\,\frac{m^*}{m_0}\right). \qquad (15.17)$$

At relatively high temperatures ($\lambda = 2\pi^2 T/\hbar\omega_H \gg 1$), only one term remains in the summation over k:

$$\widetilde{\Omega} = \frac{V}{\hbar^3}\sqrt{\frac{2}{\pi^3}}\left(\frac{e\hbar H}{c}\right)^{5/2}T\sum_m\left|\frac{\partial^2 S}{\partial p_z^2}\right|_m^{-1/2}\exp\left(-\frac{2\pi^2 T}{\hbar\omega_H}\right)\cos\left(\frac{cS_m}{e\hbar H}-\pi\pm\frac{\pi}{4}\right)\cos\left(\pi\,\frac{m^*}{m_0}\right). \qquad (15.18)$$

The factor $2\pi^2$ in the condition of validity of Eq. (15.18) makes sure that this equation is valid also at relatively low temperatures (T $\gg \hbar\omega_H/20$).

According to Eqs. (15.16)-(15.18), $\widetilde{\Omega}$ is a complex oscillatory function of the magnetic field and the intervals between the zeros of this function (its "periods") are determined by the extremal sections of the Fermi surface:

$$\Delta\left(\frac{1}{H}\right) = \frac{2\pi|e|\hbar}{cS_m}. \qquad (15.19)$$

We must draw attention to the fact that when Ω is plotted as a function of the reciprocal of the magnetic field 1/H, the periods are independent of the magnetic field. Moreover, $\Delta(1/H)$ is independent of the temperature.

If $\Delta H \ll H$, it follows that

$$\Delta H = \frac{2\pi e\hbar}{cS_m}H^2, \qquad (15.20)$$

i.e., in strong fields the period is proportional to the square of the magnetic field.

The temperature dependence of $\widetilde{\Omega}$ is determined by the effective mass $m^*(\varepsilon_F, p_z^m)$. However, we must bear in mind that the interactions of electrons with impurities, with lattice vi-

---

[4] In calculating the x and y projections of the magnetic moment by the differentiation of $\widetilde{\Omega}$, we must bear in mind that m* and S_m depend on the magnetic field direction (see §7).

brations, and with other departures from periodicity reduce the oscillation amplitude. An allowance for the scattering of electrons by impurities shows that the reduction of the oscillation amplitude can be included in the calculation if the temperature is replaced by an effective temperature defined by $T + \hbar/\tau$, where τ has the same order of magnitude as the time between two collisions of an electron [25, 26]. The formulas given in [25, 26] were derived on the assumption that the dispersion law of conduction electrons is quadratic. They provide convenient expressions for estimating the oscillation amplitude.

We must stress that, without allowance for the scattering, the magnetic susceptibility $\partial\tilde{M}_z/\partial H$ would be infinite at $T = 0$. Therefore, at sufficiently low temperatures, an allowance for the finite lifetime of the electron states is essential.[5]

We shall now obtain an expression for the component of the magnetic moment along the magnetic field. Since $cS_m/|e|\hbar H \gg 1$, we should differentiate only the cosine; the slowly varying amplitude in front of the cosine need not be differentiated:

$$\tilde{M}_z \approx -\frac{V}{\hbar^3}\frac{1}{2\pi^3\sqrt{2\pi}}\left(\frac{e\hbar}{c}\right)^{3/2}\sqrt{H}\sum_m \frac{S_m}{m^*\left|\frac{\partial^2 S}{\partial p_z^2}\right|_m^{1/2}}\sum_{k=1}^{\infty}\frac{\psi(k\lambda)}{k^{3/2}}\sin\left[k\left(\frac{cS_m}{e\hbar H}-\pi\right)\pm\frac{\pi}{4}\right]\cos\left(\pi k\frac{m^*}{m_0}\right). \quad (15.21)$$

In relatively strong fields and at low temperatures, i.e., when $\psi \approx 1$, the oscillatory component of the moment is much larger than its monotonic component. In fact, we can show [24] that in a wide range of magnetic fields ($\mu H \ll \varepsilon_F$) the monotonic component of the magnetic moment is of the same order of magnitude as the diamagnetic value of this moment, i.e.,

$M \sim \frac{V}{\hbar^3}\left(\frac{e\hbar}{c}\right)^2 \frac{S^{1/2}}{m^*} H$, and, consequently, $\frac{\tilde{M}}{M} \approx \left(\frac{cS}{e\hbar H}\right)^{1/2} \gg 1.$

It is not surprising that the oscillatory component of the moment is larger than its monotonic part because the whole moment (including its monotonic part) is of quantum-mechanical origin. Formally, this is related to the fact that each differentiation with respect to the magnetic field multiplies the function by a large factor $cS/e\hbar H$.

Figure 43a shows a typical dependence of the magnetic moment on the magnetic field. For most metals this dependence is not limited to a single harmonic but is the result of the superposition of several harmonics which may have very different periods (Fig. 43b). This means that the Fermi surface of most metals is very complex; the numerical estimates show that all metals (with the exception of those in the first group) have relatively small Fermi-surface cross sections, which are considerably less than $(\hbar/a)^2$, where a is the lattice constant $[S_0 = 4\pi(3/8\pi)^{2/3}(2\pi\hbar)^2/a^2$ is the maximum cross-sectional area of the Fermi sphere for a free-electron gas whose density is $1/a^3]$. Usually, the experimental data are interpreted on the assumption that small cross sections correspond to separate small pockets or thin necks of the Fermi surface.[6] Estimates of the volumes of such pockets show that they may contain $\sim 10^{-5}$ electrons or "holes" per atom. An estimate of the amplitude, i.e., of the factor in front of the cosine, gives very small values of the effective masses of the electrons associated with these small sections: $m^* \sim (10^{-3}\text{-}10^{-2})\,m_0$, where m_0 is the mass of a free electron.

---

[5] The role of collisions in the oscillatory effects cannot be allowed for completely in any qualitative discussion. In particular, the quantity $\partial M_z/\partial H$, which is calculated rigorously in [25], differs considerably from the same quantity obtained from formulas derived from qualitative considerations in [26].

[6] According to [27], the existence of small sections may be associated not only with split-off regions but also with small humps or depressions on a large Fermi surface.

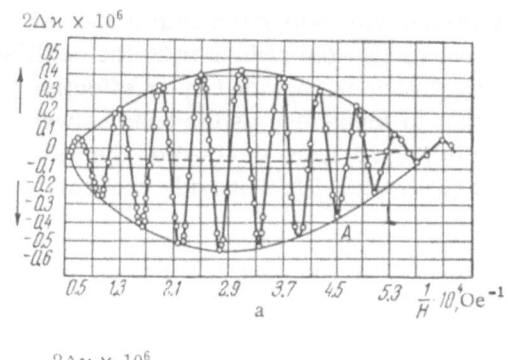

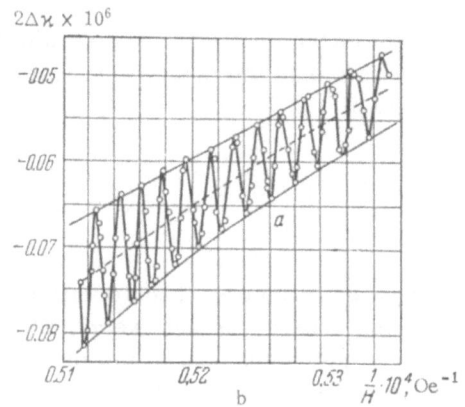

Fig. 43. a) Typical dependence of the magnetic susceptibility on the reciprocal of the magnetic field. b) Fine structure of the de Haas–van Alphen effect. Curve a in Fig. 2b represents an expanded – along both coordinates – part of curve A in Fig. 2a. A Zn crystal, $T = 4.2°K$.

The large number of extremal sections, i.e., a large number of harmonics in the dependence of the magnetic moment on the magnetic field, makes it very difficult to analyze the experimentally obtained curves. However, if we restrict our analysis to certain ranges of the magnetic field intensity and also measure carefully the angular dependences of the periods, we can interpret the curves $\widetilde{M} = \widetilde{M}(H)$ and even find the shape of the Fermi surface of some metals from the measured periods (§17).

Equations (15.16)-(15.21) make it possible to identify some of the features of the de Haas–van Alphen effect which distinguish it from other oscillatory effects:

1) the period, expressed in the reciprocal of the magnetic field, depends only on the direction of this field, and for a fixed direction it is a constant for a given metal;

2) the oscillation amplitude decreases when the temperature and the contamination of the sample are increased.

We have mentioned earlier that the de Haas–van Alphen effect can be observed only in single crystals. The mosaic structure destroys the effect because the phase of the periodic function depends very strongly on the direction, and a small change in S_m alters the phase by an amount which is of the order of π.

The formulas derived in the present section, particularly Eq. (15.16), describe only the principal contribution to the oscillatory part of the thermodynamic potential. They are derived by using the first nonvanishing terms in the expansion in powers of the small parameter $\mu H/\varepsilon_F$ or, more precisely, $e\hbar H/cS_{extr}$. An analysis of Eq. (15.14) shows that the oscillatory components are due to those values of p_z for which $S(\zeta, p_z)$ has an extremal value or $S(\varepsilon, p_z)$ – considered as a function of p_z – has a singularity. Additionally, there are special sections known as the "figure of eight," which occur near each saddle point on the Fermi surface and are observed in a wide range of directions of the magnetic field; for these sections $S(\zeta, p_z)$ has a singularity of the $\Delta p_z \ln \Delta p_z$ type (§7). Integration near such a section makes a finite contribution to the oscillatory part of the potential Ω and, according to [27], the amplitude of the oscillatory terms associated with the "figure of eight" sections is $cS/e\hbar H$ times smaller than the amplitude of the oscillations associated with the extremal sections.

We shall now consider the role of magnetic breakthrough in the de Haas–van Alphen effect [28]. When an electron moves along a quasiclassical trajectory, the probability of its transition to a nearby trajectory is usually very low. The exceptions to this rule are the cases when quasiclassical trajectories belonging to different bands are close to one another in the p space. Then, an electron subjected to relatively weak fields (the relevant estimates are given in §10) does not "see" small potential barriers and moves along a trajectory which consists of elements of several different trajectories (Fig. 44). Then, some of the harmonics in the de Haas–van Alphen effect may have periods which correspond to areas which exceed the dimensions of the Brillouin zone section even in the case of those metals which have closed Fermi surfaces. A phenomenon of this kind, reported first for magnesium in [29], is now known as the magnetic breakthrough or the magnetic breakdown.

The energy spectrum of electrons in a magnetic field capable of producing the breakthrough effect is known only for very few directions of the magnetic field (see §10 and [28]) but the available information on the motion of electrons in a magnetic field is sufficient to develop a theory of oscillatory effects which would include an allowance for the probability that an electron jumps from one classical orbit to another. The main feature of the de Haas–van Alphen effect under magnetic breakthrough conditions is the reduction in the amplitude of the isolated trajectories and the increase in the amplitude of the composite trajectories with increasing probability of the breakthrough phenomenon.

A consistent derivation of the oscillatory component of the potential $\tilde{\Omega}$, which makes allowance for magnetic breakthrough, is given in [29a].

We shall conclude this section by estimating the order of magnitude of the oscillatory component of the magnetic moment $\tilde{M}$ and of the differential magnetic susceptibility $\varkappa \approx (d\tilde{M}/\partial H)/V$. According to Eq. (15.21), at T = 0 we have

$$\tilde{M} \approx N\mu^* \left(\frac{\mu^* H}{\varepsilon_F}\right)^{1/2} \quad \left(\mu^* = \frac{e\hbar}{m^* c}, \ \varepsilon_F \approx \frac{S}{m^*}\right), \tag{15.22}$$

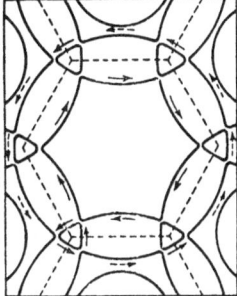

Fig. 44. Magnetic breakthrough. The continuous arrows represent the trajectory of an electron in a moderately strong magnetic field; the dashed arrows represent the corresponding trajectory in a strong field. The dashed curves represent the boundaries of the Brillouin zone.

and

$$\tilde{\varkappa} \approx \frac{\mu^{*2}N}{\varepsilon_F}\left(\frac{\varepsilon_F}{\mu^*H}\right)^{3/2}.$$

In deriving these formulas, we omitted the periodic function. Although the paramagnetic susceptibility of a metal is weak ($\mu^{*2}N/\varepsilon_F \ll 1$), the value of $\tilde{\varkappa}$ may be of the order of unity because of the factor $(\varepsilon_F/\mu^*H)^{3/2}$. This unexpected result (a diamagnetic susceptibility of the order of unity and increasing when the magnetic field is reduced) leads to very interesting consequences (§§20, 21), which affect the stability of the electron state of a metal in a magnetic field.

§16. de Haas–van Alphen Effect and the Theory of Fermi Liquids

Most of the results deduced in the preceding sections are not affected by the transition from a Fermi gas to a Fermi liquid. However, we must bear in mind that the dispersion law of elementary excitations, which we discussed in earlier sections, includes also the mutual interaction between the electrons. This is in the spirit of Landau's theory of Fermi liquids (§1). We shall now consider the effect of electron–electron collisions on the formulas describing the de Haas–van Alphen effect.

An allowance for the scattering of electrons by impurities and phonons shows that the finite nature of the mean free path of electrons does not alter the oscillation period but simply reduces the oscillation amplitude. The effect of electron–electron collisions is more complex. Collisions between electrons, which are responsible for the electron component of the resistivity, reduce somewhat the oscillation amplitude. Moreover, the electron–electron interaction considered within the framework of the theory of Fermi liquids is included in the dispersion law of quasi particles, which we have called conduction electrons. It is this dispersion law which includes the electron–electron interaction that occurs in the quantization conditions (7.3) and determines the oscillation period. It is worth noting that the interaction between electrons occurs in the expression for the period only via the dispersion law. To prove this, we shall show that the separations between the energy levels of conduction electrons interacting in a magnetic field are given (when considered in the quasiclassical approximation) by the well-known formula[1]

$$\Delta\varepsilon = \hbar\omega_H \qquad \left(\omega_H = \frac{eH}{m^*c};\ \ m^* = \frac{1}{2\pi}\frac{\partial S}{\partial\varepsilon}\right). \qquad (16.1)$$

We must stress again that the proof of Eq. (16.1) does not mean that the interaction between electrons "disappears" from the quantization condition. It means that the interaction is included in the quantization condition only via the dispersion law $\varepsilon = \varepsilon(\mathbf{p})$ which makes allowance for this interaction.

In order to prove Eq. (16.1), we shall use the fact that, in the quasiclassical approximation,

$$\varepsilon_{n+1} - \varepsilon_n = \Delta\varepsilon = \hbar\omega, \qquad (16.2)$$

---

[1] The results of the "gas" and the "liquid" treatments are identical if the spectra and the densities of states are identical. The density of states in a Fermi liquid is identical with the corresponding density in a Fermi gas and this will be proved simultaneously with Eq. (16.1).

where ω is the frequency of classical motion. Thus, we must find the frequency of classical motion of an electron in a magnetic field. We shall do this by considering the oscillations of an electron liquid in a magnetic field.

The state of electrons will be described by the distribution function $f(\mathbf{r}, \mathbf{p}, t)$ which satisfies the transport equation

$$\frac{\partial f}{\partial t} + \frac{\partial f}{\partial r}\, v + \frac{\partial f}{\partial p}\, \boldsymbol{F} = 0. \tag{16.3}$$

Here, $\mathbf{v}$ is the velocity of a particle given by

$$v = \frac{\partial \varepsilon}{\partial p} + \frac{\partial}{\partial p} \int \Phi\,(\boldsymbol{p},\,\boldsymbol{p}')\,f_1\,(\boldsymbol{p}')\,d\tau_{p'}.$$

The rest of the notation is the same as in §1. We omitted the collision integral because we are interested in the case $r_H \ll l$, which corresponds to $\omega_H \tau_0 \gg 1$, where τ_0 is the relaxation time.

In the case considered, the electrons are acted upon by the Lorentz force

$$\boldsymbol{F} = \frac{e}{c}\,[v\boldsymbol{H}]. \tag{16.4}$$

The linearization of Eq. (16.3) and the allowance for the spatial inhomogeneity of the problem shows that

$$\frac{\partial f_1}{\partial t} + \frac{e}{c}\,[v_0 \boldsymbol{H}]\,\frac{\partial f_1}{\partial p} = -\frac{\partial n_F}{\partial \varepsilon}\,\frac{e}{c} \int v_0 \left[\frac{\partial}{\partial p}\,\Phi(\boldsymbol{p},\,\boldsymbol{p}')f_1\,(\boldsymbol{p}'),\,\boldsymbol{H}\right] d\tau_{p'}, \tag{16.5}$$

where $\mathbf{v}_0 = \partial \varepsilon / \partial \mathbf{p}$ is the velocity of the particle when $f_1 \equiv 0$ and $f_1(\mathbf{p}) = f(\mathbf{p}) - n_F(\varepsilon(\mathbf{p}))$.

If we introduce a new function χ $[f_1 = -(\partial n_F / \partial \varepsilon)\chi]$ and replace the variables p_x, p_y, p_z with ε, p_z, and τ, where τ is the time of motion along a quasiclassical trajectory, we find that the transport equation of Eq. (16.5) can easily be written in the form[2]:

$$\frac{\partial \chi}{\partial t} + \frac{\partial \chi}{\partial \tau} = \int \int F\,(p_z,\, \tau;\, p_z',\, \tau')\,\chi\,(p_z',\,\tau')\,dp_z\,d\tau'. \tag{16.6}$$

The form of the function $F(p_z, \tau; p_z', \tau')$ is found by comparing it with the integral in Eq. (16.5). The integration in Eq. (16.6) is carried out over the Fermi surface because $\partial n_F / \partial \varepsilon = -\delta(\varepsilon - \varepsilon_F)$. The solution of Eq. (16.6) will be sought in the form

$$\chi\,(p_z,\, \tau;\, t) = ie^{-i\omega t} \sum_n \chi_n\,(p_z)\,e^{in\omega_H\,(p_z)\,\tau}, \tag{16.7}$$

which is justified by the periodic dependence of the distribution function on τ (we are considering only closed Fermi surfaces).

---

[2] The simplicity of the left-hand side (affected by the field) of the transport equation expressed in terms of ε, p_z, and τ is due to the fact that ε and p_z are invariant in a constant and uniform magnetic field and because τ is a canonical conjugate of ε (the same variables will be used also in the theory of galvanomagnetic effects in §27).

Substituting the expansion of Eq. (16.7) into Eq. (16.6), we obtain

$$[\omega - \omega_H(p_z)]\chi_n(p_z) = \sum_{n'} \int F_{nn'}(p_z, p_z')\chi_{n'}(p_z')\,dp_z' \tag{16.8}$$

(we shall not give the relationship between F and $F_{nn'}$). For simplicity, we shall assume that the operator $F_{nn'}$ is degenerate, i.e., we shall assume that

$$F_{nn'} = \delta_{nn'}g_n(p_z)g_n^*(p_z'). \tag{16.9}$$

We shall show later that this simplifying assumption has no influence on the basic result. It follows from Eqs. (16.9) and (16.8) that

$$[\omega - \omega_H(p_z)]\chi_n(p_z) = g_n(p_z)\langle \chi_n, g_n \rangle, \tag{16.10}$$

where

$$\langle \chi_n, g_n \rangle = \int g_n^*(p_z')\chi_n(p_z')\,dp_z' \tag{16.11}$$

is the scalar product of the functions $g_n(p_z)$ and $\chi_n(p_z)$. Since we are interested in the separation between neighboring levels, we shall consider the case $n = 1$:

$$[\omega - \omega_H(p_z)]\chi_1(p_z) = g_1\langle \chi_1, g_1 \rangle.$$

The last equation yields

$$\langle \chi_1, g_1 \rangle = \int \frac{|g_1|^2\,dp_z}{\omega - \omega_H(p_z)}\langle \chi_1, g_1 \rangle,$$

and hence the dispersion equation for the determination of the oscillation frequency ω becomes

$$1 = \int \frac{|g_1|^2\,dp_z}{\omega - \omega_H(p_z)}. \tag{16.12}$$

This equation can be analyzed conveniently by going over from continuous to discrete values of the momentum by, for example, applying the quantization due to the boundaries of a metal. Then, Eq. (16.12) becomes

$$1 = \sum \frac{|g_1|^2\,\Delta p_z}{\omega - \omega_H(p_z)}, \tag{16.13}$$

in which p_z represents a discrete set of values $p_z^{(k)}$. Figure 45 shows the graphical solution of Eq. (16.13): it is evident from this figure that the roots are located between the values $\omega_H(p_z^{(k)})$. Moreover, there may be a "split-off" root (Fig. 45). Returning to the continuous spectrum (this can be done by assuming that the sample stretches to infinity), we find that the oscillation frequencies tend to the values $\omega_H(p_z)$. Moreover, we may find a split-off frequency, which corresponds to the zeroth sound in a Fermi liquid [30]. Since the frequencies ω are always located between the values $\omega_H(p_z^{(k)})$, it is clear that the spectral density, which is the number of frequencies in the interval $(p_z, p_z + dp_z)$, of a Fermi liquid is the same as the spectral density of a Fermi gas.

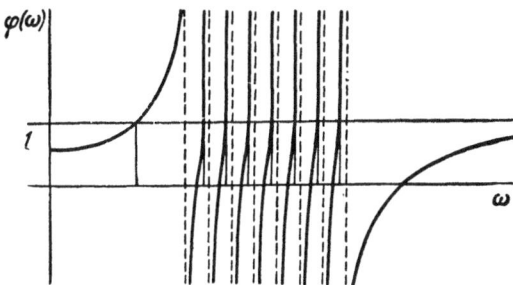

Fig. 45. Graphical solution of the dispersion equation

$$\psi(\omega) = 1 = \sum \frac{|q_1|^2 \, \Delta p_z}{\omega - \omega_H(p_z)}.$$

We note that ω_H is independent of p_z if the dispersion law is quadratic. In this case, the oscillation frequency is eH/mc. Thus, Eq. (16.1) represents quasiclassical quantization of the energy levels of a Fermi liquid in a magnetic field. We have, therefore, demonstrated that the formulas in the preceding section (this applies to the oscillation periods) are valid not only in the case of a Fermi gas but also in the case of a Fermi liquid if $\varepsilon(\mathbf{p})$ is understood to be the dispersion law which includes the interaction between electrons in the ground state.[3]

§ 17. Determination of the Electron Energy Spectrum from the de Haas–van Alphen Effect

The energy spectra of condensed systems can be determined by a great variety of methods, which include the absorption of ultrasound, cyclotron resonance, galvanomagnetic effects, anomalous skin effect, de Haas–van Alphen effect, inelastic scattering of neutrons, characteristic electron-energy losses, and so on (these methods are considered in §48).

The present section deals with the possibility of deducing the electron energy spectrum of metals (the shape of the Fermi surface and the distribution of the electron velocities on this surface) from the de Haas–van Alphen effect [36]. We shall use the principal oscillation terms in the magnetic moment, i.e., the terms which are due to the extremal sections of the Fermi surface (§15). We shall start from Eq. (15.21), which shows that the dependence of the period of each harmonic $\Delta(1/H)$ on the direction of the magnetic field determines the angular dependences of the areas of the extremal sections S_m. Measurements of the amplitudes of the harmonics and of their temperature dependences makes it possible to determine the derivative $\partial S_m/\partial \zeta = 2\pi m^*(\zeta, p_{zm})$. However, in the determination of the effective mass, we must bear in mind the comments made in §§15 and 16 on the influence of the interactions on the oscillation amplitude. Since the oscillation period can be expressed in terms of the dispersion law even when the interactions are taken into account (§16), whereas this does not apply to the amplitude, it becomes obvious that the de Haas–van Alphen effect can be used very conveniently to determine the shape of the Fermi surface. However, this effect will give less reliable data on the distribution of the velocities on the Fermi surface (the relationship between the effective mass and the velocity $\mathbf{v} = \partial \varepsilon/\partial \mathbf{p}$ will be discussed later).

Let us assume that $\boldsymbol{\xi}$ is the unit vector along the magnetic field and $\sigma(\boldsymbol{\xi}) = [S_m(\boldsymbol{\zeta})]_\xi$ is the extremal area of the section of the Fermi surface $\varepsilon(\mathbf{p}) = \zeta$ cut by a plane perpendicular to $\boldsymbol{\xi}$. Then, according to Eq. (15.21),

$$\sigma(\boldsymbol{\xi}) = \frac{2\pi e\hbar}{c\Delta_\xi}, \tag{17.1}$$

---

[3] This is confirmed by direct calculations reported in [31].

where Δ_ξ is the period (in terms of the reciprocal of the field) of the corresponding harmonic of the oscillations.

We shall assume that the Fermi surface we are seeking has a center of symmetry and that any ray drawn from this center meets the surface at only one point. If the Fermi surface consists of several closed subsurfaces, our comments and conclusions will apply to each of them separately.

The Fermi surfaces which satisfy the properties just specified have central extremal sections. We shall denote the length of the radius vector drawn from the center of the surface along $\mathbf{e}$ by $\rho(\mathbf{e})$. Then

$$\sigma(\xi) = \frac{1}{2} \int \rho^2(e)\, \delta(e\xi)\, d\Omega_e, \tag{17.2}$$

where $\delta(x)$ is the Dirac δ function and $d\Omega_e$ is an element of area of a unit sphere. In other words,

$$\sigma(\xi) = \pi \overline{\rho^2(e)}\,|_{e\xi = 0}, \tag{17.3}$$

i.e., $\sigma(\xi)$ is the average value of the radius vector $\rho(\mathbf{e})$ along the equator, where $\mathbf{e}\xi = 0$.

Thus, the plotting of the Fermi surface reduces to the calculation of the function $\pi\rho^2(\mathbf{e})$ $[\rho(-\mathbf{e}) = \rho(\mathbf{e})]$ on the basis of its average values along the equator $\mathbf{e}\xi = 0$, where ξ is arbitrary.

The solution of this problem can be obtained on the basis of the following relationship, which applies (this can be demonstrated by direct calculations) to any function $\psi(\xi) = \psi(-\xi)$ specified for a unit sphere:

$$\int \bar{\psi}(\xi)\, d\Omega - \int\limits_{z^2 > \lambda^2} \frac{\bar{\psi}(\xi)\, z}{\sqrt{z^2 - \lambda^2}}\, d\Omega = \int\limits_{z^2 > 1-\lambda^2} \psi(\xi)\, d\Omega, \quad z = (e\xi), \quad \lambda^2 < 1. \tag{17.4}$$

Here, $\overline{\psi}(\xi)$ represents the average value of $\psi(\mathbf{e})$ along the equator $\mathbf{e}\xi = 0$; the range of integration on the surface of the unit sphere is determined by the inequality specified above.

Assuming that $\psi(\xi) = \pi\rho^2(\xi)$, $\overline{\psi}(\xi) = \sigma(\xi)$ we obtain the following expression for sufficiently small values of λ

$$\pi\rho^2(e) = \left\{ \int \sigma(\xi)\, d\Omega - \int\limits_{z^2 > \lambda^2} \frac{\sigma(\xi)\, z}{\sqrt{z^2 - \lambda^2}}\, d\Omega \right\} \Big/ \pi\lambda^2, \quad z = e\xi. \tag{17.5}$$

In the limit $\lambda \to 0$, we obtain

$$\pi\rho^2(e) = \chi_e(0) - \int\limits_0^1 [\chi_e(u) - \chi_e(0)]\, \frac{du}{u^2}, \tag{17.6}$$

where

$$\chi_e(u) = \overline{\sigma(\xi)}\,|_{\xi e = u} = \frac{1}{2\pi} \int \sigma(\xi)\, \delta(\xi e - u)\, d\Omega_\xi.$$

In each case, we can select the direction $\mathbf{e}$ to be the polar axis of the coordinate system and we can introduce an angle $\cos\theta = \mathbf{e}\xi$ as well as a polar angle φ in a plane perpendicular to $\mathbf{e}$. Denoting $\sigma(\xi)$ by σ_e (cos θ, φ) and integrating by parts, we obtain

$$\pi\rho^2(e) = \sigma(e) - \frac{1}{2\pi} \int_0^{2\pi} d\varphi \int_0^1 \frac{\partial \sigma_e (\cos\theta, \varphi)}{\partial \cos\theta} \frac{d\cos\theta}{\cos\theta}. \tag{17.7}$$

The above formula is the solution to our problem. However, in numerical calculations, it is more convenient to use Eq. (17.5) by selecting a finite value $\lambda \ll 1$.

It follows from Eq. (17.6) that the solution obtained is stable. The average error in the determination of the shape of the surface is related to the error in the determination of the area $\sigma(\xi)$, i.e., of the period $\Delta(1/H)$. In terms of orders of magnitude, we have $\delta\rho^2 \sim \delta\sigma$.

The construction of the Fermi surface automatically gives the value of $\partial^2 S(\zeta, p_z)/\partial p_z^2$, which is, therefore, not an independent parameter. An additional parameter is the value of $\partial S_m/\partial \zeta$, which can be determined by measuring the oscillation amplitude or the temperature dependence of this amplitude. If we know the value of $\partial S_m/\partial \zeta$ as a function of the direction of the magnetic field ξ, we can find the electron velocities on the Fermi surface.

The electron velocity $\mathbf{v}$ is defined by $\mathbf{v} = \partial\varepsilon/\partial\mathbf{p}$. If we consider the surface $\varepsilon(\mathbf{p}) = \zeta$ as well as the neighboring surface $\varepsilon(\mathbf{p}) = \zeta + \delta\zeta$, we find that

$$v = |\nabla\varepsilon| = \frac{\delta\zeta}{\delta n},$$

where δn denotes the distance between these surfaces along the normal. On the other hand, the area of an extremal section of a neighboring surface is

$$S_m(\zeta + \delta\zeta) = S_m(\zeta) + \frac{\partial S_m}{\partial \zeta}\delta\zeta.$$

For the surfaces considered here, an extremal section coincides with the central section $p_z = 0$. Therefore, the derivative with respect to the chemical potential is basically taken at a constant value of p_z, i.e., $\partial S_m/\partial \zeta = 2\pi m^*$, or in the previous notation:

$$\delta\sigma(\xi) = \delta\zeta \left.\frac{\partial S_m}{\partial \zeta}\right|_{\xi}.$$

According to Eq. (17.6), this allows us to calculate $\delta\rho^2(\mathbf{e})$ and, therefore, also $\delta n = \delta\rho(\mathbf{en})$, which is the solution to the problem:

$$\frac{1}{v(e)} = \frac{(en)}{\rho(e)}\left\{m^*(e) - \frac{1}{2\pi}\int_0^{2\pi} d\varphi \int_0^1 \frac{\partial m^*(\cos\theta, \varphi)}{\partial \cos\theta}\frac{d\cos\theta}{\cos\theta}\right\}. \tag{17.8}$$

Knowing the shape of the surface $\mathbf{e}$ we can find the volume enclosed by it and, therefore, the number of electrons in the associated band.

Formulas (17.7) and (17.8) deduced in the present section show that, in principle, it should be possible to determine the electron spectrum simply by investigating the de Haas–van Alphen effect. However, very extensive experimental data would be required: we would have to know the period and the amplitude oscillations for all the directions of the magnetic field. Moreover, Eqs. (17.7) and (17.8) are derived on the assumption that the Fermi surface is very simple. Consequently, in actual determination of the electron energy spectrum from oscillatory effects [37] it is normal to employ test models and to determine the set of constants describing such a model from the experimental data. Recent investigations of this type have been carried out on the basis of one of the variants of the almost-free electron theory.

§18. General Theory of Oscillatory Effects

An analysis of the oscillations of the thermodynamic quantities in a magnetic field (§15) shows that they are the result of the oscillations of the density of states and that the magnetic field simply quantizes the energy levels. Obviously, a similar effect should occur in all cases when the conditions are favorable for energy quantization. We should then observe an oscillatory dependence on the parameters which determine the positions of the electron energy levels, provided these levels are separated by gaps of the order of or greater than the temperature (expressed in energy units) but considerably smaller than the Fermi energy [38].

A general theory of oscillatory effects can be developed using the well-known conditions of the quasiclassical quantization, according to which the adiabatic invariants I_i of a system in finite motion are equated to half-integral values of the Planck constant:

$$I_i = \left(n_i + \frac{1}{2}\right)\hbar \qquad n_i = 0,\ 1,\ 2,\ \ldots . \tag{18.1}$$

If the "classical" problem of the motion of an electron with an arbitrary dispersion law is solved, the electron energy can be expressed in terms of the adiabatic invariants:

$$\varepsilon = \varepsilon\,(I_1,\ I_2,\ I_3),$$

and the conditions of Eq. (18.1) can be used to quantize the energy levels [see Eq. (7.33)]:

$$\varepsilon\,(n) = \varepsilon\left[\left(n_1 + \frac{1}{2}\right)\hbar;\ \left(n_2 + \frac{1}{2}\right)\hbar;\ \left(n_3 + \frac{1}{2}\right)\hbar\right]. \tag{18.2}$$

The vector n denotes all three integers n_1, n_2, and n_3.

The use of the quasiclassical energy levels restricts the application of the theory to the case when the energy levels are separated by gaps smaller than the Fermi energy ε_F. However, this restriction is not very important because oscillations appear only if the condition just specified is obeyed.

The quantization condition of Eq. (7.3) is a special case of the conditions given by Eq. (18.2). The adiabatic invariant of an electron moving in a magnetic field is the quantity cS/eH of Eq. (4.18).

The knowledge of the quantum energy levels of Eq. (18.2) makes it possible to calculate the oscillatory component of the thermodynamic potential

$$\Omega = -\,T \sum_n \ln\left\{1 + \exp\frac{\zeta - \varepsilon\,(n)}{T}\right\}.$$

Using the three-dimensional analog of the Poisson formula, we obtain

$$\Omega = -\,T \int \ln\left\{1 + \exp\frac{\zeta - \varepsilon\,(n)}{T}\right\}dn + 2\,\mathrm{Re}\sum_k L_k + \ldots, \tag{18.3}$$

where

$$L_k = -\,T \int \ln\left\{1 + \exp\frac{\zeta - \varepsilon\,(n)}{T}\right\}\exp\,(2\pi i k n)\,dn,$$

and the vector $\mathbf{k}$ represents the three indices k_1, k_2, and k_3, all of which are integers. The summation is carried out over all the positive values of k_i ($i = 1, 2, 3$), with the exception of $k_i = 0$.

Since the integration over $\mathbf{n}$ is carried out only in the first octant ($0 \le n_i \le \infty$, $i = 1, 2, 3$) because of the discontinuity of the integrand at the limit of the integration region, we find that the final result includes three-dimensional integrals L_k, as well as two- and one-dimensional integrals whose explicit forms will not be given.

Calculations similar to those given in §15 show that the oscillatory component of L_k has the form

$$\tilde{L}_k = \frac{1}{(2\pi)^2} \frac{1}{\sqrt{k_1^2 + k_2^2 + k_3^2}} \sum_v \frac{\psi(\lambda_k) \exp\left[2\pi i k n_v(\mathbf{k}, \zeta) + i\varphi_v\right]}{\left(k \frac{\partial n_v}{\partial \zeta}\right)^2 |\nabla \varepsilon(n_v)| \sqrt{|K_v(\zeta)|}}. \tag{18.4}$$

Here, $\mathbf{n}_v = \mathbf{n}_v(\mathbf{k}, \zeta)$ are the radius vectors in the space $\mathbf{n}$ of those points on the surface $\varepsilon(\mathbf{n}) = \zeta$ at which the direction of the normal to the surface is parallel to the vector $\mathbf{k}$; $K_v(\xi)$ is the Gaussian curvature of the surface at these points and $\lambda_k = 2\pi^2 \left(k \frac{\partial n_v}{\partial \zeta}\right) T$; $\varphi_v = \pm \pi/2$, if $\mathbf{n}_v$ is an elliptic point and the minus sign corresponds to the case when the surface is convex toward $\mathbf{k}$ and the plus sign corresponds to the opposite case (for a hyperbolic point $\varphi_v = 0$). Once again, we shall omit the two- and one-dimensional terms.

Substituting the asymptotic expressions for $\tilde{L}_k$ of Eq. (18.4), as well as the omitted two- and one-dimensional terms in Eq. (18.3), we obtain the oscillatory component of the potential Ω. In order to determine the oscillation period, we must know the dependence of $\mathbf{n}_v$, i.e., of the adiabatic invariants I_i, on the parameters which determine the positions of the quantized energy levels. It is clear from Eq. (18.4) that the oscillation period is given by the condition

$$\Delta I_i^{(v)} = \hbar,$$

where the symbol ΔI_i denotes the change in the i-th adiabatic invariant due to the change in the parameters which determine the finite motion of the electron. We must bear in mind that $\tilde{\Omega}$ is the sum of a large number of terms of the same order of magnitude, each of them having its own period. Since the ratios of these periods are, in general, arbitrary, the oscillations observed can be very complex.

The conclusions of the present section may be made more specific by considering the oscillations of the thermodynamic quantities resulting from the quantization of the electron energy levels in a film. An analysis of the oscillatory components of the thermodynamic properties of an electron gas with an arbitrary dispersion law [32, 33] shows that, in this case, the oscillation period is

$$\Delta L = \frac{2\pi \hbar}{P_e^{\text{extr}}(\varepsilon_F)}, \tag{18.5}$$

where L is the thickness of the film and $P_e^{\text{extr}}(\varepsilon_F)$ is the extremal chord of the Fermi surface; this chord is parallel to $\mathbf{e}$, which is normal to the film surface (Fig. 46). At $T \gtrsim \Delta\varepsilon$, the temperature dependence of the amplitude of each harmonic is determined by the factor

$\exp\left(-2\pi^2 T/\Delta\varepsilon\right)$, where $\Delta\varepsilon = \frac{2\pi\hbar}{L}\left(\frac{1}{v_1} + \frac{1}{v_2}\right)^{-1}$, and v_j is the modulus of the projection of the

electron velocity $\mathbf{v}$ onto the chord at the point j ($j = 1, 2$). Thus, a study of the size-induced oscillatory effects allows us to determine the values of the extremal chords of the Fermi surface and of the sums of the reciprocal velocities at the points of intersections of these chords

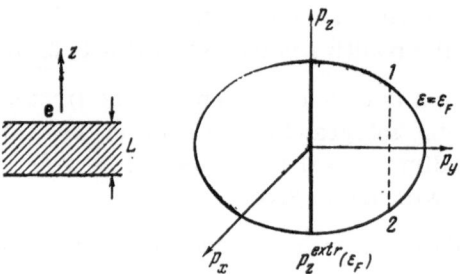

Fig. 46. Extremal chord of the Fermi
surface.

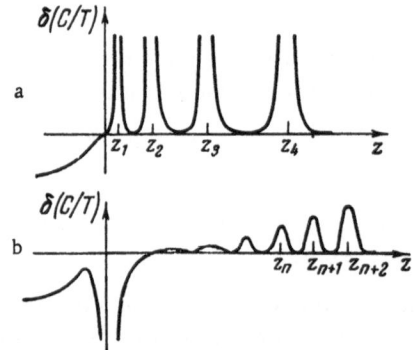

Fig. 47. Anomaly of the electron
specific heat of a metal film near a
transition point considered as a function of $z \propto \gamma (V-V_0)$: a) at T = 0;
b) at $T/\Delta \varepsilon \gtrsim 1$, $T/|z| \ll 1$.

with the surface. In the case of centrally symmetric surfaces, we can determine directly the
radius vector of the Fermi surface along a given direction, as well as the projection of the
electron velocity along this direction. If the Fermi surface consists of several sheets, the
size-induced oscillatory effects may become sensitive to the relative positions of these sheets.

An allowance for an oscillatory dependence of the density of states complicates the thermodynamic properties when they are considered as a function of the pressure near an anomaly
[32]. Instead of an "isolated singularity" (§13), we obtain a system of singularities which are
very sensitive to temperature.

Figure 47 shows an anomaly in the electron specific heat of a metal film near a transition
of type $2\frac{1}{2}$ at T = 0 and at a temperature above absolute zero.[1] The enhancement of the anomalies of the electron parameters is due to the superposition of two effects: oscillations of the density of states with the thickness of the film and a phase transition of type $2\frac{1}{2}$. Since the oscillation amplitude is not small (in the case considered, it is infinite), a situation of this type may be
called "giant oscillations."

Such giant oscillations may appear not only due to a change in the Fermi surface topology
but also when the Fermi surface has a saddle point and a plane $\mathbf{p} \cdot \mathbf{e}$ = const (corresponding to a
quantized value of $\mathbf{p} \cdot \mathbf{e}$) is in contact with the saddle point [32].

---

[1]An analysis of the effects of pressure on the behavior of various quantities and their temperature dependences are considered in [32].

Statistical thermodynamic properties of an electron gas of finite volume can be investigated also in magnetic fields [32, 34]. We shall not consider any details of this situation but note only that, in moderately strong magnetic fields, the oscillation period should depend on the dimensions and that this dependence should disappear if the extremal orbits are located entirely within a sample.

In considering the interaction of electrons with the surface of a sample, we have restricted our analysis to the case of specular reflection (§7). If the scattering of electrons by the boundary of a sample is diffuse, all the theories become more complex and the oscillatory dependences resulting from the quantization of the electron energy because of reflection from the boundaries may become less distinct or even disappear. Thus, the amplitudes of those harmonics in the de Haas–van Alphen effect which do not fit within a sample should decrease rapidly in amplitude [35]. This provides an additional possibility for the direct determination of the dimensions of sections of the Fermi surface.

§19. Strong Magnetism of Conduction Electrons. Anomalies of Thermodynamic Properties in Strong Magnetic Fields

We have frequently stressed the point that a magnetic field can be regarded as weak because the energy gaps $\Delta\varepsilon$ between the quantized levels become of the order of the atomic energy e^2/a or of the Fermi energy ε_F only in fields $H \sim 10^8$-10^9 Oe. This has allowed us to use the quasiclassical approximation in all our calculations. On the other hand, we pointed out in §15 [see Eq. (15.22)] that the magnetic susceptibility may be of the order of or greater than unity because of the oscillatory dependence of the magnetic moment. This means that metals may exhibit strong electron magnetism. The theory of this phenomenon has been developed only recently. The first problem which arises from the possibility of strong magnetism due to conduction electrons is whether the magnetic induction B or the magnetic field H should be used in the expression for the oscillatory correction to the thermodynamic potential, which is calculated in §15.

The problem of the behavior of charged quasi particles (e.g., conduction electrons in a magnetic field) is a typical problem in the field theory dealing with the behavior of a system of free charges in vacuum (in this case, the vacuum refers to quasi particles). Therefore, strictly speaking, we should consider only the microscopic magnetic field which is established at a given point by all the charges moving along orbits whose radii are of the order of the Larmor radius r_H. If the distance a between such charges is small compared with r_H, which is the usual case in metals ($a \sim 10^{-8}$ cm and $r_H \sim 10^{-3}$ cm for $B \sim 10^4$ Oe), the microscopic magnetic field at the point in question is determined by $4\pi(r/a)^2$ electrons and, therefore, it has the "self-averaging" property.

Such a self-consistent average field is, by definition, equal to the magnetic induction B. This is the only induction "seen" by each of the electrons and it determines the magnetic moment M [39] (the proof of this is given in [40]). Clearly, the relationship between M and B is nonlocal: M at the point considered is expressed in terms of the values of B at all the points separated from the one in question by the distance $4r_{max}$ (the fact that all the formulas should include B follows naturally from $\mathbf{p} = \mathbf{P} - (e/c)\mathbf{A}$, and rot $\mathbf{A} = \mathbf{B}$ since div $\mathbf{B} = 0$).

Before moving to the next step, we note that the important factor in front of the periodic function in the expression for the oscillatory component of the thermodynamic potential [see Eq. (15.16)] can be represented in the form $V \dfrac{H^2}{8\pi} \left(\dfrac{v_F}{c}\right)^2 \left(\dfrac{\hbar\omega_H}{\varepsilon_F}\right)^{1/2}$, which shows, first, that the diamagnetism of the degenerate electron gas is clearly of relativistic quantum origin, and secondly, that the oscillatory part of the potential is small compared with the magnetic field energy $VH^2/8\pi$. The strong magnetism, mentioned earlier, is not due to a change in the energy structure of the metal (such as that occurring in the ferromagnetic ordering of spins) but is simply due to the existence of oscillations.

However, a radical change in the energy spectrum of conduction electrons may occur in relatively weak magnetic fields ($H \ll 10^8$-10^9 Oe) for a completely different reason, which will be considered below.

Many metals have energy bands with an anomalously low degree of occupancy and a degeneracy energy of the order of 100°K. The electrons located in these bands have usually a small effective mass, which is of the order of $(0.1$-$0.01)\,m_0$, where m_0 is the mass of a free electron. We recall that the small effective masses and the small number of carriers are closely related (see §11 and [8]). In bismuth-type "semimetals" (Bi, Sb, As), these electrons are the only ones present. The quasiclassical approach is inapplicable to such electrons in fields of the order of 10^4-10^5 Oe.

Investigations in strong magnetic fields (quantum or extreme quantum limits)[1] are of special interest because such fields allow us to shift energy band edges, alter significantly the Fermi energy, change the number of carriers (strictly speaking, redistribute the carriers between several bands), and observe anomalies when the Fermi level crosses some critical values of the energy (§§3, 13).

We must stress once again that the electron energies in the spectrum of a metal (even a metal of the Bi type) are of the order of the energy of the Coulomb interaction between electrons and ions ($U \approx e^2/a$). Therefore, in a sense, the magnetic field is always weak provided $H \ll 10^8$-10^9 Oe. However, sometimes, the properties of conduction electrons depend strongly on the fine structure of the spectrum (a slight overlap of the bands in metals of the Bi type, low energy barriers in polyvalent metals predicted by Harrison's model, etc.). Therefore, the validity of the quasiclassical approximation is frequently determined not by the Coulomb energy but by the characteristic energy of the fine structure of the electron spectrum. Under such circumstances, even a weak magnetic field ($H \sim 10^4$-10^5 Oe) may appreciably alter the energy spectrum of conduction electrons. In some cases, a "new" dispersion law may be derived from the "old" law. This applies, for example, along certain directions of the magnetic field under magnetic breakthrough conditions (§10). However, usually this problem is too complex and cannot be solved without making some fairly rough approximation. Therefore, it would be natural and in the spirit of our treatment to assume that the quantized dispersion law of electrons in a magnetic field is known and to determine the relationship between the characteristics of this new law and the experimentally observable quantities.

It is evident from very general considerations that the quantum numbers are not affected. Since there is no dependence on the coordinates x and z if we use (to the lowest order in a/r_H) the vector potential of Eq. (4.16), it follows that there are two integrals of motion p_z and P_x. In the same approximation, the state of infinite multiplicity of the degeneracy applies to P_x. We shall assume that the classical motion in a magnetic field is finite. This ensures the existence of a discrete quantum number n. The magnetic field lifts the double degeneracy with respect to spin. Consequently, there are two branches of the spectrum, which will be denoted by the subscripts "+" and "−" (we shall assume that $\varepsilon^{(+)} > \varepsilon^{(-)}$):

$$\varepsilon = \varepsilon_n^{(\pm)}(p_z), \tag{19.1}$$

---

[1] By the extreme quantum limit we mean the situation in which the gaps between the magnetic levels are $\Delta\varepsilon \gg \varepsilon_F \gg T$ ($\Delta\varepsilon$ is the largest parameter of the problem).

which are not separated by $2\mu H$, where $\mu = e\hbar/2mc$, because the spin–orbit coupling can no longer be regarded as weak (see, for example [21] and §14).

The magnetic field is included in the dispersion law of Eq. (19.1) as a parameter. The wave functions (or the matrix elements) of an electron are also assumed to be known. The problem is to determine, first of all, which physical quantities are independent of the dispersion law (19.1) and, secondly, to find a method for deducing experimentally the law (19.1). This formulation of the problem is analogous to theoretical investigations involving "classical" dispersion laws of the $\varepsilon = \varepsilon(\mathbf{p})$ type.

We shall illustrate the solution of this problem by calculating the magnetic moment of the electron gas in the extreme quantum limit in the presence of one band [41]. Since the distance between the levels with given values of p_z is assumed to be the largest parameter of the problem, it is clear that only those states are important whose energies are close to the lowest value. This means that out of all possible branches of the spectrum described by (19.1) only one will be important and we shall denote it by $\varepsilon(p_z)$. Thus, we are faced with a problem in low-temperature thermodynamics of a "one-dimensional" gas whose dispersion law is

$$\varepsilon = \varepsilon(p_z). \tag{19.2}$$

The density of states per interval dp_z is given by Eq. (15.5) and the dispersion law (19.2) allows us to introduce the density of states per unit energy interval $\nu_H(\varepsilon)$:

$$\nu_H(\varepsilon) = \frac{2|e|HV}{(2\pi\hbar)^2 cv(\varepsilon)}, \tag{19.3}$$

where $v = |d\varepsilon/dp_z|$. The number 2 appears in this formula because the dispersion law is even: this is a consequence of the invariance of the directions along and opposite to the magnetic field.

Using the expression for $\nu_H(\varepsilon)$, we can write all the formulas in the statistical thermodynamics of the electron gas [see Eqs. (12.2)-(12.5)] but to be more specific we shall take account of the fact that the density of states (19.3) increases when the magnetic field is increased. This means that the Fermi level is located very close to the lowest energy $\varepsilon_{min} = \varepsilon_H$. Therefore, the dispersion law $\varepsilon = \varepsilon(p_z)$ can be expanded in powers of p_z and we need retain only the lowest term of the expansion:

$$\varepsilon(p_z) = \varepsilon_H + \frac{p_z^2}{2m_H}, \tag{19.4}$$

where $1/m_H = (\partial^2\varepsilon/\partial p_z^2)_{p_z=0}$ is the effective mass of an electron (deduced from the "new" dispersion law).

Equations (19.3), (19.4), and the formulas given in §12 yield an expression for the free energy of the electron gas at temperatures which are low compared with the separation between the levels

$$F = N\left\{\varepsilon_H + \frac{(2\pi\hbar)^4 c^2}{2e^2 m_H H^2}\left(\frac{N}{V}\right)^2\left[\xi - 2\int_0^\infty \frac{\sqrt{u}\,du}{e^{\frac{u-\xi}{\tau}}+1}\right]\right\}. \tag{19.5}$$

The function $\xi = \xi(\tau)$ is found from the normalization condition

$$1 = \int_0^\infty \frac{du}{\sqrt{u}\left(e^{\frac{u-\xi}{\tau}}+1\right)}, \tag{19.6}$$

where the parameter τ is the dimensionless temperature:

$$\tau = \frac{2e^2 m_H H^2}{(2\pi\hbar)^4 c^2} \left(\frac{V}{N}\right)^2 T. \qquad (19.7)$$

Thus, Eqs. (19.5)-(19.7) describe the dependences of all the thermodynamic quantities on the magnetic field and the temperature, and they are valid in a wide range of values of H and T. Naturally, we can use these equations to derive various limiting relationships. We shall not consider this point in detail. It is clear from Eq. (19.5) that in sufficiently strong magnetic fields we have $F \approx N\varepsilon_H$ and, consequently, $M \approx -N\partial\varepsilon_H/\partial H$. The quantity $-\partial\varepsilon_H/\partial H$ will be denoted by μ_H. Then,

$$M = N\mu_H. \qquad (19.8)$$

This quantity resembles the magnetic moment of a ferromagnet, i.e., we are dealing with "induced" ferromagnetism. No difficulty is encountered in separating the effects of the conduction electrons from the background of the ion-core electrons because the magnetic moment of the ions is $M_i \ll M$. Consequently, the magnetic field of the ions is weak and it follows that M_i is of the usual order of magnitude[2]: $\varkappa_i H \approx \mu N_i(\mu H/\varepsilon_a)$. Although $N_i \gg N$, we find that $\mu_H \gg \mu$ (this is a manifestation of the small effective mass, as discussed above) and, particularly, that Eq. (19.8) does not contain the very small factor $\mu H/\varepsilon_a$, where ε_a is an energy of the order of the atomic energy, i.e., $\varepsilon_a \approx e^2/a$. Thus, measurements of the magnetic moment under these conditions should make it possible to determine the dependence of the minimum energy in a band (19.4) on the magnetic field, and the temperature dependence should yield the effective mass or the equivalent quantity, which is the density of states at the bottom of the band. It is interesting to note that the chemical potential has a maximum in its temperature dependence.

The case considered (the one-band approximation) is suitable only as an illustration of the main idea. A band with an anomalously low occupancy in a pure metal is never the only one: we either have several such bands with equal numbers of electrons and holes (this applies to semimetals such as Bi) or − in addition to the bands with anomalously low occupancies − we have bands which have about one electron per atom. Naturally, a shift of the top or the bottom of an energy band in a magnetic field results in a redistribution of the electrons between the bands. This may be an increase in the number of carriers, if the band overlap increases, or it may be a loss of all carriers by some bands. At the moment when a band becomes empty, the electron characteristics of the metal should exhibit some anomalies. In the case of a semimetal, the final result is the complete filling of one band and the transformation of the semimetal into an intrinsic semiconductor.[3]

We shall now consider the anomalies in the electron characteristics of a metal with two bands (electron and hole) with equal numbers of carriers. Let us assume that the magnetic field reduces the band overlap, i.e., that the quantity $\delta\varepsilon = \varepsilon_2 - \varepsilon_1$ decreases when the magnetic

---

[2] We shall assume that $\varkappa_i$ is of the order of the usual paramagnetic susceptibility of a metal.

[3] The possibility that bismuth can be transformed into an intrinsic semiconductor in very strong magnetic fields was pointed out, in principle, in [42]. An experimental search for such a transformation in antimony was reported in [43]. A change in the number of carriers in bismuth was observed in [44] and anomalies of the electron characteristics were predicted in [45]. Recent experiments on Bi with impurities [46] indicated a dielectric−metal transition in strong magnetic fields. A rapid fall of the resistivity with increasing magnetic field (the dielectric−metal transition) was reported in [46a] and a change in the nature of the temperature dependence of the magnetoresistance was described in [46b].

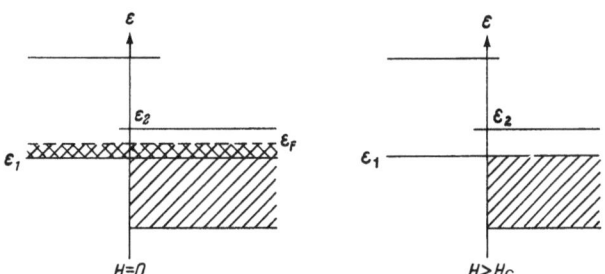

Fig. 48. Transformation of a metal into a dielectric in strong magnetic fields.

field is increased (in other words, that diamagnetism is the dominant effect) so that the quantity $\delta\varepsilon$ vanishes at $H = H_c$, and in the fields $H > H_c$ the metal becomes a dielectric at absolute zero (Fig. 48).

If $|H_c - H| \ll H_c$, we can expand all the quantities in powers of the deviation of the magnetic field from its critical value H_c. Bearing in mind the singularity of the density of states at the band edge, we obtain

$$n_1 = n_2 \sim (H_c - H)^{1/2}, \tag{19.9}$$

and the correction to the thermodynamic potentials due to electrons and holes is proportional to $(H_c - H)^{3/2}$. This applies only if the temperature is sufficiently low.

If $H > H_c$ and $T = 0$, the numbers of carriers diminish to zero and, consequently, no corrections are needed to the thermodynamic potentials. At low but finite temperatures, the numbers of carriers depend exponentially on the temperature.

It is clear that at $T = 0$ the magnetic moment $M_H = -\partial F/\partial H$ varies continuously whereas the magnetic susceptibility $\varkappa = \partial M_H/\partial H$ has an infinite discontinuity at $H = H_c$. By analogy with §13, we can show that a magnetic field causes a phase transition of the $2\frac{1}{2}$ type at $T = 0$. At finite temperatures, only strong anomalies of the electron characteristics are observed.[4]

Anomalies in the electron characteristics may occur also when the Fermi energy becomes equal to some value of the energy at which a new constant-energy surface is formed, a closed Fermi surface becomes open, etc. These anomalies resemble the singularities observed at high pressures, which were predicted in [14] and are described in §13.

If the band overlap $\delta\varepsilon$ increases with increasing magnetic field, the only possible anomalies are of the type just described. Such a situation provides a convenient means for investigating the dispersion law and its singularities in a wide range of energies.

The anomalies described above should be manifested also in the transport properties of a metal; in particular, they should be observed in the galvanomagnetic effects (see §§27, 28).

---

[4] If the number of carriers is sufficiently small, a phase transition of the first kind to a dielectric may take place (see [8] and §11). Since the electron-induced correction to $\partial P/\partial V$ may be negative, we find that, if this correction is sufficiently large, an isomorphous phase transition may take place at $\partial P/\partial V = 0$.

§20. Domains and Periodic Structures in Magnetic Fields

In pure samples at sufficiently low temperatures, when

$$\omega_H \gtrsim \frac{2\pi^2}{\tau}, \quad \frac{2\pi^2 T}{\hbar} \tag{20.1}$$

(τ is the relaxation time), the magnetic susceptibility $\varkappa$ estimated at the end of §15 (see also §19) may have values which have no upper limit. This represents strong magnetism.[1] The values $\varkappa \geq 1$ are obtained only if [provided the conditions of Eq. (20.1) are satisfied]:

$$\frac{e_F}{\hbar\omega_H} \gtrsim \left(\frac{c}{v_F}\right)^{4/3} \sim 10^3 \tag{20.2}$$

(here, c is the velocity of light).

A special feature of the conduction-electron diamagnetism is that, at $T = 0$ and $\tau = \infty$, it increases with decreasing magnetic field ($M/B \propto B^{-1/3}$, $\varkappa \propto B^{-3/2}$).

The term "diamagnetism" is used solely because the strong magnetism being discussed originates from the diamagnetic Landau quantization. The magnetic susceptibility (§15) is generally a sum of the diamagnetic and the paramagnetic (in the sense of their sign) components and may well correspond to strong diamagnetism ($-\varkappa \gtrsim 1$) as well as to strong paramagnetism ($\varkappa \gtrsim 1$). Strong paramagnetism gives rise to instability of the homogeneous state (this point will be discussed later).

The nonoscillatory term of the thermodynamic potential associated with the magnetic field can be shown to be much larger than the oscillatory term but it varies smoothly, and at low temperatures defined by Eq. (20.1) the contributions of this term to the total magnetic susceptibility $\varkappa$ and to the magnetic moment $\mathbf{M}$ are small.

If the field is sufficiently weak ($T = 0$, $\tau = \infty$), the magnetic moment M may be larger than the magnetic induction ($|\mathbf{M}| > |\mathbf{B}|$) but this is possible only at $T \lesssim 10^{-5}$–10^{-4} °K, $l \gtrsim 10$–10^3 cm, i.e., under conditions which are not yet attainable in the laboratory. Therefore, we shall assume that $|\mathbf{M}| \ll |\mathbf{B}|$, although $|\varkappa| \gtrsim 1$.

Only those states are thermodynamically stable for which $\partial H / \partial B > 0$ (see [47]), i.e., those which satisfy $4\pi\varkappa < 1$. Therefore, an increase in $\varkappa$ results in the splitting of a sample into magnetic layers with different values of $\mathbf{B}$ (Fig. 49), i.e., it gives rise to diamagnetic domains [48].[2]

Physically, this means that the magnetic field alters the density of states and, consequently, the internal energy of the electron gas; when $4\pi\varkappa_{max} > 1$ it is found that the energy considerations favor a change in the density of states which is due to a change in B associated with the splitting into domains.

Since the dependence M(B) is oscillatory (§15), it has a series of periodically repeated flat stratification regions (they are shown dashed in Fig. 49).

The homogeneous state of a magnet may be unstable also when the splitting into domains is not yet favored by the thermodynamic considerations, provided $\varkappa$ has paramagnetic as well

---

[1] Attention to this point was first drawn in [39].

[2] A calculation dealing with a domain wall is given in [49]. A direct experimental proof of the existence of diamagnetic domains is reported in [49a]: it was found that the NMR lines were split because of the variation of the magnetic moment from one domain to another. The results that follow from this model are presented and discussed in [49b].

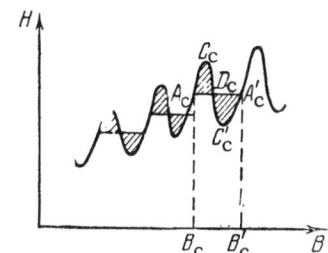

Fig. 49. Origin of diamagnetic domains: the oscillatory dependence of the magnetic field H on the induction B.

as diamagnetic components ($\varkappa = \sum_\alpha \varkappa_\alpha$, where $4\pi\varkappa < 1$). However, in the case of the paramagnetic component, a domain instability should appear even in the absence of an interaction because $4\pi\varkappa_\text{J} = \sum_\nu 4\pi\varkappa_\nu > 1$ (ν are those values of α for which $\varkappa_\nu > 0$). Actually, the splitting into domains does not occur because of a self-consistent interaction with the component for which the homogeneous state is stable. This means that the interaction results in "mixing," i.e., the interaction can be described as an effective "attraction" of the diamagnetic ($\varkappa_\alpha < 0$) to the paramagnetic ($\varkappa_\alpha > 0$) component.[3]

Thus, the instability of the components ν gives rise to a splitting tendency and the attraction between these components (negative surface energy) results in mixing. Under these conditions, we can expect phase ordering, i.e., the appearance of a self-consistent periodic structure whose characteristics are determined by the properties of the thermodynamic system. The period of such a structure is macroscopic and of the order of the "interaction radius," which is the Larmor radius r_H.

The transitions from a homogeneous to a periodic structure can be understood from some general considerations.[4] The requirement of thermodynamic stability implies the continuity of the thermodynamic potential at the transition point (for example, a positive value of the specific heat C_V means that the free energy is a continuous and monotonic function of the temperature T). This means that one of the following two events may occur at the transition point: an infini-

---

[3] The splitting into magnetic phases in the case $(\partial H/\partial B)_T < 0$ is analogous to the usual stratification of a multicomponent system into vapor and liquid when $(\partial P/\partial\rho)_T < 0$; the pressure corresponds to the magnetic field, equal to the sum of the partial fields of the components representing various extremal sections or bands, and the specific volume $V = \rho^{-1}$ corresponds to the magnetic induction. The interaction between the components formally reduces to the identity of the "specific volumes" of the various components (B obviously depends on the coordinates). Consequently, although the splitting into phases with B and B' is favorable from the thermodynamic point of view in the case of the ν "components," it is forbidden by the existence of other bands for which such specific volumes are unfavorable. However, if the induction B is nonuniform, the different values of $\varkappa_\alpha$ vary in different ways and the total susceptibility $\varkappa$ may increase, which is favored by the thermodynamic considerations (this point will be discussed later).

[4] The possibility of a periodic structure was demonstrated first in [50]. The theory of periodic structures was also given in that paper. The appearance of a periodic structure was predicted also in [51] but the solution given in that paper corresponded to $4\pi\varkappa > 1$, which would be thermodynamically absolutely unstable and therefore could not be realized physically under any conditions (this point will also be discussed later).

tesimal amount of a new phase with radically different properties, i.e., with a finite amplitude A of the spatial oscillations, may be formed, or a new state may differ infinitesimally from the old (i.e., the oscillation amplitude may be infinitesimal).

In the first case, the creation of a new phase is associated with nuclei which appear in fluctuations and which should be sufficiently large to be favored by the thermodynamic considerations (this is necessary because of the surface energy at the interfaces of different phases). Therefore, overheating and supercooling may occur and a metastable homogeneous phase may be formed. Consequently, the transition point is the point of intersection of the thermodynamic potentials of homogeneous phases and is not a singularity (see [10], §§81, 83), i.e., it is a phase transition of the first kind.

In the second case, there is no surface energy and therefore overheating or supercooling are impossible. The new state appears simultaneously throughout the whole sample and the transition point is a singularity of the thermodynamic potential of the system, i.e., we are now dealing with a phase transition of the second kind. The simultaneous change of state throughout the whole macroscopic volume should be preceded by a preliminary stage: the dimensions of the fluctuating regions of the thermodynamically unfavorable state (correlation radius r_c) should increase without any restriction on approach to the transition point. In Landau's theory, this takes place in accordance with the law (§119 in [10])

$$r_c \approx a \left| \frac{T - T_c}{T_c} \right|^{-1/2}$$

where a is the distance between particles and T_c is the transition temperature. As long as the correlation radius is small compared with the interaction radius ρ (if this is possible at all), which is the Larmor radius, the fluctuations of the oscillation amplitude can be ignored and the amplitude can be regarded as homogeneous and "authentic" (not accidental), i.e., determined by the thermodynamic equilibrium conditions. The inequality $r_c \ll r_H$ corresponds to $T \gg (a/r_H)^2 T_c$ ($a \sim 10^{-8}$ cm, $r_H \sim 10^{-3}$ cm in magnetic fields $H \sim 10^4$ Oe and $T_c \sim 10°$K – the temperature is assumed to be low because quantum oscillations are observed only at low temperatures), i.e., $|T - T_c| \gg 10^{-10}°$K. This inequality is always obeyed under experimental conditions.

If $(a/r_H)^2 \leq N^{-1/2}$ (N is the number of particles), the inequality $r_c \ll r_H$ is obeyed throughout the range of validity of the thermodynamic treatment because a change in the temperature has meaning only within the limits set by the fluctuations: $|(T_c - T)/T_c| \gg N^{-1/2}$. It is worth pointing out that $(a/r_H)^2 \propto n^{-4/3}$, where n is the density of the fermions.

If we neglect the fluctuations, we find that the magnetic-field correction to any thermodynamic potential (let us denote it by θ) depends on the "total" induction. Since an inhomogeneous correction near a phase transition point of the second kind is small compared with a homogeneous correction [$A \to 0$ when $\tau = (T_c - T)/T \to 0$], θ can be expanded as a series in terms of this correction. Assuming that the oscillation period λ is given[5] (i.e., that the period ensures a minimum of the thermodynamic potential at a given amplitude A of the spatial oscillations), we obtain an expansion of θ in powers of A^2, which is analogous to the well-known Landau expansion:

$$\theta = \theta_0 + \alpha A^2 + \beta A^4 + \gamma A^6 + \ldots .$$

$$(20.3)$$

---

[5] The period is finite at $T = T_c$. If we assume that the function $\lambda(T)$ is regular, we obtain a correction λ_c near T_c (this correction is a linear function of the temperature). A proof of this statement will be given later.

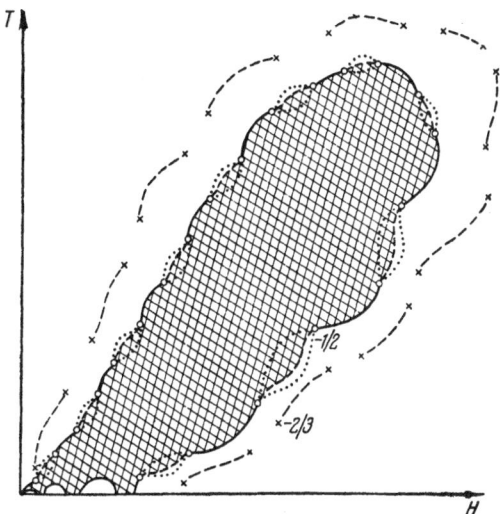

Fig. 50. Phase diagram. The continuous and the dashed curves represent phase transitions of the second and the first kind, respectively; the dotted curve is the line of absolute instability; ○ is a critical point for phase transitions of the second kind; × is a critical point; the shaded area represents the range of existence of a spatially periodic structure.

If $\alpha = 0$, $\beta > 0$, we obtain transitions of the second kind, according to Landau's classification (§§137, 138 in [10]), with a finite discontinuity in the specific heat C_H (if the transition occurs at a fixed value of the magnetic field H and the temperature is varied) or in the magnetic susceptibility (if T is fixed and H is varied). We can show that the relative magnitude of the discontinuity is of the order of $(\hbar\omega_H/\varepsilon_F)^{1/2}$.

A curve representing phase transitions of the second kind may terminate at its critical point where it intersects a curve representing phase transitions of the first kind (§140 in [10]). We then find that the coefficients in Eq. (20.3) are $\alpha(T_0, B_0) = \beta(T_0, B_0) = 0$. If $\alpha = a_1 T_1 + b_1 H_1$, $\beta = a_2 T_1 + b_2 H_1$ (where $T_1 = T - T_0$, $H_1 = H - H_0$, $|T_1/T| \ll 1$, $|H_1/H| \ll 1$), Eq. (20.3) can be reduced to its minimum value most conveniently using the coordinate system (α, β) and then converting the solution back to (T_1, H_1). If $\gamma > 0$, it is found that the curve representing transitions of the second kind continues as the absolute instability line of a homogeneous phase with A = 0 (the boundary of the absolute instability region is discussed in §21). At the point (T_0, H_0) this instability line is in contact with the line of phase transitions of the first kind (representing splitting into homogeneous, A = 0, and periodic, A ≠ 0, phases) and with the absolute instability line of a periodic phase with A ≠ 0 (Fig. 50). The specific heat C_H or the susceptibility $\varkappa$ become infinite because they are proportional to $|T_1|^{-1/2}$ or $|H_1|^{-1/2}$ (§21).

A phase transition may take place not only from a homogeneous structure to a periodic one but also when a finite or an infinitesimal perturbation with a new period, differing by a finite amount from the "old" value, appears against the background of an inhomogeneous structure.

A periodic dependence of the quantum oscillations on B^{-1} in the homogeneous case gives rise to a series of transitions periodic in B^{-1} when B is varied and T is kept fixed.

All the difficulties encountered in developing a theory of transitions of the second kind are found in the region $r_c \gtrsim \rho$, if such a region does exist.

Phase transitions associated with diamagnetic quantization occur also in the presence of a domain structure. If an external magnetic field H_0 is applied parallel to the surface of a sample so that $H_t = H_0$, the independent variable is H. In this case, a minimum is observed in the potential

$$\delta\theta_t = -\frac{1}{4\pi}\,B\,\delta H. \tag{20.4}$$

A phase transition of the first kind from B_c to B_c', known as "boiling," occurs at $H = H_c$. If the field H_0 is normal to the surface (Fig. 51), it is equal to the average value of the magnetic induction $\bar{B}$ for the whole surface of the sample:

$$H_0 = \bar{B}. \tag{20.5}$$

This relationship follows from the continuity of the integral $\int B_n \, dS$, taken over the surface shown in Fig. 51, when $l_2/l_1 \to \infty$ and $l_1 \to \infty$. It follows from Eq. (20.5) that $H_c = cB_c + (1 - c)B_c'$, where c is the concentration of the phase B_c. The points B_c and B_c' correspond to $4\pi\varkappa < 1$ and are deduced from the minimum of the thermodynamic potential. Since, according to Eq. (20.5), the value of $\bar{B}$ is fixed, the independent variable is now B and we shall seek a minimum of θ_t^* ($\delta\theta_t^* = H\,\delta B/4\pi$) on the assumption that the condition (20.5) is satisfied. This means that we shall be seeking the absolute minimum of $\theta_t = \theta_t^* + \zeta_H \int B \, dv$. Having determined the constant ζ_H from $\delta\theta_t/\delta B = 0$, we find that

$$\delta\theta_t = -\frac{1}{4\pi} B \,\delta H = -M\,\delta B + 2\pi\,(\delta M)^2 + \frac{1}{8\pi}\,(\delta H)^2. \tag{20.6}$$

Thus, θ_t has a minimum with respect to the independent variable B so that the points denoted by B_c and B_c' in Fig. 49 are determined by the equality of the areas $A_c C_c D_c$ and $D_c C_c' A_c'$. When H_0 is varied in the range $B_c \le H_0 \le B_c'$, we encounter "evaporation," which is accompanied by a discontinuity in the susceptibility and is a phase transition of the second kind. The situation is thus analogous to the usual stratification in a vapor–liquid system. [If the volume is constant, the independent variable is the average specific volume v and the function v(P) is not single-valued, so that we can vary P keeping v fixed; a minimum is exhibited by the thermodynamic potential Φ ($d\Phi = v\,dP$) considered as a function of v; the values of Φ per particle, i.e., the chemical potentials are then identical for both phases.]

Thus, transitions to periodic and domain structures may be of the first or second kind.

The phase transition to a periodic structure can be distinguished from the transition to a domain structure because of the strong dependence of a domain structure on the geometry of a bulk sample and the lack of such a dependence in the case of a periodic structure. Thus, for example, if H_0 is rotated in a plane perpendicular to a fourfold axis (the dimensions of the sample are L_1, L_2, $\sqrt{r_H L_1} \gg L \gg r_H$), the nature of the phase transitions for $H_0 \parallel x$ and $H_0 \parallel y$ will be the same for a periodic structure but will differ considerably for a domain structure (in the case of a domain structure, the transition will be of the first kind for $H_0 \parallel x$ and of the second kind for $H_0 \parallel y$).

An important property of these phase transitions is the macroscopic nature of the interaction radius. Since a sample is finite, the relative width of a transition is considerable along the temperature and the magnetic-field scales, the spread being of the order of $(r_H/D)^2$, where D represents the linear dimensions of the sample. Within this range, all the thermodynamic properties vary rapidly but are analytic functions.

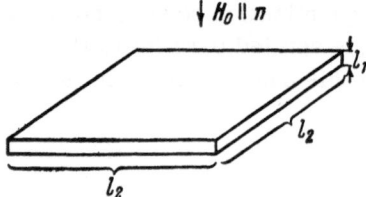

Fig. 51. External field perpendicular to the surface of a sample.

The estimates represented by Eqs. (20.1) and (20.2) show that, in pure samples, inhomogeneous structures appear in weak magnetic fields and at low temperatures (at T = 0 and $\tau = \infty$, the susceptibility tends to infinity $\varkappa \rightarrow \infty$ and B $\rightarrow$ 0).

Interference between the various periods reduces the magnetic susceptibility and widens the range of temperatures at which a transition occurs, or prevents the appearance of an inhomogeneous structure if $\varkappa$ is too small. This means that the mosaic structure of a crystal should be very weak and that the magnetic field should be stable in time and uniform in space (fluctuations in the magnetic field should be small compared with the period ΔB of the oscillations of $\varkappa$, i.e., they should be small compared with $B\hbar\omega_H/\varepsilon_F \sim B^2$).

The appearance of an inhomogeneous magnetic moment gives rise to an inhomogeneous electrostatic potential φ. According to the conditions of the thermodynamic equilibrium, the chemical potential ζ is constant over the whole system and the existence of $\varphi(\mathbf{r})$ is essential to ensure the constancy of the total electron density, which follows from the condition of electrical neutrality, satisfied to within terms of the order of $(a/r_H)^2$ (see, for example, [52]). Since our results indicate that $e\varphi(\mathbf{r}) \ll \hbar\omega_H$, $\varphi(\mathbf{r})$ may be ignored in the quantization so that if N = $N_0 + \Delta N$ [$\Delta N = -d\Delta\Omega/d\zeta$ is the component of N which oscillates when ζ is constant, $\Omega(\mathbf{r})$ is the density of the potential Ω, N_0 is the monotonic component of the electron density which is independent – in the first approximation – of the magnetic field], we obtain

$$N(\zeta) = N_0(\zeta + e\varphi) + \Delta N\{\boldsymbol{B}(\boldsymbol{r})\},$$

$$e\varphi(\boldsymbol{r}) = -\frac{1}{\nu(\varepsilon_F)}\Delta N\{\boldsymbol{B}(\boldsymbol{r})\} \sim \hbar\omega_H\left(\frac{\hbar\omega_H}{\varepsilon_F}\right)^{1/2}\exp\left(-\frac{2\pi^2 T}{\hbar\omega_H} - \frac{2\pi^2}{\omega_H\tau}\right), \quad (20.7)$$

$$\nu = \frac{dN_0}{d\varepsilon_F}.$$

The value of ζ is found by postulating that the total number of electrons in and out of the magnetic field is the same. The knowledge of the values of ζ and of $\varphi(\mathbf{r})$ allows us to determine the electron density in each band at the point r.

Using the formula just obtained and an earlier estimate of the correction to the thermodynamic potential, we can easily demonstrate that the potential φ does not affect the quantization in a magnetic field or the magnetic susceptibility in those cases which are of interest to us. Therefore, all the calculations can be carried out ignoring φ and substituting the results of the calculations in Eq. (20.7), which is used to calculate $\varphi(\mathbf{r})$.

The spatial periodic structure affects also other properties of a magnet. The propagation of electromagnetic and ultrasonic waves is different in the presence of a periodic superstructure. In particular, if the incident wave is of sufficient amplitude, we may observe singularities which are associated with phase transitions that are periodic in time. New resonances may appear because of new branches of free oscillations in the superstructure. The specific volume becomes modulated in space (magnetostriction in a periodic field).

Clearly, periodic structures, diamagnetic domains, and all the associated effects may occur also in a ferromagnet where, even in the absence of an external magnetic field, we have $B = 4\pi M_0$, where $M_0(T)$ is the spontaneous magnetic moment.

We shall now consider the thermodynamics of domain and periodic structures.

In order to simplify our discussion, we shall restrict it to the case when the magnetic field is directed along one of the principal crystallographic axes so that M ‖ B ‖ H ‖ z. We shall consider only the first approximation in respect of the mean free path $l: l = \infty$. In this case, there is no characteristic length along the z axis and all the quantities are thus independent of z. Since rot H = 0, this means that

$$H = H_z = \text{const}, \quad M = M_z(x, y), \quad B = B_z(x, y). \tag{20.8}$$

In order to determine the equation of state in the form of the dependence of M on B, which is generally nonlocal, we must calculate a correction due to the diamagnetic and the spin quantization and add this correction to the corresponding thermodynamic potential. (Since estimates show that this correction is always small, it follows that the corrections to all the thermodynamic potentials are equal when expressed in suitable variables; see [10] for details.)

Following the treatment given in §15, it is most convenient to calculate the potential Ω:

$$\Omega = -T \sum_c \ln \left(\exp \frac{\zeta - \varepsilon_c}{T} + 1 \right), \tag{20.9}$$

where the sum is taken over all the states.

In the homogeneous case the energy levels are known. According to Eq. (20.8), the inhomogeneous correction is related only to M, which is always small compared with B (as shown at the beginning of the present section). This makes it possible to find the energy levels in the inhomogeneous case by applying the perturbation theory to M; the degeneracy in respect of P_x is lifted because the position of the center of the orbit, determined by P_x, becomes important in an inhomogeneous field. In the one-dimensional case, when B depends only on one coordinate and P_x remains an integral of motion, the calculations are elementary [53]; in the two-dimensional case, the calculations are somewhat more complex because P_x is no longer conserved and it is necessary to determine the "correct" functions of the zeroth approximation in the case of infinitely multiple degeneracy [54].

The order of magnitude of the relative correction to the separation between levels can be found quite easily: it is given by $M/H \lesssim (v_F/c)^2 (\varepsilon_F/\hbar\omega_H)^{1/2}$. A small correction would not influence our earlier estimates but would affect mainly the argument of the rapidly oscillating function in which the number of levels is no longer $\varepsilon_F/\hbar\omega_H$ but $(\varepsilon_F/\hbar\omega_H)(1 + bM/H)^{-1}$, where $b \sim 1$. A small correction (of the order of $\varkappa$) to the phase appears in the approximation which is linear in respect of M. The next approximation is known to give rise to a small correction: $(v_F/c)^2 \times (\hbar\omega_H/\varepsilon_F)^{1/2}$. This means that even in the case of a periodic function it is sufficient to include the approximation linear in respect of M. It follows that we need to apply only the first approximation in the perturbation theory to the quantization rules and wave functions.[6] In general, the "correct" wave functions in the zeroth approximation and the energy levels calculated only as far as the first order are of the form:

$$\begin{aligned}
\psi^{(0)} &= \int C(P_x) \psi^{(0)}_{n p_z P_x} dP_x, \\
C(P_x) &= \exp \left\{ \frac{1}{\hbar} \int^{P_x} x(P_x') dP_x' \right\}, \\
\varepsilon &= \varepsilon^{(0)}_{n p_z} + \varepsilon',
\end{aligned} \tag{20.10}$$

where ε' is a continuous quantum number and $x(P_x)$ is found from the equation

---

[6] It follows that we can ignore the spin-orbit interaction associated with the force $\mu_0 \nabla(\sigma B)$ in an inhomogeneous field. This allows us to use the correspondence principle in the derivation of the quantization rules. The classical frequency in the inhomogeneous case can be calculated by applying the perturbation theory to M in the classical Hamiltonian–Jacobian equations: they are more convenient to use than the Newtonian equations when the period of motion is variable.

$$\varepsilon' = \frac{1}{H\frac{\partial s}{\partial \varepsilon}} \oint p_y(p_x)\, dp_x B_1\left(x, \frac{c\,(p_x - P_x)}{eH}\right),$$

$$B_1 = 4\pi M = \boldsymbol{B} - \boldsymbol{H}, \tag{20.11}$$

and $\varepsilon^{(0)}_{np_z}$, $\psi^{(0)}_{np_zP_x}$ correspond to the zeroth approximation (B = H).

If the energy levels are known, the thermodynamic potential can be calculated in the same way as in the homogeneous case (§15). The result is simplest in the one-dimensional case [54]

$$\Omega = \sum_\alpha \int dP_x \Omega^{(\alpha)} \left\{ \frac{cS_\alpha}{ehH} - \frac{2c}{ehH^2} \int_{p_{x\alpha}^{\min}}^{p_{x\alpha}^{\max}} p_y(p_x) M\left(\frac{p_x - P_x}{eH/c}\right) dp_x \right\}, \tag{20.12}$$

where $\sum_\alpha \Omega^{(\alpha)}(cS_\alpha/ehH)$ is the potential in a homogeneous magnetic field, calculated in §15 (see also [55]).

As in the homogeneous case, it is sufficient to replace the chemical potential with the Fermi energy.

The formulas obtained apply only to an ideal electron gas and they ignore, for example, an interaction of the type found in Fermi liquids (such an interaction does not affect the oscillation period and the order of magnitude of the oscillation amplitude; see §16). However, the actual form of M(r) is of little importance and the nature of this function can be deduced from very general considerations.

At the beginning of the present section, we demonstrated that, in general, all the quantities are periodic functions and their arguments are linear functions of the magnetic moment. This means that the density of any thermodynamic potential $\theta_1(\boldsymbol{r})$ of a magnet should, bearing in mind the translation symmetry, be of the form (as usual, we shall assume that B is directed along a crystallographic axis):

$$\left. \begin{array}{l} \theta_1(\boldsymbol{r}) = \sum_\alpha f_\alpha \left\{ \int K(\boldsymbol{r} - \boldsymbol{r}')\, M(\boldsymbol{r}')\, d\boldsymbol{r}' \right\}, \\[2mm] \theta = \int \theta_1(\boldsymbol{r})\, d\boldsymbol{r}, \end{array} \right\} \tag{20.13}$$

and the central symmetry of a crystal ensures that the function $K_\alpha(\boldsymbol{r})$ is even.

Knowledge of the thermodynamic potential allows us to determine the moment $\boldsymbol{M} = \delta\theta/\delta B$ and to derive the equation which is basic to the determination of B(x, y):

$$\boldsymbol{B} - 4\pi M\{B\} = \boldsymbol{H} = \text{const.} \tag{20.14}$$

However, we must bear in mind that this equation can have not only equilibrium solutions but also metastable and absolutely unstable ones. In fact, the equation of state gives us only an extremum of the thermodynamic potential and not its smallest value, and therefore such an extremum is a necessary (but not a sufficient) condition of equilibrium. The absolutely unstable solutions can be eliminated by the use of the second variational derivative of the total thermodynamic potential whose magnetic correction is associated not only with the quantization but also with the change in the fields within the magnet. We have shown that this potential is always θ ($\delta\theta = -\boldsymbol{B}\,\delta H/4\pi$, and this applies not only to domains but also to periodic structures). Since H does not vary (H = const) and $-\boldsymbol{M}\,\delta B = \delta\theta$, it follows that

$$\theta_t = \theta + 2\pi \int M^2 \, dr, \quad M = \frac{B - H}{4\pi} = -\frac{\delta\theta}{\delta B}. \tag{20.15}$$

Equations (20.15), (20.13), and (20.14) allow us, in principle, to determine completely the thermodynamic properties of the equilibrium and the metastable states of the magnets which we are considering. Since the thermodynamic potentials of these states differ by a finite amount and have a minimum in both cases, a study of the variational derivatives gives no new information: the equilibrium state corresponds to the smallest value of θ_t which can be found by comparing the values of θ_t for the various solutions substituted into Eq. (20.15).

The metastable solutions in magnets may be very stable because of the long-range nature of the interaction involved, and therefore a new phase can form only if the nuclei of this phase are macroscopic (of the order of r_H, i.e., 10^{-3}-10^{-4} cm). This should be contrasted with boiling, when atomic-size nuclei are sufficient even under weak overheating conditions. It must be stressed that the metastable states can be quenched-in only by very rapid variation of the magnetic field because the characteristic relaxation times are of electron origin and are therefore very short (even in very pure metals at helium temperatures, these relaxation times are $\tau \sim 10^{-9}$ sec).

In the next section, we shall consider the very interesting case of the formation of an inhomogeneous structure.

§21. Theory of Diamagnetic Phase Transitions

We shall first consider the case when the spatial diamagnetic structure is the result of a phase transition of the second kind.

As mentioned earlier, the basic equation in the theory of periodic structures is

$$\left.\begin{array}{l} B - 4\pi M\{B\} = H = \text{const}, \\ B = B(R), \quad R \equiv (x, y). \end{array}\right\} \tag{21.1}$$

Equation (21.1) has a homogeneous solution B_0:

$$B_0 - 4\pi M\{B_0\} = H, \tag{21.2}$$

but there may be other solutions. We shall determine the conditions for the existence of an inhomogeneous solution which is infinitesimally close to the homogeneous one:

$$B = B_0 + B_1. \tag{21.3}$$

In this case we have

$$B_1 = 4\pi \hat{\chi} B_1, \tag{21.4}$$

where $\hat{\chi}$ is a linear integral operator. It follows from the spatial homogeneity (a shift amounting to one lattice period is, in the lowest approximation, infinitesimal compared with r_H) that this operator should be of the difference type and the invariance of a crystal under inversion $(R \rightarrow -R)$ makes this operator even, so that Eq. (21.4) can be written in the form

$$B_1(r) = \int \chi(R - R')\, B_1(R')\, dR', \quad \chi(-R) = \chi(R). \tag{21.5}$$

Assuming that

$$B_1 = \text{Re}\,(A e^{-ikR}), \tag{21.6}$$

we obtain an equation for the period of the spatial oscillations:

$$4\pi \tilde{\chi}(k_0) = 1, \qquad \tilde{\chi}(k) = \int \chi(R) \cos(kR)\, dR. \tag{21.7}$$

The condition for the solubility of Eq. (21.7) is, provided $4\pi\chi < 1$, not only the necessary but also the sufficient condition for the appearance of a periodic structure.

To show this, we shall expand $\theta\{B\}$ in B_1. Bearing in mind that Eq. (20.15) should then be identical with Eq. (21.5) and using the symmetry of the crystal, we find that

$$\theta = \theta_0 + \frac{1}{8\pi} \int B_1^2(R)\, dR - \frac{1}{2} \int \chi(R - R')\, B_1(R)\, B_1(R')\, dR\, dR' + \frac{1}{12\pi} \int f(R - R';\ R - R'')\, B_1(R)\, B_1(R')\, B_1(R'') \times$$

$$\times dR\, dR'\, dR'' + \frac{1}{16\pi} \int g(R - R';\ R - R'';\ R - R''')\, B_1(R)\, B_1(R')\, B_1(R'')\, B_1(R''')\, dR\, dR'\, dR''\, dR'''. \tag{21.8}$$

The term linear in B_1 is absent from Eq. (21.8) because, in view of the translation symmetry, it should be of the form $K_0(B_0) \int B_1(R)\, dR$; and the stability in relation to homogeneous perturbations ($B_1 = \text{const}$) implies that

$$\left(\frac{\partial \theta}{\partial B_1}\right)_{B_1 = 0} = K_0(B_0) = 0 \quad \text{and} \quad \left(\frac{\partial^2 \theta}{\partial B_1^2}\right)_{B_1 = 0} = 1 - 4\pi\chi > 0$$

in accordance with our assumptions.

We shall write Eq. (21.8) in the form

$$\theta_t = \theta_0 + \frac{\pi}{2} \int (1 - 4\pi\tilde{\chi}(k))\, |\tilde{B}_1(k)|\, dk + \ldots \tag{21.9}$$

(here and later, the tilde $\sim$ denotes the Fourier components of the corresponding functions).

If Eq. (21.7) has a solution, it follows that $4\pi\tilde{\chi}_{\max} \geq 4\pi\tilde{\chi}(k_0) = 1$ and Eq. (21.9) implies that a weak inhomogeneity is stable whereas a homogeneous solution is unstable. We have mentioned earlier that the instability of the homogeneous state may be associated with the appearance of a periodic structure or the splitting into diamagnetic domains. This occurs (Fig. 45) even when $4\pi\chi < 1$ (i.e., when $\partial H/\partial B > 0$); only these values of χ are then possible. Therefore, a periodic structure may exist[1] only if $4\pi\tilde{\chi}_{\max} \geq 1$ (the maximum value is taken with respect to k) and $4\pi\tilde{\chi} = 4\pi\tilde{\chi}(0) < 1$.

We shall assume that this condition is satisfied. A phase transition point of the second kind corresponds to the first appearance of the root of Eq. (21.7), i.e., $4\pi\tilde{\chi}_{\max} = 1$, which means that this point corresponds to a multiple root of Eq. (21.7) (Fig. 52). Near this point, the range of values of k in which $1 - 4\pi\tilde{\chi} < 0$ is obviously small: $|\Delta k/k_0| \ll 1$. This means (Fig. 48) that

----

[1] We can show that, for a given band and a given section, $\chi_\mu(R)$ has a fixed sign and that

$$\text{sign}\, \chi_\mu(R) = \text{sign}\, \chi_\mu, \quad 4\pi\,|\tilde{\chi}_\mu| \leqslant 4\pi\,|\tilde{\chi}_\mu(0)| = 4\pi\,|\chi_\mu|.$$

Therefore, a periodic solution is obtained if there are several extremal sections and if $4\pi\chi_0 > 1$ and $4\pi\chi < 1$ (the definition of χ_0 is given on p. 125 in §20, where it is denoted by $\varkappa_0$).

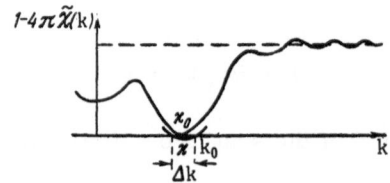

Fig. 52. Appearance of a multiple root.

if $|\Delta k| \gtrsim k_0$, where $1 - 4\pi\tilde{\chi}(k) \gtrsim 1$, the value of B_1 should be small compared with its value in the range where $1 - 4\pi\tilde{\chi} < 0$, since the term given in Eq. (21.9) predominates in the case of small values of $\tilde{B}_1$ (in accordance with our initial assumption) and the state with a small "scatter" Δk is known to be unfavorable. This means that, even when further terms of the expansion are included, B_1 near the transition point can be represented by

$$B_1(\boldsymbol{R}) = A(\boldsymbol{R}) \exp(i\varkappa\boldsymbol{R}) + A^{*}(\boldsymbol{R}) \exp(-i\varkappa\boldsymbol{R}) + C(\boldsymbol{R}), \qquad (21.10)$$

$$|C| \ll |A|,$$

$$\tilde{\chi}(\varkappa) = \tilde{\chi}_{max}, \qquad \nabla\tilde{\chi}(k) = 0, \qquad (21.11)$$

where $A(\boldsymbol{R})$ is a slowly varying function which changes significantly over distances large compared with the interaction radius in the kernels of Eq. (21.8), i.e., compared with the Larmor radius r_H. An asterisk is used for the complex conjugate and $C(\boldsymbol{R})$ may correspond to any value of k and vary in any manner provided it is small compared with $A(\boldsymbol{R})$.

Equation (21.10) allows us to apply the method of successive approximations[2] to the non-linear equation for $B_1(\boldsymbol{R})$, which follows from $\delta\theta/\delta B_1 = 0$:

$$B_1(\boldsymbol{R}) = 4\pi \int \chi(\boldsymbol{R}-\boldsymbol{R}') B_1(\boldsymbol{R}') d\boldsymbol{R}' + \int f(\boldsymbol{R}-\boldsymbol{R}'; \ \boldsymbol{R}-\boldsymbol{R}'') B_1(\boldsymbol{R}') B_1(\boldsymbol{R}'') d\boldsymbol{R}' d\boldsymbol{R}'' + \int g(\boldsymbol{R}-\boldsymbol{R}';$$

$$\boldsymbol{R}-\boldsymbol{R}''; \ \boldsymbol{R}-\boldsymbol{R}''') B_1(\boldsymbol{R}') B_1(\boldsymbol{R}'') B_1(\boldsymbol{R}''') d\boldsymbol{R}' d\boldsymbol{R}'' d\boldsymbol{R}''' + \ldots \qquad (21.12)$$

Substituting Eq. (21.10) into Eq. (21.12), we find (in the second approximation with respect to A) the value of C:

$$C = E \exp(2i\varkappa\boldsymbol{R}) + E^{*} \exp(-2i\varkappa\boldsymbol{R}) + 2D, \qquad (21.13)$$

where

$$E = \frac{f(\varkappa, -\varkappa)}{1 - 4\pi\tilde{\chi}(\varkappa)} A^2, \qquad D = \frac{f(\varkappa, -\varkappa)}{1 - 4\pi\tilde{\chi}(0)} |A|^2. \qquad (21.14)$$

The next approximation includes (in addition to the third harmonics) the first harmonics of the term of third order in B_1. The corresponding equation [this equation is, of course, iden-

---

[2] It is clear that the selection of the zeroth approximation in the form of Eq. (21.10) determines completely the nature of the successive approximations (an allowance is made for the condition of solubility of the inhomogeneous equation in the case when the homogeneous equation has a nontrivial solution). In this case, the solution of the equation $\int h(\boldsymbol{R} - \boldsymbol{R}') \Phi(\boldsymbol{R}') d\boldsymbol{R}' = F(\boldsymbol{R}) \exp(is\boldsymbol{R})$ with a slowly varying function $F(\boldsymbol{R})$ is of the form

$$\Phi(\boldsymbol{R}) = H^{-1}\left(i\frac{d}{d\boldsymbol{R}}\right) F(\boldsymbol{R}), \qquad H(\xi) = \int \exp\{i\boldsymbol{R}(\xi - s)\} h(\boldsymbol{R}) d\boldsymbol{R}.$$

tical with the condition stipulating that the "perturbing" inhomogeneous terms in Eq. (21.12) are orthogonal to the solution of the homogeneous equation] is of the form:

$$-2\pi \sum_{i=1}^{2} \frac{\partial^2 A}{\partial R_i^2} \frac{\partial^2 \tilde{\chi}}{\partial \varkappa_i^2} + \tau A - 4\beta A |A|^3 = 0,$$

$$\tau = 4\pi \tilde{\chi}(\varkappa) - 1; \tag{21.15}$$

$$\alpha_i^{-1} = -2\pi \frac{\partial^2 \tilde{\chi}}{\partial \varkappa_i^2} > 0. \tag{21.16}$$

The relatively cumbersome expression for β is not given. Here, $\alpha_i > 0$ because $\tilde{\chi}$ has a maximum at the point $\varkappa$; the directions of the axes x and y are selected in such a way that

$$\frac{\partial^2 \tilde{\chi}}{\partial \varkappa_x \partial \varkappa_y} = 0.$$

We shall introduce a point $\varkappa_0$ at which

$$4\pi \tilde{\chi}(\varkappa_0) = 1, \qquad \nabla \tilde{\chi}(\varkappa_0) = 0. \tag{21.17}$$

The two expressions in Eq. (21.17) define, in addition to $\varkappa_0$, the relationship between T and H, i.e., a curve representing phase transitions from a homogeneous to a periodic structure in the plane (T, H). If a transition occurs at a fixed value of H, we find that $\tau \propto T - T_0(H)$, where $T_0(H)$ is the transition point; however, if T is fixed, we then have $\tau \propto H - H_0(T)$.

Substituting $A = |A| \exp(i\psi)$ into Eq. (21.15) we find that, if $\psi \neq 0$, then $\psi \sim |A|^{-2}$ is a function of **R** which oscillates rapidly (because, by definition $|A|$ is small). However, we have shown that this is incorrect. Therefore, $\psi = 0$ and A is real. The substitution of Eq. (21.10) into Eq. (21.9) in the case of real A gives the expression

$$\theta_t - \theta_0 = \int \left\{ -U(A) + \sum_{i=1}^{2} \frac{1}{2\alpha_i} \left(\frac{\partial A}{\partial R_i} \right)^2 \right\} d\mathbf{R} \equiv \int \theta_1 \, d\mathbf{R}, \tag{21.18}$$

where

$$U(A) = \frac{1}{4} \tau A^2 - \frac{1}{2} \beta A^4.$$

It follows from our general discussion that a periodic structure with a period $2\pi \varkappa_i^{-1} \sim r_H$ appears at the transition point; moreover, Eq. (21.18) can be obtained directly in the form of an expansion of θ_t in terms of the slowly varying (compared with r_H) small correction A. This can be done bearing in mind that: a) the requirement of a minimum of θ_t (which is averaged out over distances of the order of r_H) with respect to A at the transition point implies that $\delta \theta_t / \delta A = 0$; b) $\delta H / \delta A \sim \tau$ because $4\pi \tilde{\chi}(\varkappa_0) = 1$ and this gives $\delta^2 \theta_t / \delta A^2 \sim \tau$; c) the requirement of a minimum of $\theta_t(A)$ at $T = T_0$ demands that $\delta^3 \theta_t / \delta A^3 = 0$, $\delta^4 \theta_t / \delta A^4 = \beta > 0$; d) an expansion in terms of the small quantity ∇A may contain only the even powers of ∇A because of the slow variation of $A(\mathbf{R})$ and the invariance under the inversion of the sign of **R**.

We shall now consider the meaning of the requirement $\beta > 0$ in the case considered. If $\beta < 0$, it follows from Eq. (21.18) that the point $\tau = 0$ is not singular at all: a periodic structure of finite amplitude already exists at this point because the transition has occurred earlier at some finite value of A (this follows from the fact that we are seeking a transition corresponding to $A \rightarrow 0$), i.e., we are dealing with a phase transition of the first kind.

Since $\alpha_i > 0$ [see Eq. (21.16)], it follows that $\theta_1 \geq -U(A) \geq U_{max}$, and the equality $\theta_1 = -U_{max}$ is satisfied by a homogeneous A. The term with the derivative in Eq. (21.18) vanishes, θ_1 assumes a form typical of Landau's transitions of the second kind [see Eq. (20.3)], and this gives rise to a change from A = 0 for $\tau < 0$ to $\pm A_\theta = \frac{1}{2}\sqrt{\tau/2\beta}$ for $\tau > 0$. The states with $\pm A_0$ differ only by a phase shift. This difference may be considerable for a finite sample in which such states are analogous to domains. The definitions of $\varkappa$ and $\varkappa_0$ can be used to find easily the dependences of the space period $2\pi\varkappa_i^{-1}$ on τ: $\varkappa - \varkappa_0 \propto \tau$. Thus, the oscillation period varies linearly with τ near the transition point and the oscillation amplitude is proportional to $\sqrt{\tau}$.

All these points become especially clear in the one-dimensional case A = A(y), where y is the time. The functional (21.18) can then be interpreted formally as the action integral for a particle of mass α moving along a coordinate A which is restricted by the condition that the particle must not escape to infinity (because B must remain finite). The transition between states with $+A$ and $-A$ is then of the domain type and occurs (for $\tau > 0$) in accordance with the law

$$A = A_0 \,\mathrm{th}\left\{ \tfrac{1}{2}\sqrt{\tau\alpha}\,(y - y_0)\right\}. \tag{21.19}$$

It is clear from the definition of Eq. (21.7) that as $k \to \infty$ the function $\tilde{\chi}(k)$ oscillates and approaches zero. This means that when the temperature (or the magnetic field) is varied, new roots of Eq. (21.11) may be obtained and new phase transitions may be encountered. Splitting into phases with different periods may become possible if any extremum of $\tilde{\chi}(k)$ is degenerate in the range k > 0.

In all the discussion so far, we have assumed that there is only one solution of the equations denoted by (21.11). Clearly, this is true only of a rectangular lattice in the x–y plane. The presence of threefold or higher symmetry axes gives rise to a "star" of the vectors $\varkappa$, which represents a larger number of solutions of the two equations in (21.11). If no three of these vectors add up to zero, all our considerations still apply but the treatment becomes more cumbersome. If any three solutions of (21.11) add up to zero, the third-order term in the expansion $\theta_t(A)$ does not vanish, a phase transition of the second kind becomes impossible, and a periodic structure appears as a result of a phase transition of the first kind.

Till now, we have assumed that the instability of the homogeneous state is unstable at $\varkappa \neq 0$. Let us now consider the opposite case when $\tilde{\chi}_{max} = \tilde{\chi}(0) \neq \chi$. [The point $\varkappa \neq 0$ must correspond to an extremum because of the central symmetry of a crystal: $\tilde{\chi}(-k) = \tilde{\chi}(k)$, $\nabla\tilde{\chi}(-k) = -\nabla\tilde{\chi}(k)$, and hence $\nabla\tilde{\chi}(0) = 0$; it follows that this extremum is a maximum because the homogeneous state is initially stable.] In this case, we can use the theory developed in the preceding section but it is simpler to point out that when $\varkappa_0 = 0$ the whole of B varies slowly and therefore $\theta_1\{B\}$ can be expanded in powers of ∇B. In the lowest approximation, θ_1 is identical with the "local" homogeneous density $\theta_1^0(B)$. In the next approximation, we have

$$\theta_1\{B\} = \theta_1^0(B) + \sum_{i=1}^{2}\frac{1}{2\alpha_i}\left(\frac{\partial B}{\partial R_i}\right)^2. \tag{21.20}$$

Following the same reasoning as before, we can show that the critical point corresponds to the splitting into diamagnetic domains and that the transitions between these domains are governed by Eq. (21.19). Within a domain wall, the derivatives in $\theta_1\{B_1\}$ vanish.

Since $4\pi\chi = 1$ and, consequently, the homogeneous case corresponds to $\partial H/\partial B = 0$, the stability of the homogeneous state on one side of the transition point implies that $\partial^2 H/\partial B^2 = 0$ and $\partial^3 H/\partial B^3 = 0$. Assuming that $B = B_0 + B_1$, $T = T_0 + T_1$, $H = H_0 + H_1$, and bearing in mind that

$$\delta\theta_t = -\, \boldsymbol{B}\, \delta \boldsymbol{H}, \tag{21.21}$$

we find that

$$\theta_t = \theta_0 + \left(a_2 T_1 - \frac{H_1}{4\pi} \right) B_1 + \frac{1}{3}\, a_3 T_1 B_1^2 + \frac{1}{4}\, a_4 B_1^4. \tag{21.22}$$

This expansion (justified in §20) is analogous to the Gibbs expansion near the usual critical point of a liquid–vapor system (see, for example, §84 in [10]) except that now the pressure p is replaced by H and the specific volume v is replaced by B. (For a given relationship between p and v experimental investigations near the usual critical point are extremely difficult. This does not apply to a domain-type critical point in an inclined magnetic field because, in this case, experimental investigations meet with no difficulties.) Therefore, all the results of the Gibbs analysis of the critical points can be applied directly to our case. Thus, by analogy with Eq. (84.10) in Landau and Lifshitz's book [10], we obtain

$$C_H \propto \left(a_3 T_1 + 3 a_4 B_1^2 \right)^{-1}. \tag{21.23}$$

In particular, on the equilibrium curve where $B_1 \propto |T_1|^{1/2}$, the specific heat is $C_H \propto T_1^{-1}$, whereas in a "critical" magnetic field ($H_1 = 0$), when a minimum of θ_t (i.e., $\partial\theta_t/\partial B_1 = 0$) implies $B_1 \propto T_1^{-1/3}$, Eq. (21.33) yields $C_H \propto T_1^{-2/3}$.

The shapes and the dimensions of diamagnetic domains are of special interest. The most interesting and important is that case in which the characteristic linear dimensions of a sample L are large compared with the Larmor radius r_H so that the domain size g is considerably greater than the thickness of a domain wall d. In this case, the solution of the problem can be split into "microscopic" and "macroscopic" parts.

The "microscopic" problem is to determine the nature of a domain wall (i.e., the distribution of the magnetic induction) whose thickness is of the order of d. Since d ≪ g, it follows that to the lowest order in d/g such a wall may be regarded as planar and the problem as one-dimensional so that, in accordance with Maxwell's equations, B_n = const, H_t = const (n is the normal whose direction coincides with the y axis and t is a direction parallel to the wall).

Then, according to Eq. (21.1),

$$\left. \begin{aligned} H_t\{B_t(y),\ B_y\} &= H_t, \\ H_y\{B_t(y),\ B_y\} &= H_y(y). \end{aligned} \right\} \tag{21.24}$$

The two equations denoted by (21.24) are completely analogous to the equations of state near the boundaries of coexisting phases except that B_y should be regarded as the given parameter because near the surface of a sample B_n is not, generally speaking, equal to the bulk equilibrium value. Therefore, the conditions for the saturation of B_t and H_y obtained in the limit $y \rightarrow \pm\infty$ (i.e., for y ≫ d, provided y ≪ g) are identical with the usual condition of the equality of the areas under the curve $H_t = H_t(B_t)$ for a given value of B_y, which is analogous to the curve in Fig. 49 where H should be replaced with H_t and B with B_t. Consequently, at infinity (y → ± ∞), we obtained a definite relationship between H_t and B_y:

$$H_t = H_c(B_y). \tag{21.25}$$

This condition is especially clear in the isotropic case when $H_y = H_y(B_y) = 0$, H_t is independent of B_y, and the condition $H_t = H_c$ corresponds to the usual equality of the areas under the curve $H_t = H_t(B_t)$.

The actual form of the dependences $B_t(y)$ and $H_y(y)$ can be found near a critical point when the approximation (21.20) is applicable. The corresponding formula for the isotropic case was first derived in [49]. It is naturally analogous to the formula (21.19) (in particular, the thickness of the domain wall is of the order of $r_H\mu^{-1/2}$) and it allows us to find the surface energy Δ at the boundary of a domain:

$$\Delta = \frac{r_H \mu^{1/4} (B - B')^2}{24 \sqrt{2\pi}},$$
(21.26)

where $\mu = 1 - 4\pi\chi$ and r_H is the maximum value of the Larmor radius.

The "macroscopic" problem is to determine the nature and the dimensions of a domain in the lowest approximation with respect to $d/g \ll 1$, i.e., when $d = 0$. We first specify the period D of the domain structure and then solve the magnetostatic problem. Then, Eq. (21.25) provides an additional boundary condition for the domain walls and determines the shape of these walls (this shape has been found for the limiting cases in [56]).

Equation (20.5) allows us to express g_1 and g_2 in terms of D and H_0:

$$g_1 = \frac{(H_0 - B') D}{B - B'}, \qquad g_2 = D - g_1 = \frac{(B - H_0) D}{B - B'}.$$
(21.27)

Finally, we can determine the period D by calculating the total thermodynamic potential for a known structure of the fields. We include in this potential the surface energy of the domain walls and minimize the potential with respect to D. We then find that the period is of the order of (see [56]):

$$D \approx \sqrt{\mu r_H L}.$$
(21.28)

Equations (21.26)-(21.28) justify the approximations

$$\frac{d}{g} \approx \frac{1}{\mu} \sqrt{\frac{r_H}{L}} \ll 1, \qquad \frac{g}{L} \approx \sqrt{\frac{r_H}{L\mu}} \ll 1.$$

Far from a critical point, where $\mu_A \gg 1$ (and correspondingly $\chi_{max} \ll 1/4\pi$), the calculations are much more complicated [56]. A domain wall then has a very complex shape: B(y) oscillates from A_c to A_c' and has many gradually decreasing narrow high maxima, as well as corresponding deep minima. The order of magnitude of the surface energy and the domain size can still be found from Eqs. (21.26) and (21.28) but because $\mu \gg 1$, we must replace $\mu^{1/2}$ with $\mu^{1/3}$.

The validity of the expansion of the thermodynamic potential by analogy with the expansion used by Landau in his theory of phase transitions is related to the long-range effects at distances $r \gg a$. This expansion makes it possible to study a singularity in the absolute instability curve for which $\partial H/\partial B = 0$ in the homogeneous case. Let us assume that, for given values of H and $T = T_0$, the equation for B, which can be written symbolically as

$$\hat{L}\{H, T_0; B\} = 0,$$
(21.29)

has a partial solution $B_0(R)$, so that

$$\hat{L}\{H, T_0; B_0(\boldsymbol{R})\} = 0.$$
(21.30)

Let us now consider the nature and the stability of the solution for $T = T_0 + T_1$. We shall assume that $B = B_0 + \psi$. Then,

$$\hat{L}_1\psi + T_1\hat{L}_2\{B_0(\boldsymbol{R})\} + \tfrac{1}{2}\hat{L}_3\psi^2 + \ldots = 0, \tag{21.31}$$

where

$$\hat{L}_2\{B_0(\boldsymbol{R})\} = \frac{\partial}{\partial T_0}\hat{L}\{H,\ T_0;\ B_0(\boldsymbol{R})\},\ \hat{L}_3 = \frac{\delta^2\hat{L}}{\delta B_0^2}. \tag{21.32}$$

The value of T is varied and H is fixed only to make the case more specific: the analysis is exactly the same when T is fixed and H is varied.

The solution and the analysis of Eq. (21.31) follow the same pattern as the treatment of Eq. (21.12) and the nature of the solution is again determined by the presence or the absence of a solution of the homogeneous equation $\hat{L}_1\psi = 0$. The values of T corresponding to the first appearance of such a solution define the curve $T_0 = T_0(H)$. The only difference is that now the cubic terms in the thermodynamic potential may either vanish (because of symmetry or at isolated values of H) or they may be retained. This is because, in contrast to Eq. (21.12), the kernels in Eq. (21.31) need not be of the difference type since the inhomogeneity of $B_0(\boldsymbol{R})$ violates the translation symmetry of the system. In this case, the expansion of the thermodynamic potential in terms of $A_1 = A - A_0$ (A_0 is the amplitude of the periodic structure on the absolute instability curve) and $T_1 = T - T_0$ is of the form:

$$\theta_t = \theta_0 + \alpha T_1 A_1 + \tfrac{1}{6}\beta A_1^3. \tag{21.33}$$

If $\alpha T_1/\beta < 0$, there is no minimum of θ_t but if $\alpha T_1/\beta > 0$, there is a relative minimum at

$$A_1 = (-2\alpha T_1/\beta)^{1/2} \text{ and } \theta_t - \theta_0 \sim |T_1|^{3/2}.$$

This means that when the absolute instability curve is approached in a fixed external field H_0, the specific heat approaches infinity as $|T_1|^{-1/2}$. In the isothermal approximation to this curve ($H_0 = H + H_1$), the susceptibility tends to infinity as $|H_1|^{-1/2}$. The singularity is similar to that at a critical point in a phase transition of the second kind.

We shall now consider the general form of the phase diagram in a magnetic field, bearing in mind the possibility of the appearance of periodic and domain structures. At high temperatures $T \gg (1/2\pi^2)\hbar\omega_H$, the susceptibility associated with the Landau diamagnetism and the Pauli paramagnetism is weak at any temperature, and an inhomogeneous structure cannot exist even in the metastable state (this can be shown quite easily by means of the perturbation theory).

A local minimum of the inhomogeneous field appears for the first time at some lower temperature. In principle, we can distinguish several possible cases.

A local minimum may appear first at some finite amplitude of the inhomogeneity so that the smallest value of θ_t is again due to a homogeneous magnetic induction B. In this case, the function $\theta(A)$ has at least two minima which are always separated by a relative maximum. Thus, the function $\theta_t(A)$ acquires (in addition to a minimum corresponding to the homogeneous induction B, i.e., corresponding to A = 0) a relative maximum and a relative minimum, so that the temperature T_c at which a minimum appears corresponds to a triply degenerate solution of $\partial\theta_t/\partial A = \partial^2\theta_t/\partial A^2 = 0$; this condition determines the temperature T_c and the amplitude at this temperature (Fig. 53a). The appearance of a local minimum means that the corresponding phase may exist although it may be metastable, and the disappearance of such a minimum means that this phase is absolutely unstable. The curve corresponding to the appearance of

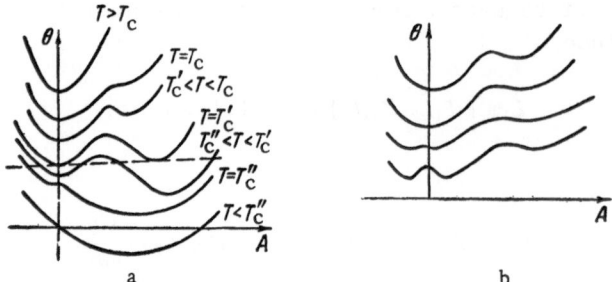

Fig. 53. Dependences of θ on A in the presence
of a phase transition of the first (a) and the
second (b) kind.

minima in the T–H plane thus defines the absolute instability region of a given phase [it is analogous to the curve $(\partial p/\partial v)_T = 0$ associated with the stratification in a liquid–vapor system].

Next, when a minimum of θ_t becomes deeper and reaches, at $T = T'_c$, the same value as in the preceding absolute minimum at $A = 0$, we find that a phase transition of the first kind results in splitting of a metal into a homogeneous ($A = 0$) and an inhomogeneous ($A \neq 0$) phase.[3] At $T < T'_c$, the inhomogeneous phase becomes stable and the homogeneous phase becomes unstable in the range $T''_c < T < T'_c$ and absolutely unstable at $T < T''_c$ (in the latter case, the singularity near $T = T_c$ is of the same kind as that discussed earlier; it must be stressed that we regard H_0 as fixed and T as variable only to make the case more specific).

However, it may be found that the homogeneous phase may become unstable at $A = 0$ before it splits into a homogeneous and an inhomogeneous phase (such splitting may not occur at all, since a minimum at $A \neq 0$ may begin to move upward beyond a certain temperature). This corresponds to a phase transition of the second kind (Fig. 53b) considered earlier, because— as already demonstrated – the expansion $\theta_t(A)$ contains only the even powers of A. The minimum at $\pm A$ corresponds to phases with the same periods and different "points of origin." If the minima of $\theta_t(A)$ coincide at different values of $A \neq 0$, we find that splitting results in structures with different periods.

When the external conditions are altered, the minima in the $\theta_t(A)$ curves can move "upward" or "downward," and therefore we can have various combinations of the cases considered.[4] In particular, a new structure may appear as a result of a transition of the first kind at one point only (T_0, H_{00}), as illustrated in Fig. 54, which shows the behavior of two structures I and II.

All the conclusions reached so far were taken into account in the phase diagram of a magnet plotted in Fig. 50 using the two variables T and H. The transition to the variables T and H_0 can be made using the relationship between H_0 and H established earlier in our analysis. Figure 50 was drawn bearing in mind that the phase transition curve cannot terminate in the case of a periodic structure but can terminate at a critical point for a domain structure. The nature

---

[3] It follows, therefore, that a periodic solution with a finite amplitude always appears first as a metastable state, which is "unfavorable" compared with the homogeneous phase.

[4] When the magnetic field is varied keeping the temperature T constant, the minima in the $\theta_t(A)$ curves move periodically. The number of periods at temperatures $T < \hbar\omega_H/2\pi^2$ is of the order of $(1/4)(\varepsilon_F/\hbar\omega_H)(v_F/c)^{4/3}$. We recall that the amplitude of the spatial oscillations is of the order of the amplitude of the quantum oscillations of the magnetic moment, i.e., $B(\hbar\Omega/\varepsilon_F)^{1/2}$.

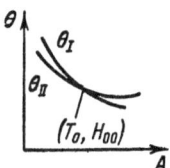

Fig. 54. Dependence of θ on A for two structures near a phase transition of the first kind at T_0, H_{00}.

of the transitions to a domain structure is the same along the whole curve and depends only on the orientation of H_0 (Fig. 50 shows the specific case when H_0 is perpendicular to the plane of the sample). We also took into account (in addition to the periodicity of χ_α in B^{-1}) the fact that, in accordance with the limitations imposed on the magnetic field, the transition curve has upper and lower bounds in respect of B and an upper bound in respect of the temperature. The nature of the phase diagram would be much more complicated in the next approximation with respect to a/r_H (a is the atomic spacing), in which the comparable values of a^2 and ehH/c are important (see also [57]).

§ 22. Emission Properties of Metals

The free-electron model can explain the basic relationships which govern the emission properties of metals (the exponential temperature dependence of the thermionic emission current, the characteristic dependence of the cold-emission current on the electric field, the threshold nature of the photoelectric emission, etc.). Investigations in which an allowance was made for the complex nature the dispersion law of conduction electrons [58-61] have confirmed the principal conclusions obtained from the free-electron theory. However, the theory of electrons with an arbitrary dispersion law predicts several additional effects, the main of which are: a) the anisotropy of the work function; b) the difference between the work functions of the various effects.

Although investigations of the emission properties of metals cannot be used to determine the dispersion law of the conduction electrons, they can often yield information which is difficult to obtain from other experiments; for example, in some cases it is possible to determine the position of the Fermi surface in the crystal momentum space [60].

The application of a relatively strong (quantizing) magnetic field should give rise to a dependence of the emission current on this field. The greatest interest lies in the oscillations of the emission current, which are due to the oscillations of the chemical potential of conduction electrons (§15), and in the magnetic-field dependence of the transmission coefficient of electrons crossing the potential barrier at the boundary of a metal [62, 63].

The quantization of the electron energy in a magnetic field gives rise to a magnetic-field dependence of the contact potential difference between two metals or between two faces of a metal single crystal [63]. This phenomenon has been confirmed experimentally [64].

Surface phenomena can also be investigated by measuring the current in a tunnel diode as a function of an applied potential φ. The tunnel current [65] oscillates not only when the magnetic field is varied (the nature of these oscillations is the same as in the de Haas—van Alphen effect) but also when the potential φ is altered. The oscillation period in respect of φ is then:

$$\Delta\varphi = 2\hbar\omega_H/e, \qquad \omega_H = eH/m^*c. \tag{22.1}$$

This means that oscillations of the tunnel current can be used to determine simultaneously the extremal sections of the Fermi surface (they can be deduced from the periodic dependence on the magnetic field) and the effective masses corresponding to these sections (the masses can be found from the periodic dependence on the potential difference).

The determination of the effective masses from $\Delta\varphi$ on the basis of Eq. (22.1) does not have the disadvantage which is encountered in the derivation of m* from the temperature dependence of the amplitude of the de Haas–van Alphen effect (§§15, 17) because the period $\Delta\varphi$ is insensitive to the mean free path of electrons. Basically, the tunnel diode provides a resonance method for measuring the effective mass (Chap. IV).

CHAPTER III

TRANSPORT PROPERTIES OF ELECTRONS IN METALS

This chapter deals with the transport properties of metals, particularly those at low temperatures. As elsewhere in the book, our principal concern will be the effects and the properties which are sensitive to the energy spectrum of conduction electrons. The mathematical treatment will be based on Boltzmann's transport equation.

The treatment will be limited to the static or the quasistatic properties. This will make it possible (§23) to use the "gas" approximation because the interaction encountered in Fermi liquids disappears from the final formulas.

Most of the transport phenomena are very sensitive to the nature of the interaction between conduction electrons and phonons. In particular, this interaction governs the temperature dependences of the transport coefficients. However, no attempt will be made to present the extensive theoretical material relating to the electron–phonon interaction but the most important results will be used in the present chapter. Special attention will be paid to those relationships and properties which are more or less independent of the nature of the interaction. In some cases, we shall use the τ approximation, i.e., we shall replace the integral collision operator with the operator corresponding to multiplication by an associated phenomenological constant (the relaxation time τ); in all such cases, the use of the approximation will be mentioned specially. We shall restrict our treatment to the problems associated with the electrical conductivity (resistivity), the thermal conductivity, and the thermoelectric properties of a metal, and we shall pay special attention to the effects of magnetic fields. Initially, we shall ignore the energy quantization in a magnetic field. This problem will be discussed in the last section of the chapter (§31). Most of the quantum transport phenomena in a magnetic field are of the same nature as the de Haas–van Alphen effect (§§15-17). However, in contrast to the de Haas–van Alphen effect, the quantum transport phenomena in fields $\mu H \ll \varepsilon_F$ are manifested by most metals in the form of weak oscillations superimposed on a relatively smooth curve representing the dependence of a given transport coefficient on the magnetic field. This smooth dependence can be explained by the classical theory. Consequently, we can divide the calculation of the transport coefficients into its classical and quantum parts, and we can ignore the quantization of the energy by magnetic fields in the classical treatment.

A situation leading to nonlinear effects is usually difficult to establish in a good metal because of the high carrier density. One should bear in mind that when the electrical resistivity depends strongly on the magnetic field, an indirect nonlinear effect may appear in the form of the dependence of the resistivity on the magnetic field of the current flowing through the metal. We shall not consider these problems and in all cases we shall restrict our treatment to the linear approximation.

§23. Boltzmann's Transport Equation

Boltzmann's transport equation applies to those cases when the motion of a conduction electron in a crystal lattice can be regarded as "free,"[1] or as the motion under the action of external forces interrupted by few collisions. The criterion of such motion is the mean free path l, defined as the mean distance between collisions. This path is determined by the properties of the electrons in question (in particular, by their dispersion law) and especially by various imperfections in the periodic structure of a crystal, in the form of chemical or physical inhomogeneities, phonons, electron–electron interactions, etc.

The collision integral, which is the term in Boltzmann's equation which describes rare collisions of electrons, is of fundamental importance in the transport equation.

The derivation of an expression for the collision integral is related to the problem of the scattering and it requires knowledge of the laws of interaction between electrons and phonons, impurities, and other electrons. However, the recent developments in the electron theory of metals have demonstrated that there are many nonequilibrium transport properties of metals which depend only weakly on the detailed structure of the collision integral and which are determined mainly by the dispersion law of conduction electrons. As a rule, these properties are sensitive to the structure of the electron energy spectrum. Therefore, we can concentrate on the field part of the transport equation and ignore almost completely the structure of the collision integral.

We shall ignore initially the quantization of the electron energy (for example, in a magnetic field). The quantum nature of the problem will be reflected in the dispersion law and in the statistics of conduction electrons. We shall assume that, between collisions, an electron moves along a classical trajectory. The restrictions imposed by our ignoring of the quantum nature of the motion of electrons are discussed in detail in §1. We recall that the classical approach allows us to neglect band–band transitions caused by external fields. This effect is of quantum origin. However, the many-band nature of the electron energy spectrum is manifested in the summation over states and in the calculation of the probabilities of various collisions. This must be allowed for in the formulation of the collision integral.

The state of electrons in a metal will be described by the electron distribution functions $f_s(\mathbf{p}, \mathbf{r}, t)$, each of which represents the density of electrons in the s-th band, where $[2f_s/(2\pi\hbar)^3]dp_x dp_y dp_z dx dy dz$ is the number of electrons of the s-th band in an element of the phase space[2] $dp_x dp_y dp_z dx dy dz$; $\mathbf{r}$ is the coordinate; t is the time; and $\mathbf{p}$ is the kinematic momentum. In the absence of a magnetic field, the kinematic momentum $\mathbf{p}$ is identical with the momentum $\mathbf{P}$, which is canonically conjugate with the coordinate.[3] However, in the presence of a magnetic field we have

$$P = p + \frac{e}{c} A, \tag{23.1}$$

where $\mathbf{A}$ is the vector potential of the magnetic field $\mathbf{H}$ (rot $\mathbf{A} = \mathbf{H}$).

All the quantities which will be of interest to us (the current density $\mathbf{j}$, the energy flux density $\mathbf{Q}$, etc.) can be calculated if we know the distribution function f_s. Thus,

---

[1] To avoid any misunderstanding, we must stress that we are speaking of electrons which have a complex dispersion law and whose mechanics is as described in Chap. I.

[2] The neglect of the quantum effects allows us to ignore the spin variable: each state of electron is assumed to be doubly degenerate.

[3] We recall that in the classical approach the concepts of the crystal momentum and the true momentum are identical.

$$j = \frac{2e}{(2\pi\hbar)^3} \sum_s \int v_s f_s(\boldsymbol{p},\ \boldsymbol{r},\ t)\,d\boldsymbol{p} \equiv e \int vf\,d\Gamma, \tag{23.2}$$

$$Q = \frac{2}{(2\pi\hbar)^3} \sum_s \int \varepsilon_s v_s f_s(\boldsymbol{p},\ \boldsymbol{r},\ t)\,d\boldsymbol{p} \equiv \int \varepsilon vf\,d\Gamma, \tag{23.3}$$

where $\mathbf{v}_s = \partial\varepsilon_s/\partial\mathbf{p}$ is the velocity of an electron of energy $\varepsilon_s(\mathbf{p})$ and the integration with respect to $d\Gamma$ includes the summation over all the partly filled energy bands of a metal.

According to Liouville's theorem, in the absence of collisions the distribution function is independent of time, i.e.,

$$\frac{df_s}{dt} \equiv \frac{\partial f_s}{\partial t} + \dot{\boldsymbol{p}}\,\frac{\partial f_s}{\partial \boldsymbol{p}} + \dot{\boldsymbol{r}}\,\frac{\partial f_s}{\partial \boldsymbol{r}} = 0. \tag{23.4}$$

Equation (23.4) represents the constancy of the number of particles in an element of the phase space during the motion of electrons along phase trajectories. The quantities $\dot{\mathbf{p}}$ and $\dot{\mathbf{r}}$ are obtained from the equations of motion according to which $\dot{\mathbf{p}} = \mathbf{F}$, where $\mathbf{F}$ is an external force acting on an electron and $\dot{\mathbf{r}} = \mathbf{v}_s$.

Since $\mathbf{F} = -\partial\varepsilon_s/\partial\mathbf{r}$ (here, ε_s is the total energy including that of the external fields), df_s/dt vanishes if f_s is an arbitrary function of the energy.

If an electron is acted upon by an electric field $\mathbf{E}$ and a magnetic field $\mathbf{H}$,[4] we find that

$$F = e\left(E + \frac{1}{c}\,[vH]\right). \tag{23.5}$$

We must stress once again the classical nature of the expression just given. In particular, we have ignored the interaction of the magnetic moment of the electron with the magnetic field. This is fully justified in the great majority of the most interesting cases (in a uniform magnetic field the force acting on the magnetic moment is zero).

In some cases the external force acting on a conduction electron cannot be expressed in terms of the macroscopic fields $\mathbf{E}$ and $\mathbf{H}$. Thus, for example, when an acoustic wave is traveling across a metal, a conduction electron is subjected to the Lorentz force of Eq. (23.5), as well as to the force due to the deformation interaction of the electron with the lattice.

Collisions violate the condition stated in Eq. (23.4). The measure of violation is known as the collision integral

$$\frac{df_s}{dt} = \mathcal{L}_c\,\{f_s\}. \tag{23.6}$$

The collision integral is a complex nonlinear functional of the distribution functions. The structure and the actual form of the collision integral are determined by the interaction of electrons with impurities, with other electrons, and with quasi particles in general. In the case of their interaction with quasi particles, Eq. (23.6) should be supplemented by the transport equations for the quasi particles in question (for example, phonons). The transport equations can be

---

[4] The Lorentz force includes the magnetic induction $\mathbf{B}$ (the average magnetic field acting on an electron) and not the microscopic field $\mathbf{H}$. In calculations of the oscillatory effects, it may be necessary to distinguish $\mathbf{B}$ from $\mathbf{H}$ (§§19-21). Following the traditional treatments, we shall always use $\mathbf{H}$ and not $\mathbf{B}$, with the exception of those cases where it might lead to a misunderstanding.

written in the form of Eq. (23.6) only in those cases when the motion of a particle can be divided into the motion along phase trajectories and collisions, which are regarded as sudden changes in the momentum of a particle without a significant (from the macroscopic point of view) change of the coordinate. Hence, it is clear that the collision integral is an operator which is related to the dependence of the distribution function on the momentum and not on the space coordinates or on time. Some very general properties of the collision integral are discussed in §24. Here, we shall mention only that the collision integral vanishes for an equilibrium Fermi function with arbitrary values of the temperature T and the chemical potential ζ; T and ζ may depend on the space coordinates and on time.[5]

Boltzmann's transport equations expressed in the form of Eq. (23.6) represent a system of complex nonlinear integrodifferential equations which determine uniquely the state of a solid provided the boundary and the initial conditions are specified in a rigorous manner. The system cannot be solved in its general form and it is usual to make many simplifications, which are determined by the physical nature of the problem.

Since an external electric field which is applied directly to a metal or which appears as a result of some external force (for example, the field due to the passage of an acoustic wave) is normally very weak (compared with the internal electric field), the deviation of the system of electrons from the equilibrium state is usually slight. This permits us to linearize the system of equations (23.6) by replacing the distribution function f_s with the sum

$$f_s = n_F + f_1, \tag{23.7}$$

where n_F is the equilibrium Fermi function (zeroth approximation) and the function f_1 (first approximation) is small because the external forces are weak. In other words, the function f_s is proportional to those external forces which have disturbed the equilibrium of the system. For example, the function f_1 is proportional to the electric field when a current flows along a conductor kept under isothermal conditions, or to the temperature gradient when heat flows through a metal. The selection of the zeroth approximation, or, more exactly, the selection of the parameters T and ζ in the Fermi function, is determined by the formulation of the problem. Usually, the most natural approach is to assume local equilibrium by postulating that the parameters in the Fermi function are selected so that $T = T(\mathbf{r})$ determines the temperature at a point $\mathbf{r}$ and $\zeta = \zeta(\mathbf{r})$ is the chemical potential. This means that the density of electrons at the point $\mathbf{r}$ and the average electron energy are determined by the zeroth-approximation function n_F and

$$\int f_1\, d\Gamma = 0, \quad \int \varepsilon f_1\, d\Gamma = 0. \tag{23.8}$$

Moreover, the energy $\varepsilon_s(\mathbf{p})$ does not include the energy of an external electric field.

In the present chapter, we shall be concerned mainly with the calculation of the electrical and the thermal conductivity tensors and of the thermoelectric coefficients. Therefore, we shall consider those cases when the equilibrium of a system is disturbed by an electric field $\mathbf{E}$ and a temperature gradient ∇T, which will be assumed to be so weak that the linearization is justified.[6]

---

[5] If the collision integral describes collisions with other quasi particles (for example, with phonons), it vanishes only if all the equilibrium functions are substituted: the Fermi function for electrons, the Bose function for phonons, etc. (the temperature is assumed to be constant).

[6] The linearization is permissible provided $l\,|\nabla T| \ll T$ and $eEl \ll T$. In practice, these two conditions do not restrict the values of the electric field and the temperature gradient.

Substituting Eq. (23.7) into Eq. (23.6), using the expression for the Lorentz force given by Eq. (23.5), and neglecting the quadratic terms, we obtain

$$\frac{\partial f_1}{\partial t} + \frac{\partial f_1}{\partial r}\,v + \frac{\partial f_1}{\partial p}\,\frac{e}{c}\,[vH] = -\frac{\partial n_F}{\partial \varepsilon}\,v\,(eE - \nabla\zeta) - \frac{\partial n_F}{\partial T}\,v\nabla T + \left(\frac{\partial f_1}{\partial t}\right)_c .\qquad (23.9)$$

Here, $(\partial f_1/\partial t)_c$ is the linearized form of the collision integral:

$$\left(\frac{\partial f_1}{\partial t}\right)_c = -\,\hat{W} f_1,$$

where $\hat{W}$ is the linear collision operator which is equal to

$$-\left[\frac{\delta\hat{\mathscr{L}}_c\{f\}}{\delta f}\right]_{f=n_F} .$$

The intensity of an electric field in a conductor E', i.e., the force acting on a unit charge, is the sum of the field due to the applied external potential difference ($E = -\nabla\varphi$) and the quantity $-(1/e)\nabla\zeta$, where ζ is the chemical potential of electrons (see, for example, §25 in [1]), i.e.,

$$E' = E - \frac{1}{e}\,\nabla\zeta.$$

From this point onwards, we shall omit the primed notation for the electric field but we must remember that the derivative of the Fermi function with respect to the temperature is calculated on the assumption that the chemical potential is constant, i.e., $\partial n_F/\partial T \equiv (\partial n_F/\partial T)_\zeta$.

Thus, the transport equation for the linear (in respect of perturbations) correction to the distribution function is of the form

$$\frac{\partial f_1}{\partial t} + \frac{\partial f_1}{\partial r}\,v + \frac{\partial f_1}{\partial p}\,\frac{e}{c}\,[vH] + \hat{W}\{f_1\} = -\frac{\partial n_F}{\partial \varepsilon}\,evE - \frac{\partial n_F}{\partial T}\,v\nabla T.\qquad (23.10)$$

In some cases, estimates are made and even solutions obtained (if the problem is relatively complex) by replacing the linear collision operator with the operator corresponding to multiplication by a phenomenologically introduced constant τ^{-1}, i.e., by assuming that

$$\hat{W}\{f_1\} = \frac{1}{\tau}\,f_1.\qquad (23.11)$$

The positive[7] quantity τ has the dimensions of time and is known as the relaxation time. The associated quantity $l = v\tau$ is known as the mean free path and Eq. (23.11) is called the τ approximation. We must bear in mind that, in the case of an anisotropic dispersion law, the τ approximation cannot be justified theoretically and therefore the substitution involved in Eq. (23.11) can be used only in estimates or in those cases when the final result is independent of the nature of the collision integral.

Let us consider the first two terms of Eq. (23.10). The derivation of the distribution function with respect to time $\partial f_1/\partial t$ takes account of the time dispersion of the transport coefficients, i.e., of the effects associated with the delay of the reaction of an electron gas to an external stimulus. If the characteristic frequency of the external field is ω, it follows that $\partial f_1/\partial t \sim \omega f_1$. This term is important at frequencies of the order of or greater than $\nu = 1/\tau$.

---

[7] The positive nature of τ ensures that the law of entropy increase is obeyed.

If $\omega \ll \nu$, it can be omitted. The relaxation time varies within very wide limits ranging from 10^{-14} sec at room temperature to 10^{-9} sec for extremely pure samples of metals at liquid helium temperature (T < 4.2°K).

The derivatives with respect to the coordinates, represented by the term $(\partial f_1/\partial \mathbf{r})\,\mathbf{v}$ in Eq. (23.10), are responsible for the effects of the spatial dispersion of the transport coefficients. If the characteristic distance over which the distribution function changes significantly is of the order of d, we find that $(\partial f_1/\partial \mathbf{r})\,\mathbf{v} \sim (f_1/d)\,\mathbf{v}$. This term is important if v/d $\geqslant \nu$, i.e., $l \geqslant$ d. If $l \ll$ d, it can be omitted. Since, in the case of the purest samples, the mean free path l does not exceed 10^{-3}-10^{-1} cm, the derivatives with respect to the spatial coordinates can be omitted from calculations of the transport coefficients of bulk samples of metals.[8] We shall show in Chap. IV that, under some conditions, the spatial dispersion must be included in the calculations of the high-frequency conductivity and the absorption of sound.

An inhomogeneity in a sample (the presence of boundaries, etc.) may be manifested not only by an inhomogeneity in the function f_1 but also in the distribution of conduction electrons (for example, in an open-circuited conductor). However, in contrast to semiconductors, this frequently occurring inhomogeneity does not give rise to macroscopic effects in metals because the Debye–Hückel radius r_D (which is a measure of the inhomogeneity of the distribution of charged particles) is very small for a degenerate electron gas ($r_D \leqslant 10^{-8}$ cm).[9] This allows us to calculate the transport coefficients on the assumption that the Fermi function which occurs in Eq. (23.10) is homogeneous (independent of the coordinates).

It will be useful to introduce the following notation:

$$\frac{d}{dt} \equiv \frac{\partial}{\partial t} + \boldsymbol{v}\,\frac{\partial}{\partial \boldsymbol{r}} + \frac{e}{c}\,[\boldsymbol{v}\boldsymbol{H}]\frac{\partial}{\partial \boldsymbol{p}}, \tag{23.12}$$

$$\widehat{W}_p = \left(\frac{\partial n_F}{\partial \varepsilon}\right)^{-1} \widehat{W}\left\{\frac{\partial n_F}{\partial \varepsilon}\ldots\right\}, \tag{23.13}$$

$$\widehat{W}_\varepsilon = \left(\frac{\partial n_F}{\partial T}\right)^{-1} \widehat{W}\left\{\frac{\partial n_F}{\partial T}\ldots\right\}, \tag{23.14}$$

and to replace the function f_1 (for homogeneous distributions of $\mathbf{E}$ and ∇T) by two vector functions ψ and φ:

$$f_1 = -e\boldsymbol{E}\psi\,\frac{\partial n_F}{\partial \varepsilon} - \nabla T \cdot \boldsymbol{\varphi}\,\frac{\partial n_F}{\partial T}, \tag{23.15}$$

which can be described by the following easily derived and very compact expressions:

$$\frac{d\psi}{dt} + \widehat{W}_p\psi = \boldsymbol{v}, \tag{23.16}$$

$$\frac{d\varphi}{dt} + \widehat{W}_\varepsilon\varphi = \boldsymbol{v}. \tag{23.17}$$

On comparing Eqs. (23.12) and (23.4), we find that the "time" which is involved in the differentiation in Eqs. (23.16) and (23.17) is the time of motion of an electron along a phase trajectory in a magnetic field (§4).

---

[8] At low temperatures, the size effects may become important even in the case of relatively thick samples (d $\leqslant$ 1 mm).

[9] By definition, $e^2/r_D = \hbar\omega_0$, where ω_0 is the plasma frequency of an electron gas ($\omega_0^2 = 4\pi n e^2/m$, n is the electron density, and m is the electron mass). Since n $\sim 1/a^3$ (a is the atomic spacing), it follows that $r_D = a\sqrt{U_c/\varepsilon_F}$, where U_c is the Coulomb interaction energy of electrons.

Substituting the distribution function of Eq. (23.15), expressed in terms of the vector functions ψ and φ, into Eqs. (23.2) and (23.3) and noting that the current density j and the energy flux Q vanish when $f = n_F$, we obtain

$$j_i = -e^2 \int \frac{\partial n_F}{\partial \varepsilon}\, v_i \psi_k\, d\Gamma E_k - e \int \frac{\partial n_F}{\partial T}\, v_i \varphi_k\, d\Gamma \nabla_k T, \qquad (23.18)$$

$$Q_i = -e \int \frac{\partial n_F}{\partial \varepsilon}\, \varepsilon v_i \psi_k\, d\Gamma E_k - \int \frac{\partial n_F}{\partial T}\, \varepsilon v_i \varphi_k\, d\Gamma \nabla_k T. \qquad (23.19)$$

In some cases, we shall also use

$$-\int \frac{\partial n_F}{\partial \varepsilon}\, \chi(p)\, \eta(p)\, d\Gamma \equiv \langle \chi \eta \rangle \qquad (23.20)$$

and we shall consider integrals of this type as the scalar products of χ and η. We can easily show that such integrals have all the necessary properties of the scalar products [2].

Using such integrals, we obtain

$$j_i = e^2 \langle v_i \psi_k \rangle E_k - \frac{e}{T} \langle (\varepsilon - \zeta)\, v_i \varphi_k \rangle \nabla_k T, \qquad (23.21)$$

$$Q_i = e \langle \varepsilon v_i \psi_k \rangle E_k - \frac{1}{T} \langle \varepsilon(\varepsilon - \zeta)\, v_i \varphi_k \rangle \nabla_k T. \qquad (23.22)$$

Examination of specific scattering mechanisms shows [3] that the operator $\hat{W}_p$ has the following important properties. First, it is Hermitian:

$$\langle \chi \hat{W}_p \eta \rangle = \langle \eta \hat{W}_p \chi \rangle, \qquad (23.23)$$

and, secondly, it is positive:

$$\langle \chi \hat{W}_p \chi \rangle > 0. \qquad (23.24)$$

Similar properties are exhibited by the operator $\hat{W}_\varepsilon$.

In the treatment given so far we have assumed implicitly that conduction electrons form an almost ideal quasi particle gas. In other words, we have ignored the fact that the energy of a given quasi particle depends on the states of the whole system, i.e., on the distribution function of the system.[10] This dependence is allowed for in the theory of Fermi liquids [4, 5].

Let us consider what changes are involved in Boltzmann's transport equation when the Fermi-liquid interaction between electrons is included.

The transport equation for the distribution function in the theory of Fermi liquids is derived in exactly the same way as in the gas model, i.e., Eq. (23.6) is used. All that is necessary is to take into account the fact that the energy ε of a single quasi particle (a conduction electron) is determined not only by the dispersion law $\varepsilon_0(p)$ but also by the distribution function $f(p)$. For near-equilibrium states, we have

---

[10] We recall that the dispersion law of elementary excitations (electrons) considered in the present and preceding chapters includes the mutual interaction of the electrons (see §§1 and 16).

$$\left. \begin{array}{l} \varepsilon(p) = \varepsilon_0(p) + \displaystyle\int \Phi(p,\, p')\, f_1(p')\, d\Gamma' \equiv \varepsilon_0(p) + \eta, \\[2mm] f_1(p) = f(p) - n_F(\varepsilon_0), \qquad |f_1(p)| \ll n_F(\varepsilon_0). \end{array} \right\} \tag{23.25}$$

Here, $\varepsilon_0(\mathbf{p})$ is the energy of an electron whose momentum is $\mathbf{p}$ in the equilibrium state which can be described by the Fermi function; $\Phi(\mathbf{p},\,\mathbf{p}')$ is the correlation function (the principal characteristic of the interaction between electrons in the theory of Fermi liquids). In the microscopic theory, the correlation function $\Phi(\mathbf{p},\,\mathbf{p}')$ is related to the amplitude of the electron-electron scattering process [6]. An experimental determination of this quantity is an important task in the physics of metals. We shall show later that quasistatic transport properties cannot be used to determine this quantity.

The energy of a quasi particle $\varepsilon(\mathbf{p})$ defined in our treatment is naturally a function of the temperature. There are some restrictions which must be imposed (the temperature should be considerably less than the Fermi value) because the lifetime of one-particle excitations decreases rapidly away from the Fermi surface. At high temperatures, all the transport phenomena are affected by the participation of electrons whose energies differ considerably from the Fermi value.

Let us now return to the derivation of the linearized transport equation. Using Eq. (23.25), we can easily show that

$$\dot{p} = eE + \frac{e}{c}\, [vH] - \frac{\partial \eta}{\partial r}, \tag{23.26}$$

$$\dot{r} = v + \frac{\partial \eta}{\partial p}, \qquad \dot{\varepsilon} = eE\, \frac{\partial \varepsilon}{\partial p},$$

where $\mathbf{v} = \partial\varepsilon_0/\partial\mathbf{p}$ is the velocity of an electron under thermodynamic equilibrium conditions. It is convenient to introduce, in addition to the function f_1 which describes the departure of a system from the state of thermodynamic equilibrium, a different function f_1^*, which is defined by

$$f = n_F(\varepsilon_0) + f_1 \equiv n_F(\varepsilon) + f_1^*. \tag{23.27}$$

In the approximation linear in f_1, we obtain

$$f_1^* = f_1 - \frac{\partial n_F}{\partial \varepsilon}\, \eta. \tag{23.27a}$$

Using Eq. (23.26) and linearizing consistently the left-hand part of the transport equation (23.6), we obtain[11]:

$$\frac{df}{dt} \approx \frac{\partial n_F(\varepsilon_0)}{\partial \varepsilon_0}\, evE + \frac{\partial n_F(\varepsilon_0)}{\partial T}\, v\, \nabla T + \frac{\partial f_1}{\partial t} + \frac{\partial f_1^*}{\partial r}\, v + \frac{\partial f_1^*}{\partial p}\, \frac{e}{c}\, [vH]. \tag{23.28}$$

In considering the collision integral, which is the right-hand part of the transport equation (23.6), we should note that the dependence on the distribution function is included in the probabilities of collision processes (because of the statistical properties of electrons) and in the δ function which describes the law of conservation of the electron energy. This dependence

We recall that the linearization involves the electric field $\mathbf{E}$ (including the gradient of the chemical potential) and the temperature gradient ∇T. In the zeroth approximation with respect to these quantities, the values of ε and ε_0 become equal and the velocity of an electron $\partial\varepsilon/\partial\mathbf{p}$ becomes $\mathbf{v} = \partial\varepsilon/\partial\mathbf{p}$.

is the consequence of the Fermi-liquid effects [see Eq. (23.25)]. Both dependences can be included by the use of

$$\hat{\mathscr{L}}_c = \hat{\mathscr{L}}_c \{f, \varepsilon\}. \tag{23.29}$$

The substitution of the equilibrium distribution function with an arbitrary dispersion law into the collision integral makes it vanish, i.e.,

$$\hat{\mathscr{L}}_c \{n_F(\varepsilon), \varepsilon(p)\} = 0. \tag{23.30}$$

We note that $\varepsilon(\mathbf{p})$ includes the Fermi-liquid correction η, i.e., it is a complex functional of the distribution function.

Substituting into Eq. (23.29) the second of the expansions in Eq. (23.27) and bearing in mind that the approximation which is linear in respect of perturbations implies that

$$\left\{ \frac{\delta \hat{\mathscr{L}}_c}{\delta f} \right\}_{\substack{f=n_F(\varepsilon) \\ \varepsilon=\varepsilon(p)}} f \approx \left\{ \frac{\delta \hat{\mathscr{L}}_c}{\delta f} \right\}_{\substack{f=n_F(\varepsilon_0) \\ \varepsilon=\varepsilon_0(p)}} f_1^*, \tag{23.31}$$

we conclude that the inclusion of the Fermi-liquid effects in the collision integral simply results in the replacement of the function f_1 by the function f_1^* and the transport equation of Eq. (23.6) can finally be written in the form

$$\frac{\partial f_1}{\partial t} + \frac{\partial f_1^*}{\partial r} \mathbf{v} + \frac{\partial f_1^*}{\partial p} \frac{e}{c} [\mathbf{v}\mathbf{H}] - \left(\frac{\partial f_1^*}{\partial t} \right)_c = - \frac{\partial n_F(\varepsilon_0)}{\partial \varepsilon} e\mathbf{v}\mathbf{E} - \frac{\partial n_F(\varepsilon_0)}{\partial T} \mathbf{v}\nabla T. \tag{23.32}$$

It follows that in all those quasistatic cases when the term $\partial f_1/\partial t$ can be neglected, i.e., when $\omega\tau \ll 1$, the transport equation can be used in the "gas" form by introducing a new distribution function f_1^*. The correlation function $\Phi(\mathbf{p}, \mathbf{p'})$ then drops out of Boltzmann's equation. If we add that the linear approximations for the current density $\mathbf{j}$ and the energy density $\mathbf{Q}$ can also be expressed in terms of the function f_1^* [12]:

$$\mathbf{j} = e \int \mathbf{v}f_1^* \, d\Gamma, \qquad \mathbf{Q} = \int \varepsilon\mathbf{v}f_1^* \, d\Gamma, \tag{23.33}$$

it becomes obvious that in dealing with quasistatic problems we can ignore completely the Fermi-liquid interaction between electrons. The correlation between electrons appears only at fairly high frequencies ($\omega\tau \gtrsim 1$).

§24. Electrical Conductivity. Ohm's Law

We shall now consider the flow of a constant current through a metal in the absence of a magnetic field or any temperature gradient.

The results deduced in the preceding section allow us to ignore the Fermi-liquid interaction between electrons. The results obtained (the values of the transport coefficients) can be formulated in terms of the "gas" theory. However, we must remember that the basic characteristic of a conduction electron — its dispersion law $\varepsilon_0(\mathbf{p})$ — depends on the electron–electron

---

[12] The expressions in Eq. (23.33) follow directly from Eqs. (23.2) and (23.3) if we bear in mind that

$$\int \frac{\partial \varepsilon}{\partial p} n_F(\varepsilon) \, d\Gamma = \int \varepsilon \frac{\partial \varepsilon}{\partial p} n_F(\varepsilon) \, d\Gamma = 0.$$

correlation. An analogous situation is encountered in the de Haas–van Alphen effect (§16): the oscillation periods are determined solely by the shape of the Fermi surface $\varepsilon_0(\mathbf{p}) = \varepsilon_F$.

In the case we shall consider, the transport equation simplifies considerably: we can use Eq. (23.16), dropping the term $d\psi_i/dt$. Thus,

$$\hat{W}_{\boldsymbol{p}}\{\psi_i\} = v_i. \tag{24.1}$$

Dropping of the terms with the spatial (coordinate) derivatives implies that the distance over which the electric field or the distribution function varies significantly is large compared with the mean free path of electrons. Dropping of the time derivative implies that the field frequency ω is considerably lower than the collision frequency ν (if only an estimate is required, we can use the τ approximation).

Introducing the operator $\hat{W}_{\mathbf{p}}^{-1}$, which is the reciprocal of the collision operator, we find that Eq. (24.1) yields

$$\psi_i = \hat{W}_{\boldsymbol{p}}^{-1} v_i. \tag{24.2}$$

If Eq. (24.1) is supplemented by the linearized transport equations for the distribution functions of phonons or other quasi particles with which electrons may be colliding, the operator $\hat{W}_{\mathbf{p}}^{-1}$ will be understood to be that which is obtained after excluding all the distribution functions with the exception of that applicable to electrons.

It follows from Eqs. (23.18), (23.21), and (24.2) that

$$j_i = -e^2 \int \frac{\partial n_F}{\partial \varepsilon} v_i \hat{W}_{\boldsymbol{p}}^{-1} v_k \, d\Gamma E_k = e^2 \langle v_i \hat{W}_{\boldsymbol{p}}^{-1} v_k \rangle E_k. \tag{24.3}$$

Comparing Eq. (24.3) with Ohm's law

$$j_i = \sigma_{ik} E_k, \tag{24.4}$$

we find the formal expression for the electrical conductivity tensor:

$$\sigma_{ik} = -e^2 \int \frac{\partial n_F}{\partial \varepsilon} v_i \hat{W}_{\boldsymbol{p}}^{-1} v_k \, d\Gamma = e^2 \langle v_i \hat{W}_{\boldsymbol{p}}^{-1} v_k \rangle. \tag{24.5}$$

The properties of the operator $\hat{W}_{\mathbf{p}}$ given by (23.23) and (23.24) are possessed also by the operator $\hat{W}_{\mathbf{p}}^{-1}$ (this fact can be shown easily) and, therefore, the tensor σ_{ik} is symmetric and its principal values are positive.

We note that the symmetry of the tensor σ_{ik} ($\sigma_{ik} = \sigma_{ki}$) is the manifestation of a general principle of nonequilibrium thermodynamics known as the principle of symmetry of the transport coefficients. The positive values of the principal values of the tensor σ_{ik} ensure that the law of entropy increase is obeyed.

The number of independent components of the tensor σ_{ik} is determined by the symmetry class of a given crystal. Most metals have either the cubic or hexagonal symmetry. In the first case, the tensor degenerates to a scalar; in the second case, it has two identical principal values. Some metals (for example, Mg) have three different principal values of the electrical conductivity tensor because they belong to the orthorhombic system.

The absence of anisotropy of the electrical conductivity tensor is not evidence of isotropy of the dispersion law of conduction electrons. For example, gold, silver, copper, and other metals have cubic lattices but the Fermi surfaces of these metals are strongly nonspherical.

If $\hat{W}_p^{-1}v_k$ is a sufficiently smooth function of the energy, which is the case in all the situations when the change in the energy of an electron resulting from a collision is small compared with its initial energy,[1] we can replace the derivative of the Fermi function in the first of the formulas in Eq. (24.5) with the δ function. Integrating with respect to the energy, we obtain

$$\sigma_{ik} = \frac{2e^2}{(2\pi\hbar)^3} \sum_s \oint_{\varepsilon_s(p)=\varepsilon_F} \frac{v_i^{(s)} \hat{W}_p^{-1} v_k^{(s)}}{v^{(s)}} \, dS_s. \tag{24.6}$$

The integration in this expression is carried out over the Fermi surface and dS_s is an element of the area of the s-th sheet of the Fermi surface.

The temperature dependences of the components of the σ_{ik} tensor are governed by the temperature dependence of the factor $\hat{W}_p^{-1}v_k$ in the integrand.

Equation (24.6) must be modified if we introduce the "mean free path operator," which acts on a unit vector along the normal to the Fermi surface:

$$\hat{l}_p\{n_i\} \equiv \hat{W}_p^{-1}\{vn_i\}. \tag{24.7}$$

Then,

$$\sigma_{ik} = \frac{2e^2}{(2\pi\hbar)^3} \sum_s \int_{\varepsilon_s(p)=\varepsilon_F} n_i^{(s)} \hat{l}_p\{n_k^{(s)}\} \, dS_s. \tag{24.8}$$

This form of the electrical conductivity tensor makes it clear that the symmetry of σ_{ik} is a consequence of the Hermitian nature of the mean free path operator.

In the case of a cubic crystal, it is convenient to introduce the mean free path l_p by means of

$$\frac{1}{3} l_p \delta_{ik} = \frac{1}{S_F} \oint_{\varepsilon(p)=\varepsilon_F} n_i \hat{l}_p\{n_k\} \, dS,$$

where S_F is the Fermi surface area and the summation over s will now be omitted for the sake of simplicity. The electrical conductivity σ ($\sigma_{ik} = \sigma\delta_{ik}$) can now be represented in a very compact form:

$$\sigma = \frac{2e^2 S_F l_p}{3(2\pi\hbar)^3}, \tag{24.9}$$

or

$$\sigma \,(\text{sec}^{-1}) \approx 0.6 \cdot 10^{10} S_F \left(\frac{\text{g} \cdot \text{cm}^2}{\text{sec}^2} \right) l_p \,(\text{cm}). \tag{24.10}$$

---

[1] In the case of collisions with phonons (this point will be treated later), the condition $\Delta\varepsilon/\varepsilon \ll 1$ ($\Delta\varepsilon$ is the change in the energy resulting from a collision) implies that $\theta/\varepsilon_F \ll 1$: This condition may not be satisfied by poor metals such as bismuth or graphite.

These expressions for the tensor σ_{ik} are very general and are subject to very few restrictions (for example, it is assumed that the electron gas is degenerate). It is usual to introduce an anisotropic relaxation time by means of the relationship

$$\widehat{W}_p^{-1} v_k = \tau_p(\boldsymbol{p}) v_k, \tag{24.11}$$

which should be regarded as the definition of the relaxation time $\tau_{\mathbf{p}}(\mathbf{p})$. Then, Eq. (24.6) can be rewritten in the form

$$\sigma_{ik} = \frac{2e^2}{(2\pi\hbar)^3} \oint_{\varepsilon(p)=\varepsilon_F} \tau_p(\boldsymbol{p}) \frac{v_i v_k}{v} dS \equiv \frac{2e^2}{(2\pi\hbar)^3} \oint_{\varepsilon(p)=\varepsilon_F} \tau_p(\boldsymbol{p}) v_i dS_k, \tag{24.12}$$

$$(d\boldsymbol{S} = \boldsymbol{n}\, dS, \quad \boldsymbol{n} = \boldsymbol{v}/v).$$

In the case of an isotropic dispersion law, the relaxation time $\tau_{\mathbf{p}}(\mathbf{p})$ depends only on the modulus of the crystal momentum $\tau_{\mathbf{p}}(\mathbf{p}) \equiv \tau_{\mathbf{p}}(\mathbf{p})$ and Eq. (24.12) yields the usual expression for the electrical conductivity:

$$\sigma = \frac{ne^2 l_p}{p_F}, \tag{24.13}$$

where n is the electron density and $l_{\mathbf{p}} = \tau_{\mathbf{p}}(p_F) v_F$ is the mean free path.

The formulas given in the present section are somewhat formal and semiphenomenological: the interactions of electrons which make the conductivity finite are included in the mean free path $l_{\mathbf{p}}$. The calculation of the mean free path is one of the basic problems in the theory of metals. Much work has been done on this problem and it is still attracting attention. A fairly comprehensive review of the work on the mean free path can be found, for example, in [7].

In the introduction to the present chapter, we pointed out that detailed studies of the mechanisms of resistance are outside our scope. We shall make only some general comments.

The problem of the calculation of the mean free path of conduction electrons cannot yet be regarded as fully solved but it is clear that we know already all the principal resistance mechanisms: 1) collisions of electrons with phonons; 2) collisions of electrons with one another; 3) collisions of electrons with impurity atoms and other static defects in the crystal lattice.[2] The first two mechanisms are encountered even in ideal crystals and they define the "ideal resistance," which vanishes at absolute zero. The third mechanism is encountered only in crystals with defects and is the cause of the residual resistance, i.e., of the resistance which is still observed at absolute zero. The value of the residual resistance may differ considerably from one sample to another (for the same metal).

We shall start by considering the first mechanism, i.e., the electron–phonon collisions. All the modern ideas on the energy structure of metals are based on the assumption that conduction electrons and phonons represent two relatively weakly coupled subsystems. The interaction between conduction electrons and phonons (lattice vibrations) is weak because the principal interaction between electrons and the lattice is included in the dispersion laws of the electrons and the phonons.[3] This weakness of the electron–phonon interaction allows us to

---

[2] We shall not consider the additional scattering mechanisms in metals having some special properties. For example, the collisions of electrons with spin waves are known to occur in ferromagnetic and antiferromagnetic metals and the scattering by long-range-order fluctuations may be encountered in orderable alloys.

[3] See [8] for details.

limit our analysis to one-phonon processes in the form of the absorption and the emission of phonons by electrons. Since the velocity of those electrons whose energies are of the order of the Fermi energy is much greater than the velocity of sound, such processes are allowed by the laws of conservation.[4]

If we assume that the phonon gas is in equilibrium, i.e., if we ignore the drag of phonons by electrons, we can show that the scattering of electrons by phonons results in the following temperature dependence of the resistivity:

$$\frac{\rho_{eff}(T)}{\rho(\Theta)} \approx \begin{cases} \left(\dfrac{T}{\Theta}\right)^5 & (T \ll \Theta), \\ \dfrac{T}{\Theta} & (T \gg \Theta). \end{cases} \tag{24.14}$$

Here, $\rho(\Theta)$ is the resistivity of a metal at the Debye temperature Θ: $\rho(\Theta) \approx m\Theta/ne^2\hbar$. Although these results are now regarded as classical, the assumptions on which their derivations are based have not yet been fully established. In particular, the role of the drag of phonons by electrons at very low temperatures is still an open question [11].

A consistent derivation of the expressions in Eq. (24.14) can be found in any detailed treatment of the electron theory of metals (it is often referred to as the Bloch solution of Boltzmann's equation). This derivation is usually given for the isotropic model of a metal. However, it is easy to show that the nature of the temperature dependence of the resistivity is simply related to the fact that the number of phonons (N_{ph}) is proportional to T^3 at $T \ll \Theta$ and to T at $T \gg \Theta$. Moreover, at $T \ll \Theta$ a collision between an electron and a phonon is accompanied by a very small change in the electron momentum: $|\Delta p| = |\overline{\hbar k}| \sim (\hbar/a) T/\Theta$ (the bar denotes averaging over the equilibrium phonons, k is the phonon wave vector, and a is the lattice constant). Since $p_F \sim \hbar/a$, an electron must undergo many collisions in order to lose its momentum $\simeq p_F$ at temperatures in the range $T \ll \Theta$. This means that the dissipation of momentum at $T \ll \Theta$ can be described as the diffusion of an electron on the Fermi surface. An estimate of the relaxation time τ can then be obtained easily from the expression for the diffusion coefficient in the momentum space $D_p = \overline{\Delta p^2} \nu_{ph}$ ($\nu_{ph} \propto N_{ph} \propto T^3$ is the frequency of electron collisions with phonons) and from Einstein's relationship which, in this case, assumes the form $p_F^2 \approx D_p \tau$. Hence, it follows directly that $\tau \propto T^{-5}$, i.e., $\rho_{eff} \propto T^5$ (at $T \ll \Theta$). At high temperatures, the most likely collisions are those between electrons and phonons whose energy is Θ and whose momentum $\approx \hbar/a$. Therefore, only a few collisions are needed to lose the electron momentum $\approx p_F$ ($p_F \approx \hbar/a$) and the temperature dependence of the resistivity is determined solely by the temperature dependence of the number of phonons. Thus, in the range $T \gg \Theta$, $\rho_{eff} \propto T$.

The basic temperature dependence of the phonon component of the resistivity [see Eq. (24.14)] is independent of the dispersion law of conduction electrons. However, the values of all the transport coefficients are naturally affected strongly by the parameters of the energy spectrum. Recent numerical calculations of the electron energy spectra of metals have yielded the temperature dependences of the resistivity at arbitrary temperatures and not only in the limiting cases.

---

[4] The weak coupling between electrons and phonons is manifested, in particular, by situations in which each of the two subsystems is described by its own temperature. Under these conditions, the relaxation time is equal to the time necessary to equalize the electron and the phonon temperatures [9]. The difference between the electron and the phonon temperatures is one of the causes of the dependence of the resistance on the current (the departures from Ohm's law), which has been observed experimentally for bismuth at anomalously high current densities ($j \sim 10^6$ A/cm^2) [10].

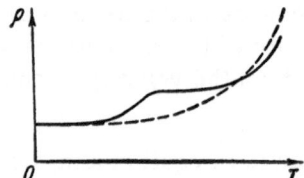

Fig. 55. Temperature dependence of the resistivity (in the low-temperature range) of metals with non-spherical Fermi surfaces.

In some cases, an allowance for the shape of the Fermi surface makes it possible to understand details of the temperature dependence of the resistivity (and to obtain analytic expressions) in the intermediate range of temperatures. In particular, it is shown in [12] that if the dispersion law is not isotropic a transition from the residual to the phonon component of the resistivity is more complex than in the case of metals with spherical Fermi surfaces (Fig. 55).

Collisions of electrons with one another are the cause of the resistance only in those cases when these collisions are accompanied by spin flip. Since the Fermi surface usually has some regions located close to the boundaries of the Brillouin zone, collisions involving spin flip can occur relatively easily. According to [13], the electron–electron component of the resistivity is proportional to the square of the temperature:

$$\rho_{e-e} \approx \frac{m}{ne^2} \frac{\varepsilon_F}{\hbar} \left(\frac{T}{\varepsilon_F} \right)^2 . \qquad (24.15)$$

A comparison of Eqs. (24.14) and (24.15) shows that the electron–electron component of the resistivity of all metals (with the exception of the transition elements) may be observed only at very low temperatures and even then it may be masked completely by the residual resistivity. A comparison of the experimental data on the resistivity at low temperatures (when the electron–electron interaction may be important) with theoretical calculations [we are not speaking here of rough estimates represented by Eq. (24.14) or (24.15) but of rigorous calculations [7]] shows that there is a systematic disagreement between the experimental and the theoretical results. The theory overestimates the values of the electron–electron component of the resistivity.

The residual resistivity is determined entirely by the purity of a metal and the quality of a single crystal (the presence of stresses, dislocations, etc.). It is now possible to prepare samples of such purity that the mean free path of electrons reaches several millimeters (for example, $l = 1$ cm for the purest samples of tungsten).

A very interesting resistance mechanism is predicted in [14]: in this mechanism the resistance is due to the isotopic inhomogeneity of a metal. Estimates of the mean free path based on this mechanism give values of 0.1-1.0 cm. A rigorous calculation of the electrical conductivity of isotopically inhomogeneous metals is given in [15]. We have mentioned this resistance mechanism specially because its origin is far from trivial. We must remember that electron shells of the isotopes of a given metal are identical (the slight difference resulting from the hyperfine structure of the terms can be ignored). The scattering is solely due to the inhomogeneity. of the zero-point vibrations of the lattice.

In most theoretical calculations of the temperature dependence of the resistivity, it is assumed that the temperature of the electron-gas degeneracy is considerably higher than the characteristic temperature of the quasi particles undergoing scattering (for example, phonons). This makes it possible to use the limiting value of the Fermi function: $-\partial n_F/\partial \varepsilon = \delta(\varepsilon - \varepsilon_F)$.

This has been done in the derivation of Eqs. (24.6) and (24.9)-(24.13). The temperature dependence of the conductivity tensor is then determined by the collision integral. However, we must bear in mind that the degeneracy criterion for a gas of conduction electrons is much more rigorous than the degeneracy criterion for a gas of free electrons (§12). For those metals whose Fermi surfaces are located close to singular points in the **p** space (Bi, As, Sb, graphite, etc.), the temperature dependence of the resistivity depends strongly on the factor $\partial n_F/\partial \varepsilon$ in the integrand of Eq. (24.5), and to calculate this dependence it is necessary to know in detail the structure of the electron energy spectrum of the metal being considered.

In the first approximation, we may assume that the various scattering mechanisms are additive (Matthiessen's rule). This conclusion can be "proved" using Eq. (24.13) by noting that $1/l_p$ is the probability that an electron will be scattered in a unit length of its path. If there are several independent causes of the scattering, it follows from the theorem on additive probabilities that

$$1/l = \sum_j 1/l_j, \tag{24.16}$$

where the subscript j labels the various scattering mechanisms.

It follows from Eq. (24.16) that[5]

$$\rho = \sum_j \rho_j. \tag{24.17}$$

Matthiessen's rule should not be regarded as absolute. It is quite clear that its derivation ignores the correlation between the various scattering processes. However, even if we ignore this correlation, i.e., if we assume that the probability of scattering is the sum of the probabilities of independent processes, we still find that an anisotropic dispersion law gives rise to a more complex dependence of the resistivity on the individual scattering mechanisms than that given by Eq. (24.16).

If one of the mechanisms makes only a small contribution compared with the other mechanisms, we can deduce a rule which is analogous to Matthiessen's but much weaker.

Let us assume that the collision operator $\hat{W}_p$ includes a small additive correction. We shall denote this correction by $\hat{W}_1$:

$$\hat{W}_p = \hat{W}_0 + \hat{W}_1, \tag{24.18}$$

where $|\hat{W}_1\{f\}| \ll |\hat{W}_0\{f\}|$. Then, the solution of the transport equation (24.1) can be found by the method of successive approximations. The conductivity and the resistivity tensors can be calculated in the approximation which is linear with respect to $\hat{W}_1$:

$$\rho_{ik} = \rho_{ik}^{(0)} + \Delta\rho_{ik}, \tag{24.19}$$

where

$$\Delta\rho_{ik} = \rho_{il}^{(0)} \Delta\sigma_{lm}\rho_{mk}^{(0)}, \qquad \Delta\sigma_{ik} = -\int v_i \hat{W}_0^{-2}\hat{W}_1 v_k \frac{\partial n_F}{\partial \varepsilon} d\Gamma. \tag{24.20}$$

The resistivity tensor $\rho_{ik}^{(0)}$ is related to the principal collision operator $\hat{W}_0$.

---

[5] The more usual formulation of Matthiessen's rule is that the resistivity of a metal is the sum of the temperature-independent residual component and the temperature-dependent ideal resistivity. This formulation follows directly from Eq. (24.17).

At high temperatures (T $\gg$ Θ, where Θ is the Debye temperature), electrons are scattered mainly by phonons. In this range of temperatures, $\rho_{ik}^{(0)} \propto T$ (this is true to within terms $\Theta/T \ll$ 1), and it follows from the relationships in Eq. (24.20) that the correction to the resistivity resulting from the scattering by impurities (described by the operator $\hat{W}_1$) is completely independent of the temperature (in the first approximation with respect to Θ/T). At low temperatures, the principal resistance mechanism is the scattering by impurities and other static inhomogeneities (in the case of very pure samples, this applies at very low temperatures). In this case, the operator $\hat{W}_0$ describes impurity scattering. The interaction with phonons is now simply a small correction (the operator $\hat{W}_1$). At low temperatures, the phonon term is proportional to T^5 and the coefficient in front of T^5 is independent of the total number of impurity atoms [see Eq. (24.14)]. However, this coefficient does depend on the nature of the scattering of electrons by impurities and, therefore, may vary from one sample to another. This less restrictive form of Matthiessen's rule is not always satisfied and the most important deviations are associated (as established by recent investigations) with the scattering of electrons by quasilocal and local vibrations of a crystal, i.e., with the inelasticity of the collisions between electrons and impurities and with changes in the phonon spectrum under the influence of impurities. The detailed analysis in [16] explains the main experimental observations relating to the dependence of the resistivity on the mass of impurities, their concentration, etc.

The inelasticity of the collisions is particularly important in the interaction of electrons with paramagnetic impurities. It is shown in [17] that the departure from the Born type of scattering gives rise to a minimum in the temperature dependence of the resistivity, which has been observed for several metals.

In some cases, it is important to make allowance for the collective effects resulting from the interaction between a Fermi gas of electrons and localized spins [17a].

At low temperatures, the mean free path in pure metals may be comparable with or even considerably greater than the thickness of a film d or the radius of a wire R. Under these conditions, the collisions of electrons with the boundaries of a sample are of considerable importance. In the limiting case $l \gg R$, the average value of the conductivity over the cross section of a cylindrical wire or a plane-parallel plate can be calculated without making any special assumptions as to the dispersion law of conduction electrons [17b]. In this case, the order of magnitude of the electrical conductivity can be obtained by replacing the mean free path in the usual formula [for example, in Eq. (24.13)] with the radius R in the case of a cylinder or the quantity d ln (l/d) in the case of a plate. The appearance of a large logarithmic term is related to the contribution of electrons moving parallel to the surface and not colliding with it.

We have already pointed out that the direct cause of the resistance of a metal is the dissipation of the crystal momentum in collisions and, therefore, it is necessary to include the umklapp processes in the derivation of Eq. (24.15). However, there are some situations in which the umklapp processes are hindered and the collisions in which the crystal momentum is conserved are relatively frequent. If we use l_N for the mean free path in the case of collisions which conserve crystal momentum, and l_U for the mean free path in the case of collisions accompanied by the umklapp processes, the situations we are considering can be described by $l_U \gg l_N$. If the dimensions of a wire or a film are intermediate, $l_U \gg d \gg l_N$, the dissipation of the momentum occurs at the boundaries of the sample (because of the viscosity of the electron gas) and the electrons reach the boundaries as a result of diffusion. We can easily see that the order of magnitude of the time involved is $d^2/l_N v_F$ [17c]. Although the Fermi surfaces of some metals have a structure which should ensure a considerable difference between l_U and l_N ($l_U \gg l_N$), the viscous resistance mechanism has not yet been observed experimentally.

Equation (24.6) and all that follows it show that the electrical conductivity depends strongly on the dynamic properties of electrons and especially on the nature of their dispersion law. We can show that this conclusion, which is supported by all the experimental data on the conductivity of metals, is in conflict with the classical experiments of Stewart and Tolman, according to whom the ratio e/m for electrons in a metal is equal to e/m_0 for free electrons. This apparent contradiction can best be resolved as follows. If an electron is acted upon by an electric field as well as by an inertial force, the transport equation can be written in the following way [18]:

$$\left(\frac{\partial f_1}{\partial t}\right)_c = -\frac{\partial n_F}{\partial \varepsilon}\, v\,(e\mathbf{E} - m_0\mathbf{a}),\qquad(24.21)$$

where $\mathbf{a}$ is the acceleration. This expression for the inertial force $(-m_0\mathbf{a})$ can be justified in two ways: either we can use the principle of equivalence and replace the force of gravity by acceleration, or we can consider the Schrödinger equation for an electron in a periodic field in a noninertial system of coordinates (a detailed calculation is given in [18]). In both cases, one must bear in mind that the number of conduction electrons (the number of quasi particles) is equal to the number of free electrons. Consequently, the Fermi-liquid interaction does not alter the expression for the inertial force. It follows from Eq. (24.21) that the field which appears in an open-circuited conductor (this corresponds to one of the variants of the experiment performed by Stewart and Tolman) is equal to $m_0\mathbf{a}/e$ and we can see that the measured quantity is determined by the ratio m_0/e for a free electron.

In a different variant of their experiment, Stewart and Tolman determined the ratio of the density of the momentum[6] $\mathbf{p}$ to the current density j. We can easily show that

$$p = \frac{m_0}{e}\, j.$$

According to the general principles of relativistic mechanics, the momentum of a particle is $\varepsilon v/c^2$, where ε is its total energy, including its energy at rest. Since the energy of interaction of an electron with the lattice and with other electrons is considerably less than its energy at rest, the momentum of an electron is given quite accurately by $m_0 v$ (m_0 is the mass of a free electron). The relationship given above follows directly from these considerations.

§ 25. Thermal Conductivity. Wiedemann–Franz Law. Thermoelectric Effects

Electrons participate not only in the transport of charge but also in the transport of heat — in fact in good metals, heat is transferred mainly by electrons. Thermoelectric effects are solely due to free electrons.

The transport coefficients which relate the appropriate fluxes (the heat flux $\mathbf{q}$ and the current density j) and forces (the electric field $\mathbf{E}$ and the temperature gradient ∇T) can be calculated by means of the transport equation we formulated earlier in this chapter.

We shall start with a brief phenomenological description of the conduction of heat and of the thermoelectric effects.

Restricting our analysis to the linear approximation and bearing in mind the principle of symmetry of the transport coefficients (see, for example, §25 in [1]), we obtain the following

---

[6] The density of the momentum is equal to the density of the mass flow.

relationship between the electric field $\mathbf{E}$,[1] the temperature gradient ∇T, the current density $\mathbf{j}$, and the heat flux $\mathbf{q}$:

$$\left.\begin{aligned} E_i &= \rho_{ik} j_k + \alpha_{ik} \frac{\partial T}{\partial x_k}, \\ q_i &= T\alpha_{ki} j_k - \varkappa_{ik} \frac{\partial T}{\partial x_k}. \end{aligned}\right\} \tag{25.1}$$

Here, ρ_{ik} is the resistivity tensor; $\varkappa_{ik}$ is the thermal conductivity tensor; α_{ik} is the tensor representing the thermoelectric properties of a metal. The components of the tensor α_{ik} can be used to derive the Thomson and the Peltier coefficients as well as the thermoelectric power [1]. We note that Onsager's relationships (the principle of symmetry of the transport coefficients), which postulate that the tensors ρ_{ik} and $\varkappa_{ik}$ are symmetric, allow for the possibility of the existence of conductors with an asymmetric tensor of the thermoelectric coefficients ($\alpha_{ik} \neq \alpha_{ki}$).

We shall now consider calculations of the thermal conductivity and the thermoelectric coefficients. It will be convenient to rewrite the transport equation by deriving the equation for the phonon distribution function χ.

We recall that collisions with phonons are almost always important. Moreover, in all solids (including metals), phonons take part in the transport of heat. If the nonequilibrium correction to the Fermi function is denoted by f_1 and the equilibrium phonon distribution function by χ_1 [this is the Bose function $N_B = (e^{\hbar\omega/T} - 1)^{-1}$, where $\hbar\omega = \hbar\omega(\mathbf{p})$ is the energy of a phonon whose crystal momentum is $\mathbf{p}$], the system of Boltzmann's transport equations can be written in the following way (for reasons stated earlier, we shall omit the terms with the space and time derivatives of the distribution functions; moreover, we shall omit the terms representing the effect of a magnetic field):

$$\left.\begin{aligned} \left(\frac{\partial f_1}{\partial t}\right)^{\text{e-e}}_{\text{st}} + \left(\frac{\partial \chi_1}{\partial t}\right)^{\text{e-ph}}_{\text{st}} &= \frac{\partial n_F}{\partial \varepsilon}\, e v \mathbf{E} + \frac{\partial n_F}{\partial T}\, v\,\nabla T, \\ \left(\frac{\partial f_1}{\partial t}\right)^{\text{ph-e}}_{\text{st}} + \left(\frac{\partial \chi_1}{\partial t}\right)^{\text{ph-ph}}_{\text{st}} &= \frac{\partial N_B}{\partial T}\, \mathbf{u}\,\nabla T. \end{aligned}\right\} \tag{25.2}$$

Here, $\mathbf{u} = \partial\hbar\omega/\partial\mathbf{p}$ is the velocity of a phonon whose crystal momentum is $\mathbf{p}$. The linearized collision integrals in the above equations are:

$$\left.\begin{aligned} \left(\frac{\partial f_1}{\partial t}\right)^{\text{e-e}}_{\text{st}} &= -\,\widehat{W}_{\text{e-e}}\{f_1\}, \\ \left(\frac{\partial f_1}{\partial t}\right)^{\text{ph-e}}_{\text{st}} &= -\,\widehat{W}_{\text{ph-e}}\{f_1\}, \end{aligned}\right\} \tag{25.3}$$

and so on, where

$$\left.\begin{aligned} \widehat{W}_{\text{e-e}} &= -\left\{\frac{\delta \mathscr{L}^{\text{e}}_{\text{st}}}{\delta f}\right\}_{\substack{f=n_F \\ \chi=N_B}}, \\ \widehat{W}_{\text{ph-e}} &= -\left\{\frac{\delta \mathscr{L}^{\text{ph}}_{\text{st}}}{\delta f}\right\}_{\substack{f=n_F \\ \chi=N_B}}, \end{aligned}\right\} \tag{25.4}$$

---

[1] The intensity of the electric field in a conductor is the sum of the field resulting from the applied potential difference and of the quantity $-(1/e)\nabla\zeta$, where ζ is the chemical potential of electrons [1]. This means that the change in the chemical potential resulting from the temperature gradient is included in $\mathbf{E}$.

and so on, are the linear operators acting on functions f_1 and χ_1; $\hat{\mathscr{L}}_{st}^{e}$ and $\hat{\mathscr{L}}_{st}^{ph}$ are the electron and the phonon collision operators, which include the interaction of electrons not only with phonons but also with all static inhomogeneities (impurities, dislocations, etc.). The existence of several electron and phonon bands is postulated and allowed for in the same manner as in §23 [see Eqs. (23.2) and (23.3)].

By analogy with Eq. (23.15), we introduce the vector functions ψ, φ, μ, and ν so that

$$\left.\begin{array}{l} f_1 = - eE_i \dfrac{\partial n_F}{\partial \varepsilon} \psi_i - \nabla_i T \dfrac{\partial n_F}{\partial T} \varphi_i, \\[2mm] \chi_1 = - eE_i \dfrac{\partial N_B}{\partial (\hbar\omega)} \mu_i - \nabla_i T \dfrac{\partial N_B}{\partial (\hbar\omega)} \nu_i. \end{array}\right\} \tag{25.5}$$

Boltzmann's transport equations can then be written as a system of vector equations for the functions ψ_i, φ_i, μ_i, and ν_i:

$$\left.\begin{array}{ll} \hat{W}_p^{e-e}\ \psi_i + \hat{W}_p^{e-ph}\ \mu_i = v_i, & \hat{W}_\varepsilon^{e-e}\ \varphi_i + \hat{W}_\varepsilon^{e-ph}\ \nu_i = v_i, \\[2mm] \hat{W}_p^{ph-e}\ \psi_i + \hat{W}_p^{ph-ph}\ \mu_i = 0, & \hat{W}_\varepsilon^{ph-e}\ \varphi_i + \hat{W}_\varepsilon^{ph-ph}\ \nu_i = u_i, \end{array}\right\} \tag{25.6}$$

and the linearized collision integrals $\hat{W}$ are given by formulas such as Eqs. (23.13) and (23.14).

Using the operators introduced so far, we can easily derive the operators $\hat{W}_p$ and $\hat{W}_\varepsilon$ from the results given in §23; for example:

$$\hat{W}_p = \hat{W}_p^{e-e} - \hat{W}_p^{e-ph} \left(\hat{W}_p^{ph-ph}\right)^{-1} \hat{W}_p^{ph-e}. \tag{25.7}$$

The generalized forces in the transport equations are the electric field **E** and the temperature gradient ∇T. According to the principles of the thermodynamics of nonequilibrium processes [19], the fluxes corresponding to these forces should be defined in such a way that the time derivative of the entropy $\dot{S}$ (the rate of change of the entropy) is given by the following equation:

$$\dot{S} = \boldsymbol{j}^* \boldsymbol{E} + \boldsymbol{q}^* (- \nabla T). \tag{25.8}$$

We can easily show that, in this case, the fluxes **j*** and **q*** are of the form:

$$\boldsymbol{j}^* = \frac{e}{T} \int \boldsymbol{v} f_1\, d\Gamma_e, \tag{25.9}$$

$$\boldsymbol{q}^* = \frac{1}{T^2} \left\{ \int \boldsymbol{u}\hbar\omega\,(p)\,\chi_1\, d\Gamma_{ph} + \int \boldsymbol{v}\,(\varepsilon - \zeta) f_1\, d\Gamma_e \right\}. \tag{25.10}$$

Here, the symbol $\int \ldots d\Gamma_e$ denotes integration over all the electron states and the symbol $\int \ldots d\Gamma_{ph}$ denotes integration over all the phonons. Naturally, the procedure must include summation over the various bands of the phonon spectrum.

Usually, the fluxes **j*** and **q*** are replaced with the current density $\boldsymbol{j} = T\boldsymbol{j}^*$ and the heat flux $\boldsymbol{q} = T^2\boldsymbol{q}^*$ which obey the phenomenological relationships of Eq. (25.1).

If the generalized forces are $-\nabla\varphi$ and $-\nabla T$, the corresponding fluxes are **j*** and **q**'*, where the latter differs from that given by Eq. (25.10) because ζ is replaced with $W = - T^2\partial/\partial T(\zeta/T)$ (W is the thermal function per one electron; $W \approx \zeta$ and $T \ll \varepsilon_F$).

Using Eqs. (25.5), (25.9), and (25.10), we can write the expressions for **j** and **q** in the following way:

$$j_i = \sigma_{ik} E_k + b_{ik} \frac{\partial T}{\partial x_k}, \left.\begin{array}{c} \\ \\ \end{array}\right\}$$
$$q_i = c_{ik} E_k + d_{ik} \frac{\partial T}{\partial x_k}, \left.\begin{array}{c} \\ \\ \end{array}\right\} \qquad (25.11)$$

where

$$\sigma_{ik} = e^2 \langle v_i \psi_k \rangle_e, \qquad b_{ik} = \frac{e}{T} \langle v_i \varphi_k (\varepsilon - \zeta) \rangle_e,$$
$$c_{ik} = - e \{ \langle v_i \psi_k (\varepsilon - \zeta) \rangle_e + \langle u_i \mu_k \hbar \omega \rangle_{\text{ph}} \}, \qquad (25.12)$$
$$d_{ik} = - \frac{1}{T} \{ \langle v_i \varphi_k (\varepsilon - \zeta)^2 \rangle_e + \langle u_i v_k (\hbar \omega)^2 \rangle_{\text{ph}} \},$$

and

$$- \int \frac{\partial n_F}{\partial \varepsilon} \chi(p) \eta(p) \, d\Gamma_e = \langle \chi \eta \rangle_e,$$
$$- \int \frac{\partial N_B}{\partial (\hbar \omega)} \chi(p) \eta(p) \, d\Gamma_{\text{ph}} = \langle \chi \eta \rangle_{\text{ph}}. \qquad (25.13)$$

As before (§23), integrals of this type can be regarded as scalar products of the functions $\chi(p)$ and $\eta(p)$.

When the fluxes are written in the form given by Eq. (25.11), Onsager's principle of symmetry of the transport coefficients requires that the components of the tensors σ_{ik}, b_{ik}, c_{ik}, and d_{ik} should be related in the following way:

$$\sigma_{ik} = \sigma_{ki}, \quad d_{ik} = d_{ki}, \quad c_{ik} = - T b_{ik}. \qquad (25.14)$$

These relationships are the consequences of certain properties of the collision integral. We shall not deal with them in detail but recall only that in order to prove relationships of the type given by Eq. (25.14) it is sufficient to use the Hermitian nature of the operator $\hat{W}_p$ of Eq. (23.23) [see also Eq. (25.22)].

Comparing Eqs. (25.1) and (25.11), we can easily show that Onsager's relationships follow from Eq. (25.14), that these have been allowed for in Eq. (25.1), and that, moreover,

$$\rho_{ik} = \sigma_{ik}^{-1}, \qquad \alpha_{ik} = - \sigma_{il}^{-1} b_{lk},$$
$$\varkappa_{ik} = - d_{ik} + c_{im} \sigma_{ml}^{-1} b_{lk}. \qquad (25.15)$$

If we use the last relationship in Eq. (25.14) and replace the tensor c_{im} in the formula for the thermal conductivity tensor $\varkappa_{ik}$ with the equivalent tensor $-T b_{mi}$, we find that – in accordance with the formulas of Eq. (25.12) – the thermal conductivity splits into the sum of the electron and the phonon components:

$$\varkappa_{ik} = \varkappa_{ik}^e + \varkappa_{ik}^{\text{ph}}, \qquad (25.16)$$

where

$$\varkappa_{ik}^e = - d_{ik}^e - T b_{mi} \sigma_{ml}^{-1} b_{lk}, \qquad d_{ik}^e = - \frac{1}{T} \langle v_i \varphi_k (\varepsilon - \zeta)^2 \rangle_e, \qquad (25.17)$$

and

$$\varkappa_{ik}^{\text{ph}} = \frac{1}{T} \langle u_i v_k (\hbar \omega)^2 \rangle_{\text{ph}}. \qquad (25.18)$$

This division into components does not mean that there is no interaction between the electron and the phonon subsystems. In a wide range of temperatures, phonons are the principal .

cause of the scattering of electrons and this is responsible (at practically all temperatures) for the main part of the thermal resistance to the phonon flux. The mean free path of phonons is short and the role they play in heat conduction in metals is relatively unimportant, largely because they are scattered by electrons. Formally, this allows us to drop the second terms in Eq. (25.16) and in the expression for c_{ik}.

The Wiedemann–Franz law relates the electron component of the thermal conductivity to the electrical conductivity.

Thus, the electron component of the thermal conductivity of a metal and the thermoelectric coefficients can be described in terms of b_{ik}, d_{ik}, and c_{ik}. We demonstrated in §24 that the electrical conductivity can be expressed in the form of an integral over the Fermi surface [see Eq. (24.9) and the subsequent formulas]. The formulas in Eq. (25.15) can be expressed in a similar manner by expanding the equations obtained in powers of the temperature (more precisely, in powers of $T/|\varepsilon_F - \varepsilon_k|$); the coefficients of such an expansion are also functions of the temperature because of the temperature dependence of the collision operator. All the tensors can be written in such a form that the integration with respect to the energy is explicit:

$$\sigma_{ik} = -e^2 \int \frac{\partial n_F}{\partial \varepsilon} \psi_{ik}(\varepsilon)\, d\varepsilon, \qquad b_{ik} = \frac{e}{T} \int \frac{\partial n_F}{\partial \varepsilon} (\varepsilon - \zeta) \varphi_{ik}(\varepsilon)\, d\varepsilon, \\ c_{ik} = -e \int \frac{\partial n_F}{\partial \varepsilon} (\varepsilon - \zeta) \psi_{ik}(\varepsilon)\, d\varepsilon, \quad d_{ik} = \frac{1}{T} \int \frac{\partial n_F}{\partial \varepsilon} (\varepsilon - \zeta)^2 \varphi_{ik}(\varepsilon)\, d\varepsilon, \tag{25.19}$$

where

$$\psi_{ik} = \frac{2}{(2\pi\hbar)^3} \oint_{\varepsilon(p)=\varepsilon} \frac{v_i \psi_k}{v}\, dS, \quad \varphi_{ik} = \frac{2}{(2\pi\hbar)^3} \oint_{\varepsilon(p)=\varepsilon} \frac{v_i \varphi_k}{v}\, dS, \tag{25.20}$$

and the integration is carried out over the constant-energy surface $\varepsilon(\mathbf{p}) = \varepsilon$.

Using the well-known properties of the Fermi function [see, for example, Eq. (12.7)], we can easily show that

$$\sigma_{ik} \approx e^2 \psi_{ik}(\varepsilon_F), \qquad b_{ik} \approx -\frac{\pi^2}{3} eT \left(\frac{d\varphi_{ik}}{d\varepsilon}\right)_{\varepsilon=\varepsilon_F}, \\ c_{ik} \approx \frac{\pi^2}{3} eT^2 \left(\frac{d\psi_{ik}}{d\varepsilon}\right)_{\varepsilon=\varepsilon_F}, \quad d_{ik} \approx -\frac{\pi^2}{3} T\varphi_{ik}(\varepsilon_F). \tag{25.21}$$

In this approximation, Onsager's relationships (25.14) imply that

$$\left(\frac{d\varphi_{ik}}{d\varepsilon}\right)_{\varepsilon=\varepsilon_F} = \left(\frac{d\psi_{ki}}{d\varepsilon}\right)_{\varepsilon=\varepsilon_F}. \tag{25.22}$$

Substituting the expressions in Eq. (25.21) into the formulas of Eq. (25.15) and retaining only the terms containing the lowest power of the ratio $T/|\varepsilon_F - \varepsilon_k|$, we obtain:

$$\rho_{ik} \approx \frac{1}{e^2} \psi_{ik}^{-1}(\varepsilon_F), \\ \alpha_{ik} \approx \frac{\pi^2}{3} \frac{T}{e} \psi_{il}^{-1}(\varepsilon_F) \varphi_{lk}'(\varepsilon_F) = \frac{\pi^2}{3} \frac{T}{e} \psi_{il}^{-1}(\varepsilon_F) \psi_{kl}'(\varepsilon_F), \\ \varkappa_{ik} \approx \frac{\pi^2}{3} T\varphi_{ik}(\varepsilon_F). \tag{25.23}$$

As in §24, we find it convenient to introduce mean free path operators acting on a unit vector

$n_i = v_i/v$. If we employ the form of the collision operators used in §24, we find that

$$\left. \begin{aligned} \psi_i &= \hat{W}_p^{-1}\{v_i\} \equiv \hat{l}_p\{n_i\}, \\ \varphi_i &= \hat{W}_\varepsilon^{-1}\{v_i\} \equiv \hat{l}_\varepsilon\{n_i\}. \end{aligned} \right\} \tag{25.24}$$

The subscripts p and ε stress that, in the first case, the change in the momentum is the most important aspect of the collision, whereas, in the second case, the change in the energy is the dominant effect. In other words, the order of magnitude is $|\psi| \approx l_p$, where l_p is the distance in which the electron momentum is dissipated, and $|\varphi| \approx l_\varepsilon$, where l_ε is the distance in which the energy (more correctly, $\varepsilon - \zeta$) is lost. At low temperatures ($T \ll \Theta$), when the collisions of electrons with long-wavelength phonons play the dominant role, these distances differ considerably $[l_p \approx (\Theta/T)^2 l_\varepsilon]$. This gives rise to the experimentally observed departures from the Wiedemann–Franz law. In those cases where inelastic collisions can be neglected,[2] there is no difference between l_ε and l_p, which is to be expected since, in this case, the relaxation involves the loss of anisotropy in the motion of electrons (in particular, the difference between l_p and l_ε vanishes in those rare cases to which the τ approximation can be applied).

The average values of the free paths are convenient to use in the case of a cubic crystal. These average values are functions of the electron energy and are given by the following formula:

$$\frac{1}{S(\varepsilon)} \oint n_i \hat{l}_{p\,(\varepsilon)}\{n_k\}\,dS = \frac{1}{3} l_{p\,(\varepsilon)}(\varepsilon)\,\delta_{ik}, \tag{25.25}$$

where the integration is carried out over the constant-energy surface $\varepsilon(p) = \varepsilon$ whose area is $S(\varepsilon)$.

The mean free path l_p, introduced in §24, is identical with the average path if the energy ε is replaced with the Fermi energy: $l_p = l_p(\varepsilon_F)$ and, correspondingly, $l_\varepsilon = l_\varepsilon(\varepsilon_F)$.

Using Eqs. (25.20), (25.23)-(25.25), we obtain

$$\left. \begin{aligned} \sigma_{ik} &= \sigma\delta_{ik}, \quad \sigma = \frac{2e^2}{3\,(2\pi\hbar)^3} S(\varepsilon_F)\,l_p, \\ \varkappa_{ik} &= \varkappa\delta_{ik}, \quad \varkappa = \frac{2\pi^2 T}{9\,(2\pi\hbar)^3} S(\varepsilon_F)\,l_\varepsilon; \\ \alpha_{ik} &= \alpha\delta_{ik}, \quad \alpha = \frac{\pi^2 T}{3e} \left[\frac{d}{d\varepsilon}\ln\left(S(\varepsilon)\,l_p(\varepsilon)\right)\right]_{\varepsilon=\varepsilon_F}. \end{aligned} \right\} \tag{25.26}$$

We note that Onsager's relationships of Eq. (25.14) or (25.22) can be written in a very compact form if the new notation is employed:

$$\frac{d}{d\varepsilon}(Sl_p)_{\varepsilon=\varepsilon_F} = \frac{d}{d\varepsilon}(Sl_\varepsilon)_{\varepsilon=\varepsilon_F}. \tag{25.27}$$

If $l_p = l_\varepsilon$, it follows that

$$\varkappa = \frac{\pi^2 T}{3e^2}\,\sigma, \tag{25.28}$$

_____

[2] This means that not only can we assume $\mu_i = \nu_i = 0$ [see Eq. (25.5)] but (and this is the more important point) we can ignore the transfer of the energy from electrons to equilibrium phonons in each collision event. We recall that at high temperatures $T \gg \Theta$ the collisions with phonons are quasielastic. This is why the Wiedemann–Franz law is satisfied at high temperatures.

i.e., the Wiedemann–Franz law is satisfied. It follows from the derivation of Eq. (25.28) that this law is determined solely by the nature of the scattering process and is, for example, quite independent of the dispersion law. In the case of a noncubic crystal, this law is obeyed (for $\varphi_i = \psi_i$) by each of the components of the tensors σ_{ik} and $\varkappa_{ik}$.

We shall not consider the theories which explain (or, more accurately, try to explain) the experimental observations relating to the thermoelectric power and the thermal conductivity (this point is dealt with in several sections in [11]). We wish to point out only that the reversal of the sign[3] of the coefficient α (i.e., of the thermoelectric power) of some metals can be explained by the hole nature of the Fermi surface: if this surface is located near a point in the p space where the energy reaches its maximum value, the surface area decreases (the Fermi surface contracts) with increasing energy. Moreover, the velocity decreases on the approach to an extremal point: the slow electrons are scattered more frequently and, therefore, the free path becomes shorter as the energy increases.

§26. Galvanomagnetic Effects (Introduction)

The influence of a magnetic field on the conductivity of a metal is due to its effect on the motion of electrons (§§4, 5). One must remember that, in contrast to the equilibrium thermodynamic properties, the transport characteristics (the electrical resistivity, the thermal conductivity, etc.) depend strongly on the magnetic field even in the classical approximation. In other words, the dependence on the magnetic field is observed even if we ignore the quantization of the energy of electrons in this field (see §1). A characteristic dimensionless parameter which determines the role of the magnetic field is the ratio r_H/l, where r_H is the radius of an electron orbit and l is the path traveled by an electron. Since the orbit radius r_H is inversely proportional to the magnetic field, it is usual to speak of weak fields if $r_H \gg l$ and of strong fields if the opposite inequality is satisfied. The radius of a free-electron orbit r_H is equal to cp/eH, where p represents here the radius of the Fermi sphere, whereas, in general, $p = \sqrt{S}$, where S is some special (for example, extremal) section of the Fermi surface.

The role of the quantum effects is determined by two parameters: $\hbar\omega/|\varepsilon_F - \varepsilon_k|$ and $\hbar\omega/T$ (§8). The first of these parameters is usually very small and this is why we can ignore the quantum effects. The second parameter, $\hbar\omega/T$, can vary within a wide range when the field and the temperature are varied. If $\hbar\omega \gtrsim T$, we can observe quantum oscillations of the galvanomagnetic effects in the form of the Shubnikov–de Haas and similar effects, which are analogous to the de Haas–van Alphen effect (§15). For most metals, these quantum oscillations are superimposed in the form of fine ripples onto the main dependence of a given galvanomagnetic property on the magnetic field. This allows us to consider first the classical effect and then to include the quantum effects in the form of corrections.[1]

The simplest variant of the electron theory of transport phenomena (the τ approximation for one group of carriers with an isotropic and quadratic dispersion law) cannot even qualitatively explain the dependence of the resistivity on the magnetic field although an estimate of the Hall coefficient R obtained in this way is often correct (R = 1/nec, where n is the electron den-

---

[3] For free electrons, we have $\alpha < 0$.

[1] The quantum corrections always contain the factor $(\hbar\omega/|\varepsilon_F - \varepsilon_k|)^n$, where n > 0. The quantum effects in some metals, for example, Bi, can alter significantly the dependence of the resistivity, and particularly of the Hall coefficient R, on the magnetic field. Therefore, the results obtained in the present section should be applied to these metals with some caution. The theory of the quantum oscillations of the galvanomagnetic effects is presented in §31.

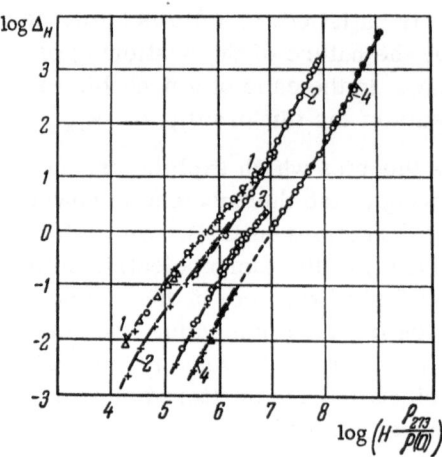

Fig. 56. Dependence of the increase in the resistivity in a magnetic field on the effective field $H\rho_{273}/\rho(0)$: 1) Mg; 2) Cd; 3) Cu; 4) Pb; ●) $T = 2°K$; ○) $T = 22°K$; ▲) $T = 14°K$; ×) $T = 20.35°K$; △) $T = 78°K$.

sity). Recent investigations have demonstrated that in strong fields the galvanomagnetic properties are very sensitive to the structure of the electron energy spectra. Such investigations provide one of the reliable methods for the determination of the Fermi surface topology.

We shall not attempt to summarize the experimental data but we shall merely use the experimental results to illustrate some theoretical conclusions. We shall start by listing some well-known facts.

1. Polycrystalline samples and some single crystals satisfy approximately Köhler's rule [20], which states that

$$\Delta_H = \frac{\rho(H) - \rho(0)}{\rho(0)} \tag{26.1}$$

is a function of the effective field

$$H_{eff} = H \frac{\rho_{273}}{\rho(0)},$$

where $\rho(H)$ is the resistivity in a magnetic field H, whereas ρ_{273} is the resistivity at 273°K in $H = 0$. This rule (Fig. 56) supports our assumption that the parameter which determines the change in the resistivity is r_H/l. The considerable deviations from Köhler's rule [21], which can be explained in a natural manner by developing a rigorous theory (this will be considered in later sections), have often been explained by various other "rules" containing more adjustable parameters than Köhler's rule. One of the recent attempts [22] was made on the basis of the results obtained for the resistivity of Al in a magnetic field [23].

2. All the metals investigated so far can be divided into three groups in accordance with the dependences of their resistivity on the magnetic field. The first group consists of metals whose resistivity tends to saturation as $H \to \infty$, irrespective of the direction of the magnetic field. The increase in the resistivity (the resistivity always increases in strong fields) of these metals can reach a few hundreds of percent (Δ_∞ may amount to several units). The second group comprises those metals whose resistivity increases quadratically with the magnetic field ($r_H \ll l$), again irrespective of the direction of the field. The resistivity of these metals can

increase by a factor of a few millions (for example, Bi); essentially, a metal of this kind behaves as a dielectric (insulator) in strong magnetic fields. The third group has intermediate properties: along some directions of the magnetic field, the resistivity increases quadratically with the field; along other directions, it tends to saturation.

We have been speaking so far of the transverse galvanomagnetic effect, i.e., of the case when the current is perpendicular to the magnetic field. This effect is known as the transverse magnetoresistance. The longitudinal magnetoresistance is relatively weak: the longitudinal resistivity of all metals tends to saturation. For most metals, there is a wide range of the transverse magnetoresistance ($j \perp H$) in which the resistivity is a linear function of the magnetic field (Kapitza's law). For some of these metals, the linear region separates two quadratic dependences, or is located between a quadratic region and a saturation level. Others show no deviation from the linear law even in strong fields.

Most of the experimental observations can be explained by the modern transport theory of galvanomagnetic effects. Some of the observations have made it necessary to refine this theory on the basis of the electron energy spectrum of a particular metal.[2]

We shall begin the theory of galvanomagnetic effects from a phenomenological description [1]. The relationship between the electric field and the current is linear in the region of Ohm's law:

$$j_i = \sigma_{ik} E_k, \quad E_i = \rho_{ik} j_k, \quad \sigma_{ik}^{-1} = \rho_{ik}. \tag{26.2}$$

In a magnetic field, the tensors σ_{ik} and ρ_{ik} become functions of this field and are no longer symmetric. The principle of symmetry of the transport coefficients predicts a more complex relationship between the components of the conductivity tensor:

$$\sigma_{ik}(H) = \sigma_{ki}(-H). \tag{26.3}$$

The tensor σ_{ik}, like any other tensor of the second rank, can be represented in the form of a sum of symmetric s_{ik} and antisymmetric a_{ik} tensors:

$$\sigma_{ik} = s_{ik} + a_{ik}. \tag{26.4}$$

Using Onsager's relationships of Eq. (26.3), we can easily show that the components of the tensor s_{ik} are even and those of a_{ik} are odd functions of the magnetic field:

$$s_{ik}(H) = s_{ik}(-H), \quad a_{ik}(H) = -a_{ik}(-H). \tag{26.5}$$

The components of the antisymmetric tensor a_{ik} can be used to construct a dual vector $\boldsymbol{a}$:

$$a_x = a_{yz}, \quad a_y = a_{zx}, \quad a_z = a_{xy}$$

or

$$a_i = \varepsilon_{ikl} a_{kl},$$

---

[2] In particular, it has been found recently that magnetic breakthrough (breakdown) alters considerably the dependence of the resistivity on the magnetic field (this point will be discussed later).

where ε_{ikl} is a unit antisymmetrical tensor of the third rank.

An analogous expansion can be performed also in the case of the resistivity tensor ρ_{ik}, which is the reciprocal of the tensor σ_{ik}:

$$\rho_{ik} = \rho_{ik}^s + \rho_{ik}^a, \qquad \rho_{ik}^s(\boldsymbol{H}) = \rho_{ik}^s(-\boldsymbol{H}), \qquad \rho_{ik}^a(\boldsymbol{H}) = -\rho_{ik}^a(-\boldsymbol{H}).$$

The components of the tensors ρ_{ik}^s and ρ_{ik}^a can be expressed in terms of the components of the tensors s_{ik} and a_{ik}:

$$\rho_{ik}^s = \frac{|s|(s^{-1})_{ik} + a_i a_k}{|s| + (a\hat{s}a)}, \qquad b_l = \frac{(\hat{s}a)_l}{|s| + (a\hat{s}a)}. \tag{26.6}$$

Here, $|s|$ is a determinant consisting of components of the tensor s_{ik}; $a\hat{s}a = a_i s_{ik} a_k$; b is a vector which is dual to the antisymmetric tensor ρ_{ik}^a: $b_i = \varepsilon_{ikl}\rho_{kl}^a$.

Using the quantities introduced in the preceding paragraphs, we find that Ohm's law for a conductor subjected to a magnetic field can be written in the following way:

$$\boldsymbol{E} = \hat{\rho}^s \boldsymbol{j} + [\boldsymbol{b}\boldsymbol{j}]. \tag{26.7}$$

The vector $\hat{\boldsymbol{\rho}}^s \boldsymbol{j}$ has the components $\rho_{ik}^s j_k$; the second term in the above equation is known as the Hall field. We note that this field is perpendicular to the current and that its sign reverses with the sign of the magnetic field. The Joule heat evolved in a conductor is determined only by the tensor ρ_{ik}^s

$$Q = \rho_{ik}^s j_i j_k; \tag{26.8}$$

therefore, the tensor ρ_{ik}^s (i.e., the symmetric part of the resistivity tensor) is frequently simply called the resistivity tensor.

It also follows from Eq. (26.7) that a complete description of the galvanomagnetic properties is possible if we know, for all values of the magnetic field, the three principal values of the tensor ρ_{ik}^s, the directions of its principal axes, and the three components of the vector $\boldsymbol{b}$. In general, the principal directions of the tensor ρ_{ik}^s do not coincide with crystallographic axes but are functions of the magnitude and the direction of the magnetic field.

The calculation of the density of the current and of the electric field direction in a sample of finite dimensions is a complex mathematical problem which has been solved only for a few special cases,[3] and, therefore, one usually employs long cylindrical samples (wires) in which the directions of the lines of the current are set by the geometry of the sample.

The components of the tensor ρ_{ik}^s and of the vector $\boldsymbol{b}$ depend in a complex manner on the magnetic field. If the magnetic field is weak, it can be expanded in powers of this field. It follows from Eq. (26.5) that such expansions are of the form:

$$\rho_{ik}^s(\boldsymbol{H}) = \rho_{ik}(0) + \lambda_{iklm}H_l H_m, \quad b_i = R_{ik}H_k. \tag{26.9}$$

---

[3] We are speaking of sufficiently large samples so that we can ignore the surface phenomena and use the macroscopic description (see, however, §29). Under these conditions, the mathematical formulation of the problem of the calculation of the current density $\boldsymbol{j}$ and the field intensity $\boldsymbol{E}$ is as follows: div $\boldsymbol{j} = 0$, rot $\boldsymbol{E} = 0$, and $\boldsymbol{j}\boldsymbol{\nu} = 0$ at all points of the boundary except those to which current is supplied; at the latter points $\boldsymbol{j}\boldsymbol{\nu}$ is continuous ($\boldsymbol{\nu}$ is the normal to the surface of the sample under consideration).

We shall retain only the first nonvanishing terms containing the magnetic field; $\rho_{ik}(0)$ is the resistivity tensor for $\mathbf{H} = 0$.

The number of independent components of the tensors λ_{iklm} and R_{ik} is determined by the symmetry class of a metal. Therefore, the angular dependence (the dependence on the directions of the magnetic field and of the current) of the resistivity and the Hall field is completely determined by the symmetry class if the magnetic field is weak. For example, in the case of a cubic crystal as well as for all isotropic bodies, we have $R_{ik} = R\delta_{ik}$. The quantity R is known as the Hall coefficient. For most metals, this coefficient is not affected greatly by the temperature, the purity of a sample, etc. However, it is more convenient to use the high-field Hall coefficient R_∞, which is defined in §27 (this cannot be done for all metals).

In the case of an isotropic metal (a polycrystalline sample), the only special direction is that of the magnetic field. The component of the tensor along the field will be denoted by $\rho_\parallel^s$ and that in a plane perpendicular to the field by $\rho_\perp^s$. It follows from Eq. (26.9) that

$$\left.\begin{aligned}
\rho_{ik}^s &= \left(\rho\left(0\right) + \lambda_1 H^2\right)\delta_{ik} + \lambda_2 H_i H_k, \\
\rho_\parallel^s &= \rho\left(0\right) + \lambda_\parallel H^2 \qquad \left(\lambda_\parallel = \lambda_1 + \lambda_2\right), \\
\rho_\perp^s &= \rho\left(0\right) + \lambda_\perp H^2 \qquad \left(\lambda_\perp = \lambda_1\right).
\end{aligned}\right\} \tag{26.10}$$

Usually, the constants $\lambda_\parallel$ and $\lambda_\perp$ are positive. The only exceptions to this rule are ferromagnets, for which the negative sign of the magnetoresistance (the reduction of the resistivity in weak magnetic fields) requires a special explanation [24].

Equations (26.9) and (26.10) are phenomenological. The task of the microscopic theory is to calculate the tensors λ_{iklm} and R_{ik}.

Usually, the formulas obtained for λ_{iklm} and R_{ik} are very complex and cannot be used to draw any clear conclusions. However, it can be shown [25] that Onsager's relationship (26.3) is a consequence of the general properties of the collision operator formulated in §23 [see Eqs. (23.23) and (23.24)]. Similar considerations can be used to show that the principal values of the tensor $\rho_{ik}(\mathbf{H})$ are larger than the principal values of the tensor $\rho_{ik}(0)$. In other words, the resistivity increases when the magnetic field is increased. (We must bear in mind that this conclusion is a consequence of the classical theory. The quantum effects, resulting from quantization of the energy in a magnetic field, may reduce the longitudinal resistivity. This effect has been observed in semiconductors [26].)

The dependences of the components of the resistivity tensor on the magnitude and the direction of the magnetic field cannot be determined for an arbitrary value of the field. These dependences vary with the details of the scattering process and they are affected considerably by the dynamic properties of conduction electrons. However, the asymptotic behavior of the components of the resistivity tensor ρ_{ik} (for $r_H \ll l$) are determined mainly by the structure of the Fermi surface of a metal and depend little on the nature of the scattering process. The dependences of the components of ρ_{ik} on the magnetic field in the case $r_H \ll l$ can be derived rigorously by analyzing the solutions of the transport equation. This will be done in the next section. However, the principal galvanomagnetic properties of metals and their relationship to the shape of the Fermi surface (in the case $r_H \ll l$) can be established from very general considerations using the "diffusion" approximation, i.e., by considering the motion of an electron in mutually perpendicular electric and magnetic fields as "jumps" from one orbit to another.

Since we are interested in weak electric fields (the ohmic region), we shall consider the diffusion in a magnetic field and calculate the electrical conductivity from Einstein's relationship between the diffusion coefficient D and the electrical conductivity σ, which is of the following form for a degenerate Fermi gas[4]

$$\sigma = \frac{e^2 D}{\partial \zeta / \partial n} \approx \frac{Dne^2}{\varepsilon_F}.$$
(26.11)

The problem thus reduces to estimating the diffusion coefficient. We can use two formulas which are valid in two opposite limiting cases:

$$D \approx l_p v_F \quad \text{or} \quad D \approx r_H^2 W \qquad \left(W \approx \frac{1}{\tau_p} \right).$$
(26.12)

The first of the formulas is valid in those cases when the motion along a given direction is limited by the mean free path: this applies to weak magnetic fields ($r_H \gg l$) or to infinite motion in a magnetic field (motion along a magnetic field and along an open trajectory in a plane perpendicular to this field).[5] The second formula in Eq. (26.12) applies to finite motion in a plane perpendicular to the magnetic field: this includes motion along a closed trajectory or an open trajectory at right-angles to an open direction ($r_H \approx cp_F/eH$). Using Eqs. (26.11) and (26.12), we easily find that

$$\sigma \approx \frac{ne^2 l}{p_F} \quad \text{or} \quad \sigma \approx \frac{ne^2 l}{p_F} \left(\frac{r_H}{l} \right)^2.$$
(26.13)

The validity of the two formulas in Eq. (26.13) is restricted in the same way as the validity of the formulas in Eq. (26.12). The components of σ are not identified but the derivation implies that they are the diagonal (dissipative) components.

In the case of metals with a closed Fermi surface, we have $\sigma_{zz} \approx \sigma$, and $\sigma_{xx} \approx \sigma_{yy} \approx \sigma (r_H/l)^2$. For metals with an open Fermi surface and a particular orientation of the magnetic field, we have $\sigma_{xx} \approx \sigma$, and $\sigma_{yy} \sim \sigma (r_H/l)^2$ (σ_{zz} is always $\approx \sigma$).

It follows from our derivation that the increase in the resistivity in a magnetic field is always associated with the finite nature of the motion of an electron and with the fact that diffusion "jumps" from one trajectory to a neighboring one are relatively rare: $W = 1/\tau_p \ll \omega_H$.

The Hall component of σ_{xy} cannot be obtained in this way. Its value is not determined by diffusion but by purely mechanical motion in the form of the drift of electrons along a direction perpendicular to the electric and the magnetic fields.

Since, according to Eq. (4.26), the drift velocity is c/H^2 [**EH**], the Hall current is given by the simple formula

$$j_x = \frac{ec}{H^2} [\textbf{EH}] (n_1 - n_2),$$
(26.14)

where n_1 is the number of occupied electron states characterized by a positive effective mass and n_2 is the number of vacant states characterized by a negative effective mass. Therefore, the Hall component of the electrical conductivity is given by $\sigma_{xy} = [(n_1 - n_2) ec]/H$ (the magnetic field is, as usual, directed along the z axis; for details see the next section). If we know the components of σ_{ik}, we can calculate the components of $\rho_{ik} = \sigma_{ik}^{-1}$. This will be done in §27 after a rigorous derivation of the asymptotic form of $\sigma_{ik}(\textbf{H})$.

---

[4] We shall omit factors of the order of unity because the formulas are approximate.
[5] We recall that, when an open direction coincides with the p_x axis, infinite motion takes place along the y axis in the **r** space.

The formulas obtained so far are sufficient to demonstrate clearly the basic difference between the behavior of a bulk conductor in and out of a magnetic field. When H = 0, all the components of the conductivity tensor tend to infinity with increasing mean free path because $\sigma_{ik} \propto l$. In a magnetic field (H ≠ 0) when $l \to \infty$ (in this case, the magnetic field is naturally strong: $r_H \ll l \to \infty$), the components σ_{xx}, σ_{yy} and, when $n_1 = n_2$, the components σ_{xy}, σ_{yx} all tend to zero because $\sigma (r_H/l)^2 \propto 1/l$, in accordance with the formulas derived earlier in the section. The difference between the cases H ≠ 0 and H = 0 is particularly striking when $n_1 = n_2$: when H = 0, a sample with an infinite mean free path is a perfect conductor (its resistivity $\rho = 0$), whereas in H ≠ 0, such a sample is a perfect dielectric in a plane perpendicular to the magnetic field (the conductivity $\sigma = 0$).

Moreover, our analysis has made it clear that there is a relationship between the energy structure of a metal and its galvanomagnetic properties.

§27. Galvanomagnetic Effects in Strong Fields (Closed Trajectories)

The development of a microscopic theory of the galvanomagnetic effects, i.e., the calculation of the resistivity tensor as a function of the magnetic field, is based on the solution of the linearized Boltzmann transport equation, which should be used in a form somewhat different from that given in §23.

The nature of the motion of an electron in a constant and uniform magnetic field (§4) shows that, in order to describe the position of an electron in the momentum space, it is convenient to replace (in the presence of a magnetic field) the Cartesian coordinates p_x, p_y, p_z with the coordinates linked to the electron trajectory in the momentum space. The position of an electron can be specified by giving its trajectory in the momentum space, i.e., its energy ε, the projection of the momentum along the magnetic field p_z (a uniform and constant magnetic field H is directed, as usual, along the z axis), and its position on the trajectory. The position on the trajectory can be specified either by the length of an arc s, measured from some point, or (this is more convenient) the time t that the electron has taken to travel from some fixed point to that being considered. Thus, the position of an electron may be described by three quantities (coordinates): ε, p_z, and t. The formulas for the transformation of the coordinates (from p_x, p_y, p_z to ε, p_z, t) are provided by the dispersion law ε = ε (p_x, p_y, p_z) and the equation of motion of an electron in a magnetic field

$$\frac{dp}{dt} = \frac{e}{c}[vH], \quad v = \frac{\partial \varepsilon}{\partial p},$$ (27.1)

according to which

$$dt = \frac{c}{eH}\frac{ds}{v_\perp},$$

where $v_\perp$ is the projection of the vector **v** onto a plane perpendicular to the magnetic field.

Boltzmann's transport equation (23.6) can now be rewritten in terms of these new variables[1]:

$$\frac{\partial f}{\partial t} \dot{t} + \frac{\partial f}{\partial p_z} \dot{p}_z + \frac{\partial f}{\partial \varepsilon} \dot{\varepsilon} = \left(\frac{\partial f}{\partial t}\right)_{st},$$ (27.2)

---

[1] Here and in the succeeding sections, we shall consider only the spatially homogeneous case, i.e., we shall ignore the presence of the boundaries of a sample. This means that the results obtained are applicable to fairly large samples. Suitable estimates are given in §23 (see also [27] and §29). Moreover, we shall deal only with the static case.

where the generalized forces ($\dot{t}$, $\dot{p}_z$, and $\dot{\varepsilon}$) describe the change in the state of an electron caused by the application of external fields (electric field **E** and magnetic field H). These generalized forces should be calculated using the equation of motion

$$\dot{p} = eE + \frac{e}{c}\,[vH]. \tag{27.3}$$

It follows from Eqs. (27.1) and (27.3) that

$$\dot{\varepsilon} = evE, \qquad \dot{p}_z = eE_z, \qquad \dot{t} = 1 + \frac{dt}{dp}\,eE. \tag{27.4}$$

The vector dt/d**p** is defined by Eq. (27.1).

Using the relationships in Eq. (27.4), we can easily linearize Eq. (27.2) with respect to the electric field so that it assumes the following simple form:

$$\frac{\partial f_1}{\partial t} - \left(\frac{\partial f_1}{\partial t}\right)_{\text{st}} = -\,evE\,\frac{\partial n_F}{\partial \varepsilon}. \tag{27.5}$$

As in §23, we shall replace the function f_1 with the vector function ψ:

$$f_1 = -\,\frac{\partial n_F}{\partial \varepsilon}\,e\psi E. \tag{27.6}$$

This vector function ψ satisfies the following equation [see Eq. (23.16)]:

$$\frac{\partial \psi_l}{\partial t} + \hat{W}_p\,\{\psi_i\} = v_l. \tag{27.7}$$

The operator $\hat{W}_p$ is defined in §23 [Eq. (23.13)].

Equation (27.7) is not only formally but physically identical with Eq. (23.16) because, in the absence of an allowance for the space and the time inhomogeneities, the quantity df/dt [Eq. (23.12)] describes the change in the distribution of electrons solely because of their motion in a constant and uniform magnetic field. It is clear that Eq. (27.7) is a differential equation in t. We shall now consider the expressions which will be used as the boundary conditions. If, for given values of ε and p_z, the trajectory of an electron in a magnetic field is closed, the boundary condition is that of periodicity

$$\psi_i\,(t + T) = \psi_i\,(t), \tag{27.8}$$

where T = T(ε, p_z) is the Larmor precession period. According to Eq. (4.11):

$$T\,(\varepsilon,\ p_z) = -\,\frac{c}{eH}\,\frac{\partial S}{\partial \varepsilon} = -\,\frac{2\pi m^* c}{eH} = \frac{2\pi}{\omega_H},$$

where S is the area of the section of the constant-energy surface ε(**p**) = ε which is cut by a plane p_z = const.

If the trajectory is open, the boundary condition requires that ψ should be finite at t = ± ∞.

The presence of the factor $-\partial n_F/\partial \varepsilon$ in the expression (27.6) indicates that electrons whose energy is of the order of the Fermi value play the dominant role. Consequently, we can deal separately with closed and open surfaces (§2).

In the present section we shall consider only the closed surfaces (the open surfaces will be dealt with in the next section). More specifically, we shall restrict our treatment to the case of closed trajectories. This will be done because there are some cases when an open surface has no open sections, and because open surfaces may have closed sections along certain directions of the magnetic field (§2).

Averaging Eq. (27.7) over t and using the boundary condition (27.8), we obtain

$$\overline{\hat{W}_p\{\psi_i\}} = \bar{v}_i,$$
(27.9)

where the bar represents the averaging over one period:

$$\bar{u} = \frac{1}{T(\varepsilon, p_z)} \int\limits_0^T u\, dt.$$
(27.10)

We shall regard Eq. (27.9) as the boundary condition which applies to ψ_i.

The high intensity of the magnetic field is manifested by the fact that the term $\partial \psi_e/\partial t \approx \psi_i/T$ in Eq. (27.7) is considerably larger than the "collision" term $\hat{W}_p\{\psi_i\} \approx \psi_i/\tau_p$. This is valid when $T \ll \tau_p$ or $r_H \ll l$ (see preceding section) and it allows us to use the method of successive approximations which corresponds to an expansion in terms of powers of the reciprocal of the magnetic field [28]. We can show that the solution of Eq. (27.7) which satisfies the condition (27.9) can be represented in the form

$$\psi_i = \overline{\hat{W}_p^{-1}}\bar{v}_i + \frac{\hat{q}}{1 - \hat{q}}\left(\overline{\hat{W}_p^{-1}}\bar{v}_i - \hat{W}_p^{-1}v_i\right),$$
(27.11)

where the operator $\hat{q}$ is given by the following expression[2]:

$$\hat{q}\varphi = \overline{\hat{W}_p^{-1}}\overline{\hat{W}_p \int\limits_{-\infty}^t \overline{\hat{W}_p}\varphi\, dt} - \int\limits_{-\infty}^t \hat{W}_p\varphi\, dt.$$
(27.12)

Equations (27.11) and (27.12) represent an algorithm for the calculation of the functions ψ_i in the form of a series in powers of the reciprocal of the magnetic field. We need to know the explicit form of the operators $\hat{W}_p$ and $\hat{W}_p^{-1}$ in order to carry out this calculation. However, some important conclusions, particularly the dependences of the various components of the resistivity tensor on the strong magnetic field, can be obtained knowing only the topology of the Fermi surface and of nearby constant-energy surfaces.

In the case of closed trajectories, it follows from the equation of motion (27.1) that

$$\bar{v}_\alpha = 0, \quad \bar{v}_z \neq 0 \qquad (\alpha = x,\, y).$$
(27.13)

Therefore, the expansion of ψ_α begins from terms proportional to $1/H$ and that of ψ_z from a term which is independent of the magnetic field. Using Eqs. (27.11) and (27.12) together with the equation of motion (27.1), we easily obtain

$$\left.\begin{aligned}
\psi_x &= -\frac{c}{eH}\left(p_y - \overline{\hat{W}_p^{-1}}\overline{\hat{W}_p p_y}\right) + \dots, \\
\psi_y &= \frac{c}{eH}\left(p_x - \overline{\hat{W}_p^{-1}}\overline{\hat{W}_p p_x}\right) + \dots, \\
\psi_z &= \overline{\hat{W}_p^{-1}}v_z + \dots.
\end{aligned}\right\}
$$
(27.14)

---

[2] Equation (27.12) has meaning if the function φ satisfies the condition $\overline{\hat{W}_p\varphi} = 0$.

The electrical conductivity tensor σ_{ik} can be calculated by substituting the value of ψ_i into Eq. (23.21) and going over from the integration with respect to p_x, p_y, p_z to the integration with respect to ε, p_z, and t. The Jacobian of this transformation can be calculated most easily employing the equation of motion (27.1). If Eq. (27.1) is written in the form

$$\frac{\partial(p_x, \varepsilon, p_z)}{\partial(t, \varepsilon, p_z)} = \frac{eH}{c} \frac{\partial(\varepsilon, p_x, p_z)}{\partial(p_y, p_x, p_z)}, \qquad \frac{\partial(p_y, \varepsilon, p_z)}{\partial(t, \varepsilon, p_z)} = -\frac{eH}{c} \frac{\partial(\varepsilon, p_y, p_z)}{\partial(p_x, p_y, p_z)}$$

and the numerators are canceled, we obtain

$$\left| \frac{\partial(p_x, p_y, p_z)}{\partial(\varepsilon, p_z, t)} \right| = \left| \frac{eH}{c} \right|. \tag{27.15}$$

Thus,

$$\sigma_{ik} = -\frac{2e^2}{(2\pi\hbar)^3} \left| \frac{eH}{c} \right| \int \int \int \frac{\partial n_F}{\partial \varepsilon} v_i \psi_k \, d\varepsilon \, dp_z \, dt. \tag{27.16}$$

Using this form of the electrical conductivity tensor, we can show that Onsager's relationship in a magnetic field (26.3) does not impose any other restrictions on the collision integral apart from those formulated in §23 (see [25]).

We shall now substitute the expansion (27.14) into Eq. (27.16). We shall begin by calculating the components σ_{xy} and σ_{yx}. We shall replace v_x, v_y with their values from the equation of motion (27.1):

$$\sigma_{xy} = -\frac{2e^2}{(2\pi\hbar)^3} \int \frac{\partial n_F}{\partial \varepsilon} \, d\varepsilon \int dp_z \oint \frac{\partial p_y}{\partial t} [p_x - \overline{\hat{W}_p^{-1}} \, \overline{\hat{W}_p p_x}] \frac{c}{eH} \, dt + \ldots .$$

It follows from this expression[3] that the term containing $\overline{\hat{W}_p^{-1}} \, \overline{\hat{W}_p p_x}$ always drops out after integration with respect to t and

$$\oint \frac{\partial p_y}{\partial t} p_x \, dt = \int\limits_{(\varepsilon, p_z)} p_x \, dp_y = \pm S(\varepsilon, p_z). \tag{27.17}$$

Here, $S(\varepsilon, p_z)$ is the area of the section of the constant-energy surface $\varepsilon(\mathbf{p}) = \varepsilon$ cut by a plane $p_z = $ const. The sign of the area is given by the direction in which its circumference is described, i.e., by the sign of the effective mass m^* (§4). The asymptotic expression for strong fields is thus

$$\sigma_{xy} = -\frac{ec}{H} \frac{2}{(2\pi\hbar)^3} \int \frac{\partial n_F}{\partial \varepsilon} \, d\varepsilon \int dp_z \{S_1(\varepsilon, p_z) - S_2(\varepsilon, p_z)\}, \tag{27.18}$$

where the first integral is taken over those parts of the energy bands where $m^* > 0$ and the second over those parts where $m^* < 0$. Moreover, it is assumed that the summation is carried out over all the partly filled bands.

------

[3] We have not given here the self-evident limits of integration. We recall that, by definition, we are dealing with closed trajectories.

If $-\partial n_F / \partial \varepsilon = \delta(\varepsilon - \varepsilon_F)$, it follows that

$$\sigma_{xy} = \frac{ec}{H} \frac{2}{(2\pi\hbar)^3} \int \{S_1(\varepsilon, \, p_z) - S_2(\varepsilon, \, p_z)\} \, dp_z, \tag{27.19}$$

and, finally,

$$\sigma_{xy} = \frac{2ec}{H(2\pi\hbar)^3} \{V_1(\varepsilon_F) - V_2(\varepsilon_F)\}. \tag{27.20}$$

Here, $V_1(\varepsilon_F)$ is the volume bounded by that part of a multiple-sheet Fermi surface within which the energy is less than the Fermi energy ($m^* > 0$) and $V_2(\varepsilon_F)$ is the volume bounded by that part of the Fermi surface within which the energy is greater than the Fermi energy ($m^* < 0$).

If the closed trajectories are located on an open Fermi surface, we shall define V_1 and V_2 as the volume bounded by the Fermi surface and the boundaries of a unit cell in the reciprocal lattice.

At absolute zero, electrons occupy all the states with energies less than ε_F and, therefore,

$$\sigma_{xy} = \frac{ec(n_1 - n_2)}{H}, \tag{27.21}$$

where n_1 is the number of occupied electron states with a positive effective mass and n_2 is the number of vacant electron states with a negative effective mass. The quantity n_1 can be called the number of "electrons" and n_2 the number of "holes."

Thus, if the Fermi surface is closed (or if only closed sections are obtained for all the values of p_z in the case of an open surface), the asymptotic form of σ_{xy} is given by Eq. (27.21). We must stress that, in this case, σ_{xy} is independent of the direction of the magnetic field and of the nature of the collisions involving conduction electrons, but is determined simply by the numbers of electrons and holes. The other component is given by $\sigma_{yx} = -\sigma_{xy}$. This can be demonstrated by direct calculations and it follows from Onsager's relationship (we shall show that the symmetric part of the tensor is much smaller in the asymptotic limit: it is proportional to $1/H^2$).

The case which deserves special consideration is that in which the number of electrons is equal to the number of holes ($n_1 = n_2$). Here, the expansion of the components σ_{xy}, σ_{yx} begins with terms proportional to $1/H^2$. The expansion of the antisymmetric part then begins from the terms $\sim 1/H^3$. We can select the x and y axes in such a way that the symmetric part of the tensor σ_{ik} (denoted by s_{ik}) is diagonal; in such a system of coordinates, the expansions of σ_{xy} and σ_{yx} begin from terms which are proportional to $1/H^3$.

We note that the equal number of electrons and holes are not exceptional: all metals with an even number of electrons per unit cell have conduction electrons only because the energy bands of these metals overlap. The number of states which become vacant in the lower band (the number of holes) is equal to the number of occupied states (electrons) in the upper band (Fig. 57).

The concepts "electron" and "hole" are defined uniquely if the Fermi surface is closed. In the case of an open surface, we should speak of electron or hole orbits, depending on the direction of motion along these orbits. If the overlap of the energy bands is slight (Fig. 57), the constant-energy surfaces are usually closed (§2).

We shall now continue our calculations of the components of the tensor σ_{ik}. Proceeding as before, we find that the expansions of σ_{xx} and σ_{yy} begin from terms $\sim 1/H^2$, whereas σ_{xz}, σ_{yz} begin from terms $\sim 1/H$; σ_{zz} tends to a constant value when the field approaches infinity.

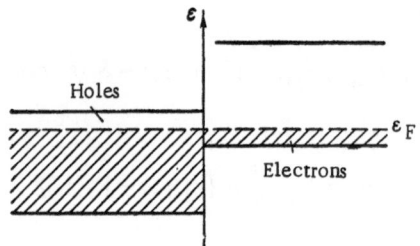

Fig. 57. The number of electrons
is equal to the number of holes for
metals with an even number of electrons per unit cell.

Thus, in strong fields, the electrical conductivity tensor ($n_1 \neq n_2$) is

$$(\sigma_{ik}) = \begin{vmatrix} \dfrac{a_{xx}}{H^2} & \dfrac{ec\,(n_1 - n_2)}{H} & \dfrac{a_{xz}}{H} \\[2mm] -\dfrac{ec\,(n_1 - n_2)}{H} & \dfrac{a_{yy}}{H^2} & \dfrac{a_{yz}}{H} \\[2mm] \dfrac{a_{zx}}{H} & \dfrac{a_{zy}}{H} & a_{zz} \end{vmatrix}, \tag{27.22}$$

and the components of the matrix a_{ik} tend to constant values. In specially symmetric cases, some of the components of this matrix may vanish. For example, the tensor σ_{ik} for an iso-tropic sample and an arbitrary value of the magnetic field has the following form in the τ ap-proximation:

$$(\sigma_{ik}) = \begin{vmatrix} \dfrac{\sigma}{1 + \omega_H^2\tau^2} & \pm\,\sigma\,\dfrac{\omega_H\tau}{1 + \omega_H^2\tau^2} & 0 \\[2mm] \mp\,\sigma\,\dfrac{\omega_H\tau}{1 + \omega_H^2\tau^2} & \dfrac{\sigma}{1 + \omega_H^2\tau^2} & 0 \\[2mm] 0 & 0 & \sigma \end{vmatrix}. \tag{27.23}$$

Here, $\omega_H = eH/m^*c$ is, as usual, the Larmor precession frequency; the plus sign corresponds to an electron band; the minus sign corresponds to a hole band; $\sigma = ne^2\tau/|m^*|$. If electrons be-longing to several bands participate in the conduction process, the electrical conductivity is the sum of the corresponding tensors.

It is evident from Eq. (27.22) that, in those cases when the Fermi surface is closed, all the components of the tensors σ_{ik} (except σ_{zz}) vanish when the magnetic field tends to infinity. The nature of this approach to zero depends on the actual component.

An allowance for the "spreading" of the Fermi step may alter the results obtained but the correction would normally be negligibly small. In some cases it is convenient to use a more general expression for the electrical conductivity tensor, which is valid throughout the whole range of magnetic fields. If we restrict our treatment to the τ approximation $[\hat{W}_p\{\psi\} = (1/\tau)\psi]$, Eq. (27.7) combined with the condition (27.8) shows that

$$\psi_i = \int_0^\infty v_i\,(t - t')\,e^{-t'/\tau}\,dt'. \tag{27.24}$$

We note that this solution is valid also in the case of open trajectories. It can be used, for ex-ample, to calculate the dependence of the transverse resistivity of noble metals on the mag-netic field [29].

There is also a method [28] for solving Eq. (27.7) which is valid for any collision integral and any value of the magnetic field provided we restrict the treatment to the closed trajectories. The solutions are obtained in the form of a series.

Substituting Eq. (27.24) into Eq. (27.16) for σ_{ik}, we obtain[4]

$$\sigma_{ik} = -e^2 \int \varphi_{ik}(\varepsilon,\, p_z)\, \frac{\partial n_F}{\partial \varepsilon}\, d\Gamma, \qquad (27.25)$$

where

$$\varphi_{ik} = \int_0^\infty \overline{v_i(t)\, v_k(t-t')}^{\,t}\, e^{-t'/\tau}\, dt'. \qquad (27.26)$$

The bar represents averaging over one period in the case of closed trajectories [Eq. (27.10)]. In the case of open trajectories, the averaging should be taken in the sense of the limit corresponding to $T \to \infty$.

The galvanomagnetic effects are usually investigated by passing a current of known magnitude and direction, and measuring the components of the electric field along three noncoplanar directions (if possible, along three orthogonal directions; see, for example [31]). This means that the measured quantity is the resistivity tensor in a magnetic field ρ_{ik}, which is the reciprocal of the conductivity tensor ($\rho_{ik} = \sigma_{ik}^{-1}$).

Using the asymptotic expressions for the conductivity tensor $\sigma_{ik}(\mathbf{H})$, we can easily derive the asymptotic form of the resistivity tensor in strong fields.

If the number of electrons is not equal to the number of holes ($n_1 \neq n_2$), it follows from Eq. (27.22) that

$$(\rho_{ik}(\mathbf{H})) = \begin{pmatrix} b_{xx} & \dfrac{H}{(n_1-n_2)\, ec} & b_{xz} \\[2mm] -\dfrac{H}{(n_1-n_2)\, ec} & b_{yy} & b_{yz} \\[2mm] b_{zx} & b_{zy} & b_{zz} \end{pmatrix}, \qquad (27.27)$$

where the components of the matrix b_{ik} tend to constant values as $H \to \infty$. These components can be determined from the asymptotic values of the components of the matrix a_{ik}. If the current density $\mathbf{j}$ is perpendicular to the magnetic field, the x axis can be taken along the direction of the current. The resistivity ρ (the ratio of the electric field along the current to the current density) is then ρ_{xx} (in the general case, $\rho = \rho_{ik} j_i j_k / j^2$) and the quantity ρ_{yx}/H is identical with the Hall coefficient $R = E_y/Hj$. Thus, in this case (closed trajectories, $n_1 \neq n_2$), the resistivity always tends to a saturation value irrespective of the directions of the magnetic field and the current. The asymptotic value of the resistivity naturally depends on the directions of the magnetic field and the current because of the anisotropy of the components of the matrix b_{ik}, which is related to the anisotropy of the dispersion law and of the scattering processes. As mentioned earlier, we can show that $\rho\,(H \to \infty)$ is always greater than $\rho\,(H = 0)$ and that this conclusion is based solely on the observation that the collision operator is independent of the magnetic field [25]. This does not apply in the quantum case [26].

The asymptotic value of the Hall coefficient R_∞ is an important characteristic of the electron energy spectrum of a metal. If the Fermi surface is closed (or if all the sections are

---

[4] A similar form of the solution of the transport equation has been used in [30].

closed for all values of p_z in the case of an open surface), we find that this value is

$$R_\infty = \frac{1}{(n_1 - n_2)\, ec} \qquad (27.28)$$

and is independent of the direction of the magnetic field.

Recent investigations have demonstrated that some metals have closed Fermi surfaces with unequal numbers of electrons and holes: these metals are In, Al, K, Na [23, 32].

If the Fermi surface is closed and the number of electrons is equal to the number of holes ($n_1 = n_2$), all the transverse (in relation to the magnetic field) components of the tensor ρ_{ik} tend to infinity because $\sigma_{xy} \propto 1/H^2$ and

$$\rho_{\alpha\beta} \propto H^2, \quad \rho_{\alpha z} \propto H \qquad (\alpha, \beta = x, y), \qquad (27.29)$$

whereas the longitudinal component ρ_{zz} tends to a constant value.

It follows that, in this case, the resistivity along any direction of the magnetic field tends to an infinite value if the current has a transverse component in relation to the field.

The Hall field for $n_1 = n_2$ is not the main component in the asymptotic expression for the electric field perpendicular to the current. If the direction is selected in an arbitrary manner, we find a term which is quadratic in the magnetic field and this term is associated with the resistance anisotropy (i.e., with the difference between the principal values ρ_{xx} and ρ_{yy}).

The high-field Hall coefficient R_∞ (the limiting value of the quantity $[\rho_{xy}(\mathbf{H}) - \rho_{xy}(-\mathbf{H})]/2H$ for $H \to \infty$) depends on the magnetic field direction and is determined not only by the energy spectrum but also by the nature of the scattering of conduction electrons. If $n_1 \neq n_2$, the Hall vector denoted by $\boldsymbol{b}$ in §26 is practically parallel to the magnetic field and if $n_1 = n_2$ all the components of the Hall vector are of the same order of magnitude.

The transverse resistivity of some metals (Bi, As, Sb) increases irrespective of the direction of the magnetic field. This result is in excellent agreement with our ideas on the nature of the energy spectra of these metals [33]. However, the resistivity usually increases a little less rapidly than H^2.

Certain model assumptions about conduction electrons allow us to derive compact formulas which describe the dependence of the resistivity on the magnetic field in a wide range of values of this field. Thus, if we assume that there are only two bands with quadratic but anisotropic dispersion laws, that the mobility tensors are proportional to each other ($u_{ik}^{(1)} = k u_{ik}^{(2)}$), and that the number of electrons is equal to the number of holes ($n_1 = n_2$), we find that elementary calculations yield the following relationships:

$$\rho = \frac{\sigma_2 \cos^2 \alpha + \sigma_1 \sin^2 \alpha}{\sigma_1 \sigma_2}\left(1 + \frac{H^2}{H_0^2}\right), \quad R = -\frac{1}{nec}\frac{1-k}{1+k}. \qquad (27.30)$$

Here, $H_0 = (c/e)\,[u_1^{(1)} u_2^{(1)}]^{-1/2}$; σ_1 and σ_2 are the principal values of the conductivity tensor $\sigma_{ik} = ne^2 u_{ik}(1+k)$; the magnetic field is directed along one of the principal axes (the third); the current is perpendicular to the magnetic field; α is the angle between the first axis and the current. This simple model yields a unique quadratic dependence for the resistivity and no dependence of the Hall coefficient on the field. This is a characteristic of a two-band model with a quadratic dispersion law. The introduction of a third band alters the position radically. We can show (making the same assumptions about the dispersion law) [34] that the coefficients in front of H^2 in weak ($r_H \gg l$) and strong ($r_H \ll l$) fields are different and that, in agreement with the experimental data [35], the coefficient in front of H^2 in weak fields is always greater than the corresponding coefficient in strong fields. The Hall coefficients are also different in weak and strong fields.

The galvanomagnetic properties (more correctly, their asymptotic values in strong fields) are very sensitive to the nature of the electron energy spectrum and, as we shall show later, provide excellent means for the determination of the Fermi surface topology.

In the case of metals with closed Fermi surfaces, the galvanomagnetic properties can be used to establish that the Fermi surface is, in fact, closed and to determine the difference between the numbers of electrons and holes ($n_1 - n_2$).

The most important information on the nature of the electron energy spectrum can be obtained by investigating metals with open Fermi surfaces, which are considered in the next section.

§ 28. Galvanomagnetic Effects in Strong Fields (Open Trajectories)

In the preceding section, we considered in detail the galvanomagnetic effects in those metals whose Fermi surfaces are closed. In the present section, we shall discuss open Fermi surfaces. The most important distinguishing characteristic of the galvanomagnetic properties of metals with open Fermi surfaces is the strong anisotropy of these properties [28, 38]. Along some directions of the magnetic field, the high-field ($l \gg r_H$) transverse resistivity tends to saturation, whereas it increases quadratically along other directions. This property can be used to investigate the Fermi surface topology.

The strong anisotropy of the resistivity is a consequence of the different nature of the motion of electrons along different directions in a plane perpendicular to the magnetic field (§26), which is a characteristic of the open Fermi surfaces. This anisotropy can be demonstrated by considering the simplest case of such a surface: a metal whose Fermi surface and the neighboring constant-energy surfaces are corrugated cylinders (Fig. 58).

If a magnetic field (directed, as usual, along the z axis) is not perpendicular to the cylinder axis, all the sections are closed. This is the case we discussed in the preceding section.

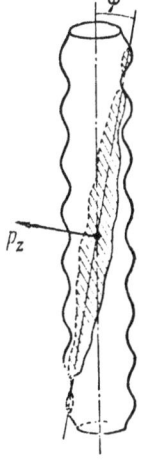

Fig. 58. Constant-energy surface of the corrugated-cylinder type. A section by a plane p_z = const, perpendicular to a magnetic field ($\varphi = \theta - \pi/2$), is shown shaded.

Special treatment is required when the magnetic field is almost perpendicular to the open direction and the electron trajectories are closed but highly elongated so that they are considerably larger than a unit cell in the reciprocal lattice. We shall consider this situation below.

A number of features have to be taken into account in solving the transport equation in the case of open surfaces. First, the characteristic time of motion [needed to determine the order of magnitude of $|\partial\psi_i/\partial t|$ in Eq. (27.7)] is that time during which the electron momentum changes by an amount of the order of $|2\pi\hbar b|$, where b is a reciprocal lattice vector along the open direction (in our case, along the x axis). The magnetic field is regarded as strong if the relaxation time is much longer than this characteristic time [it follows from the equation of motion (27.1) that this time is inversely proportional to the magnetic field and its order of magnitude is the same as that of the period of revolution of an electron T]. Secondly, the boundary conditions must be refined. It is mentioned in §27 that, in the case of open trajectories, the boundary condition is the requirement that the function ψ_i must be finite when the value of the "time" t is large (it must remain finite as $t \to \pm\infty$). If we generalize the definition of the average value of a function by assuming that

$$\bar{u} = \lim_{T \to \infty} \frac{1}{T} \int_0^T u(t)\, dt, \tag{28.1}$$

we find that the boundary condition for the function ψ_i can be written formally in the same way as in the case of closed trajectories:

$$\overline{\hat{W}_p\{\psi_i\}} = \bar{v}_i. \tag{28.2}$$

Thirdly, the average values of the transverse (in relation to the magnetic field) components of the velocity do not vanish. These components (in the selected coordinate system) are given by

$$\bar{v}_y = -\frac{c}{eH} \overline{\frac{\partial p_x}{\partial t}} \neq 0, \tag{28.3}$$

$$\bar{v}_x = \frac{c}{eH} \overline{\frac{\partial p_y}{\partial t}} = 0. \tag{28.4}$$

We shall now refine the statement that there are some values of p_z ($|p_z| < p_1$, Fig. 59) for which the sections are open and $\bar{v}_y \neq 0$.

The formal solution (27.11) of Eq. (27.7) is valid also in the case of a corrugated cylinder and the expansion of ψ_i in powers of the reciprocal of the magnetic field is given by [compare it

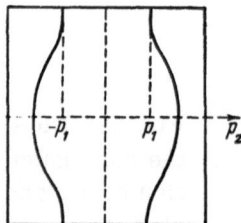

Fig. 59. Open trajectories for $|p_z| < p_1$, but closed for $|p_z| > p_1$.

with Eq. (27.14)]:

$$\psi_x = -\frac{c}{eH}\left(p_y - \overline{\widehat{W}_p^{-1}}\,\overline{\widehat{W}_p p_y}\right) + \dots,$$
$$\psi_z = \overline{\widehat{W}_p^{-1}}\,\bar{v}_z + \dots,$$
$$\psi_y = \begin{cases} \overline{\widehat{W}_p^{-1}}\,\bar{v}_y + \dots & (|p_z| < p_1), \\[2mm] \frac{c}{eH}\left(p_x - \overline{\widehat{W}_p^{-1}}\,\overline{\widehat{W}_p p_x}\right) + \dots & (|p_z| > p_1). \end{cases} \tag{28.5}$$

These asymptotic values of ψ_i can be used to deduce the asymptotic expressions for the components of the electrical conductivity tensor:

$$(\sigma_{ik}) = \begin{pmatrix} \dfrac{a'_{xx}}{H^2} & \dfrac{a'_{xy}}{H} & \dfrac{a'_{xz}}{H} \\[3mm] \dfrac{a'_{yx}}{H} & a'_{yy} & a'_{yz} \\[3mm] \dfrac{a'_{zx}}{H} & a'_{zy} & a'_{zz} \end{pmatrix}, \tag{28.6}$$

where the expansion of the elements of the matrix a_{ik} in powers of the reciprocal field begin from the zeroth term. Inversion gives us the asymptotic values of the resistivity tensor:

$$(\rho_{ik}) = \begin{pmatrix} H^2 b'_{xx} & H b'_{xy} & H b'_{xz} \\[2mm] H b'_{yx} & b'_{yy} & b'_{yz} \\[2mm] H b'_{zx} & b'_{zy} & b'_{zz} \end{pmatrix}. \tag{28.7}$$

The elements of the matrix b'_{ik} are defined in terms of the elements of the matrix a'_{ik}.

Since $\bar{v}_y \neq 0$, b'_{xy} and a'_{xy} normally depend not only on the energy spectrum but also on the nature of the collision integral [see the derivation of Eq. (27.21)]. Let us now analyze the results obtained.

One of the components of the transverse resistivity (ρ_{xx}) increases quadratically with magnetic field, whereas the other (ρ_{yy}) tends to saturation. We note that the resistivity along the open direction (along the x axis)[1] increases when the magnetic field is increased. The asymptotic difference between ρ_{xx} and ρ_{yy} shows that the resistivity depends strongly on the direction of the current j (even when j ⊥ H). The resistivity $\rho = \rho_{ik} j_i j_k / j^2$ for an arbitrary direction of the current increases quadratically but if the current is directed along the y axis, the resistivity tends to saturation.

Let us consider a metal whose Fermi surface is a corrugated cylinder and analyze the dependence of its resistivity on the direction of a strong magnetic field ($r_H \ll l_p$) for an arbitrary direction of the current. In this case, we should observe the following behavior: the resistivity along an arbitrary direction of the magnetic field should tend to saturation and it should increase quadratically along some of the directions (those directions for which the magnetic field is perpendicular to the open direction). An analysis of the equation for the function ψ_i shows [38] that the transition from the one to the other dependence occurs in a narrow range of angles $\Delta\theta$, of the order of $H_0/H \ll 1$ (here, H_0 is the magnetic field in which the ratio $2\pi\hbar bc/eH$ is equal to the mean free path). This follows from the observation that the period

---

[1] In the momentum space, the open direction would coincide (in the present case) with the direction of the p_y axis.

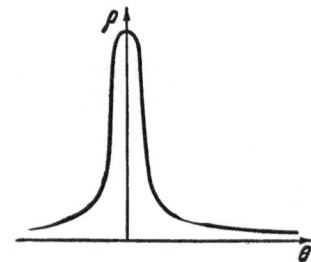

Fig. 60. Dependence of the resistivity on the direction of a magnetic field applied to a metal with a Fermi surface of the corrugated-cylinder type.

of revolution of an electron along elongated trajectories is inversely proportional to the angle θ between the magnetic field and a plane perpendicular to the open direction.

We note that, in addition to this strong angular dependence of the resistivity and other galvanomagnetic properties, there is also a smooth dependence on the magnetic field direction. The latter can be explained only by the numerical solution of the transport equation but this point will not be discussed here.

An analytic dependence of the resistivity ρ on the magnitude and the direction of the magnetic field is reported in [38]. This dependence is of the form

$$\rho = \frac{\beta H^2 \cos^2 \alpha}{\theta^2 H^2 + \lambda^2 H_0^2} \, C(\eta) + A. \qquad (28.8)$$

Here, β, λ, and A are smooth functions of the angle; $\lambda \sim 1$; $\eta = H/\theta H_0$; $C(\eta)$ is a smooth function of its argument, and $C(0) = C(\infty) = 1$; α is the angle between the direction of the current and the x axis. Figure 60 shows schematically the dependence of the resistivity on θ, which is plotted in accordance with Eq. (28.8).

In experimental investigations, the direction of the current is practically always determined by the geometry of the sample (its axis) and the magnetic field is rotated in a plane perpendicular to the current. If we know the direction of the current in relation to the principal axes of a crystal and the direction of the magnetic field along which the resistivity increases quadratically, we can find the plane on which the open direction lies. Therefore, to determine the actual open direction we need to carry out experiments on at least two samples with different directions of the current. In practice, the problem of finding the open directions is more complex because one needs to know not only these directions but also the nature of the Fermi surface.

Some comments should be made about the Hall coefficient. If the magnetic field is directed in an arbitrary manner and does not lie close to the yz plane, which is perpendicular to the open direction, the Hall coefficient is isotropic and is governed by the volume V' bounded by that part of the Fermi surface which is located in a unit cell of the reciprocal lattice (to make the case specific, we shall assume that the size of the corrugated cylinder increases with increasing energy):

$$R_\infty = \frac{1}{n'ec}, \quad n' = \frac{2V'}{(2\pi\hbar)^3}.$$

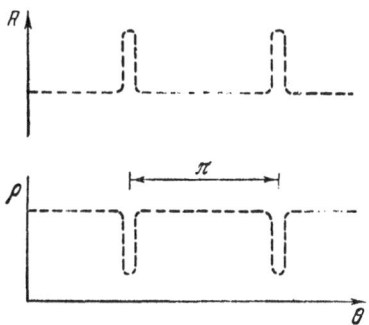

Fig. 61. Schematic representation of the dependences of the resistivity and the Hall coefficient on the direction of a magnetic field applied to a metal with a Fermi surface of the corrugated-cylinder type.

Fig. 62. Stereographic projection of the special directions for a Fermi surface of the corrugated-cylinder type.

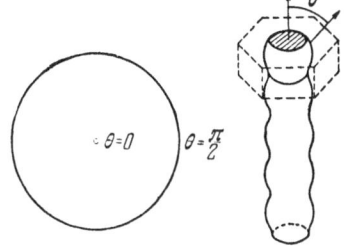

When the magnetic field is directed close to the yz plane, the value of the Hall coefficient may change very considerably. It depends on the collision integral (this is because $\overline{\partial p_x / \partial t} \neq 0$ and, therefore, $\overline{\hat{W}_p^{-1} \hat{W}_p p_y}$ does not drop out of the expression for σ_{xy}). Figure 61 shows the angular dependences of the resistivity and the Hall coefficient for metals whose Fermi surface is a corrugated cylinder.

The dependence of the resistivity on the magnetic field direction described in the preceding paragraphs is obtained also when the Fermi surface can be divided into a corrugated cylinder and an arbitrary number of closed regions, provided there is no volume compensation, i.e., provided the expansion of σ_{xy} for an arbitrary direction of the magnetic field does not begin with a term of the order of $1/H^2$. In this case, the resistivity increases quadratically irrespective of the direction of the magnetic field (just as in the case of a metal in which the number of electrons is equal to the number of holes).

Our analysis of the galvanomagnetic properties of metals whose Fermi surface is a corrugated cylinder has demonstrated that strong dependences of these properties on the magnetic field direction are observed in those cases when the field alters the topology of the electron trajectory. In the case considered, the special directions are those for which the magnetic field is perpendicular to the corrugated cylinder axis. In a stereographic projection,[2] this cor-

[2] See footnote 2 on p. 21.

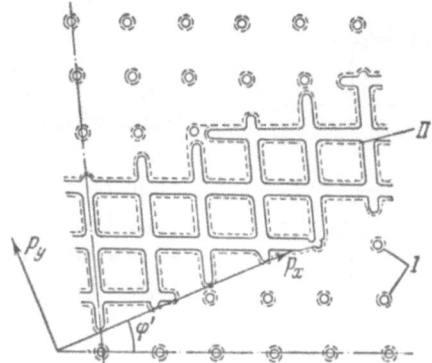

Fig. 63. Continuous curves show a section, by a plane p_z = const, of a constant-energy surface $\varepsilon = \varepsilon_0$ of the three-dimensional network type. The dashed curves represent a section of a neighboring constant-energy surface $\varepsilon = \varepsilon_0 + \delta\varepsilon$. Closed type I trajectories surround regions where the energy is $\varepsilon < \varepsilon_0$, whereas type II trajectories surround regions where the energy is $\varepsilon > \varepsilon_0$. These two types of trajectory are followed in opposite directions. Open trajectories divide these two types of curve. The direction of an open trajectory is given by an angle φ', which satisfies $\cos \varphi' = \cos \varphi \, (1 + \sin^2 \varphi \, \mathrm{tg}^2 \theta)^{-1/2}$.

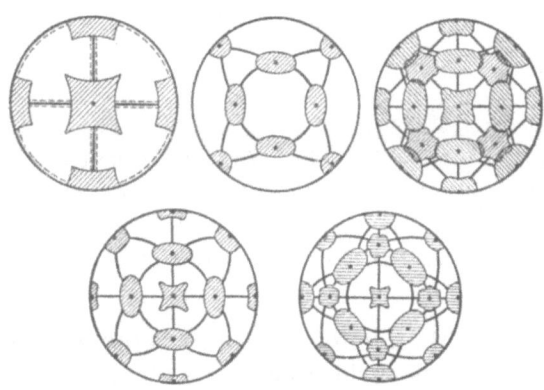

Fig. 64. Stereographic projection of the magnetic field directions (shaded regions and continuous curves) which give rise to open trajectories in various forms of Fermi surface of the three-dimensional network type.

responds to a circle of unit radius (Fig. 62, where the polar axis coincides with the open direction).

We shall now consider an open surface in the form of a three-dimensional network (Fig. 2). In this case, open trajectories are encountered very frequently although they are difficult to represent (Fig. 63). Figure 64 shows stereographic projections of the magnetic field directions (shaded regions and continuous curves) corresponding to open trajectories ($\varepsilon = \varepsilon_F$, r_z = const). The different parts of the figure correspond to three-dimensional networks oriented in various ways.[3]

---

[3] This is only a summary of the principal results. The derivation of the formulas given in the present section can be found in [38, 40].

An examination of the stereographic projections in Fig. 64 shows that there are different ranges of directions of the magnetic field characterized by: a) the absence of open trajectories (unshaded regions); b) the presence of layers of open trajectories with a single open direction (shaded regions); c) the presence of layers of open trajectories with different open directions (overlapping shaded regions). Additionally, there are at least four different types of special direction of the magnetic field.

1. There are directions of the magnetic field for which a layer of open trajectories forms a one-dimensional set. This corresponds, in particular, to the case of single directions of open trajectories. Examples of such cases are the directions perpendicular to the axis in the case of a corrugated cylindrical surface, and the thick continuous curves in the stereographic projections in the case of three-dimensional network surfaces (Fig. 64).

2. The boundaries of the two-dimensional regions (solid angles) which enclose open trajectories (they are the boundaries of the shaded regions in Fig. 64).

3. A single direction within a region of open trajectories in which a layer of such trajectories degenerates into isolated sections (denoted by points in Fig. 64). Usually, this direction coincides with a threefold, fourfold, or sixfold axis.

4. The boundary of a region of those directions of the magnetic field which correspond to layers of open trajectories with different open directions (these are the boundaries of the overlapping shaded regions in Fig. 64).

When the magnetic field does not coincide with any of the special directions, we observe the following dependences on this field: a) in the unshaded regions (the range of closed sections), the resistivity tends to a constant value (saturation) whereas the Hall coefficient is determined by the volume which is either occupied by electrons or is free of holes (§27); b) in the shaded regions (layers of open trajectories with a common open direction), the resistivity increases quadratically with the magnetic field (with the exception of the case when the current is perpendicular to the magnetic field and to the open direction) whereas the Hall coefficient depends in a complex manner on the collision integral (this point has been considered earlier); c) in the overlapping (doubly shaded) regions, corresponding to layers of open trajectories with different open directions, the resistivity tends to a constant value (saturation) whereas the Hall coefficient depends on the collision integral and is very small (it tends to zero as H → ∞).

Let us now consider the dependences of the resistivity on the magnitude and the direction of the magnetic field near the special directions.

1. In the case of directions represented by the thick curves in Fig. 64 the situation is fully analogous to that which applies to a Fermi surface of the corrugated-cylinder type near a plane perpendicular to the cylinder axis. In particular, Eq. (28.8) can be used. The angle θ is a measure of the "distance" to the thick lines in the stereographic projections.

2. An analysis of the behavior of the resistivity near the boundaries of the two-dimensional regions representing open trajectories must take account of the fact that an unshaded region contains only closed trajectories, whereas a shaded region contains closed and open trajectories and the number of the latter (the layer thickness) vanishes at the boundary of the region. In view of this, the standard dependence of the resistivity on the magnetic field in the form of saturation is observed outside the shaded regions right up to the boundaries of these regions, and when these boundaries are approached from the shaded side, the dependence of the resistivity on the magnitude and the direction of the magnetic field is [39]:

$$\rho = \theta\beta \left(\frac{H}{H_0}\right)^2 \cos^2\alpha + A \qquad (28.9)$$

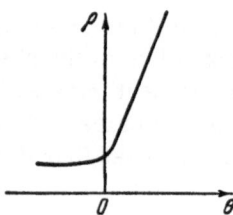

Fig. 65. Dependence of the resistivity on the magnetic field direction on transition to the region of open trajectories (this region is represented by $\theta > 0$).

[the notation is the same as in Eq. (28.8)]. Figure 65 shows the dependence of the resistivity on the inclination of the magnetic field near these special directions.

3. A similar situation occurs also to those single directions for which layers of open trajectories degenerate into isolated sections (these are represented by points in Fig. 64). In particular, Eq. (28.9) is still valid and the situation is as shown in Fig. 65.

When the magnetic field is rotated in a plane, we can sometimes obtain a resistivity rosette (the dependence of the resistivity on the direction of the magnetic field for a fixed value of this field), which has three types of singularity.

The dependences considered are characterized by the fact that, for a certain direction of the current ($\alpha = \pi/2$), the anomaly in the dependence of the resistivity on the magnetic field disappears.

4. At the boundaries of the overlapping (doubly shaded) regions, the field dependence of the resistivity changes from saturation (within these regions) to the quadratic dependence (outside these regions). An analytic expression for the resistivity near the boundary of such a region is

$$\rho = \frac{\rho'}{\theta A + (H_0/H)^2},$$
(28.10)

where ρ' and A are smooth functions of the angle, and $A \sim 1$.

This discussion shows that a great variety of situations may be encountered in investigations of open Fermi surfaces of the three-dimensional network type, and that a comparison of the asymptotic dependences of the resistivity on the magnitude of the magnetic field (obtained for different directions of the field) with the theoretical "patterns," i.e., with stereographic projections of the type shown in Fig. 64, allows us to determine not only the general nature of the Fermi surface but also quite a few details of its shape.

Similar analyses can be carried out also in the case of surfaces of other types. For example, the dependence of the resistivity on the magnitude and the direction of the magnetic field of metals whose Fermi surfaces are of the two-dimensional network type is reported in [41]. The results of this investigation are presented in Fig. 66.

In all cases, the characteristic properties of metals with open Fermi surfaces are, first, the strong anisotropy of the transverse (relative to the magnetic field) resistivity and, secondly, the strong dependence of the resistivity on the direction of the electric current (in the case when $\mathbf{j} \perp \mathbf{H}$).

We have discussed so far the cases in which the resistivity tends to a constant value in the unshaded regions, i.e., along those directions of the magnetic field for which there are no

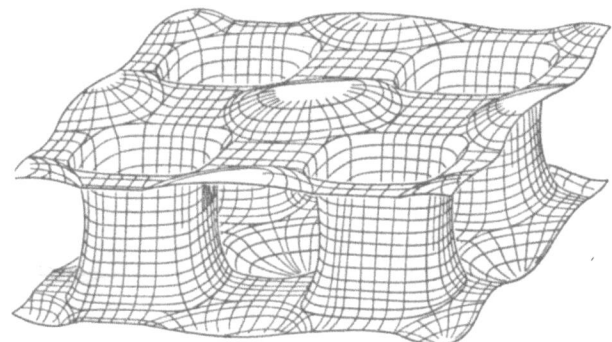

Fig. 66. Fermi surface of tin based on the
results reported in [41].

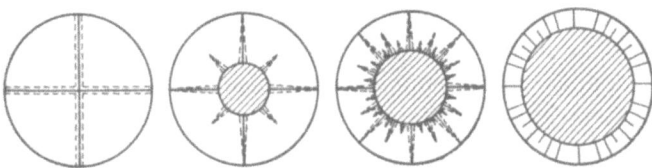

Fig. 67. Stereographic projections of the special
directions of the magnetic field for constant-energy
surfaces of the two-dimensional network type.

open sections. In such cases there is no compensation of the hole and the electron volumes.
If compensation does occur, we should observe a quadratic rise of the resistivity in the unshaded and the shaded regions. At the boundaries of these regions, the resistivity should exhibit a kink. Such kinks should be observed also for all the special directions. This case has
been considered in detail in connection with the galvanomagnetic properties of tin,[4] which are
described in [41]. Typical stereographic projections of a Fermi surface of this type are shown
in Fig. 67.

We must also mention that, as in the case of closed surfaces, the compensation of the hole
and the electron volumes is not accidental. It should occur whenever a unit cell of a crystal
contains an even number of electrons.

The strong dependence of the resistivity on the magnetic field direction, observed in
some cases, can be used to explain the linear rise of the resistivity of polycrystalline samples
when the magnetic field is increased (Kapitza's law), whereas the quadratic rise ($\rho \propto H^2$) is observed in a narrow range of angles $\Delta\theta \sim H_0/H$ [see Eq. (28.8)] and any averaging of the resistivity over the angles in the interval $\delta\theta \gg H_0/H$ (which includes $\Delta\theta$) results in a linear rise in
the resistivity with increasing field. It must be stressed that the averaging procedure is of
little importance because the maximum is narrow: in particular, the averaging of the conductivity and the subsequent transition to the resistivity gives the same result, i.e., a linear de-

---

[4] The Fermi surface of tin consists of an open surface of the two-dimensional network type
(Fig. 66) and a number of closed regions whose shape cannot be determined by the methods
described here. The shape of these closed regions can be determined by means of oscillatory, resonance, magnetoacoustic, and other methods (§§17 and 48).

pendence of the resistivity on the magnetic field [38]. It is quite likely that this is not the only cause of the linear rise of the resistivity. It is shown in [42] that the linear dependence is frequently observed in the transition region (where H ~H_0) from a quadratic dependence in weak fields (H $\ll$ H_0) to a saturation region, or to a second quadratic dependence in strong fields (H $\gg$ H_0). Therefore, in some cases, a linear dependence may mean that the measured values of the resistivity are not asymptotic.

Investigations of the galvanomagnetic properties in strong magnetic fields (particularly investigations of the resistivity) have become one of the principal methods for investigating the electron energy spectra [43]. About thirty metals have been studied in this way. It should be stressed that open Fermi surfaces are encountered much more frequently than closed surfaces. Although the dependence of the component of the tensor ρ_{ik} on the magnetic field direction is related not only to the shape of the Fermi surface but also to the nature of the scattering processes, all the singularities in this dependence are governed entirely by the geometry of the constant-energy surfaces. Therefore, angular dependences of the type shown in Fig. 68 can be used not only to reconstruct the general shape of the Fermi surface but also to calculate some of its detailed features such as constrictions, necks, etc. Naturally, one has to use some trial models. The Harrison model [44] is of great help in comparisons of the theory with the experimental data: according to this model, the principal features of the Fermi surface can be deduced from the almost-free electron approximation (§11).

Most of the formulas given in this section and in §27 are derived on the basis of the most general assumptions about the nature of the electron energy spectra. However, the range of the validity of these formulas should be determined more accurately. If the Fermi surface is very simple (which is the case for metals of the first group in the periodic table), the inequalities mentioned in our discussion ($r_H \ll l_p$) need no special explanation. However, in the case of complex Fermi surfaces, which are encountered for most other metals, some refinement is necessary. Let us consider a simple example. It is shown in §27 that if $n_1 \neq n_2$ (closed Fermi surfaces) the resistivity tends to saturation in strong magnetic fields. The question arises as to what happens if $|n_1 - n_2| \ll n_1 + n_2$. The next question is what are the values of the magnetic fields in which saturation does actually occur. The calculations reported in [45] show that it occurs only if H $\gg$ $H_0 n/\Delta n$ (n = $n_1 + n_2$, $\Delta n = |n_1 - n_2|$) where H_0 is the field in which $r_H = l_p$.

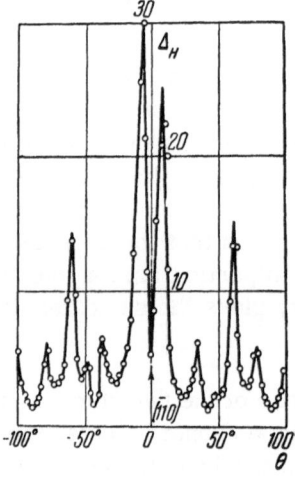

Fig. 68. Typical dependence of the rise of the resistivity of gold in a strong magnetic field on the direction of this field.

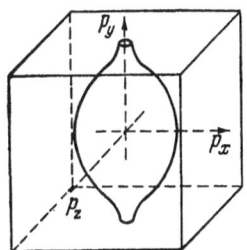

Fig. 69. Simplest example of an open surface with very narrow necks.

In weaker fields, a metal of this kind behaves like a substance with equal numbers of electrons and holes.[5]

We have already shown that metals with a Fermi surface of the corrugated-cylinder type may exhibit a quadratic dependence of the resistivity on the magnetic field ($r_H \ll l$) for certain directions of this field. However, if the corrugations are such that the necks are very narrow (Fig. 69), the quadratic rise of the resistivity appears in fields exceeding $H_0/\sqrt{\nu}$, where $\nu = \Delta p/p_F$ and Δp is the thickness of the neck. The role of a narrow layer of open trajectories will be discussed in detail later (see also [47]).

In comparing the experimental data on the galvanomagnetic effects with the theoretical conclusions, we should take into account an additional point not mentioned so far. When the magnetic field is increased sufficiently, magnetic breakthrough (§10) may alter the topology of the electron trajectory and this will naturally affect the dependence of the components of the resistivity tensor on the magnetic field. The influence of magnetic breakthrough will be considered later.

We shall now consider several special points in the theory of galvanomagnetic effects. We shall begin with a discussion of the role played by a narrow layer of open trajectories [47].

Such a layer appears in very strong magnetic fields and gives rise to a quadratic dependence of the corresponding component of the resistivity tensor on the magnetic field. This conclusion follows from the expression for σ_{yy} (the open direction along p_x), which itself follows from the diffusion estimates given in §26:

$$\sigma_{xx} = \left[\frac{\Delta p}{p_F} \frac{l_{\text{eff}}}{l} + \left(\frac{r_H}{l} \right)^2 \right] \sigma, \qquad (28.11)$$

where Δp is the thickness of a layer of open trajectories (it is assumed that $\Delta p/p_F \ll 1$; see Fig. 69).

The effective mean free path l_{eff} is introduced in order to make allowance for the following point: the probability of the escape of an electron from a narrow layer of open trajectories is considerably higher than the probability that an electron will enter such a layer. If the momentum of an electron changes only slightly in each collision (as in the case of collisions with long-wavelength phonons), a single collision will be sufficient to knock out an electron from a narrow layer of open trajectories. This naturally alters significantly the temperature dependence of l_{eff} (for example, in the case of the phonon mechanism of the resistivity and when $\Delta p \le T/s$, where s is the velocity of sound, the effective mean free path is $l_{\text{eff}} \ll l$ and $l_{\text{eff}} \propto T^{-3}$).

---

[5] This behavior is exhibited by Bi with few impurities [46]. Impurity atoms act as sources of electrons or holes and this disturbs the compensation of the electron and the hole volumes.

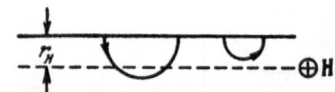

Fig. 70. Examples of trajectories of electrons which are scattered diffusely by the surface of a metal.

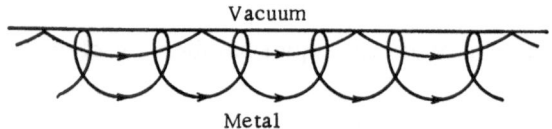

Vacuum

Metal

Fig. 71. Examples of trajectories of electrons which are reflected specularly by the metal–vacuum boundary in the presence of a magnetic field.

A layer of open trajectories may result from magnetic breakthrough. In this case, the thickness of such a layer depends strongly on the magnetic field in a certain range of the fields. This is manifested as a complex dependence of l_{eff} on the temperature and the magnetic field: when the magnetic field is varied, we may observe a transition from the "normal" dependence ($l \propto T^{-5}$) to the anomalous dependence ($l \propto T^{-3}$). All the mechanisms considered so far reduce that part of the effective mean free path[6] which depends on the scattering through small angles. This means that the role played by these mechanisms (in particular, the phonon mechanism) becomes greater and the transition to the residual resistivity case occurs earlier (at much lower temperatures) than in the absence of a magnetic field.

The complex dependence of the resistivity of Al on the magnetic field and the temperature [48] (in particular, the violation of Köhler's rule) suggests that a layer of open trajectories plays an important role in Al subjected to strong magnetic fields.

The diffusion approach helps one to understand the very interesting phenomenon [49, 50] known as the static skin effect, which is the concentration of a direct current near the surface of a metal subjected to a strong magnetic field. This effect can be explained as follows. It has been shown that if $l \gg r_H$, the transverse components of the conductivity tensor for a closed trajectory are very small compared with their value in H = 0 [$\sigma_{xx} \propto \sigma_{yy} \propto \sigma (r_H/l)^2$] because an electron in a plane perpendicular to the magnetic field is localized in an orbit whose dimensions are r_H and collisions occur at a frequency equal to $1/\tau_p = v_F/l$. If electrons are scattered diffusely by the boundary of the metal, momentum is lost in each collision with the boundary (Fig. 70). Therefore, the collision frequency W of electrons in a layer of thickness r_H – this frequency occurs in the second of the formulas in Eq. (26.12) – is equal to $\omega_H \gg 1/\tau_p$ (for $r_H \ll l$). Consequently, in the surface layer, we have

$$\sigma_{xx}^{surf} \approx \sigma \frac{r_H}{l} \gg \sigma \left(\frac{r_H}{l}\right)^2 = \sigma_{xx}. \qquad (28.12)$$

In the case of specular reflection from the boundary (Fig. 71), the electrons in the boundary layer follow infinite paths and, therefore, the electrical conductivity is of the order of the conductivity of a metal in zero magnetic field.

Clearly, a considerable rise in the conductivity of the boundary layer may result in the concentration of the lines of flow of the current near the surface. A rigorous theory, based on

---

[6] We have in mind here a formula of the type (24.16).

the solution of Boltzmann's equation [51], predicts a complex dependence of the current density on the distance from the surface (this point is discussed in the next section).

We shall conclude this section by considering the role of magnetic breakthrough in galvanomagnetic effects. This breakthrough appears most clearly when it causes a change in the topology of the electron trajectories, i.e., when closed trajectories change into open trajectories, or conversely.[7]

If we assume that a jump from one section of a trajectory to another is accompanied by a phase change in the wave functions, we can use the diffusion approximation to demonstrate easily [52] that the order of magnitude of the conductivity along an open direction is

$$\sigma_{xx} \approx \sigma \frac{r_H}{l} \left(\frac{r_H}{l} + p \right), \tag{28.13}$$

where p is the probability of breakthrough (§10). This estimate is in qualitative agreement with the results of the calculations reported in [53, 54]. However, the formula just given is valid only as long as $p \ll r_H/l$. If $p \gg r_H/l$, we must develop a theory of the galvanomagnetic effects by postulating the existence of "new" quasi particles which undergo infinite motion in a magnetic field. The electrical conductivity is then limited by the mean free path. If $r_H/l \ll p \ll 1$, we then find $\sigma_{xx} \approx p^2\sigma$ (σ is the electrical conductivity in the absence of a magnetic field). A comprehensive theory of the galvanomagnetic effects which would include magnetic breakthrough has yet to be developed.

§29. Theory of the Static Skin Effect

In the preceding section, we pointed out that the scattering of electrons by the boundary of a sample may alter significantly the values of the components of the conductivity tensor: in a bulk sample a given component may be proportional to $\sigma(r_H/l)^2$, whereas near a boundary it may be proportional to $\sigma(r_H/l)$ or even σ (in the case of specular reflection). We note also that discompensation of the electron and the hole volumes may occur in the boundary layer, which is typical of metals with an equal number of electrons and holes since the drift velocity in an inhomogeneous field depends naturally on the local characteristics of the carrier orbits.

If a metal is sufficiently pure and the temperature sufficiently low, the condition $r_H \ll l$ (strong fields) may be satisfied by relatively large (macroscopic) orbit radii ($r_H \lesssim 10^{-1}$ cm). Therefore, a redistribution of the lines of flow of the current near the boundary of a sample is a macroscopic effect which can be investigated experimentally [50]. If the sample is thin ($d \ll l$), the principal contribution to the total current is made by a surface layer whose thickness is of the order of r_H. This is the static skin effect which is manifested by the dependence of the resistivity on the magnetic field and on the thickness of the sample, and by the lack of a dependence of the total current on the thickness.

We shall now develop a consistent theory of the static skin effect.

Our prime target will be to calculate the resistivity of thin samples ($d \ll l$) in a magnetic field, taking account of the scattering by a boundary. We must draw attention to the fact that

---

[7] If we use Harrison's model (see §10 and also [44]), we find that electrons behave as if they were free provided that the magnetic field is sufficiently strong (but weaker than the atomic fields). This means that both transverse components of the conductivity tensor should be inversely proportional to the square of the magnetic field and the resistivity should tend to saturation.

the mathematical formulation of the problem is different from that in the case of the usual macroscopic effect.

When we consider the geometrical effects and their influence on the distribution of the current and the field in a weakly inhomogeneous case (when the geometrical dimensions or the distances of interest to us are large compared with r_H and l), we find that the relationship between the current density and the electric field $\mathbf{E} = -\nabla\varphi$ (because rot $\mathbf{E} = 0$) remains the same as in an infinite sample of a metal:

$$\mathbf{j} = \hat{\sigma}\mathbf{E} = -\hat{\sigma}\nabla\varphi, \tag{29.1}$$

where σ_{ik} is independent of the coordinates.

The basic equation which describes the distribution of the field in a sample is a second-order equation with respect to φ, which is the equation of continuity of charge in the static case:

$$\operatorname{div}\mathbf{j} = 0, \tag{29.2}$$

i.e.,

$$(\nabla, \hat{\sigma}\nabla\varphi) = 0, \tag{29.3}$$

The boundary conditions for this equation ensure that the equation of continuity of charge is satisfied at the surface. No charges cross the surface of the conductor, i.e.,

$$j_n = 0. \tag{29.4}$$

Here, $\mathbf{n}$ is the inward-directed normal to the surface of the conductor. In the presence of contacts $j_n = i_n$, where i is the current flowing into the sample through the contacts, i.e.,

$$(\hat{\sigma}\nabla\varphi)_n = \begin{cases} i_n & \text{at a contact} \\ 0 & \text{elsewhere on the surface.} \end{cases} \tag{29.5}$$

If φ is known, the charge density ρ' is given by[1]

$$\rho' = -\frac{1}{4\pi}\operatorname{div}\mathbf{E} = \frac{1}{4\pi}\nabla\varphi. \tag{29.6}$$

Since the density of the electron charge in a conductor is high and $\rho' \ll \rho_0$, we may assume that in the lowest approximation

$$\rho' = 0. \tag{29.7}$$

In a thin plate ($d \ll l$) and in the general microscopic theory, the relationship between $\mathbf{j}$ and φ should be determined from this theory. Since the field is inhomogeneous, $\hat{\sigma}$ is an operator which acts on the functions of the coordinates. If the microscopic theory is developed correctly, the conservation of charge is ensured automatically and, irrespective of the nature

---

[1] Here, $\rho' = \rho - \rho_0$, where ρ_0 is the equilibrium charge density. By definition, $\rho' = \frac{2e}{(2\pi\hbar)^3}\int f_1\,d\mathbf{p}$ and f_1 can be found from the transport equation which is made linear in respect of the electric field.

of $\varphi(\mathbf{r})$, the operator $\hat{\sigma}$ should be such that Eq. (29.2) is always satisfied as a consequence of the nature of $\hat{\sigma}$.

In the microscopic theory, the condition for the reflection of charges from the boundaries of a conductor should be such as automatically to ensure that the condition for the conservation of charge at the surface is satisfied, i.e., that Eq. (29.5) is always obeyed.

In the case of Eq. (29.7), the microscopic theory gives not only the relationship between $\mathbf{j}$ and φ but also the integral relationship between ρ' and φ. Consequently, Eq. (29.7) is an integral equation which can be used to determine φ and which, therefore, does not require any additional boundary conditions.

In view of the fact that Eq. (29.7) is the basic equation in the determination of ρ', we shall have to justify more thoroughly the replacement of Eq. (29.6) with Eq. (29.7). It is clear from the definition of ρ' that $\rho' = e^2\nu\hat{L}\varphi$, where $\nu = n/\varepsilon_F$ is a characteristic charge density, and $\hat{L}$ is a linear integral operator of the order of unity: $\hat{L}\varphi \approx \varphi$. If φ changes very significantly over distances of the order of λ, it follows from Eq. (29.6) that $\hat{L}\varphi \sim (\lambda_0/\lambda)^2\varphi$, where λ_0 is the Debye–Hückel radius for metals, which is equal to $(l\sqrt{4\pi\nu})^{-1}$. In good metals, for which the number of electrons is of the order of one per atom, the radius λ_0 is of the order of the atomic spacing (10^{-8} cm), whereas in semimetals such as bismuth, $\lambda \approx 10^{-7}$ cm ($\lambda_0 \propto n^{1/6}$, i.e., it depends very weakly on the carrier density).

This means that the retention of the right-hand side in Eq. (29.6) is an unnecessary refinement even in the region where the boundary condition for the electrons scattered by the surface is applicable. This is because Eq. (29.6) can be reduced to $\hat{L}\varphi = 0$, which is formally equivalent to Eq. (29.7). In alternating fields, the replacement of Eq. (29.6) with Eq. (29.7) is equivalent to the dropping of the displacement current. It should be stressed that ρ' differs from the extremely small quantity $\hat{L}\varphi$ by a dimensional factor and, therefore, it is necessary to use Eq. (29.6) in the determination of ρ' after the calculation of φ by means of Eq. (29.7). This represents the next approximation in the solution of Eq. (29.6).

It is clear from Eq. (29.6) that the Debye–Hückel radius λ_0 determines the depth at which external fields are damped out. This means that we can speak of the resistivity as an internal characteristic of a conductor, which is independent of external electric fields, with an accuracy which is determined by the ratio λ_0/λ. For example, if a current-carrying conductor is placed between the electrodes of a capacitor and the potentials of these electrodes are varied, we can alter the Joule heat evolution in the conductor with the accuracy just specified (this applies to a fixed value of the current and specified contacts).

Having determined the bulk value of ρ' as a function of φ and of the external currents on the surface, we find that mathematical considerations lead, following from Eq. (29.6), to a Poisson-type equation which has a single-valued solution only if we know the surface potential. Thus, a complete solution can be obtained only if we know the current density and the potential on the surface of a conductor, or if we go outside the conductor and consider the external conditions.[2]

We shall now consider the solution using the lowest approximation with respect to d/l, i.e., for $l = \infty$, in the case of a plate defined by $0 \leq \xi \leq d$ (Fig. 72). In this case, it will be convenient to use a modified form of the nonequilibrium distribution function.

The application of an external field increases the original energy of a quasi particle $\varepsilon(\mathbf{p})$ by $\Delta\varepsilon(\mathbf{p})$. The number of quasi particles whose energies are now ε is equal to the number of

---

[2] In the absence of external fields (apart from an applied emf), the value of φ can be determined in the $\mathbf{D} = \mathbf{E}$ case provided φ vanishes at infinity. In this case, $\lim_{\lambda_0 \to 0} \lambda_0 \partial\varphi/\partial n$ is finite.

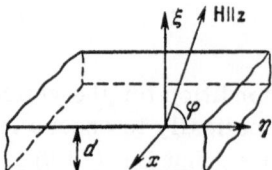

Fig. 72.   Plate in an
inclined magnetic field.

quasi particles whose initial energy was $\varepsilon - \Delta\varepsilon$, i.e.,

$$f(\boldsymbol{p}) = n_F(\varepsilon(\boldsymbol{p}) - \Delta\varepsilon(\boldsymbol{p})),$$

or, in the linear approximation with respect to the field,

$$f(\boldsymbol{p}) \approx n_F(\varepsilon) - \frac{\partial n_F}{\partial \varepsilon} \Delta\varepsilon.$$

In the equilibrium state (i.e., in the absence of an external electric field), there is no current or uncompensated charge in a normal metal and, therefore,

$$\left. \begin{aligned} \boldsymbol{j} &= -\frac{2e}{(2\pi\hbar)^3} \int \boldsymbol{v}\, \frac{\partial n_F}{\partial \varepsilon} \Delta\varepsilon(\boldsymbol{p})\, d\boldsymbol{p}, \\ \rho' &= -\frac{2e}{(2\pi\hbar)^3} \int \frac{\partial n_F}{\partial \varepsilon} \Delta\varepsilon(\boldsymbol{p})\, d\boldsymbol{p}. \end{aligned} \right\} \tag{29.8}$$

Since $-\partial n_F/\partial \varepsilon = \delta(\varepsilon - \varepsilon_F)$, we can use Eq. (15.5) for the density of states to obtain

$$\left. \begin{aligned} \boldsymbol{j} &= \frac{2e}{(2\pi\hbar)^3} \left| \frac{eH}{c} \right| \sum \int\limits_{\varepsilon(\boldsymbol{p})=\varepsilon_F} \boldsymbol{v}\, \Delta\varepsilon\, dp_z\, dt, \\ \rho' &= \frac{2e}{(2\pi\hbar)^3} \left| \frac{eH}{c} \right| \sum \int\limits_{\varepsilon(\boldsymbol{p})=\varepsilon_F} \Delta\varepsilon\, dp_z\, dt. \end{aligned} \right\} \tag{29.9}$$

The summation is carried out over different bands if they overlap. It follows from $\dot{\varepsilon} = e\mathbf{E}\boldsymbol{v}$ (§4) that

$$\Delta\varepsilon = e\boldsymbol{E}\, \Delta\boldsymbol{r}. \tag{29.10}$$

The first of the two expressions in Eq. (29.9), modified by the substitution of Eq. (29.10), yields a formula which is suitable for making estimates:

$$\sigma_{ik} = \sum \overline{\langle v_i\, \Delta r_k \rangle}. \tag{29.11}$$

Here, $\langle \ldots \rangle$ denotes the averaging whose meaning follows from Eq. (29.9) and the bar represents averaging over collisions, i.e., the replacement of the path traveled by the mean free path between collisions. In order to find this path, we must determine the nature of the scattering of electrons by the surface. We shall assume that the scattering is diffuse, i.e., that after its collision with a surface, a particle "forgets" its previous history, and that a Fermi distribution is established. [If the magnetic field is not too weak (§7) so that the de Broglie wavelength of electrons is short compared with the characteristic dimensions of surface irregularities, this assumption is fully justified.] However, we must bear in mind that the corresponding distribution function is not, in general, of the equilibrium type because the chemical potential is differ-

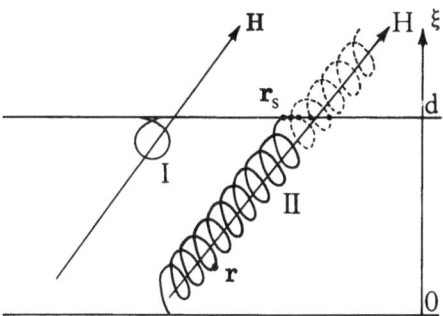

Fig. 73. Trajectories of electrons in a plate subjected to a magnetic field.

ent. This potential is determined by the density of reflected particles which is given by the law of conservation of charge on the surface. This means that the fluxes of the incident and the reflected particles (but not the densities of these particles) are equal.

In the case of diffuse reflection, the energy $\delta\varepsilon$ (additional to $\varepsilon = \varepsilon_F$) which an electron carries away from the point $\mathbf{r}_s$ on the surface is isotropic over the angles and, since $\varepsilon = \varepsilon_F$, it can depend only on $\mathbf{r}_s$. In the absence of collisions ($l = \infty$), the energy acquired between $\mathbf{r}_s$ and $\mathbf{r}$ (Fig. 73) is, according to Eq. (29.10), $-e\{\varphi(\mathbf{r}) - \varphi(\mathbf{r}_s)\}$, so that

$$\Delta\varepsilon = \delta\varepsilon\,(r_s) + e\,\{\varphi\,(r_s) - \varphi\,(r)\} = h\,(r_s) - \varphi\,(r), \tag{29.12}$$

where $h(\mathbf{r}_s)$ is an unknown function which should be determined from the condition of conservation of charge on the surface, i.e., from the boundary condition which ensures that Eq. (29.4) is satisfied.

If, at a given moment of the "orbital" time, an electron is at a point $\mathbf{r}$ and has a momentum p_z (t and p_z together with $\varepsilon = \varepsilon_F$ determine completely the state of an electron, i.e., its crystal momentum $\mathbf{p}$, in the zeroth approximation with respect to $\mathbf{E}$), we find that $\mathbf{r}_s = \mathbf{r}_s(\mathbf{r}, t, p_z)$ which can be calculated by equating the distance to the surface and the path traveled by the electron:

$$r - r_s = r\,(t) - r\,(\lambda) = \int_{\lambda}^{t} v\,(t')\,dt', \tag{29.13}$$

where λ is the moment of collision with the surface. Since a formal continuation of the orbit beyond the surface may intersect the surface at several points (Fig. 73), we find that λ is the root of Eq. (29.13) that is closest to but smaller than t; clearly a collision takes place with that surface for which $v_n > 0$ ($\mathbf{n}$ is the inward-directed normal to the surface).[3] In the case of a plate, Eq. (29.13) reduces to

$$\xi\,(\lambda) = \xi\,(t) - \xi + \xi_s \qquad (\lambda < t,\; \lambda = \lambda_{min}),$$
$$\xi_s = \begin{cases} 0 & \text{for } v_\xi\,(\lambda) > 0, \\ d & \text{for } v_\xi\,(\lambda) < 0. \end{cases} \tag{29.14}$$

---

[3] The three equations denoted by (29.13), combined with the equation of the surface of a conductor $G(\mathbf{r}_s) = 0$, determine, in general, the four quantities $\mathbf{r}_s$ and λ.

The function $h(\mathbf{r}_s)$ yields a single value of the current density and of the potential φ. The substitution of Eqs. (29.12) and (29.13) into Eq. (29.9) gives

$$j = \sum \langle v h(\mathbf{r}_s) \rangle, \tag{29.15}$$

and Eq. (29.7) yields

$$\varphi(r) = \frac{\sum \langle h(\mathbf{r}_s) \rangle}{\sum \langle 1 \rangle}. \tag{29.16}$$

Since rot $\mathbf{E} = 0$, it follows that E_x and E_η (the axes are shown clearly in Fig. 72) are independent of ξ, so that

$$\varphi(r) = -E_x x - E_\eta \eta - \varphi_1(\xi).$$

Therefore, using Eq. (29.16), we can conveniently find

$$h(\mathbf{r}_s) = -E_x x - E_\eta \eta + h(\xi_s). \tag{29.17}$$

We shall now determine $h(\xi_s)$. It follows from Eq. (29.4) that

$$\left. \begin{array}{l} j_n^s = \sum \langle v_n h \rangle^s = \sum \langle v_n h \rangle_+^s + \sum \langle v_n h \rangle_-^s = 0, \\[2mm] \langle f \rangle = \dfrac{2e}{(2\pi\hbar)^3} \left| \dfrac{eH}{c} \right| \int f \, dp_z \, dt, \end{array} \right\} \tag{29.18}$$

where the superscript s indicates that we are considering a point on a particular surface and the plus and minus subscripts refer, respectively, to electrons reflected from this surface (type I electrons in Fig. 73) and to electrons reflected from the opposite surface (type II electrons). Substituting Eq. (29.17) into Eq. (19.18), we find that

$$\begin{aligned} h(0) - h(d) = \frac{1}{\sum \langle v_\xi \rangle} \Big\{ & E_x \sum \langle v_\xi(t)(x(t) - x(\lambda_0)) \rangle + \\ & + E_\eta \sum \langle v_\xi(t)(\eta(t) - \eta(\lambda_0)) \rangle \Big\}, \end{aligned} \tag{29.19}$$

where the electrons colliding with the surface $\xi = 0$ are described by

$$\xi(\lambda_0) = \xi(t), \qquad \lambda < t, \tag{29.14a}$$

and the electrons colliding with the surface $\xi = d$ are described by

$$\xi(\lambda_0) = \xi(t) + d, \qquad \lambda < t. \tag{29.14b}$$

It is evident from Eq. (29.14) that, in the monotonic range of $\lambda(t)$, we have

$$v_\xi(t) \, dt = v_\xi(\lambda) \, d\lambda, \tag{29.20}$$

which allows us to simplify considerably Eq. (29.19). We shall avoid cumbersome calculations by considering the simplest case when not only the Fermi surface is closed but also $\lambda = \lambda(\xi, t, p_z)$ is a monotonic function of t for any value of p_z, i.e., $v_\xi \neq 0$ for any value of t and p_z. This is an exceptional situation realized, for example, when a plate is subjected to a magnetic field normal to its surface and the carriers obey a quadratic dispersion law. However, an analysis of the general case does not yield anything new but the calculations are much more complex.

In our very simple case when $\bar{v}_\xi > 0$, we obviously have $\lambda_0 = t$ and when $\bar{v}_\xi < 0$, we have

$$J = \int_0^T v_\xi(t)\, g(\lambda_0)\, dt = \int_{\lambda_0(0)}^{\lambda_0(T)} v_\xi(\lambda_0)\, g(\lambda_0)\, d\lambda_0 = \int_{\lambda_0(0)}^{\lambda_0(0)+T} v_\xi(\lambda_0)\, g(\lambda_0)\, d\lambda_0, \qquad (29.21)$$

and if $g(\lambda_0)$ is a function periodic in λ_0, for example

$$x(t) = x_0 - c p_y(t)/eH \quad \text{and} \quad y(t) = y_0 + c p_x(t)/eH),$$

we find that

$$J = \int_0^T v_\xi(t)\, g(t)\, dt.$$

The only function in (29.19) which is aperiodic in λ_0 is $\eta(t)$, which is (Fig. 72)

$$\eta(t) = \frac{y(t)}{\sin\varphi} + \xi(t)\cot\varphi.$$

Since $y(t)$ is periodic, it remains for us to calculate

$$J = \int_{\lambda_0(0)}^{\lambda_0(0)+T} v_\xi(t)\, \xi(t)\, dt = \xi^2(t)\Big|_{\lambda_0(0)}^{\lambda_0(0)+T} = 2\bar{v}_\xi\left(\xi(0) + d + \frac{1}{2}\bar{v}_\xi\right).$$

Bearing in mind that

$$\int_0^T v_\xi(t)\, \xi(t)\, dt = 2\bar{v}_\xi\left(\xi(0) + \frac{1}{2}\bar{v}_\xi\right),$$

we finally obtain Eq. (29.19) in the form (where the minus sign now corresponds to $\bar{v}_\xi = \bar{v}_z \sin\varphi < 0$, for example, $p_z < 0$)

$$h(0) - h(d) = dE_\eta \cot\varphi. \qquad (29.22)$$

This is sufficient for the determination of the potential difference from Eq. (29.16) and of the current density from Eq. (29.15) because these quantities are not affected by a change in h; the quantity $h(\mathbf{r})$, which is needed only when $\xi = 0$ or $\xi = d$, can be written in the form

$$h = -E_x x - E_\eta \eta + E_\eta \xi \cot\varphi = -E_x x - E_\eta y/\sin\varphi. \qquad (29.23)$$

Equations (29.15), (29.16), and (29.23) give an exact solution of the problem for any magnitude and direction of the magnetic field. The only exception is the case when the field is exactly parallel to the surface ($\varphi = 0$) because then the solution obtained loses its physical meaning. This is to be expected because there are some electrons whose orbits fit within the plate and which move along the surface without colliding with it (these electrons would give rise to an infinite conductivity).

Substituting Eq. (29.23) in Eq. (29.15) and bearing in mind that

$$\xi(\lambda) = \begin{cases} \xi(t) - \xi & (\bar{v}_z > 0, \ \lambda = \lambda^+), \\ \xi(t) - \xi + d & (\bar{v}_z < 0, \ \lambda = \lambda^-), \end{cases} \qquad (29.24)$$

and that

$$r_s = r + r(\lambda) - r(t),$$

we obtain:

$$j = \sum \langle v(t)(K(t) - K(\lambda)) \rangle, \qquad (29.25)$$

where

$$K(t) = -E_x x(t) - E_\eta y(t)/\sin\varphi;$$

K(t) is a periodic function of t (period T). As expected, it follows automatically from Eqs. (29.25) and (29.22) that $j_\xi(\xi) = 0$.

Equation (29.25) yields

$$j_\alpha(\xi) = \sigma_{\alpha\beta}(\xi) E_\beta, \qquad (29.26)$$

where $\sigma_{\alpha\beta}(\xi) \sim r_H$. Here, α and β take the values of x and η.

The current density can be calculated from Eq. (29.25) using the fact that K(λ) is a periodic function of λ (period T) and, in accordance with Eq. (29.24), λ changes by T when ξ changes by $\bar{v}_\xi$. We shall expand $K(\lambda^\pm)$ as a Fourier series in ξ:

$$K(\lambda^\pm) = \sum_{n=-\infty}^{\infty} x_n^\pm \exp(2\pi in\xi/\bar{v}_\xi). \qquad (29.27)$$

We thus obtain

$$j_\alpha = \sum_{m=-\infty}^{\infty} j_m^\alpha,$$

where

$$j_0^\alpha = [\overline{v_\alpha K} - \bar{v}_{_\cdot}\overline{v_{_\cdot}K}/\bar{v}_\xi], \qquad (29.28)$$

$$j_m^\alpha = \frac{1}{2\pi im}\left[T \exp\left\{\frac{2\pi im}{|\bar{v}_\xi|}\left[\frac{d}{2} + \text{sign}\,\bar{v}_\xi\left(\frac{d}{2}+\xi\right)\right]\right\}\overline{v_\alpha(t)\exp\left(-\frac{2\pi im\xi(t)}{\bar{v}_\xi}\right)}\,\overline{K(t')\exp\left(\frac{2\pi im\xi(t')}{\bar{v}_\xi}\right)}\right]. \quad (29.29)$$

We shall now consider in detail the case of strong magnetic fields $r_H \ll d$.

Noting that $\overline{v_x x} = \overline{v_y y} = 0$ and $\sum\langle v_y y\rangle = -\sum\langle v_y x\rangle \sim n_1 - n_2$, we find that, if $d \gg r_H$, then

$$j_0^{x,\,\eta} = \mp\frac{(n_1 - n_2)ec}{H\sin\varphi}E_{\eta,\,x}. \qquad (29.30)$$

Let us consider the physical meaning of this result. If the number of holes in a metal is equal to the number of electrons, it follows that $j_0^x = 0$ and the current at any depth is determined by terms of the type given by Eq. (29.29). However, because of the integration with respect to p_z, these terms oscillate and they rapidly decrease away from the surface of the plate (because $d \gg r_H$). Moreover, the principal contribution to the total current is made only by the surface layer whose thickness is of the order of r_H (this can be demonstrated quite simply by calculating the total current in the region far from the surface of the plate). In the lowest approximation with respect to r_H (i.e., when $d \to \infty$), the current density on the surface is zero [this can be demonstrated easily by means of Eq. (29.25)] but at a depth of the order of r_H it increases to a value of the order $necE/H$, whereas at greater depths $\xi \gg r_H$ the current density performs damped oscillations proportional to $(r_0/\xi) \cos(\xi/r_0)$, $r_0 = (\bar{v}_\xi)_{max}$ (Fig. 74). Because of this, the total current in the case $n_1 = n_2$ is independent of the plate thickness and is proportional to r_H^2. The current in deep layers of the metal exhibits only weak Sondheimer oscillations [55], which are considered in detail in [56]; the exact formula for these oscillations can be derived from Eqs. (29.28) and (29.29). [If $d \gg r_H$, no problems are encountered in an analysis of such oscillations: the principal contribution to Eq. (29.29) is made by a region near the limits of integration with respect to p_z, i.e., near the limiting points p_0.]

The damping of the current in the interior, shown in Fig. 74, is the static skin effect. It is important to note that this effect is observed only when $n_1 = n_2$ and that it does not alter the dependence of the resistivity on the magnetic field. We shall now determine the resistance of the plate. To do this, we must bear in mind that although the plate extends far along a direction μ (even compared with d) it is still bounded and the total current can flow only along the direction $\zeta \perp \mu$, ξ and $J_\mu = \int j_\mu \, d\mu = 0$ (this is the equation of continuity in its integral form, which ensures that there is no flow of charge across the plate along the direction μ). Since

$$J_{x,\eta} = \mp \frac{(n_1 - n_2)\,ec}{H \sin \varphi}\, dE_{\eta,x} + \frac{ne^2}{mv}\, r_H^2 a_\beta^{x,\eta}\, E_\beta, \quad a^{x,\eta} \approx 1,$$

it follows that $R = E_\zeta/J_\zeta$ is of the form

$$R(H, d) \approx \begin{cases} R(0, d) \propto d^{-2} & (n_1 \neq n_2), \\ R(0, d)\left(\dfrac{d}{r_H}\right)^2 \propto r_H^{-2} & (n_1 = n_2), \end{cases}$$

and the total resistance is independent of the thickness if $n_1 = n_2$.

We have considered so far the one-dimensional problem (a film or a wire with the lines of flow of the current parallel to the boundaries). However, it is evident that the method used is applicable to a sample of arbitrary shape with arbitrary positions and dimensions of the contacts supplying the current [57]. This problem is of interest because it is associated with the

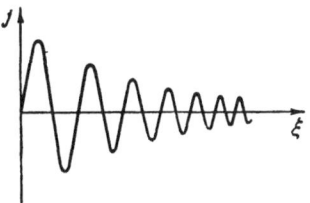

Fig. 74. Dependence of the current density on the depth in a plate.

focusing of the current in strong magnetic fields. It is found that the application of a magnetic field alters considerably the resistance of a sample, particularly when the dimensions of the contacts are small compared with r_H; in this case, the resistance is inversely proportional to the contact area [57].

The scattering of electrons by the surface is usually assumed to be diffuse. However, in semimetals, the scattering may be almost specular (because of the large de Broglie wavelength of electrons). This gives rise to a linear dependence of the resistivity on the magnetic field [58], which is known as Kapitza's law. In good metals, the reflection is almost specular for those electrons which meet the surface at low angles (§7). This makes it possible to determine the dependence of the reflection coefficient of electrons on their angle of incidence from the dependence of the resistivity on the magnetic field.

In this section, we have discussed only the closed Fermi surfaces. The open surfaces also give rise to the static skin effect; the theory of this case is developed in [59]. Experimental investigations of the static skin effect have been reported for cadmium [50], tungsten [60], indium and aluminum [61], and bismuth [61, 62].

§30. Thermal Conductivity and Thermoelectric Effects in Strong Magnetic Fields

The magnetic field not only alters the electrical resistivity of a metal but also its thermal conductivity, the Thomson and the Peltier coefficients, as well as other thermoelectric effects [63]. This is quite natural because all these phenomena are governed by the motion of electrons which changes appreciably when a magnetic field is applied.

The dependences of the thermal conductivity and the thermoelectric effects on the magnetic field are frequently called the thermomagnetic effects. Like the galvanomagnetic effects, they can be divided into transverse and longitudinal, odd and even. Some of them are analogous to the Hall effect. It follows from the symmetry considerations that the dependences of the thermomagnetic coefficients on a weak magnetic field can be represented by expansions in powers of the magnetic field. The number of independent components of the tensors (coefficients of proportionality) obtained in this way is determined by the symmetry class of a crystal. Traditionally, the thermomagnetic investigations have been used to determine the carrier scattering mechanisms (this is true of semiconductors rather than of metals). The thermomagnetic effects have not yet been used to determine the parameters of the electron energy spectra of metals.

Theoretical investigations [64, 65] and an analysis which will be given later all show that studies of the thermomagnetic properties in strong fields should make it possible to determine certain characteristics of electrons which would be very useful in the determination of their energy spectra.

We shall not repeat all the postulates on which the quasiclassical approach is based but simply use the methods developed in the investigation of the galvanomagnetic effects. We shall apply these methods to the determination of the dependences of the thermomagnetic coefficients (see beginning of §27) on the magnitude and the direction of a strong magnetic field ($r_H \ll l$).

All the thermomagnetic effects can be described by the magnetic-field dependences of the coefficients α_{ik} and $\varkappa_{ik}$ in Eq. (25.1). These coefficients are related by the formulas of Eq. (25.15) to σ_{ik}, b_{ik}, d_{ik}, c_{ik}, and, in accordance with the principle of symmetry of the transport coefficients, we have

$$\sigma_{ik}(\boldsymbol{H}) = \sigma_{ki}(-\boldsymbol{H}), \quad d_{ik}(\boldsymbol{H}) = d_{ki}(-\boldsymbol{H}), \quad -Tb_{ik}(\boldsymbol{H}) = c_{ki}(-\boldsymbol{H}). \tag{30.1}$$

The tensors d_{ik} and b_{ik} are given by the formulas in Eq. (25.12), into which we must substitute the solution of the transport equation

$$\frac{\partial \varphi_i}{\partial t} + \hat{W}_\varepsilon \{\varphi_i\} = v_i. \tag{30.2}$$

Equation (30.2) is derived in the same way as Eq. (27.7) and it differs from the latter only in respect of the collision operator. In deriving this equation the distribution function of phonons is not specified, i.e., the drag of phonons by electrons is ignored (we shall consider this point later).

Since the explicit form of the collision operator has no influence on the asymptotic (high-field) behavior of the solution, we can use the results of the analysis given in §§27 and 28. The dependences of the components of the thermal conductivity $\varkappa_{ik}$ and the thermoelectric α_{ik} tensors on a strong magnetic field can also be determined using Eq. (25.21), bearing in mind that $\varkappa_{ik} \approx -d_{ik}$ [see Eq. (25.23)].

With the exception of some special cases (these will be discussed later), the dependences of the components of the tensors b_{ik}, c_{ik}, and d_{ik} are of the same nature as the magnetic-field dependences of the components of the electrical conductivity tensor. This allows us to use directly the results obtained in §§27 and 28. One of these special cases is the situation when the number of electrons equals that of holes. The special nature of this situation is associated with the fact that, because of the compensation ($n_1 = n_2$), the expansion of the component σ_{xy} (the z axis is, as usual, directed along the magnetic field) begins with the quadratic term ($\sigma_{xy} \propto 1/H^2$). This leads to a quadratic rise of the transverse (relative to the magnetic field) resistivity. However, the equality of the numbers of electrons and holes does not mean that $\left(\frac{dV_1}{d\varepsilon}\right)_{\varepsilon=\varepsilon_F}$ and $\left(\frac{dV_2}{d\varepsilon}\right)_{\varepsilon=\varepsilon_F}$, are equal, where $V_{1(2)}(\varepsilon)$ is the volume inside the electron part or outside the hole part of the surface $\varepsilon(\mathbf{p}) = \varepsilon$. Consequently, even if $n_1 = n_2$ [$n_{1(2)} = 2V_{1(2)}(\varepsilon_F)/(2\pi\hbar)^3$], the expansions of b_{xy} and c_{xy} begin with a term proportional to $1/H$ [see Eq. (27.21)].

We shall now list some of the main results (details are given in [64, 65]). If the Fermi surface is closed and $n_1 \neq n_2$, we have

$$\varkappa_{yx} = \frac{\pi^2}{3} T \frac{c}{eH} (n_1 - n_2). \tag{30.3}$$

This means that the Righi–Leduc effect (like the Hall effect) can be used to determine the difference between the numbers of electrons and holes $n_1 - n_2$. Under these conditions, the asymptotic form of the matrix α_{ik} is:

$$(a_{ik}) = \begin{pmatrix} \nu_{xx} & (1/H)\nu_{xy} & \nu_{xz} \\ (1/H)\nu_{yx} & \nu_{yy} & \nu_{yz} \\ (1/H)\nu_{zx} & (1/H)\nu_{zy} & \nu_{zz} \end{pmatrix}. \tag{30.4}$$

The expansions of all the matrix components ν_{ik} begin with terms which are of zeroth power in the reciprocal of the magnetic field, and all the components ν_{ik} (with the exception of ν_{xx} and ν_{yy}) depend on the collision integral, whereas the two exceptions are given by

$$\nu_{xx} \approx \nu_{yy} \approx \frac{\pi^2}{3} \frac{T}{e} \left[\frac{d}{d\varepsilon} \ln (n_1 - n_2) \right]_{\varepsilon=\varepsilon_F}. \tag{30.5}$$

Consequently, an investigation of the Nernst–Ettingshausen effect in a strong magnetic field can be used to determine $(d/d\varepsilon) \ln(n_1 - n_2)$, which is a characteristic of the electron energy spectrum that cannot be determined from the galvanomagnetic or any other effects. If the dispersion law is quadratic and isotropic ($\varepsilon = p_2/2m$), and if $n_2 = 0$, we have

$$\frac{d}{d\varepsilon} \ln(n_1 - n_2)\Big|_{\varepsilon = \varepsilon_F} = \frac{3}{2\varepsilon_F}. \tag{30.6}$$

If the number of electrons n_1 is equal to the number of holes n_2 and the Fermi surface is closed, all the components of the thermal conductivity tensor $\varkappa_{ik}$ depend on the nature of the scattering processes, and the asymptotic dependence on the magnetic field is naturally identical with the asymptotic dependence of the components of the electrical conductivity tensor. In particular, the transverse components $\varkappa_{xx}$, $\varkappa_{yy}$ decrease rapidly with increasing magnetic field ($\varkappa_{xx}$, $\varkappa_{yy} \propto 1/H^2$). This makes it possible to determine separately the phonon (lattice) contribution to the transport of heat in metals. The asymptotic form of the tensor α_{ik} is

$$(\alpha_{ik}) = \begin{pmatrix} H\nu_{xx} & H\nu_{xy} & H\nu_{xz} \\ H\nu_{yx} & H\nu_{yy} & H\nu_{yz} \\ \nu_{zx} & \nu_{zy} & \nu_{zz} \end{pmatrix}. \tag{30.7}$$

In this case, all the components of the matrix ν_{ik} depend on the angles of inclination of the vector $\mathbf{H}$ with respect to the crystallographic axes, and the actual form of these components is determined by the collision integral. We must point out that metals with $n_1 = n_2$ should differ from the others by a relatively high thermoelectric power in strong magnetic fields.

An analysis of various types of open Fermi surface shows that, as in the case of the galvanomagnetic effects, the most typical characteristic which distinguishes such surfaces from closed ones is the strong anisotropy of the thermomagnetic effects: the asymptotic forms of the tensors $\varkappa_{ik}$ and α_{ik} change greatly on approach to some special directions and the asymptotic form of the vector $\varkappa_{ik}$ becomes completely analogous to the asymptotic form of the tensor σ_{ik} (§28).

The drag of phonons by electrons in strong magnetic fields may alter the asymptotic forms of the thermomagnetic coefficients expressed in terms of the reciprocal of the magnetic field [66]. This is because the role played by the drag reduces (as shown in [66]) to the appearance of an additional "force" on the right-hand side of Eq. (30.2) and the average values of all the components of this force do not vanish (we recall that in some cases $\bar{v}_x = \bar{v}_y = 0$). Because of this, the expansions of all the components of the function φ_i begin with terms which are independent of the magnetic field. We shall not give the final results obtained in [66] but we shall point out that the drag effect is most important at moderately low temperatures [when $T \gg T_0$, where $T_0 = \Theta(\Theta/\varepsilon_F)^{1/2}$], which is to be expected because the number of phonons is proportional to T^3, i.e., it decreases when the temperature is lowered.

§31. Quantum Oscillations of the Resistivity of Metals (Shubnikov–de Haas Effect)

The quantization of the energy levels of conduction electrons results in oscillatory dependences not only of the thermodynamic quantities (§§15–18) but also of the transport effects. The most widely investigated example of oscillations of a transport coefficient is the Shubnikov–de Haas effect, which is a nonmonotonic oscillatory dependence of the electrical resistivity on the magnetic field.

In addition to the oscillatory dependence, which is a consequence of oscillations of the density of states (in complete analogy with the de Haas–van Alphen effect), the resistivity may exhibit also special oscillations which are due to the periodic dependences of the cross sections of various interactions of electrons with other quasi particles. These special oscillations, investigated in detail for semimetals [67], will not be considered, i.e., we shall deal only with the "classical" Shubnikov–de Haas effect. We shall show that the quantization cannot be ignored completely in an analysis of the collision term. Although we shall still confine ourselves to the quasiclassical approximation (which describes well the electronic properties of good metals), the classical form of Boltzmann's equation (23.6) will no longer be valid. The natural approach in the calculation of the oscillatory terms of the transport coefficients is the use of the quantum transport equation for the density matrix [68].

However, in static fields, the quantization affects strongly the nature of the collision integral, even in the quasiclassical case, and it is necessary to find the specific form of this integral for each separate case (this should be compared with the classical theory presented in the preceding sections and with the quantum theory of the effects in high-frequency fields, which is given in §§39, 42, and 43). The graph technique [69] provides the most convenient approach in the quasiclassical, quantum, and extreme quantum cases. This approach is outside the scope of our treatment and we shall restrict ourselves to a brief consideration of the results applicable to the simplest case, especially as the number of possible variants increases rapidly in the quantum and the extreme quantum limits.

The conditions necessary for the observation of the de Haas–van Alphen effect were considered in detail in the earlier sections. The Shubnikov–de Haas effect is observed under the following conditions: a single crystal must be sufficiently pure, whereas the temperature and the magnetic field should obey the following moderately stringent conditions:

$$\omega_H \tau \gg 1, \qquad \hbar \omega_H \gtrsim T.$$

We shall first consider (quite formally and solely for the sake of simplicity) the τ approximation. In this case, the quantum transport equation for the density matrix

$$\hat{f} = \hat{f}_0 + \hat{f}_1 \tag{31.1}$$

$[\hat{f} = f_0(\hat{\mathcal{E}}_0)$ is the equilibrium density matrix and $\hat{\mathcal{E}}_0$ is the Hamiltonian in the absence of an electric field] assumes the following form in the approximation which is linear in the electric field:

$$\frac{i}{\hbar}[\hat{e}_0, \hat{f}_1] + \frac{1}{\tau}\hat{f}_1 = \frac{i}{\hbar}[e\boldsymbol{E}\hat{r}, \hat{f}_0]. \tag{31.2}$$

Going over to the matrix elements, using the diagonality of $\hat{\mathcal{E}}_0$ and $\hat{f}_0$, and the fact that

$$\frac{i}{\hbar}[\hat{r}, \hat{f}_0]_{nn'} = \frac{i}{\hbar}\frac{f_0(e_{n'}) - f_0(e_n)}{e_{n'} - e_n}[\hat{r}, \hat{e}_0]_{nn'} =$$

$$= -\frac{f_0(e_n) - f_0(e_n')}{e_n - e_{n'}}(\dot{\hat{r}})_{nn'} = -\frac{f_0(e_n) - f_0(e_n')}{e_n - e_n'}v_{nn'}, \tag{31.3}$$

we obtain from Eq. (31.2)

$$\left\{\frac{i}{\hbar}(e_n - e_{n'}) + \frac{1}{\tau}\right\}f'_{nn'} = -e\frac{f_0(e_n) - f_0(e_{n'})}{e_n - e_{n'}}v_{nn'}\boldsymbol{E}, \tag{31.4}$$

and hence

$$f'_{nn'} = -\frac{f_0(e_{n'}) - f_0(e_n)}{e_{n'} - e_n}\frac{e\boldsymbol{E}v_{nn'}}{\frac{i}{\hbar}(e_n - e_{n'}) + \frac{1}{\tau}}. \tag{31.5}$$

Since $\varepsilon_{n'} - \varepsilon_n \approx (a' - n)\hbar\omega_H$, it follows from the conditions of observation of the Shubnikov-de Haas effect that the first and the second factors in Eq. (31.5) have a sharp maximum at n' = n.

In the quasiclassical approximation, the diagonal matrix elements are equal to the time averages of the classical quantities. It is shown in §15 that the quantum oscillations are determined by extremal (in respect of the area) sections of the Fermi surface. For these sections, the average electron velocity is zero. Thus, the term which should make the principal contribution to the quantum oscillations drops out because of the "homogeneous" averaging over the whole orbit, i.e., because of the homogeneity in the coordinate space.

Therefore, it is clear that the terms in Eq. (31.5) which are inhomogeneous in the coordinate space may dominate the resistivity oscillations even if they are small.

Such terms do occur in the transport equation. Our discussion shows they must be retained in this equation.

We shall consider the specific case of the elastic scattering by impurities. In the classical approximation, the conservation of the electron energy in collisions is ensured by the delta function $\delta(\varepsilon - \varepsilon')$, and in the lowest approximation we can use the expression for the energy in the absence of an electric field. In the quantum approach, when the application of crossed electric and magnetic fields $\mathbf{E} \perp \mathbf{H}$ gives rise to stationary states which correspond — for a given p_z (z ‖ H) — to discrete levels, we must use

$$\delta(\hat{\varepsilon} - eE_y\hat{y} - \hat{\varepsilon}' + eE_y\hat{y}').$$

We then obtain terms which are inhomogeneous in the classical case and whose contribution must be estimated.[1] The corresponding terms on the right-hand side of Eq. (31.4) are of the order of $(eEr_H/\hbar\omega_H)(f_0/\tau)$ because $y \sim r_H$. On the other hand, the right-hand part of Eq. (31.4) is of the order of $eEv(f_0/\hbar\omega_H)(\Delta f/f_0) \sim eEv(f_0/\hbar\omega_H)(\hbar\omega_H/\varepsilon_0)$. Since r_H replaces l in the nondiagonal components of the conductivity in a homogeneous field, i.e., the contribution to the current is made only by the next approximation in respect of $(\omega_H\tau)^{-1}$, the contribution of the "usual" right-hand part of Eq. (31.4) to the current is represented by the term of the order of $eEv(f_0/\hbar\omega_H)(\hbar\omega_H/\varepsilon_0)(1/\omega_H\tau)$. Consequently, the role played by the "additional" inhomogeneous terms is $(\varepsilon_0/\omega_H\hbar)$ times greater than the role played by the "normal" terms.

We must stress that the result obtained is entirely due to the homogeneous nature of the field. We can easily show that in a strongly inhomogeneous field (in particular, in the anomalous skin effect discussed in §§33 and 34) the "additional" dependence of the collision integral on the electric field may be ignored (this is confirmed by direct calculations); this dependence makes a contribution to the next approximation in respect of $(\omega_H\tau)^{-1}$. In the inhomogeneous case, the quantum oscillations contain neither $\bar{\mathbf{v}}$ (since $\mathbf{v}$ can be averaged out over an orbit in an inhomogeneous field) nor $(\omega_H\tau)^{-1}$ (because the dominant factor is then the skin depth and the equal number of holes and electrons does not represent a special case).

We shall now give without proof the results of a consistent treatment of the elastic scattering of electrons by impurities. The quantum transport equation can be written in the following form [70]:

---

[1] It may be found that no allowance for the electric field in the collision integral is necessary in the linearization of the quantum transport equation because the collision operator is "local." However, because the scattering of an electron may result in a jump from one orbit to another and the centers of these orbits are separated by a distance of the order of r_H, a suitable allowance for the electric field is essential [56]. Such an allowance reduces to a change in the argument of the δ function, as shown rigorously in [70] by means of the Bogolyubov method [71].

$$\frac{i}{\hbar}[\hat{e}_0,\ \hat{f}_1] + eE\hat{g}\hat{v}_y + D\{\hat{f}_1\} = eED\left\{\left(\frac{cp_x}{eH}\right)\frac{df_0}{d\varepsilon} - \xi g\right\} + \frac{i}{\hbar}[eE\hat{g},\ \hat{F}_0],\tag{31.6}$$

where

$$g_{\mu\nu} = \frac{f_{c\mu} - f_{0\nu}}{\varepsilon_\mu - \varepsilon_\nu},\tag{31.7}$$

$$\{D_\rho\}^{\mu\nu} = N\frac{2\pi}{\hbar}\sum_{l,\,l',\,s}\left\{\delta_+\left(\varepsilon_l - \varepsilon_\mu\right)\left[\rho^{\mu l'}v_s^{l'l}v_{-s}^{l\nu} - v_s^{\mu l'}\rho^{l'l}v_{-s}^{ls}\right] - \right.$$

$$\left. - \delta_+\left(\varepsilon_\nu - \varepsilon_l\right)\left[v_s^{\mu l}\rho^{ll'}v_{-s}^{l'\nu} - v_s^{\mu l}v_{-s}^{ll'}\rho^{l'\nu}\right]\right\},$$

$$F_0^{\mu\nu} = N\sum_{l,\,s} v_s^{\mu l}v_{-s}^{l\nu}(g^{\mu\nu} - g^{\nu l})/(\varepsilon_\mu - \varepsilon_\nu),$$

$$\delta_+(x) = \frac{1}{2}\,\delta(x) + \frac{i}{2\pi x},$$

$$y = y_0 + \xi,\qquad y_0 = -\frac{cP_x}{eH},\qquad \xi = \frac{cp_x(t)}{eH},$$

N is the number of scattering centers, $\hat{v}$ determines the probability of a transition in a collision, and $\hat{g}\hat{v}_y$ is understood to be a simple product of the matrices: $(\hat{g}\hat{v}_y)^{\mu\nu} = g^{\mu\nu}v_y^{\mu\nu}$. Equation (31.6) is analogous to an equation derived by Adams and Holstein [56], who were the first to draw attention to the importance of the electric field in the collision integral and who derived the correct formula for the quantum oscillations of the conductivity.[2]

The solution (31.6) can be obtained in the form of an expansion of $\hat{f}_2$ as a series of powers $(\omega_H\tau)^{-1}$:

$$\hat{f}_1 = \hat{f}_{01} + \hat{f}_{11}.\tag{31.8}$$

The first term, which is independent of the relaxation time, is of the form

$$\hat{f}_{01} = -eE\hat{\xi}\hat{g},\tag{31.9}$$

and the matrix elements of $\hat{f}_{11}$ include the probabilities of transitions from one electron state to another:

$$\frac{i}{\hbar}[\mathscr{H}_0,\ \hat{f}_{01}] = eED\left\{\left(\frac{cp_x}{eH}\right)\frac{df_0}{d\varepsilon}\right\} + \frac{i}{\hbar}[eE_y,\ F_0].\tag{31.10}$$

In the classical approximation, Eqs. (31.9) and (31.10) reduce to

$$f_{01} = -eE\xi\frac{df_0}{d\varepsilon}\tag{31.11}$$

and

$$\frac{df_{11}}{dt} = e\omega_H ED_0\left\{\left(\frac{cp_x}{eH}\right)\frac{df_0}{d\varepsilon}\right\},\tag{31.12}$$

where t is the time of revolution of an electron along a classical orbit. The definition of $D_0\{\rho\}$ remains formally the same as before but the quasiclassical matrix elements which are included in this quantity are naturally replaced by the Fourier components (see also [73]) and

---

[2] The first quantum theory of the magnetoresistance was given in [72].

the functions $\delta_+(x)$ are replaced with $\frac{1}{2}\delta(x)$. Equations (31.9) and (31.10) give the first terms in the expansions in powers of $(\omega_H\tau)^{-1}$.

The same equations can be used to determine the current which is transverse to the magnetic field:

$$j_a = e \operatorname{Sp}(\hat{f}_1,\ \vartheta_a) \qquad (a = x,\ y). \tag{31.13}$$

It follows directly from Eqs. (31.13) and (31.7) that when the elements of the electrical conductivity tensor $\sigma_{\alpha y}$ are expanded, the terms proportional to $1/H$ are of the following form in the case of closed Fermi surfaces:

$$\sigma_{yy}^{(0)} = 0, \qquad \sigma_{xy}^{(0)} = \frac{(n_1 - n_2)\,ec}{H}, \tag{31.14}$$

where n_1 is the number of electrons and n_2 is the number of holes.

The expression for $\sigma_{xy}^{(0)}$ in Eq. (31.14) includes the current along the x axis due to the last term in Eq. (31.10). The contribution to σ_{xy} is

$$\Delta\sigma_{xy}^{(0)} = \frac{ec}{H} \operatorname{Sp} \hat{F}_0. \tag{31.15}$$

The meaning of the diagonal elements of the matrix $F_0^{\mu\nu}$ shows clearly that the correction (31.15) may be included in the renormalized chemical potential of the electron gas and, therefore, it is not given explicitly in Eq. (31.14).

That part of $\sigma_{\alpha y}$ which is determined by the collision integral can be represented in the form

$$\sigma_{ay}^{(1)} = -2 \sum_{\varepsilon_n} \int \frac{df_0}{d\varepsilon}\, \chi_q^{ay} m^* \,\Delta\varepsilon_n\, dp_z. \tag{31.16}$$

The quantities $\chi_q^{\alpha y}$ which occur in Eq. (31.16) are defined by means of Eqs. (31.14) and (31.10).

If we introduce the average mobility tensor u_{ik}, the oscillatory correction to the transverse components of the electrical conductivity tensor can be written in the form:

$$\tilde{\sigma}^{ay} \sim -H^2 u_{extr}^{ay}(\varepsilon_F) \frac{d\ln S}{d\varepsilon_p} \frac{d\tilde{M}_z}{dH},$$

where S is the area of a section of the Fermi surface cut by a plane perpendicular to H; $\tilde{M}_z$ is the oscillatory component of the magnetic moment. Usually, there is no need to make allowance for oscillations of the chemical potential because these oscillations make only a small contribution (§15).

Other elements of the tensor $\tilde{\sigma}_{ik}$ can be calculated in a similar manner.[3]

The de Haas–van Alphen effect influences the Shubnikov–de Haas effect since, strictly speaking, in the classical formulas for σ_{ik} (§§27–30) we must use B (defined in §§19–21) and not H. In some cases (extremely pure metals and sufficiently weak fields), this may be the dominant influence.

A quantum theory of thermomagnetic effects can be developed, to a considerable degree, by analogy with the quantum theory of galvanomagnetic effects but care is needed in the introduction of the relevant fluxes [74].

---

[3] The conductivity tensor in a magnetic field is calculated in [69] by the consistent application of the graph technique to the quasiclassical as well as to the quantum and the extreme quantum cases.

CHAPTER IV

HIGH-FREQUENCY PROPERTIES OF METALS

The high-frequency properties of metals are much more complex than those considered in the preceding chapters. This is due to the existence of the skin effect, i.e., to the fact that, at high frequencies, the electromagnetic field is damped out in a relatively short distance from the surface so that, in most cases, one cannot restrict the treatment to the spatially homogeneous case. Therefore, it is necessary to consider (even in the simplest cases) a semi-infinite sample and to determine the behavior of conduction electrons at the metal-vacuum boundary. Moreover, a considerable number of different cases arises at high frequencies, particularly when magnetic fields are applied.

The high-frequency properties of metals are traditionally divided into two groups – radio-frequency and optical – although, of course, the boundary between them is not very sharply defined. We shall assume that the frequency which divides the radio from the optical region is of the order of $\varepsilon_F/\hbar$ and, in most cases, we shall assume that $\hbar\omega \ll \varepsilon_F$, i.e., we shall confine our treatment to the radio-frequency range. This allows us to ignore the direct absorption of photons by electrons (such absorption is allowed only because of the band structure of the electron energy spectrum) which has a frequency threshold comparable with or even higher than $\varepsilon_F/\hbar$ (optical range).

The optical properties of metals will be considered separately in §47. Experimental investigations of metals in the high-frequency range always involve measurements of various components of the surface impedance tensor. Therefore, the basic problem in the theory of high-frequency properties is the calculation of the components of the surface impedance tensor as a function of the frequency, the temperature, the external magnetic field, etc. We shall pay special attention to these properties and to the cases when the impedance depends strongly on the structure of the electron energy spectrum. We shall be particularly interested in the possibility of using the high-frequency properties of metals in the determination of the shape of the Fermi surface and of the distribution of velocities on this surface.

Boltzmann's transport equation, or its quantum analog, will be the basis of our analysis of high-frequency properties. However, because in the most interesting cases we can quite rigorously introduce the relaxation time in place of the collision operator, we can employ a more direct method of calculation of the current. This will be done in most cases.

For reasons of space, we shall refer only briefly to the experimental data. It is worth mentioning here that much progress has recently been made in experimental studies of the high-frequency properties of metals.

The present chapter consists basically of two parts: one deals with the classical approach (§§32-41, 47) and the other with the quantum approach (§§42-46).

§32. General Relationships Governing the Behavior of Metals in High-Frequency Fields

The principal characteristic of metals from the point of view of their electron properties is the high electron density. Consequently, the conductivity is the largest parameter among those which have the dimensions of frequency.

Bearing in mind that, at frequencies which are compared with the electron collision frequency ν, the conductivity determines the natural oscillations of the electron plasma in a metal, we shall assume that the frequency ω obeys

$$\omega \ll \omega_0, \tag{32.1}$$

where ω_0 is the plasma frequency whose order of magnitude is the same as that of the Fermi energy. If the characteristic electron velocity is v and the average distance between any two electrons is a, the characteristic natural frequency should be of the order of v/a (this follows also from an estimate of the energy of the zero-point vibrations: $\hbar\omega_0 \approx \hbar^2/2ma^2 \approx \hbar p/am \approx \hbar v/a$). Therefore, the condition stated in Eq. (32.1) is equivalent to

$$a \ll \frac{v}{\omega}, \tag{32.2}$$

i.e., the condition that the electrons form a gas and are not "localized" in relation to the incident high-frequency wave: the path traveled by an electron during one period of the field is large compared with the average distance between any two electrons.

The inequality state in Eq. (32.1) does not restrict greatly the frequency ω because in "good" metals, in which the number of electrons is of the order of one per atom, we have $\omega_0 \sim 10^{15}$-10^{16} sec^{-1}, whereas in "poor" metals (such as bismuth) we have $\hbar\omega_0 \approx 100°K \sim 10^{13}$ sec^{-1}. At the frequencies usually employed in radio engineering (up to $\omega = 10^{10}$-10^{11} sec^{-1}), the condition (32.1) is satisfied with a large margin and the lowest approximation in respect of ω/ω_0 is quite sufficient because higher approximations usually give results to an accuracy beyond that of the currently available experimental apparatus. However, this does not apply to metal optics.

By ignoring the frequency compared with the conductivity, we are effectively dropping the displacement current, which is then assumed to be much smaller than the conduction current in Maxwell's equation

$$\operatorname{rot} \boldsymbol{H} = \frac{4\pi}{c}\boldsymbol{j} + \frac{1}{c}\frac{\partial \boldsymbol{D}}{\partial t} \tag{32.3}$$

so that this equation can be written in the form

$$\operatorname{rot} \boldsymbol{H} = \frac{4\pi}{c}\boldsymbol{j}. \tag{32.4}$$

It follows from Eq. (32.4) that div $\boldsymbol{j} = 0$ and, therefore, the use of this equation corresponds to neglecting the uncompensated charge (which appears in a metal in external fields), which is quite natural in view of the high density of electrons in a metal. In vacuum (or in a dielectric), where the conduction current is negligibly small, Eq. (32.3) becomes

$$\operatorname{rot} \boldsymbol{H} = \frac{1}{c}\frac{\partial \boldsymbol{D}}{\partial t}. \tag{32.5}$$

The difference between Eqs. (32.4) and (32.5) is the basic difference between the electrodynamics of "good" and "poor" conductors: the displacement current can be ignored in the first case and the conduction current in the second.

The other Maxwell equation

$$\text{rot } E = - \frac{1}{c} \frac{\partial B}{\partial t} \tag{32.6}$$

is the same in all cases.

The difference between **B** and **H** in nonferromagnetic metals is of importance only when allowance is made for the quantization of the electron energy levels (this point will be discussed later). In the classical approach, Eq. (32.6) can be rewritten in the following way:

$$\text{rot } E = - \frac{1}{c} \frac{\partial H}{\partial t}. \tag{32.6a}$$

The wavelength in a metal[1] determines the conduction current; in a vacuum or in a dielectric, it determines the displacement current. Therefore, the wavelength in a metal is significantly shorter than in vacuum. This means that, for any angle of incidence of a wave on a metal, the problem becomes basically one-dimensional provided we are close to the surface of the metal and all the quantities depend only on the distance from the surface. The nature of the reflection from and the transmission by a metal is independent, at least in the lowest approximation, of the parameters of the field and particularly of the angle of incidence of the wave.

Any problem involving conditions external to a metal can be solved if we know the values of the tangential components of the electric E_t and the magnetic H_t fields [it is clear from Eq. (32.6a) that this means that we must know the electric field and its derivative on the surface, which occur in the second-order equation for the components of the electric field]. The properties of a metal affect only the relationship between these two quantities:

$$E_t^s = \hat{\zeta}[Hn]^s, \quad \hat{\zeta} = \hat{\zeta}' + i\hat{\zeta}'' \tag{32.7}$$

[n is a unit vector normal to the surface; ζ is the two-dimensional surface impedance tensor; the quantity $(4\pi/c)\zeta$ is frequently referred to as the surface impedance]. The continuity of two projections of the field and their derivatives on the surface of a metal give us, together with Eq. (32.7), six equations with six unknowns: two projections of the field intensities in the reflected wave and two projections of the fields and their derivatives on the surface of the metal. (We are assuming that the largest dimensions involved are the size and the curvature of the metal sample, so that it is sufficient to investigate the incidence of a wave on a semi-infinite sample. In general, we must know the impedance tensor at all points on the surface in order to reduce the problem to a form purely external to the metal. In the case of a thin sample, the impedance naturally depends on the thickness of the sample.) It follows that the surface impedance tensor depends weakly on the nature of the field. This makes it convenient to use this tensor.

The high conductivity of a metal thus allows us to determine just a two-dimensional tensor of the second rank, which is the surface impedance tensor ζ. This tensor also governs (in the lowest approximation with respect to ω/ω_0) the impedance of the metal as a whole since, according to Eqs. (32.7) and (32.4),

---

[1] The wavelength in a plasma (a metal can be regarded as a plasma if $\omega \gg \nu$) is of the order of c/ω_0. However, if $\omega \ll \nu$, the role of the wavelength is taken over by the skin depth $\delta \approx (c/\omega_0)(\nu/\omega)^{1/2}$.

$$E_t^s = \frac{4\pi}{c}\,\hat{\zeta}I,$$
(32.8)

where I is the total current flowing through the metal [the problem is one-dimensional and, therefore, it follows from Eq. (32.4) that $j_n = 0$]. The real and the imaginary components of the impedance allow us to determine the Joule heat dissipated in a metal and the phase shift of the field resulting from reflection by the metal.

It will be useful later to write the surface impedance for a monochromatic wave of frequency ω in a different way. If we use Eq. (32.6a), we find that

$$E_t^s = \frac{ic}{\omega}\,\hat{\zeta}E_t'^s.$$
(32.8a)

The determination of the electrodynamics of a metal in a high-frequency field reduces basically to the solution of two problems. We must first find the current density j in given external fields and then – having solved Eqs. (32.4) and (32.6a) – we have to determine the surface impedance. As pointed our earlier, it is sufficient to consider simply the one-dimensional case of normal incidence of a wave on a semi-infinite space filled by a metal. Since there is no current if $E = 0$, the equations are homogeneous. The condition for the existence of a non-trivial solution of homogeneous equations determines the tensor ζ.

Before we tackle the first problem mentioned in the preceding paragraph, we must analyze the structure of a high-frequency field in a metal.

In principle, we can have two limiting cases. If the characteristic depth δ in which the field alters significantly is the largest parameter of the dimensions of length, it follows that over microscopic distances (such as the mean free path of electrons l or the radius r_H of the electron orbits in a static magnetic field) the field does not change greatly and in the lowest approximation the relationship between j and E is local, i.e., the value of the current density at a given point is determined by the field intensity at the same point. In good metals, the condition $\delta \gg l$, r_H automatically ensures that $\omega \ll \nu$ (because if $\omega \gg \nu$, $\delta = c/\omega_0 \approx 10^{-6}$ cm, i.e., the value of δ cannot be large compared with the mean free path even at room temperature). This allows us to use, in the lowest approximation with respect to ω/ν, the relationship between j and E obtained earlier for the static case. We may then find that the value of ζ in an external static magnetic field is finite and does not vanish even when $l \to \infty$ (i.e., when formally $l/r_H \to \infty$, although $l \ll \delta$; naturally, this is possible only if $r_H \ll \delta$) because the components of the conductivity tensor (in which r_H replaces l) associated with the Hall current remain finite and do not vanish. Since $l = \infty$ corresponds to the absence of energy dissipation, this case represents the propagation of an undamped helicon wave in a metal (this is true of the approximation considered), which represents the natural oscillations of the electron plasma in a metal subjected to a magnetic field (see §39 for details).

It must be stressed that since the reflection of electrons from the surface of a metal is of importance at distances of the order of or shorter than l and since $l \ll \delta$, it follows that, in the lowest approximation with respect to l/δ, the nature of the scattering of an electron by the surface is of little importance.[2]

---

[2] The nature of such scattering is important in the next approximation. We must then bear in mind that purely specular reflection is a very special case: even when the fraction of non-specularly-reflected electrons is only of the order of $q \sim r_H/l \ll 1$, the reflection process differs considerably from the purely specular case.

The case of the extreme anomalous skin effect[3] may be realized in metals at relatively low frequencies. This effect is observed when δ is the smallest parameter of the dimensions of length[4]:

$$\delta \ll l, \ r_H, \ v/\omega. \tag{32.9}$$

Knowledge that this condition is satisfied is sufficient to draw some important physical conclusions. We must first find which electrons make the principal contribution to the current density. In the quasiclassical case, we have

$$j = \frac{2e}{(2\pi\hbar)^3} \int v f \, dp \qquad \left(v = \frac{\partial \varepsilon}{\partial p}\right), \tag{32.10}$$

$$\rho' = \frac{2e}{(2\pi\hbar)^3} \int (f - n_F) \, dp, \tag{32.10a}$$

where ρ' is the density of the charge not compensated by ions. The relationship between f and the equilibrium distribution function $n_F(\varepsilon)$ can be easily established for noninteracting quasi particles (the role of Fermi-liquid effects will be considered in §40). A quasi particle, whose crystal momentum is $\mathbf{p}$ and whose energy is $\varepsilon = \varepsilon(\mathbf{p})$, acquires an additional energy $\Delta\varepsilon(\mathbf{p})$ from the electromagnetic field. Clearly, the number of particles whose energy at a given moment is ε is, in the linear approximation with respect to the field, equal to the number of particles whose initial energy has been $\varepsilon - \Delta\varepsilon$, i.e., $f(\mathbf{p}) = n_F(\varepsilon - \Delta\varepsilon)$ or, still in the same linear approximation with respect to the field (which is quite sufficient for our purpose), $f(\mathbf{p}) = n_F(\varepsilon) - (\partial n_F/\partial\varepsilon)\Delta\varepsilon$. In the absence of external fields, there is no current or uncompensated charge in a normal metal which is in equilibrium and, therefore,

$$j = -\frac{2e}{(2\pi\hbar)^3} \int v \frac{\partial n_F}{\partial \varepsilon} \Delta\varepsilon(p) \, dp, \tag{32.11}$$

$$\rho' = -\frac{2e}{(2\pi\hbar)^3} \int \frac{\partial n_F}{\partial \varepsilon} \Delta\varepsilon(p) \, dp. \tag{32.11a}$$

The term with $n_F(\varepsilon)$ in j vanishes because of the central symmetry of the function $\varepsilon(\mathbf{p})$: $\varepsilon(-\mathbf{p}) = \varepsilon(\mathbf{p})$.

In the absence of collisions, we have

$$\dot{p} = eE + \frac{e}{c}[vH] \tag{32.12}$$

and, therefore,

$$\dot{\varepsilon} = \frac{\partial \varepsilon}{\partial p} \dot{p} = v\dot{p} = eEv. \tag{32.13}$$

If the average time that a quasi particle is subjected to a field between two successive collisions is τ_f (we are considering only orders of magnitude), it follows that

$$\Delta\varepsilon \approx eEv\tau_f. \tag{32.14}$$

---

[3] The anomalous skin effect was discovered experimentally by London [1] and the theory of this effect in the simplest case of free electrons and the absence of a static magnetic field is given in [1, 3].

[4] At high frequencies, when $v/\omega \gtrsim \delta$, i.e., $\omega \gtrsim \omega_0(v/c)$, the inequality (32.9) is not obeyed and the treatment given here is invalid. Such frequencies are discussed in §47.

We shall now take account of the extreme anomalous skin effect defined by Eq. (32.9). Almost all the electrons (except for that small fraction which suffers collisions before escaping from the skin layer) obey

$$\tau_f \approx \frac{\delta}{v_n} \tag{32.15}$$

and

$$\Delta\varepsilon \approx eEv\delta/v_n, \tag{32.16}$$

i.e., the energy acquired between collisions increases in inverse proportion to v_n as $v_n \to 0$ ($v_n = 0$ corresponds to electrons traveling parallel to the surface of a metal). Then j_t of Eq. (32.11) diverges at the point $v_n = 0$. (There is no divergence in j_n.) Consequently, only those few electrons which are reflected at glancing angles from the surface are of importance in the anomalous skin effect (we are still speaking of the lowest approximation). This circumstance is most important in the development of the theory of the anomalous skin effect. We shall now consider its consequences.

Since there is no divergence in j_n, and E_n is found, in accordance with Eq. (32.4), from $j_n = 0$ (we recall that it is sufficient to consider only the one-dimensional case), it follows that

$$E_n \approx E_t.$$

Since the term $\mathbf{E}_t\mathbf{v}_t$ in Eq. (32.16) leads to a divergence of j in Eq. (32.11) (when $l = \infty$) and the term $\mathbf{E}_n\mathbf{v}_n$ does not, it is sufficient to include only $\mathbf{E}_t$ in the definition of j_t (which is required for the determination of ζ) by assuming formally that $E_n = 0$ and to use

$$\dot{\varepsilon} = e\mathbf{E}_t\mathbf{v}_t. \tag{32.13a}$$

The normal component of the field can be neglected because the field E_n does practically no work on the electrons which are reflected at glancing angles. Since Eq. (32.13a) is linear in $\mathbf{E}$ and we are only interested in the linear approximation, all the other quantities can be calculated for $E = 0$. In the foregoing discussion, we have assumed implicitly that the "anomalous" parameter is the smallest in the problem being considered and it is therefore sufficient to make allowance for this smallness. To be specific, we shall consider only this particular case, which is of the greatest practical importance. It is easy to extend the theory to the case of, for example, an extremely sharp resonance when the largest parameter is ω/ν and E_n may have to be included (§36). Mathematical proofs of all these assertions are given in [4].

The fact that only the "glancing" electrons make a significant contribution to the current density is due to the fact that the nonequilibrium correction to the electron distribution function has a sharp maximum (which gives rise to a divergence at $l = \infty$ when integrated) located close to $v_n = 0$. Naturally, the probability that a collision causes an electron to escape from a narrow region near $v_n = 0$ to some other point in the crystal momentum space is considerably higher than the probability that this electron is scattered into this narrow region. This means that only the last term need be retained in the collision integral, which describes the change in the distribution function as a result of collisions (the changes include the gain of electrons by a given state as well as the loss from that state). This term is proportional to the "excess" (relative to the equilibrium value) number of electrons in a given state so that

$$\left(\frac{\partial f}{\partial t}\right)_c = \nu\,(\boldsymbol{p})\,(f\,(\boldsymbol{p}) - n_F\,(\varepsilon)), \tag{32.17}$$

$$\nu\,(\boldsymbol{p}) = \int W\,(\boldsymbol{p},\ \boldsymbol{p}')\,d\boldsymbol{p}', \tag{32.18}$$

where $\nu(\mathbf{p})$ is the number of collisions per unit time and $W(\mathbf{p}, \mathbf{p}')$ is the probability of a transition from $\mathbf{p}$ to $\mathbf{p}'$. [The integral term $\int W(\mathbf{p}', \mathbf{p})(f(\mathbf{p}') - n_F(\varepsilon(\mathbf{p}')))d\mathbf{p}'$ is small compared with Eq. (32.17) because integration reduces the "sharpness" of the singularity.]

The anomalous nature of the skin effect thus permits us to introduce, in a consistent manner, the concept of the "mean free time," i.e., the time between collisions of electrons at any temperature (in static problems, this is possible only in exceptional cases).

Our analysis is valid only if the probability of a transition $W(\mathbf{p}, \mathbf{p}')$ is itself a smooth function and has no sharp maxima in the narrow region between $\mathbf{p}$ and $\mathbf{p}'$. This is not true of electron–phonon collisions at low temperatures ($T \ll \Theta$): in this case, the most probable scattering is through small angles of the order of T/Θ. Therefore, we need consider only those electrons which "escape" from the region close to $v_n = 0$ provided the following inequalities are satisfied:

$$\delta/l, \quad \delta/r_H, \quad \omega\delta/v \ll T/\Theta. \tag{32.18a}$$

When Eq. (32.18a) is satisfied, the scattering through an angle of the order of T/Θ knocks an electron out of the skin layer, i.e., out of the region where the electric field is not low, and, therefore, such scattering is as equally important as that through a large angle. The number of collisions ν_{eff} is then found to be proportional to the number of phonons at low temperatures, i.e., it is of the order of $(\Theta/\hbar)(T/\Theta)^3$; the additional factor $(T/\Theta)^2$ (§24). The "effective" electron collision frequency at temperatures $T/\Theta \ll 1$ increases rapidly with increasing "anomaly" of the skin effect [when the inequalities of Eq. (32.18a) are satisfied by a margin which is a factor of $(\Theta/T)^2$] and becomes significant even at very low temperatures (at these temperatures the residual resistance is important in the static case). If $\hbar\omega > T$, the temperature is effectively described by $\hbar\omega$; this quantity then occurs in all the expressions given above, so that

$$\nu_{eff} \approx \frac{\Theta}{\hbar}\left(\frac{T + \hbar\omega}{\Theta}\right)^3. \tag{32.18b}$$

If the nonequilibrium correction to the electron distribution function has a sharp minimum in an even narrower region of the momentum space (for example, if electrons are in resonance along certain selected directions of $\mathbf{p}$, see §§35, 36), the relative number of electrons escaping from this region becomes even more important compared with the number arriving in the region and the restrictions imposed by the inequalities (32.18a) on the anomaly of the skin effect become weaker [under resonance conditions an additional small factor $(\nu/\omega)^{1/2}$ appears on the left-hand side].

For similar reasons, the anomaly of the skin effect makes the Fermi-liquid interaction unimportant: this interaction is related to a correction to the energy in the gas approximation and is expressed by an integral with respect to $f - n_F$. The exception to this rule is the case of a very sharp resonance, when this interaction is responsible for the broadening of the resonance curve and its finite amplitude (for details, see [4] and §40).

We shall also show that the nature of the reflection of electrons by the surface of a metal is not of great importance in the extreme anomalous skin effect [5]. This very important point enables us to obtain a solution in a closed form since the "elimination" of the surface of a metal from our considerations makes it possible to consider the problem over the whole space and to solve it by the standard Fourier method.

We shall extend formally the electric field as an even function to the region outside the metal.

If we now ignore the collisions of electrons with the surface, we find that even the fraction of the "glancing" electrons originally scattered by the surface acquires an energy twice as great as before from the field extended outside the metal. However, doubling of the conductivity increases the effective skin depth δ and reduces the surface impedance only by a factor $\sqrt[3]{2} \approx 1.2$ (this follows from dimensional analysis if we bear in mind that the relationship between j and $\mathbf{E}$ is of an integral nature: since l, r_H, $v/\omega \gg \delta$, the current at a given point is determined by the field intensity in the whole skin layer of thickness δ which is crossed by the electron, i.e., $j \approx \sigma(\delta/l)E$ and $\delta \propto \sqrt[3]{\sigma}$; see also the beginning of §33).

This means that, to within a small real numerical factor which differs little from unity,[5] we can solve the problem over the whole space (a rigorous proof of this assertion is given in [7]). The electric field is then an even function which, generally speaking, has a discontinuity in its derivative at the origin of the coordinates [if $\mathbf{E}_t(\xi) = \mathbf{E}_t(-\xi)$, where ξ is the coordinate along the normal to the surface, it follows that $\mathbf{E}_t'(+0) = -\mathbf{E}_t'(-0)$, and if $\mathbf{E}_t'(+) \neq 0$, a discontinuity is unavoidable; the dependence of $\mathbf{E}_t'(+0)$ on $\mathbf{E}_t(0)$ determines – in accordance with Eqs. (32.7) and (32.6) – the value of the surface impedance].

In the normal skin effect, the nature of reflection from a boundary is, in the lowest approximation, quite unimportant and, therefore, this procedure is always permissible irrespective of the nature of the skin effect.

We shall now summarize the present section.

1. It is always permissible to ignore the displacement current and the uncompensated charge density in a metal.

2. Any problem external to a metal can be solved simply by ascertaining the value of the surface impedance, i.e., the relationship between the tangential components of the electric and the magnetic fields on the surface of a metal [Eq. (32.7)].

3. The surface impedance can be determined, to within an unimportant real numerical factor close to unity, by considering a metal to be subjected to a field $\mathbf{E}(-\xi) = \mathbf{E}(\xi)$ which has a discontinuity in its derivative at the origin of the coordinates, where $\xi = 0$ is the plane surface of the metal (the exception to this general rule is discussed in §41).

4. When the smallest parameter is that related to the anomaly of the skin effect, we are permitted

a) to introduce an electron relaxation time [in the case of electron–phonon collisions at low temperatures $T \ll \Theta$ this time constant is of the order of $(\hbar/\Theta)[\Theta/(T+\hbar\omega)]^3$;

b) to disregard the dependence $E_\xi(\xi)$;

c) to ignore the Fermi-liquid effects, i.e., to consider, in accordance with 4a above, the one-electron problem in the "mean free time" approximation $1/\nu(\mathbf{p})$.

The decisive simplifications, which permit us to obtain a solution in a closed form, are those of 3 and 4a.

Finally, we must mention that, because of $\hbar\omega_0 \gg T$, we have

$$-\frac{\partial n_F}{\partial \varepsilon} \approx \delta(\varepsilon - \varepsilon_F), \qquad (32.19)$$

where ε_F is the Fermi energy and n_F is the equilibrium Fermi function.

---

[5] The special case is that of a scattered electron returning to the surface after fully specular reflection (see Footnote 2 in the present section), which happens when the magnetic field is exactly parallel to the surface (see §§22, 41, and 46, as well as [6]).

We shall not give the theory of the normal effect because it can be found in any textbook dealing with the electrodynamics of continuous media. We shall give only the expression for the principal values of the surface impedance tensor:

$$\zeta_{1,2} = \sqrt{\frac{\omega \rho_{1,2}}{4\pi i}}. \tag{32.20}$$

Here, $\rho_{1,2}$ are the principal values of the resistivity tensor in the plane representing the boundary of the metal. The application of a magnetic field, even at relatively low frequencies, may alter the impedance quite considerably (this is considered in one of the later sections).

§33. Anomalous Skin Effect in the Absence of a Static Magnetic Field

Before we consider the quantitative theory of the anomalous skin effect, we shall use the conclusions drawn in the preceding section to give a physical picture of the effect.

Since the principal contribution to the current density is made by the "glancing" electrons which move in the skin layer, i.e., in a region where the electric field is everywhere of approximately the same order of magnitude, the relationship between the current density and the field intensity can be expressed approximately in the form of a "local" Ohm's law. We must bear in mind that the current density is proportional to the number of charges participating in the current and that the relative number of electrons moving at an angle $\leqslant \delta/l$ (δ is the skin depth) is of the order of δ/l. Therefore,

$$j = \sigma_{\text{eff}} E, \quad \sigma_{\text{eff}} = a\sigma \frac{\delta}{l}, \quad a \approx 1. \tag{33.1}$$

We may expect Eq. (33.1) to remain valid even if δ is complex (the complex form of δ represents not only the decrease in amplitude of the field with depth but also the change in its phase).

If a monochromatic wave of frequency ω is incident normally on a semi-infinite metal which satisfies the "normal" form of Ohm's law, we find that Eqs. (32.4) and (32.6a), representing the electric components $\mathbf{E}_t$, become

$$\mathbf{E}_t'' = \frac{4\pi i \omega}{c^2} \sigma_{\text{eff}} \mathbf{E}_t = \frac{4\pi i \omega a \, \sigma \delta}{c^2 l} \mathbf{E}_t,$$

and hence the complex skin depth is

$$\frac{1}{\delta^2} = \frac{4\pi i \omega \sigma_{\text{eff}}}{c^2} = \frac{4\pi i \omega \sigma \, \delta a}{c^2 l} \tag{33.1a}$$

and

$$\delta = \sqrt[3]{\frac{c^2 l}{4\pi i \omega a \sigma}}, \tag{33.2}$$

which gives [see Eq. (32.8a)]

$$\zeta = \zeta' + i\zeta'' = \frac{1 + i\sqrt{3}}{2} \left(\frac{\omega^2 l}{4\pi c a \sigma} \right)^{1/3}. \tag{33.3}$$

A rigorous calculation (to be given later) confirms this formula, which is derived in a similar manner in [2], and yields $a = 3^5 \sqrt{3}\, \pi/2^8$. It is important to note that Eq. (33.3) de-

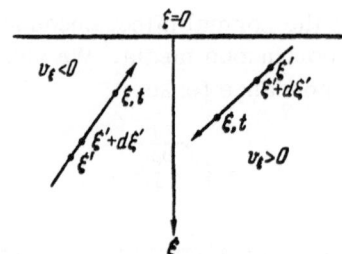

Fig. 75. Electrons moving toward the surface of a metal $(v_\xi < 0)$ and away from it $(v_\xi > 0)$.

scribes correctly the principal features of the impedance in the anomalous skin effect: the independence of the mean free path (and, consequently, of the temperature) because $\sigma \propto l$; the dependence on the frequency in the form $\zeta \propto \omega^{2/3}$ [in the normal skin effect this dependence is — in accordance with Eq. (32.20) — given by $\zeta \propto \omega^{1/2}$]; the complex nature of ζ ($\zeta''/\zeta' = \sqrt{3}$, whereas in the normal skin effect $\zeta'' = \zeta'$).

We shall now develop a consistent theory of the anomalous skin effect.

According to Eq. (32.11), it is necessary to calculate the energy acquired by an electron whose crystal momentum is p up to the moment t at a point denoted by ξ in Fig. 75. This can be done using Eq. (32.13a), assuming that the electron velocity **v** is constant. The path from ξ' to $\xi' + d\xi'$ is traveled by the electron in a time $|d\xi/v_\xi|$ and in this time the electron acquires the following energy from the field $\mathbf{E}_t \exp(i\omega t)$ (the field amplitude is not indicated by any index because there can be no misunderstanding):

$$d\varepsilon' = e\mathbf{E}_t(\xi')\,v_t\left|\frac{d\xi'}{v_\xi}\right|\exp\left\{i\omega\left(t - \left|\frac{\xi - \xi'}{v_\xi}\right|\right)\right\}.$$

The last factor in the above expression represents the change in the phase of the field with time. The probability that an electron retains this energy when it reaches the point ξ, i.e., when it does not suffer a collision on the way from ξ' to ξ (in a time $\left|\frac{\xi - \xi'}{v_\xi}\right|$), is given by

$$\exp\left\{-\left|\frac{\xi - \xi'}{v_\xi}\right|\nu(p)\right\} \equiv \exp\left\{-\left|\frac{\xi - \xi'}{v_\xi \tau(p)}\right|\right\}.$$

Consequently, the total energy $\Delta\varepsilon$ is:

$$\Delta\varepsilon = \begin{cases} \displaystyle\int_{-\infty}^{\xi} e\mathbf{E}_t(\xi')\,v_t\,\frac{d\xi'}{v_\xi}\exp\left\{-\frac{|\xi - \xi'|}{v_\xi \tau^*}\right\}\, v_\xi > 0, & \mathbf{E}(-\xi) = \mathbf{E}(\xi), \\[4mm] \displaystyle\int_{\infty}^{\xi} e\mathbf{E}_t(\xi')\,v_t\,\frac{d\xi'}{v_\xi}\exp\left\{-\frac{|\xi - \xi'|}{v_\xi \tau^*}\right\} & v_\xi < 0; \end{cases} \tag{33.4}$$

$$\frac{1}{\tau^*} = \nu(p) + i\omega \equiv \frac{1}{\tau(p)} + i\omega. \tag{33.4a}$$

We shall substitute Eq. (33.4) into Eq. (32.11) and reduce the integration over all values of v_ξ to integration over $v_\xi > 0$, using the central symmetry of the function $\varepsilon(\mathbf{p})$: $\varepsilon(-\mathbf{p}) = \varepsilon(\mathbf{p})$. We shall then obtain the required expression for $\mathbf{j}_t$:

$$j_\alpha(\xi) = \int_{-\infty}^{\infty} K_{\alpha\beta}(|\xi - \xi'|)\,E_\beta(\xi')\,d\xi', \tag{33.5}$$

where

$$K_{\alpha\beta}(W) = -\frac{2e^2}{(2\pi\hbar)^3} \int \frac{\partial n_F}{\partial\varepsilon} \, dp \exp\left(-\frac{W}{|v_\xi|\tau^*}\right)\frac{v_\alpha v_\beta}{|v_\xi|}. \tag{33.6}$$

It is worth noting that this method of finding the relationship between **j** and **E**, which uses the ideas presented in [8], is equivalent to the solution of the transport equation. It follows from Eqs. (32.4) and (32.6a) that, in the case of a monochromatic wave $\mathbf{E}_t \exp(i\omega t)$, we obtain:

$$E_\alpha'' = \frac{4\pi i\omega}{c^2} \int\limits_{-\infty}^{\infty} K_{\alpha\beta}(|\xi - \xi'|) E_\beta(\xi') \, d\xi'. \tag{33.7}$$

We must draw attention to the fact that Eq. (33.7) allows an extension of the field as an even function: $E_\alpha(-\xi) = E_\alpha(\xi)$.

This is a natural procedure because we have calculated the contribution from the electrons moving away from the boundary, using the fact that $\mathbf{E}_t(-\xi) = \mathbf{E}_t(\xi)$. Strictly speaking, Eq. (33.5) is valid only for the specular reflection of electrons from the boundary of a sample [3]. If the reflection is diffuse, the expression for the kernel of the integral relationship between the field and the current is still of the form given by Eq. (33.6) but the integration in a formula analogous to Eq. (33.5) is carried out from zero to infinity, which makes it very difficult to obtain a closed expression for the impedance. However, it is shown in the preceding section that, in the case of the extreme anomalous skin effect, the impedance is not greatly affected by the nature of the reflection of electrons from the surface.

The equation obtained in this way can be solved directly by taking the Fourier components on both sides of the equality sign. We must bear in mind that, in accordance with §32, $E_\alpha'(\xi)$ has a discontinuity at $\xi = 0$ and, therefore, integration by parts gives

$$\int\limits_{-\infty}^{\infty} e^{-ik\xi}E_\alpha''(\xi) \, d\xi = \int\limits_{-\infty}^{-0} + \int\limits_{+0}^{\infty} = -2E_\alpha'(+0) - k^2\varepsilon_\alpha(k), \tag{33.8}$$

where

$$\varepsilon_\alpha(k) = \int\limits_{-\infty}^{\infty} e^{-ik\xi}E_\alpha(\xi) \, d\xi = 2\int\limits_{0}^{\infty} \cos k\xi \cdot E_\alpha(\xi) \, d\xi,$$

$$E_\alpha(\xi) = \frac{1}{\pi}\int\limits_{0}^{\infty} e^{ik\xi}\varepsilon_\alpha(k) \, dk, \tag{33.9}$$

and (assuming that $\xi = \xi' + \xi''$), we obtain

$$\int\limits_{-\infty}^{\infty} e^{-ik\xi} \, d\xi \int\limits_{-\infty}^{\infty} K_{\alpha\beta}(|\xi - \xi'|) E_\beta(\xi') \, d\xi' = \int\limits_{-\infty}^{\infty} d\xi' E_\beta(\xi') e^{-ik\xi'} \int\limits_{-\infty}^{\infty} e^{-ik\xi} K_{\alpha\beta}(|\xi - \xi'|) \, d\xi = \varepsilon_\beta(k) K_{\alpha\beta}(k), \tag{33.10}$$

$$K_{\alpha\beta}(k) = \int\limits_{-\infty}^{\infty} K_{\alpha\beta}(|\xi|) e^{-ik\xi} \, d\xi = 2\int\limits_{0}^{\infty} K_{\alpha\beta}(\xi) \cos k\xi \, d\xi. \tag{33.11}$$

Using Eqs. (33.8)–(33.11), we find from Eq. (33.7) that

$$- k^2 e_\alpha(k) - 2E'_\alpha(0) = \frac{4\pi i \omega}{c^2} K_{\alpha\beta}(k)\, e_\beta(k), \qquad (33.12)$$

where, in accordance with Eqs. (33.6) and (33.11),

$$K_{\alpha\beta} = \frac{4e^2}{(2\pi\hbar)^3} \oiint\limits_{\substack{(\varepsilon = \varepsilon_F \\ v_\xi > 0)}} \frac{\tau^* v_\alpha v_\beta}{v}\, \frac{ds}{1 + (kv_\xi \tau^*)^2}. \qquad (33.13)$$

In writing down Eq. (33.13), we have used $-\partial n_F/\partial\varepsilon \approx \delta(\varepsilon - \varepsilon_F)$ – in accordance with Eq. (32.19) – and the fact that $dp = d\varepsilon\, ds/v$, where ds is an element of area of the surface $\varepsilon(\mathbf{p}) = \varepsilon$. Naturally, if there are several branches of the spectrum corresponding to different types of quasi particle, we must sum Eq. (33.12) over all these branches.

It follows from Eq. (33.12) that

$$E_\alpha(\xi) = -\frac{2}{\pi} \int\limits_0^\infty \cos k\xi\, dk \left(k^2 + \frac{4\pi i \omega}{c^2} \hat{K}(k)\right)^{-1}_{\alpha\beta} E'_\beta(0),$$

so that, in accordance with Eq. (32.8a),

$$\zeta_{\alpha\beta} = \frac{2i\omega}{\pi c} \int\limits_0^\infty dk \left(k^2 + \frac{4\pi i \omega}{c^2} \hat{K}(k)\right)^{-1}_{\alpha\beta}. \qquad (33.14)$$

Using the condition of the extreme anomaly of the skin effect, we can easily show that only the terms $|kv\tau^*| \gg 1$ are important and, therefore, the velocities which are important in Eq. (33.13) are those (as shown in §32) which satisfy $v_\xi \ll 1$. Consequently, we can determine ζ simply using the asymptotic expression (33.13). Employing the fact that $ds = do/\varkappa$ (do is an element of a solid angle in the velocity space), we find that a calculation of the asymptotic expression yields

$$K_{\alpha\beta} = \frac{2e^2}{(2\pi\hbar)^3}\, \frac{\pi}{k} \int\limits_0^{2\pi} d\varphi\, \frac{n_\alpha(\varphi)\, n_\beta(\varphi)}{\varkappa(\varphi)} \equiv \frac{3\pi}{4}\, \frac{B_{\alpha\beta}}{k}, \qquad n = \frac{v}{v}, \qquad (33.15)$$

where $\varkappa$ is the Gaussian curvature and φ is the angle measured over the strip $\varepsilon(\mathbf{p}) = \varepsilon_F$, $v_\xi(\mathbf{p}) = 0$.

Reducing the tensor $B_{\alpha\beta}$ to its principal axes, which are the coordinate axes in the plane $\xi = 0$, and writing the impedance in terms of these axes, we obtain:

$$\zeta_\alpha = \left(1 + i\sqrt{3}\right) \frac{2}{9} \left(\frac{\sqrt{3}\, \omega^2}{\pi^2 c B_\alpha}\right)^{1/3}. \qquad (33.16)$$

In the extreme anomalous skin effect, we can obtain exact formulas for the impedance in the case of purely specular and completely diffuse scattering of electrons by the surface [3, 9]. It is shown in §32 that the nature of the reflection has little effect on the impedance: in the specular reflection case, the impedance formula is identical with Eq. (33.16), whereas in the diffuse case, it differs from Eq. (33.16) by a factor $^9/_8$.

It follows from Eqs. (33.15) and (33.16) that measurements of the surface impedance allow us to determine directly a very important characteristic of the electron spectrum, which is the average value of the reciprocal of the Gaussian curvature of the Fermi surface in the strip defined by $\varepsilon(\mathbf{p}) = \varepsilon_F$, $v_\xi = 0$.

The anisotropy can be determined using the results of measurements on different faces of a single-crystal sample. The measurements of the impedance of copper in the anomalous skin effect can be used to determine the Fermi surface [10].

If the dispersion law is isotropic (as in the case of potassium, sodium, and other metals), Eq. (33.16) can be expressed in the form

$$\zeta = \left(1 + i\,\sqrt{3}\right)\frac{2\sqrt{3}}{9}\left(\frac{\sqrt{3}\,\omega^2 l}{3\pi^2 c\sigma}\right)^{1/3}, \tag{33.17}$$

which makes it possible to use the measurements of the impedance under the anomalous skin-effect conditions to determine the mean free path of electrons.

§ 34. Anomalous Skin Effect in a Static Magnetic Field

As in the preceding section, we shall begin by considering the physical nature of the effect. We shall determine the dependence of the impedance on the frequency of the electromagnetic field and on the mean free path. We shall then consider the most interesting regions and develop a quantitative theory for these regions. Finally, we shall examine the results in some detail.

It is shown at the beginning of the preceding section that the calculation of the impedance reduces to the calculation of the effective conductivity σ_{eff} under appropriate conditions. For the sake of simplicity, we shall restrict our treatment to the case of closed orbits when an electron subjected to a static magnetic field **H** moves along a helix whose axis is parallel to the magnetic field. The radius of this helix is $r_H \approx dp_\perp / eH$, where $p_\perp$ is the projection of the electron momentum onto a plane perpendicular to **H**. As usual, we shall consider only the lowest approximation.

1. Let us assume that the frequency of the alternating field is relatively low

$$\omega \ll \nu, \tag{34.1}$$

so that the alternating field does not change significantly during the "mean free time" (this is true only in the lowest approximation). The nonstationary nature of the field simply has the effect of reducing strongly the thickness of the skin layer.

Different electrons traverse different paths in the skin layer, depending on the angle at which they enter this layer (Fig. 76). The longest path (AB) in the skin layer during one revo-

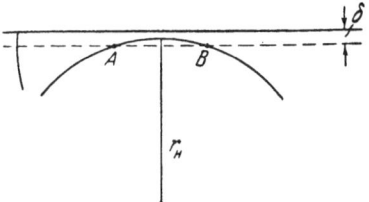

Fig. 76. Electrons traveling along different paths in the skin layer, depending on the angle at which they enter this layer.

lution is of the order of $\sqrt{r_H \delta}$; this is applicable to the "glancing" electrons, which are the only ones of importance in the anomalous skin effect (§32).

If

$$\sqrt{r_H \delta} \gg l, \quad \text{i.e.,} \quad r_H \gg \frac{l^2}{\delta} = r_1, \tag{34.2}$$

where $r_H = cp_F/eH$ is the radius of the orbit, we find that in the lowest approximation with respect to r_1/r_H all the "glancing" electrons traverse the greatest possible distance (of the order of l) in the skin layer without suffering collisions, and that the magnetic field does not affect the impedance.

The magnetic field shortens the path in the skin layer only in the case of those electrons which satisfy $cp_\perp/eH \lesssim l^2/\delta$ [Eq. (34.2)]; these electrons "escape" earlier from the skin layer. Since the only electrons of importance are those on the Fermi surface [Eqs. (32.11) and (32.19)] and the electrons of interest to us are represented by a sector of radius $eHl^2/c\delta$ on the surface whose radius is $p_F \sim eHr_H/c$, it follows that the number of these electrons and their relative contribution to the impedance are of the order of

$$\Delta\zeta/\zeta \sim (l^2/\delta r_H)^2. \tag{34.3}$$

These electrons make a considerable contribution to the impedance even in very weak fields characterized by $r_H \sim r_1$. An estimate shows that in very pure metals this contribution is appreciable in fields up to 0.1-1.0 Oe. When the magnetic field is increased so that

$$\sqrt{r_H \delta} \ll l, \tag{34.4}$$

although we still have

$$r_H \gg l, \tag{34.5}$$

the "glancing" electrons traverse a path of the order of $\sqrt{r_H \delta} \ll l$ in the skin layer and this reduces the conductivity by a factor of $\sqrt{r_H \delta}/l$ (because the static conductivity σ is proportional to l). The maximum path is traversed in the skin layer by those electrons which enter or leave the layer at an angle of the order of $\delta/\sqrt{r_H \delta}$. Since all angles are equiprobable, this expression determines the relative number of the "glancing" electrons. Thus,

$$\sigma_{\text{eff}} \approx \sigma \frac{\sqrt{r_H \delta}}{l} \frac{\delta}{\sqrt{r_H \delta}} = \sigma \frac{\delta}{l}, \tag{34.6}$$

i.e., the effective conductivity is of the same order as that in the absence of a magnetic field and — in the lowest approximation — the value of σ_{eff} is independent of the magnetic field. Simple qualitative calculations cannot be carried out in the range $r_H \approx r_1$; the change of the impedance in this range may be nonmonotonic, as reported in [11, 12] (this point is discussed in §§22 and 46).

If the magnetic field is inclined at a fairly large angle to the surface of the metal, the situation is still the same if

$$r_H \ll l, \tag{34.7}$$

because electrons escape from the skin layer in a time $\delta/v \ll \Omega^{-1}$ which is short compared with the period of one revolution. (If $\delta \gg r_\mathrm{H}$, the situation is much more complex, as discussed in §41.)

However, if the magnetic field is exactly parallel to the surface (or if the angle of inclination is small compared with r_H/l), the effective conductivity is larger than that given by Eq. (34.6) by a factor equal to the number of times that an electron returns to the skin layer in its "mean free time," i.e., by a factor of $\sim l/2\pi r_\mathrm{H}$:

$$\sigma_\mathrm{eff} \approx a\sigma \frac{\delta}{2\pi r_H}, \quad a \approx 1. \tag{34.8}$$

Consequently [Eq. (33.1)], the impedance of Eq. (33.3) has to be multiplied by $(l/2\pi r_\mathrm{H})^{-1/3}$:

$$\zeta(H) \approx \zeta(0) \left(\frac{l}{2\pi r_H}\right)^{-1/3} \sim (lH)^{-1/3}. \tag{34.9}$$

(This result was deduced rigorously for the first time in [13].)

2. We shall now consider the case of relatively high frequencies of the alternating field

$$\omega \gg \nu, \tag{34.10}$$

so that an electron usually acquires energy from the field during a time interval of the order of the period of the field. Since there is no dissipation of energy for $\nu = 0$ in the absence of a surface, the quantity ν is now replaced by $i\omega$. In general, the application of an alternating field means that ν has to be replaced with $\nu + i\omega$ [see, for example, Eq. (33.4a)], i.e., τ has to be replaced with τ^*. All the results can now be rewritten (this applies also to $\sigma \sim \tau$) making this replacement.

Moreover, in the low-frequency case, an electron entering repeatedly the skin layer finds that the field is practically constant. In the high-frequency case, if

$$r_H \ll l, \tag{34.11}$$

but

$$\Omega \lesssim \omega, \tag{34.12}$$

we may use Eq. (34.8): this formula applies only if the electron is accelerated "synchronously" by the field, i.e., under resonance conditions. Under other conditions, an electron may be accelerated only during one revolution so that Eq. (34.6) still applies. Since the resonance case is of special interest we shall consider it separately in the next section.

There are two further points worth mentioning. We have considered only one type of quasi particle. The presence of several types of quasi particle does not alter the results even when the number of electrons equals the number of holes (this should be contrasted with the case of the galvanomagnetic effects under static field conditions, which are discussed in §27). This follows from the fact that the important feature is the detailed structure of the orbits in the thin skin layer and this structure is naturally different for holes and electrons.

We have not considered bounded samples. The presence of a second boundary in the case of a magnetic field parallel to the surface has a very interesting effect. If the thickness of a plate D is small compared with the mean free path l but large compared with the skin depth δ:

$$\delta \ll D \ll l, \tag{34.13}$$

we find that irrespective of the value of ω/ν the application of a field H exceeding the field H_D, known to within δ/D and given by

$$H_D = \frac{c p_s^{max}}{eD} \tag{34.14}$$

(S is the direction which lies on the surface of the plate and is perpendicular to the magnetic field), confines all the orbits to the plate so that all the electrons return repeatedly to the skin layer. Consequently, a kink in the derivative of the surface impedance with respect to the magnetic field is observed in $H = H_D$ [14].

This property is known as the size effect and provides an important method for the determination of the shape of the Fermi surface [15].

§35. Physical Aspects of Cyclotron Resonance

It is well known that a free electron subjected to a static magnetic field **H** moves along a helix whose axis is aligned along this field. In a plane perpendicular to the magnetic field, the motion of a free electron represents uniform rotation in a circle and the frequency of this rotation is $\Omega = eH/mc$, which is independent of the magnitude and the direction of the electron velocity. This frequency of rotation, known as the cyclotron frequency, is independent of the electron velocity in the more general case of a quadratic dispersion law when the constant-energy surface in the momentum space is an ellipsoid (§4). This is the case which is frequently encountered in a semiconductor when a band is almost empty or almost filled. It is natural to expect a resonance when a semiconductor is subjected to a static magnetic field and a perpendicular (to the magnetic field) circularly polarized electromagnetic field whose frequency is $\omega \sim \Omega$. This resonance was predicted independently by Dorfman [16] and by Dingle [17] and observed since in many semiconductors. It played an important role in establishing the dispersion laws of semiconductors (see, for example, [18]).

A very important point in the Dorfman–Dingle resonance is that the relatively low carrier density in semiconductors allows us to regard the electric field in such materials as homogeneous (we shall show later that this is not always true even of semiconductors).

These considerations are inapplicable to metals for two reasons.

As stressed in §32, a very high carrier density in a metal gives rise to a strong inhomogeneity of the electromagnetic field over distances which are of the order of one orbit radius (extreme anomalous skin effect). In fact, appreciable resonance can be observed only if an electron makes at least several revolutions between two collisions (these collisions are with impurities, lattice imperfections, phonons, or other electrons). This means that we must satisfy the inequalities $\omega\tau < 1$, $l > r_H$. In this case, the skin depth δ is independent of τ and is of the order of c/ω_0 (§32), where ω_0 is the plasma frequency given by $\omega_0 \approx (ne^2/m)^{1/2}$. The usual electron density in metals is $n \sim 10^{22}$ cm^{-3} and, therefore, the skin depth is $\delta \approx 10^{-5}$-10^{-6} cm, whereas the orbit radius is $r_H \approx v/\omega \approx 10^{-3}$ cm (even at centimeter wavelengths), which corresponds to resonance magnetic fields of $\sim 10^4$ Oe.

The strong inhomogeneity of the high-frequency field alters drastically the mechanism by which an electron acquires energy from the field, and makes this mechanism strongly dependent on the orientation of the static magnetic field relative to the surface of the metal (§34). If the magnetic field is inclined to the surface of the metal, practically all the electrons (with the exception of a small fraction, δ/r_H, of electrons whose velocities along the magnetic field are

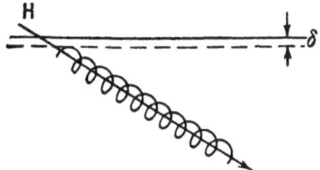

Fig. 77. Motion of an electron in a magnetic field inclined to the surface of a metal.

approximately equal to $\omega\delta$) escape from the skin layer in the first revolution and are accelerated only over a short arc (Fig. 77). Clearly, in this case the influence of a magnetic field on the surface impedance of a metal is weak (this is demonstrated rigorously in [7, 19-21]).

If the field is exactly parallel to the surface, there are some electrons which do not collide with the surface and return to the skin layer after each revolution. In this case, the skin layer performs a role completely analogous to the accelerating gap in a cyclotron. If the return of an electron to the skin layer is synchronized with the external high-frequency (hf) field and the frequency ω is equal to or a multiple of the frequency Ω (the period of one revolution is equal to or an integral multiple of the period of the hf field), it follows that (§34) electrons are accelerated in the skin layer by a factor $l/2\pi r_H$. This gives rise to a special type of cyclotron resonance.

The second factor which complicates cyclotron resonance in metals is also due to one of their basic properties, which is the high carrier density resulting in the filling of a finite fraction of the conduction band (the exceptions are semimetals such as bismuth, which will be discussed below). Since the electrons participating in the conduction processes are not concentrated at the band edge, we can see that there are no grounds for expecting their dispersion law to be nearly quadratic.[1]

If the dispersion law is nonquadratic, the trajectory of an electron in the momentum space is described (§4) by the equation

$$\varepsilon(p) = \varepsilon, \; p_z = \text{const.}$$

It also follows from the equation of motion that the trajectory projected in the coordinate space onto an xy plane perpendicular to the magnetic field is similar to the trajectory in the momentum space rotated through 90°, the conversion coefficient being c/eH (§4).

Electrons moving along orbits which are open in the bulk of a metal (i.e., electrons which follow infinite paths) have an infinite period of rotation and they do not return to the skin layer. Obviously, these electrons can never take part in resonance. However, resonance may occur in an inclined magnetic field if an "open" direction is parallel to the surface of the metal (see [22]).

We shall consider only closed orbits, which are in practice encountered for all directions of the magnetic field even if the Fermi surface is not closed. Electrons moving along closed orbits execute periodic motion but the frequency now depends on p_z because

$$\Omega = eH/m^*c \quad \text{and} \quad m^* = (1/2\pi)\,\partial S(\varepsilon, p_z)/\partial\varepsilon.$$

The effective mass m* depends also on the energy but this dependence can be ignored because

---

[1] This does not apply to alkali metals (K, Na, Rb, Sc) whose Fermi surfaces are nearly spherical.

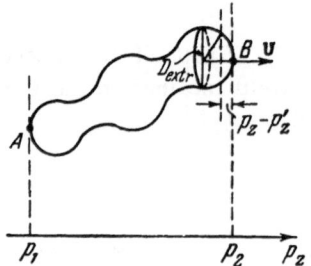

Fig. 78. Fermi surface showing the points p_1 and p_2 which
are the values of p_z corresponding to the limiting points on the
Fermi surface for $\mathbf{v}\,(p_{1,2}) \parallel \mathbf{H}$.

only those electrons whose energies are close to the Fermi value ε_F are of importance in metals.

Therefore, the resonance conditions apply rigorously[2] only to electrons associated with the sections for which $\Omega(p_z) = \omega$. This does not imply that resonance may occur irrespective of the nature of the dispersion law, and the existence of resonance must be discussed in greater detail.

However, even general considerations are sufficient to predict which electrons are in the more "privileged" position.

When p_z is close to p_0, the frequency Ω usually varies linearly with $(p_z - p_0)$. However, if p_0 corresponds to an extremal value of the cyclotron frequency $[\Omega'(p_0) = 0]$, the frequency Ω varies much more slowly with $(p_z - p_0)$. Therefore, near such a section a considerably larger number of electrons has a frequency of revolution close to $\Omega(p_0)$ and, therefore, one can expect resonance at this frequency.

The "special" frequencies are, as usual, those which correspond to the highest and the lowest values of p_z in the continuous spectrum, i.e., the frequencies corresponding to the limiting points on the Fermi surface (points A and B in Fig. 78, at which the velocity is parallel to the field). It is natural to expect resonance at these frequencies.

It is worth considering the depth and the half-width of the resonance minimum of the impedance (the impedance has a minimum because the conductance increases strongly at the resonance points). Far from the resonance point the electrons returning to the skin layer are out of phase with the electric field and are either accelerated or slowed down, and the energy acquired by the electrons from the field is of the same order of magnitude as that in the case of a single passage through the skin layer. If Ω is independent of p_z (this is the case, for example, for a quadratic dispersion law) and the frequency satisfies the resonance condition ($\omega = q\Omega$, where q is an integer), electrons are accelerated synchronously by the field each time they cross the skin layer, so that the energy acquired from the field (and, therefore, the conductivity) increases proportionally to the number of revolutions in the skin layer between two collisions, i.e., it increases by a factor $l/2\pi r_H \sim \Omega\tau/2\pi$. Consequently, the relative depth of the resonance minimum of the impedance ζ, which is proportional to $\delta \propto (\sigma_{eff}/\sigma)^{-1/3}$ [Eq. (33.1a)], is of the order of $(\Omega\tau/2\pi)^{-1/3}$. Because the attenuation depth and the oscillation period of a wave in a metal change considerably near the resonance point, an even deeper minimum is encountered when the conditions deviate slightly from resonance (see §37).

---

[2] Here and later, we shall assume that, although a resonance at a definite frequency is analyzed, the same considerations apply also to resonances at higher harmonics.

If Ω depends strongly on p_z and resonance corresponds to an extremal value of Ω, synchronous acceleration by the field is experienced by those electrons for which $|\Omega(p_z) - \omega/q| \sim 1/\tau$, i.e., $\Delta p_z/p_0 \sim 1/\sqrt{\Omega\tau}$ (p_0 is a characteristic value of p_z such as, for example, $p_2 - p_1$; see Fig. 78). Since the number of revolutions of these electrons is $l/2\pi r_H \sim \Omega\tau/2\pi$, they are the ones which make the principal contribution to the conductivity and are responsible for its resonance rise by a factor of $(1/\sqrt{\Omega\tau})(\Omega\tau/2\pi) = \sqrt{\Omega\tau}/2\pi$. Thus, the relative depth of the minimum of the impedance ζ is of the order of $(2\pi/\sqrt{\Omega\tau})^{1/3}$. Once again, we may encounter an even deeper minimum at a frequency close to resonance (§37).

Finally, if Ω depends strongly on p_z and resonance occurs at a frequency corresponding to a limiting point [for example, p_1, where $\Delta\Omega(p_z) = \Omega'(p_1)(p_z - p_1) \sim \Omega\Delta p_z/p_0$], electrons are accelerated synchronously by a factor $\Omega\tau$ when $\Delta\Omega \sim 1/\tau$.

In the anomalous skin effect (which, as we have stressed repeatedly, is of practical importance only in metals), the radius of the orbit near a limiting point is small (at the limiting point, the radius vanishes) and, therefore, the relative path traveled in the skin layer increases (Fig. 78), compared with the paths traveled by other electrons, by a factor proportional to $r_0/r(p_z) \sim \sqrt{p_0/\Delta p_z} \sim \sqrt{\Omega\tau}$ (r_0 is the characteristic Larmor radius). Consequently, in the anomalous skin effect, the rise in the conductivity under resonance conditions corresponding to a limiting point is once again due to those electrons which make the largest number of revolutions in the skin layer. Since the relative number of these electrons is of the order of $\Delta p_z/p_0 \sim (\Omega\tau)^{-1}$, the relative rise of the conductivity is of the order $\Omega\tau \cdot (\Omega\tau)\sqrt{\Omega\tau} = \sqrt{\Omega\tau}$ and the relative depth of resonance is the same as that in the case of resonance at an extremal frequency. We note that in the normal skin effect the smallness of the orbit radius near a limiting point is unimportant and the resonance at the limiting frequency corresponding to this point is of the logarithmic type.

The half-width $\Delta\Omega$ of the resonance curve is in all cases determined by the broadening of the resonance frequency resulting from collisions and the value of the half-width is of the order of $1/\tau$.

There is one other frequency at which cyclotron resonance may occur if we consider the next approximation for the anomalous skin effect [23]. This resonance is associated with the special nature of the attenuation of a high-frequency field in a metal which is subjected to a static magnetic field. In § 38 we shall show that an alternating field has peaks ("splashes") in the metal at depths which are multiples of the extremal (in p_z) orbit diameter D_{extr} (Fig. 78). Electrons associated with this diameter are in a "privileged" position and, therefore, resonance may occur at the corresponding resonance frequency. The half-width of this resonance (§36) is determined by $(\omega\tau)^{-1}$ as well as by the relationship between the effective skin depth and the orbit diameter $\delta_{eff}/r \sim (\omega_0 D/c)^{-2/3} \sim \rho$; the order of magnitude $\varkappa$ of the half-width is

$$\varkappa \approx (\omega\tau)^{-1} + \rho^{1/2}, \tag{35.1}$$

where ω_0 is the plasma frequency; the relative correction is of the order of $\varkappa^{-2}\rho^2$.

Cyclotron resonance in metals at extremal frequencies was predicted by Azbel' and Kaner [7] in 1956.[3]

---

[3] Cyclotron resonance in the case of a quadratic dispersion law is implied by formulas (1)-(2) given in [13], where the current density j for $\Omega\tau \gg 1$ is proportional to $\gamma/sh(\pi\gamma)$ and $\gamma = r(1 + i\omega\tau)/l = i\omega/\Omega + 1/\Omega\tau$ so that $j \sim 1/\sin(\pi\omega/\Omega - i\pi/\Omega\tau)$; resonance occurs at frequencies $\omega = q\Omega$, where $q = 1, 2, 3, \ldots$.

Cyclotron resonance can be observed not only in good but also in poor metals which have few carriers (for example, bismuth, antimony, arsenic, graphite, etc.) and in doped semiconductors with carrier densities of the order of 10^{17} cm^{-3}. This is due to the fact that the conductivity increases strongly (by a factor $\Omega\tau$ in the case of a quadratic dispersion law) under resonance conditions. At the same time, the skin depth decreases to such an extent that it may become less than the radius of the Larmor orbit for a given direction of the magnetic field. This may be why the experimental data do not agree with the conventional theory (see, for example, [24]). The theoretical prediction of cyclotron resonance was soon followed by its experimental observation in tin and — in a very weak form — in copper [25].

Since then, cyclotron resonance has been observed not only in tin [26] and copper [27], but also in lead [28], indium [29], zinc [30], aluminum [31], bismuth [32], mercury [33], gallium [34], silver [35], gold [36], sodium and potassium [37], tungsten [38], cadmium [39], antimony [40], and magnesium [41]. The effective masses of electrons were determined along the appropriate directions in all these metals and in some cases (the method will be described later) the experimental values of the Fermi surface diameter and of the velocity on the surface were also found. It must be stressed that, at present, the purpose of such investigations is not to study cyclotron resonance but to determine the structure of the electron energy spectra of various metals by the cyclotron resonance method, which is one of the most effective techniques available for this purpose.

Typical cyclotron resonance curves are shown in Fig. 79. The periodic repetition of the resonance frequencies can be seen clearly in this figure.

The experimental studies of cyclotron resonance have confirmed its high sensitivity to the angle between the static magnetic field and the surface of a metal.

We shall now consider cyclotron resonance in a magnetic field inclined to the surface of a metal.

In such a field, cyclotron resonance may occur, for example, for a central section when electrons do not escape into the bulk of the metal (because the average value of the velocity on a central section is zero). However, the relative number of electrons which do not escape from the skin layer after the n-th revolution is of the order of $\delta/r_H n l \ll 1$ and their relative contribu-

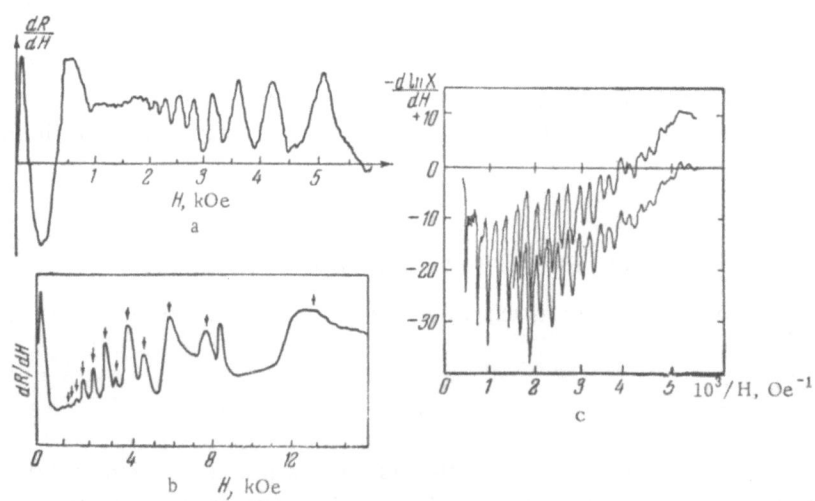

Fig. 79. Cyclotron resonance in: a), b) tin; c) copper. In all the figures, the ordinate is plotted in relative units.

Fig. 80. Effect of surface roughness on the
motion of an electron in a magnetic field.

tion to the impedance is of the order of δ/r_H and is always small (this is true under the anomalous skin effect conditions, i.e., up to frequencies $\omega \gtrsim 10^{13}$ sec^{-1}, which correspond to magnetic fields $H \gtrsim 10^6$ Oe [7]). This applies also to the resonance term for the open trajectories [43].

Quantum oscillations in an inclined field are solely due to electrons near the Fermi-surface sections which are extremal in area (this point will be discussed later) when $p_z \sim p_0(\Omega/\omega)^{1/2}$. These electrons return to the skin layer. Therefore, the amplitude of the quantum oscillations may exhibit a resonance at which this amplitude increases by a factor of $\sim (r/\delta) \times (\Omega/\omega_0)^{1/2}$; their final contribution to the impedance is then of the order of $(r_H/\delta)(\Omega/\omega_0)$. [Formally, the classical and the quantum resonance corrections in an inclined field represent the expansion of the impedance in terms of different small parameters: the "classical" parameter is δ/r_H and the "quantum" one is $(\Omega/\varepsilon_0)^{1/2}$.]

None of this applies to the special cases when $\bar{v}_z = 0$ over the full range of p_z or when cyclotron resonance is possible in a magnetic field inclined at any angle and in a range of values of p_z for which $\bar{v}_z = 0$.

The influence of the inclination of the magnetic field can be explained also in a somewhat different way. The drift of electrons along a magnetic field gives rise, on the one hand, to a Doppler-type frequency shift whose sign is different for electrons traveling toward and away from the surface of a metal. The result is the splitting of the resonance frequency into a doublet. On the other hand, the appearance of a continuous (in the wave vector) correction to the resonance frequency results in the "spreading" of the resonance and the broadening of the resonance curve. Fields inclined at small angles can thus split the resonance frequency (the initial effect may be the enhancement of the resonance peak since the "mean free time" may increase, as shown in Fig. 80, because smooth undulations of the surface may not result in collisions), and the more favorable conditions for resonance are provided by the central section for which $\bar{v}_z = 0$. A considerable inclination in the case of "noncentral" sections gives rise to a new type of resonance, which is described in the next section and which follows from the next approximation in respect of the anomaly parameter δ/r. (The experimental data on cyclotron resonance in inclined fields are reported in [35, 44].)

§ 36. Theory of Cyclotron Resonance

The calculation of the current density at a point $\mathbf{r}$ at time t reduces to the determination of the energy $\Delta\varepsilon$ acquired from an external field by an electron (whose crystal momentum is $\mathbf{p}$) arriving at this point at the specified time [Eq. (32.11)].

We shall consider the general case of the motion of an electron in an alternating electromagnetic field and a static magnetic field, the latter being oriented in an arbitrary manner in relation to the surface of a metal (Fig. 81).

The state of an electron can be represented by variables which are more convenient than the three projections of the momentum (§27). These variables are the energy ε, the projection of the momentum p_z along the static magnetic field H (this direction is taken as the z axis and the ξ axis is assumed to be perpendicular to the surface of the metal), and the time of revolu-

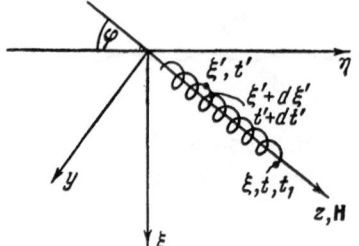

Fig. 81. Motion of an electron in a magnetic field oriented at an arbitrary angle with respect to the surface of a metal.

tion of an electron along an orbit measured from some specified point. These three quantities determine completely the crystal momentum of an electron. It is obvious that t and t + T, where $T = 2\pi/\Omega = 2\pi m^*c/eH$, correspond to the same value of **p** because all the quantities associated with the motion of an electron in the momentum space are periodic in t and the period is T.

We shall find $\Delta\varepsilon$ for an electron which arrives at a point ξ at a moment t_1 and has (in the zeroth approximation in E) an energy ε, a crystal momentum projection p_z, and is located at this moment at a point on the orbit which can be represented by a time t.

We shall determine the energy acquired in the time interval (measured along the orbit) between t' and t' + dt'.

In the time from t' to t, the electron traverses a path $\xi(t) - \xi(t') = \int_{t'}^{t} v_\xi(t')\,dt'$; the function $v_\xi(t)$ is known if we know $\varepsilon(\mathbf{p})$ and the relationship of **p** with ε, p_z, and t. Consequently, at the time t' the electron was at the point $\xi - \xi(t) + \xi(t')$. [We recall that ξ is the coordinate of a point in the coordinate space and $\xi(t)$ describes the motion of an electron in a magnetic field.] The electron was at this point earlier at the time $t_1 - (t - t')$. Therefore, Eq. (32.13a) assumes the following form (the field and its amplitude will again be denoted by the same symbol):

$$\dot{\varepsilon}(t') = ev_\alpha(t')\,E_\alpha(\xi - \xi(t) + \xi(t'))\exp\{i\omega t_2 - i\omega(t - t')\} \tag{36.1}$$

(α represents the coordinates x and η in the plane of the surface of a metal; summation is carried out whenever α is repeated), so that $d\varepsilon(t') = \dot{\varepsilon}(t')\,dt'$.

The probability that the electron retains this energy by avoiding collisions in the time t − t' is $\exp[-(t - t')/\tau]$; therefore, we finally obtain

$$\Delta\varepsilon(\xi, t_1) = \int_{-\infty}^{t} ev_\alpha(t')\,E_\alpha(\xi - \xi(t) + \xi(t'))\exp\{i\omega t_1 - i\omega(t - t')\}\,dt\,\exp\left(-\frac{t - t'}{\tau}\right). \tag{36.2}$$

Substituting Eq. (36.2) into Eq. (32.11) and replacing **p** with the new variables, we find that the amplitude of the current density $j_\alpha(\xi)$ is given by

$$j_\alpha(\xi) = \frac{2e^2}{(2\pi\hbar)^3}\left|\frac{eH}{c}\right| \int_{p_z^{min}(\varepsilon_F)}^{p_z^{max}} dp_z \int_{0}^{T(\varepsilon_F,\,p_z)} dt\,v_\alpha(t) \int_{-\infty}^{t} v_\beta(t')\,dt'\,E_\beta(\xi - \xi(t) + \xi(t'))\exp\left\{-\left(i\omega + \frac{1}{\tau}\right)(t - t')\right\}, \tag{36.3}$$

$$\xi(t) - \xi(t') = \int_{t'}^{t} v_\xi(t_1)\,dt_1.$$

The expression obtained for $j_\alpha(\xi)$ should be substituted into Maxwell's equation (32.4) and (32.6a) for a monochromatic wave of frequency ω (one-dimensional case):

$$E_\alpha''(\xi) = \frac{4\pi i \omega}{c^2} j_\alpha(\xi).$$

(36.4)

As expected, Eqs. (36.3) and (36.4) have an even solution so that the current density $j_\alpha(\xi)$ for $E_\alpha(-\xi) = E_\alpha(\xi)$ is also an even function (§33). To prove this we must use the central symmetry of $\xi(\mathbf{p})$. In the new variables, the replacement of $\mathbf{p}$ with $-\mathbf{p}$ corresponds to the substitution $p_z \to -p_z$, $t \to t + T_1$. Since the function in the integral of Eq. (36.3) is periodic in t, we find that the change of variables $p_z \to -p_z$, $t \to t + T_1$, $t' \to t' + T_1$, and proves the even nature of $j_\alpha(\xi)$.

Using the Fourier components of both parts of Eqs. (36.3) and (36.4) and bearing in mind the discontinuity of $E_\alpha'(\xi)$ at $\xi = 0$, we find that:

$$- k^2 \varepsilon_\alpha(k) - 2E_\alpha'(0) = \frac{4\pi i \omega 2 e^2}{c^2 (2\pi\hbar)^3} \left| \frac{eH}{c} \right| K_{\alpha\beta}(k) \varepsilon_\beta(k),$$

$$E_\alpha'(0) \equiv E_\alpha'(+0);$$

(36.5)

$$\varepsilon_\alpha(k) = 2 \int_0^\infty E_\alpha(\xi) \cos k\xi \, d\xi, \quad E_\alpha(\xi) = \frac{1}{\pi} \int_0^\infty \varepsilon_\alpha(k) \cos k\xi \, dk,$$

(36.6)

$$K_{\alpha\beta}(t) = \int_{p_z^{\min}(\varepsilon_F)}^{p_z^{\max}(\varepsilon_F)} dp_z \int_0^{T(p_z, \varepsilon_F)} dt \, v_\alpha(t) \exp\left\{ -ik\xi(t) - \left(i\omega + \frac{1}{\tau}\right)t \right\} \int_{-\infty}^{t} dt' v_\beta(t') \exp\left\{ ik\xi(t') + \left(i\omega + \frac{1}{\tau}\right)t' \right\},$$

(36.7)

or, reducing $\int_{-\infty}^{t}$ in Eq. (36.7) to an integral over a period, we obtain

$$K_{\alpha\beta}(k) = \int_{p_z^{\min}}^{p_z^{\max}} \left[1 - \exp\left(ik\bar{v}_z T \sin\varphi - 2\pi i \frac{\omega}{\Omega} - \frac{2\pi}{\Omega\tau} \right) \right]^{-1} dp_z \times$$

$$\times \int_0^T v_\alpha(t) \exp\left\{ -ik\xi(t) - \left(i\omega + \frac{1}{\tau}\right)t \right\} dt \int_{t-T}^t v_\beta(t') \exp\left\{ ik\xi(t') + \left(i\omega + \frac{1}{\tau}\right)t' \right\} dt',$$

(36.8)

where

$$\bar{\varphi} = \frac{1}{T} \int_0^T \varphi(t_1) \, dt_1, \quad T = 2\pi/\Omega.$$

(36.9)

In deriving Eq. (36.8), we have used $\bar{v}_x \sim \bar{p}_y = 0$, $\bar{v}_y \sim \bar{p}_z = 0$.

We note that an allowance for the dependence of τ on $\mathbf{p}$ does not alter Eq. (36.8) if $1/\tau$ is understood to be $\overline{\tau^{-1}}$. This is a natural approach since, in the lowest approximation, the orbits are represented only by the integrals of motion ε and p_z, and the degeneracy in t results in the averaging of the reciprocal of the relaxation time.

The characteristic quantity k is determined by the attenuation of the field, i.e., by the effective skin depth (this can be checked separately): $k \sim 1/\delta_{\text{eff}}$ so that $kr_H \gg 1$ and the integrals with respect to t and t' in Eq. (36.8) can be calculated by the method of steepest descent. Consequently, only "glancing" electrons with $v_\xi = 0$ are effective.

All the results deduced from the physical considerations in §35 can also be obtained from Eqs. (36.5)-(36.8). However, we shall consider only the most interesting case (in view of the associated resonance) of a magnetic field parallel to the surface of a metal, when $\varphi = 0$, $\eta \equiv z$, $\xi = y$, and $y(t) = cp_x(t)/eH$.

We must point out that the boundary condition in Eq. (36.3) is particularly simple near resonance. The electrons colliding with the surface cannot participate in cyclotron resonance if their reflection from the surface is not completely specular. Therefore, these electrons make only a small contribution to the current density. Consequently, they should simply be omitted from our considerations by substituting into Eq. (36.3) the factor $E(y - y(t) + y_{min})$, where $E(W) = 1$ for $W > 0$ and $E(W) = 0$ for $W < 0$.

It is evident from Eq. (36.8) that cyclotron resonance should occur at the frequency $\Omega = \omega$, as well as at the frequencies $\Omega = \omega/2, \omega/3, \omega/4, \ldots$. It also follows from this equation that the depth of resonance is quite different for the quadratic and the nonquadratic dispersion laws.

If the dispersion law is quadratic, the frequency Ω is independent (as demonstrated earlier) of p_z and, therefore, $j_a \sim \left[1 - \exp\left(-2\pi i\, \frac{\omega}{\Omega} - \frac{2\pi}{\Omega\tau} \right) \right]^{-1}$, so that as $\tau \to \infty$ the current density at resonance ($\omega = \Omega, 2\Omega, \ldots$) increases indefinitely and is proportional to $\Omega\tau$.

If the dispersion law is not quadratic, we can easily show that the coincidence of ω with one of the frequencies Ω, other than the extremal or limiting frequency, does not give rise to any singularity in the lowest approximation with respect to the anomaly of the skin effect.

If ω is equal to or a multiple of one of the extremal or limiting values $\Omega = \Omega_0$ ($\omega = q\Omega_0$, where $q = 1, 2, \ldots$), resonance occurs but — in contrast to the quadratic dispersion law — we now have $j \sim \sqrt{\Omega\tau}$, i.e., the amplitude of the resonance peak is much less.

The formula for the impedance can be obtained simply by using the extreme anomaly of the skin effect (the inequality $kr_H \approx r_H/\delta \gg 1$) and calculating the inner integrals in Eq. (36.8) by the method of steepest descent. The important points are, as usual (§32), those with $\dot{y} = v_y = 0$, i.e., the points corresponding to the "glancing" electrons. At these points, the smooth functions v_α and v_β are simply taken outside the integral sign. If the dispersion law is strongly nonquadratic, the only operative values of p_z are the "selected" values which ensure resonance. In this case, v_α and v_β should be taken only at the points just specified and should not be included under the integral sign. This allows us to reduce $K_{\alpha\beta}(k)$ directly to the principal axes because $K_{\alpha\beta}(k)$ is proportional to a real tensor which is independent of k.

If the dispersion law is quadratic and Ω is independent of p_z, so that the resonance factor $\left[1 - \exp\left(-2\pi i\, \frac{\Omega}{\omega} - \frac{2\pi}{\Omega\tau} \right) \right]^{-1}$ is taken outside the integral sign, the asymptotic form of the remaining expression is independent of Ω for $\Omega\tau \gg 1$ and, in general, is of the same form as in the zero magnetic field [Eq. (33.15)]. This means that once again $K_{\alpha\beta}(k)$ can be reduced directly to the principal axes.

The calculations suggested in the preceding paragraphs are straightforward but cumbersome. Therefore, we shall just quote and analyze the results. We note that the calculations are based on the anomaly of the skin effect and the results obtained are only qualitatively valid for any magnetic field [compare, for example, Eq. (36.10) with Eqs. (33.16) and (33.15); real factors of the order of unity are ignored in all cases].

We shall consider separately the quadratic and the nonquadratic dispersion laws.

Quadratic Dispersion Law.   In this case, the impedance is of the form

$$
\begin{aligned}
&\zeta_\alpha(H) \approx \zeta_\alpha(0)\left[1 - \exp\left(-2\pi i\,\frac{\omega}{\Omega} - \frac{2\pi}{\Omega\tau}\right)\right]^{-1/3}, \\
&\zeta_\alpha(0) = \left(\frac{\sqrt{3}\,\omega^2}{\pi^2 c B_\alpha}\right)^{1/3}(1 + i\sqrt{3}), \\
&B_{\alpha\beta} = \frac{8e^2}{3(2\pi\hbar)^3}\int_0^{2\pi}\left[\frac{v_\alpha v_\beta}{\varkappa v^2}\right]_{v_y=0} d\varphi,
\end{aligned}
\qquad (36.10)
$$

where B_α are the principal values of the real tensor $B_{\alpha\beta}$; $\varkappa$ is the Gaussian curvature; φ is the angle measured along the strip defined by $\varepsilon = \varepsilon_F$, $v_y = 0$. The two principal values of the surface impedance tensor are of the resonance type and the values of ζ' and ζ'' are minimal at the resonance point. Since measurements are carried out at a fixed value of the field, these minimal values correspond to an absorption minimum.

Nonquadratic Dispersion Law.   In this case, the inequalities $\Omega\tau \gg 1$ and $r_H/\delta \gg 1$ give rise to a situation in which only the electrons with $v_y = 0$, $p_z \approx p_0$, and $\Omega(p_0) = \Omega_{\text{extr}}$, are important, i.e., we need consider only those electrons which move almost parallel to the surface of a metal and have a frequency of revolution along an orbit which is close to the extremal value (with respect to φ).

The impedance can be reduced to the principal values simultaneously with the tensor $A_{\alpha\beta}$. These principal values ζ_α can be expressed in terms of the principal values A_α of the tensor $A_{\alpha\beta}$ by means of the formula:

$$
\zeta_\alpha = \frac{4}{9}\left(\frac{\sqrt{3}\,\omega^2}{\pi^2 c A_\alpha}\right)^{1/3}\exp\left(\frac{i\pi}{3}\right),
$$

$$
A_{\alpha\beta} = \frac{16 e^2}{3(2\pi\hbar)^3}\sum_{i=1}^{a}\frac{v_\alpha v_\beta}{a\varkappa v^2}\bigg|_{\varphi=\varphi_i}\int_0^\pi\left[1 - \exp\left(-2\pi i\,\frac{\omega}{\Omega} - \frac{2\pi}{\Omega\tau}\right)\right]^{-1} d\varphi.
\qquad (36.11)
$$

The variables in Eq. (36.11) are the same as those in Eq. (36.10) so that integration is carried out over an angle corresponding to a strip $v_y = 0$ on the Fermi surface (Fig. 82). Here, $\varphi_1, \varphi_2, \ldots, \varphi_\alpha$ are the points where Ω corresponds to an extremal value of φ. When the cubic root is taken in Eqs. (36.10) and (36.11), we must select that root which corresponds to $\zeta'_\alpha > 0$ (such a root always exists). Strictly speaking, Eqs. (36.10) and (36.11) should include also the nonresonance terms, particularly those associated with the electrons scattered by the surface. This is necessary because the small value of the complex denominator of a fraction does not, in general, ensure that the real and the imaginary parts of the fraction are large (for example,

$$
\lim_{a\to 0}\operatorname{Im}\frac{1}{a - ia^2} = 1\Big).
$$

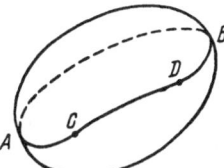

Fig. 82.   Strip on
the Fermi surface.

However, an "additional" nonresonance correction to A_α and B_α can only give rise, in the final formulas, to real factors of the order of unity. Therefore, such a correction is of no importance and there is no need to write it out explicitly.

We shall continue to assume that the principal contribution to the current density is made by the "resonance" electrons. In the case of a very complex Fermi surface, the fraction of electrons located on a given part of this surface may be numerically so small that the nonresonance electrons will dominate the experimental results corresponding to values of $\Omega\tau$ which can be realized [45]. This case can also be considered without any special difficulty.

We have confined our analysis to the lowest approximation in the anomaly of the effect, i.e., in the parameter δ/r_H. New resonance frequencies are obtained in the next approximation with respect to this parameter. For $\varphi = 0$, the new frequency represents a cyclotron resonance corresponding to extremal values of the orbit diameter D ($kD \gg 1!$). The half-width of the resonance curve is given by Eq. (35.1). If $\varphi \neq 0$, the limiting point is that for which $v_\xi = \dot{v}_\xi = 0$ and d = 0. Calculations of the integrals with respect to t and t' in Eq. (36.8) near such a point depends on the value of p_z. The procedure depends on whether p_z corresponds to a point on the orbit with $v_\xi = 0$, or whether p_z has a value for which $\xi(t)$ is a monotonic function. Consequently, the points $p_z = p_0$ become singular and make their own small contribution to $K_{\alpha\beta}$. The impedance can be calculated conveniently by expanding $(k^2\delta_{\alpha\beta} + K_{\alpha\beta})^{-1}$ as a series in terms of a small correction. In the second approximation, the terms containing the product $e^{ikd(p_0)}e^{-ikd(-p_0)} = 1$ give rise to resonance in an inclined field at frequencies

$$2\omega = n\Omega(p_0), \quad n = 1, 2, \ldots \tag{36.12}$$

We shall now consider the physical nature of resonance in an inclined field.

We shall consider the influence of the value of φ on the resonance at a limiting point. Beginning from $\varphi \sim \delta_{eff}$, electrons near a limiting point do not return to the skin layer and if $\varphi \gg \delta/r_H$, only those electrons which are sufficiently far from such a point can be scattered at glancing angles by the surface because of $v_\xi(t) = 0$. The effective electrons, i.e., those that remain for a long time in the skin layer, are those which are characterized by values close to $p_z = p_0$ (p_0 and t_0 are deduced from $v_\xi = 0$, $dv_\xi/dt = 0$). Such electrons are focused by a magnetic field at certain depths. This gives rise to peaks ("splashes") of the field and the current at depths ξ which are multiples of the "limiting" value $d_0 = d(p_0)$, where d is the path traversed in one period T:

$$d = \int_0^T v_\xi(t)\, dt.$$

These peaks are damped out only because of the collisions of electrons in the bulk of a crystal at a depth of the order of the mean free path. If 2ω is a multiple of $\Omega_0 = \Omega(p_0)$, the electrons with $p_z = -p_0$ traverse synchronously numerous field peaks. For example, if $\omega = \Omega_0/2$, the electrons with $p_z = -p_0$ find a field peak at a depth $\xi = d_0$, which has been carried to this point from the surface by electrons with $p_z = p_0$ in a time interval T_0. After the elapse of another time interval T_0, these electrons pass again through a field maximum near the surface $\xi = 0$ and this will automatically eliminate the spatial phase shift acquired by the electrons with $p_z = p_0$. This gives rise to resonance whose damping is solely due to the damping of the peaks and whose relative half-width is $\sim\gamma = T_0/\tau$. In this case, we have

$$\frac{\Delta\zeta}{\zeta} \approx \left(\frac{\delta}{r_H}\right)^{4/3} \sum_{n=1}^{\infty} \frac{1}{n^2} \exp(-2i\omega^\cdot T_0 n) \sim \left(\frac{\delta}{r_H}\right)^{4/3} A^{-1} \ln A^{-1}, \tag{36.13}$$

where

$$A = 1 - \exp(-2i\omega^* T_0).$$

This type of resonance was first predicted in [46] and the case corresponding to $\varphi \ll 1$ was considered in [47]. A physical explanation of the Doppler splitting of the frequencies for $\varphi \ll 1$ is given in [48]. The theory of this resonance is developed in [49].

This resonance is not observed if the interval $2p_0$ degenerates to a point, i.e., if p_z becomes perpendicular to a planar section $v_\xi = 0$. Under isotropic dispersion conditions, this happens when $\mathbf{H}$ is perpendicular to the surface of a metal.

Experimental investigations of this resonance can be used, in principle, to determine the effective mass m^* and the area of a cross section S of the Fermi surface as a function of p_z for any direction z [in this case we use $d = c/eH(\partial S/\partial p_z)_\varepsilon$], i.e., such investigations can give the same information as the quantum cyclotron resonance (§43). If $\varphi \ll 1$, we can use this method to find the Gaussian curvature $\varkappa$ and – using the Doppler splitting – the velocity v on the Fermi surface. If $\varphi \ll 1$, the electric field must be directed along the velocity at a limiting point and the reasons for this are the same as in the case $\varphi = 0$.

The resonance at any value of φ in the case $2\omega = n\Omega$ (n is an integer) is observed if there are sections for which d has an extremum. The resonance associated with these sections corresponds to $\Delta\zeta/\zeta \sim \delta/r_H \ln A^{-1}$ and the value of T in Eq. (36.13) is taken over these sections.

The open sections give $\Delta\zeta/\zeta \sim \delta A^{-1}/r_H$ and Ω corresponds to the diameters $D = \xi_{max} - \xi_{min}$, which are extremal in p_z (the maximum and minimum values of ξ correspond to moments t within the same period; Fig. 83). The resonance in an inclined field may be more sharply peaked than for $\varphi = 0$ since τ is not reduced by the collisions of electrons with surface irregularities. None of these results applies to special values of the angle φ, such as those for which an extremum of d disappears, or those for which $d' = 0$ and $d'' = 0$ simultaneously, or those for which the resonance frequencies corresponding to different sections are identical.

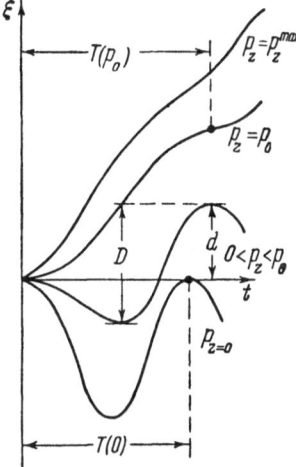

Fig. 83. Different variants
of the dependence $\xi = \xi(t)$.

§37. Investigation of Surface Impedance under Cyclotron Resonance Conditions

It is convenient to consider the surface impedance near cyclotron resonance separately for the quadratic and the nonquadratic dispersion laws.

Since the experimental measurements frequently give the derivatives of $\zeta' = \mathrm{Re}\ \zeta$ and $\zeta'' = \mathrm{Im}\ \zeta$ rather than the quantities themselves (for example, $d\zeta'/dH$, $d \ln \zeta''/dH$), we shall consider also the derivative of the impedance with respect to the magnetic field.

Quadratic Dispersion Law. If the dispersion law is quadratic, we can plot the whole resonance curve for any value of H, irrespective of whether or not it is close to a resonance value. The frequency dependences of $\zeta'(H)/\zeta'(0)$, $\zeta''(H)/\zeta''(0)$, and $\zeta''(H)/\zeta'(H)\sqrt{3}$ are plotted in Fig. 84 for $\omega\tau = 1$, 10, and 50. The maxima of ζ' and ζ'' at $\omega \approx (q + \tfrac{1}{2}\Omega)$ are not associated with resonance and as $\Omega\tau \rightarrow \infty$, the values of the impedance at these points tend to a constant nonzero limit.

It is important to note that if τ is finite the depth of the resonance minimum and the shift of its frequency relative to ω/q are quite different for ζ' and ζ'':

$$\zeta'_{\mathrm{res}} = \zeta'\left(H'_{\mathrm{res}}\right) \approx \zeta'(0)\left(\frac{2\pi q}{\omega\tau}\right)^{2/3}, \quad \frac{\omega - q\Omega'_{\mathrm{res}}}{\omega} \approx (2\pi q\omega\tau)^{-1/3},$$

$$\zeta''_{\mathrm{res}} = \zeta''\left(H''_{\mathrm{res}}\right) \approx \zeta''(0)\left(\frac{2\pi q}{\omega\tau}\right)^{1/3}, \quad \frac{\omega - q\Omega''_{\mathrm{res}}}{\omega} \approx (\omega\tau)^{-1}.$$

The frequency shifts exhibited by ζ' and ζ'' are due to different reasons. The shift of ζ'' is due to the fact that a small increase in the magnetic field, which hardly changes the resonance conditions, gives rise to a "favorable" increase in the number of electron revolutions between collisions. The frequency shift of ζ' can be understood by considering how the results are affected by the variation of the electric field phase with depth in a metal. A change in the phase disturbs the resonant synchronization and, therefore, reduces the energy acquired by an electron from the field so that the resonance is "damped." It is evident from Eq. (36.10) and from the dependence of $\zeta''(H)/\zeta'(H)\sqrt{3}$ on the magnetic field (Fig. 84) that even a very small

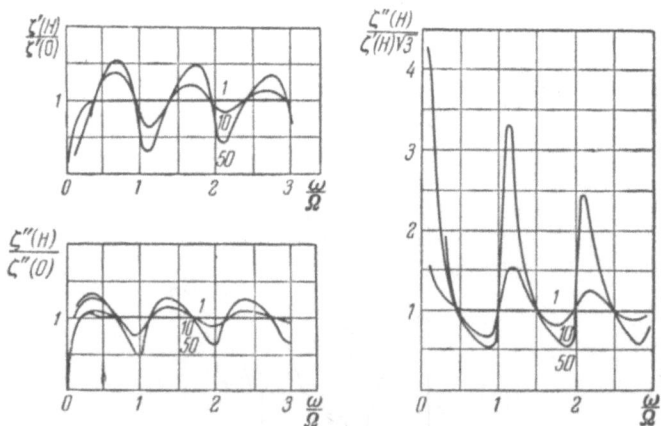

Fig. 84. Theoretically calculated resonance curves for a metal with a quadratic dispersion law (the values of $\omega\tau = 1$, 10, and 50 are given alongside the curves).

change in the magnetic field gives rise to $\zeta'' \gg \zeta'$, i.e., to a constancy of the field phase at the attenuation depth δ. This is convenient in spite of the fact that, under these conditions, an electron which makes $\sim |\omega - q\Omega'_{res}|\tau$ revolutions finds that the phase of the field near the surface is now quite different.

We shall now consider $d\zeta'/dH$ and $d\zeta''/dH$. At first sight, it might appear that at resonance, when the values of ζ' and ζ'' are minimal, their derivatives should vanish.

However, in fact, resonance does not correspond to the zero values of $d\zeta'/dH$ and $d\zeta''/dH$ but to their maximum values. This is due to the fact that when $\omega\tau = \infty$ the functions $\zeta'(H)$ and $\zeta''(H)$ do not have their minimal values but the smallest values (zero) which correspond to cusps in these functions. Thus, when $\omega\tau = \infty$, it follows from Eq. (36.10) that to the left of the resonance point (where $H < H'_{res}$) we have $\zeta'(H) \propto (H'_{res} - H)^{4/3}$ and $d\zeta'/dH \rightarrow 0$ if the resonance is approached from the left; to the right of the resonance point (where $H > H'_{res}$) we have $\zeta'(H) \propto (H - H'_{res})^{1/3}$ and $d\zeta'/dH \rightarrow \infty$ when the resonance is approached from the right (Fig. 85).

The functions $\zeta''(H)$ and $d\zeta''/dH$ behave similarly. Therefore, $(d\zeta/dH)_{res} \approx \zeta(0) \times (q^2\omega\tau)^{2/3}/H^{(1)}_{res}$ and, as $\tau \rightarrow \infty$, $[1/\zeta(0)](d\zeta/dH)$ approaches infinity and not zero and $(\omega-q\Omega_{res})/\omega \approx (\omega\tau)^{-1}$. Here, $H^{(1)}_{res}$ is the magnetic field corresponding to the fundamental resonance (the "first" harmonic $q = 1$). The relative amplitudes of the maxima of $d\zeta'/dH$ and $d\zeta''/dH$ are considerably greater than the reciprocal of the minimum value of $\zeta''(H)$ and the shift of the resonance frequency is the same as for ζ'' but considerably smaller than for ζ'.

Nonquadratic Dispersion Law. The formula (36.11) for a nonquadratic dispersion law includes the quantity $\Omega(\varphi)$ and not $\Omega(p_z)$, which we have been considering so far. Since the function $p_z(\varphi)$ evidently has extrema only at the limiting points of the Fermi surface (Fig. 78), and $d\Omega/d\varphi = (d\Omega/dp_z)(dp_z/d\varphi)$, it follows that $\Omega(\varphi)$ has extrema at the same places as $\Omega(p_z)$ as well as at extrema of $p_z(\varphi)$, i.e., at the elliptic limiting points of the Fermi surface (this has been discussed earlier). At the hyperbolic limiting points $m^* = \infty$, $\Omega = 0$, and resonance is impossible. However, since $r_H = 0$ at elliptic limiting points, the basic condition of cyclotron resonance $r_H/\delta \gg 1$ cannot be satisfied at these points. Since the important values in the vicinity of a limiting point are

$$r_H \approx \frac{v}{\Omega}(\omega\tau)^{-1/3} \approx \frac{v}{\omega}(\omega\tau)^{-1/3},$$

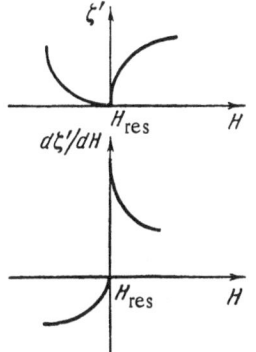

Fig. 85. Behavior of ζ' and $d\zeta'/dH$ as a function of the magnetic field H near resonance for a quadratic or a nonquadratic dispersion law in the case of the minimum effective mass m^*_{min}.

it follows that we must satisfy

$$\frac{r_H}{\delta} \approx \frac{\omega_0 \frac{v}{c}}{\omega} (\omega\tau)^{-1/2} \gg 1.$$ (37.1)

If the collisions with phonons predominate, the application of Eq. (32.18a) gives rise to the condition

$$\frac{v}{c} \frac{\omega_0}{\omega_\Theta} \left(1 + \frac{kT}{\hbar\omega}\right)^{3/2} \gg 1, \qquad \omega_\Theta = \frac{\Theta}{\hbar},$$

where Θ is the Debye temperature.

In the case of good metals, we usually have $\omega_0 v/c \sim \omega_e$ and, therefore, $\hbar\omega$ must be considerably smaller than kT. However, in practice, the condition (37.1) is very easy to satisfy because, at low temperatures we usually have $\tau \ll \tau_{eff}$, $\omega_e \approx 10^{13}$ sec^{-1}, $\omega \approx 10^{11}$ sec^{-1}, and it is sufficient to satisfy the condition $\omega\tau \ll 10^4$.

Resonance thus occurs in the central section [because the Fermi surface has a center of symmetry, it follows that $\Omega(p_z) = \Omega(-p_z)$, $\Omega'(p_z) = -\Omega'(-p_z)$, and $\Omega'(0) = 0$] at frequencies corresponding to the elliptic limiting points where, as shown in [7], $m^* = (v\sqrt{\varkappa})^{-1}$ (v and $\varkappa$ are, respectively, the carrier velocity and the Gaussian curvature at a limiting point). Resonance also occurs for "noncentral" values of Ω which are extremal in respect of p_z.

The resonance in the central section and that at the limiting points differ considerably from the resonance corresponding to other extremal values of Ω.

In the first two cases, the central symmetry of the Fermi surface makes the resonance term in $A_{\alpha\beta}$ proportional to the tensor $v_\alpha(\varphi_1) v_\beta(\varphi_1)$, one of whose principal values is zero and the other unity. Consequently, only one of the principal values of the impedance can resonate. Since the resonance values of ζ' and ζ'' are minimal and not maximal, the impedance in the case of an arbitrary polarization of the wave incident on a metal is determined by the large nonresonance principal value and the "resonance" component of the impedance represents only a small correction.

The resonance is strong only if the incident wave is polarized along the direction of the velocity v_F at the $\varepsilon = \varepsilon_F$, $v_y = 0$, $\varphi = \varphi_1$ (for a limiting point this direction coincides with that of the static magnetic field, as shown in Fig. 78).

The current density corresponding to the electric field perpendicular to v_F does not resonate because, in this case, the electric field in the skin layer is almost normal to the velocity and, consequently, does hardly any work on the electron.

The derivative of the impedance with respect to the magnetic field is determined primarily by functions which depend strongly on H and it resonates along all directions of $\mathbf{E}$ with the exception of that perpendicular to v_F.

A strong anisotropy of ζ' and ζ'', as well as of $d\zeta'/dH$, $d\zeta''/dH$, should be observed in a very wide range of angles. For example, in the case of absorption of an incident wave given by $p = \zeta'_\alpha (E_\alpha^{inc})^2 + \zeta'_\beta (E_\beta^{inc})^2$, the range of angles is determined by $[\zeta'_{res}/\zeta(0)]^{1/2}$, which usually differs little from unity.

In the resonance corresponding to a noncentral section, we have at least two centrally symmetrical sections in which resonance is possible and the two points (C and D, shown in Fig. 82) make a considerable contribution to the current density. The velocities at points C and D are, in general, not parallel and for any direction of the electric field the current density and both principal values of the impedance exhibit resonance. This is due to the fact that, in general, none of the principal values of the tensor $v_\alpha(\varphi_1) v_\beta(\varphi_1) + v_\alpha(\varphi_2) v_\beta(\varphi_2)$ vanishes.

We shall now consider only the resonance values of ζ_α.

The formulas for the resonance values of the impedance depend on whether a given section corresponds to a minimum or maximum of $\Omega = eH/m^*c$.

In the formulas for ζ_α'' this affects only the numerical constants: we find that $\zeta_\alpha''^{\,res}/\zeta_\alpha''(0) \approx (q^2/\omega\tau)^{1/6}$, $(\omega - q\Omega_{res}'')/\omega \approx (\omega\tau)^{-1}$ for m_{min}^* and m_{max}^*.

However, the formulas which give ζ_α' for m_{min}^* and m_{max}^* differ basically. This difference is due to the fact that the case of minimum effective mass $m_{min}^* = (1/2\pi)(\partial S/\partial\varepsilon)_{min}$ is analogous to the case of the quadratic dispersion law because a small change in H near resonance gives rise to $\zeta'' \gg \zeta'$ and a considerable enhancement of the resonance of ζ'. In the case of maximum effective mass m_{max}^*, the inequality $\zeta'' \gg \zeta'$ cannot be satisfied by a small shift in H [all this can be shown starting from Eq. (36.11)].

To be specific, we shall consider electrons $\partial S/\partial\varepsilon > 0$ and not holes $\partial S/\partial\varepsilon < 0$; however, the analysis is the same in both cases and the final answer includes $|\partial S/\partial\varepsilon|$. When the numbers of electrons and holes are equal, no singularity occurs (this is to be expected because the Hall field E_y does not occur in the formulas).

When the effective mass m^* has its maximum value, the real part of the impedance is equal to the imaginary part and $(\omega - q\Omega_{res})/\omega \approx (\omega\tau)^{-1}$.

When the effective mass m^* has its minimum value, the resonance amplitude of ζ_α' and the shift of the resonance frequency are given by

$$\zeta_\alpha'^{\,res}/\zeta_\alpha'(0) \approx (\omega\tau/q^2)^{-4/9}, \quad (\omega - q\Omega_{res}')/\omega \propto (q\omega\tau)^{-2/3}.$$

The curves representing $d\zeta_\alpha'/dH$ are also quite different for minimum and maximum values of m^*. For m_{min}^*, they are analogous to the curves obtained for the quadratic dispersion law (Fig. 85). In the case of m_{max}^*, the value of $d\zeta_\alpha'/dH$ to the left of the resonance point $(H = H_{res}' - 0)$ becomes $-\infty$ and to the right of the resonance point $(H = H_{res}' + 0)$ it becomes $+\infty$, and the curves are then of the type shown in Fig. 86.

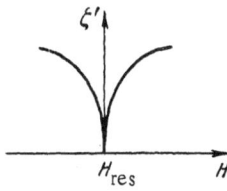

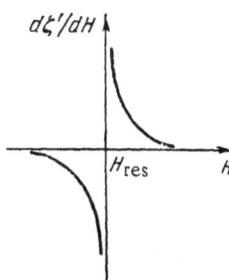

Fig. 86.   Behavior of ζ' and $d\zeta/dH$ near resonance for a nonquadratic dispersion law in the case of maximum effective mass m_{max}^*.

In both cases ($m^{\bullet}_{min}$, $m^{\bullet}_{max}$), we have $(d\zeta'_{\alpha}/dH)_{max} \approx H_1^{-1}\zeta'_{\alpha}(0)\, q^{4/3}(\omega\tau)^{5/6}$ and the frequency shift is $(\omega - q\Omega_{max})/\omega \sim (\omega\tau)^{-1}$, where H_1 is the magnetic field corresponding to resonance at the fundamental harmonic $q = 1$. It is important to note that the relative amplitude of the resonance peak of $d\zeta'_{\alpha}/dH$ is considerably higher than the reciprocal of the relative depth of the resonance of ζ'_{α} and ζ''_{α} and this means that it would be more convenient to measure $d\zeta'_{\alpha}/dH$ rather than ζ'_{α} or ζ''_{α}.

It is clear from our analysis that the behavior of ζ'_{α} and $d\zeta'_{\alpha}/dH$ at the resonance point is very sensitive to some features of the Fermi surface (large differences are observed between resonances at a limiting point, in central and noncentral sections, and for maximum and minimum values of the effective mass).

We have considered so far resonance in a bulk metal which can be regarded as a semi-infinite sample. As in the case of weak fields (§34), interesting features are observed in a plate whose thickness D satisfies the inequalities

$$\delta \ll D \ll l.$$

In this case, resonances are observed only for those harmonics whose associated orbits are smaller than the plate thickness (otherwise an electron would collide with the surface and become ineffective during its first revolution). If the n-th harmonic $n = \omega/\Omega_{extr}$ is observed experimentally but not the $(n+1)$-th (such experiments should be compared with those carried out on a bulk sample), it means that the diameter $d^{(1)}_{res}$ of the orbit corresponding to the first harmonic $\omega = q\Omega_{extr}$ lies between the limits

$$\frac{D}{n+1} < d^{(1)}_{res} = \frac{cp_x^{res}}{eH_1} < \frac{D}{n}.$$

In the case of large values of n, this method for "cutting off" the resonance orbits [50] can give accurate values of the Fermi surface diameters corresponding to the extremal effective masses (calculations are also reported in [51] and experiments on tungsten are described in [52]).

Cyclotron resonance thus makes it possible to determine directly the effective mass and the extremal diameters of the Fermi surface for the conduction electrons and is thus an effective way (in combination with other methods) for finding the shape of the Fermi surface.

§38. Damping of a High-Frequency Field in a Metal

We have not considered so far the details of the penetration of a high-frequency field into a metal. We have simply determined the characteristic depth of penetration (which can be related, for example, to the surface impedance) and compared it with the microscopic characteristics of the motion of an electron, such as the mean free path, the Larmor orbit radius, the distance traveled by an electron during one period of the high-frequency field. We have considered also the normal and the anomalous skin effects.

In the normal skin effect, the field in a metal can be described completely by the complex expression for the skin depth.

In the anomalous skin effect, this is impossible. In the presence of a magnetic field, an electron that has acquired energy from the field in a skin layer carries away this energy into the interior of the metal to a distance of the order of the mean free path and generates a high-

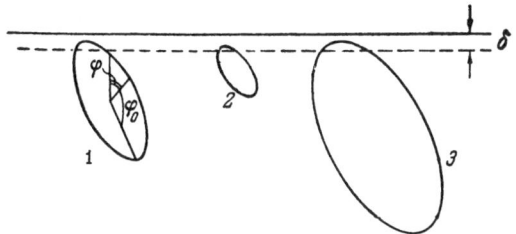

Fig. 87. Motion of electrons in a magnetic field which is exactly parallel to the surface of a metal.

frequency field on the way. Since the electrons escaping into the interior acquire much smaller amounts of energy from the field than the "glancing" electrons which do not escape into the interior (§32) and since the current representing the first type of electrons is spread over a layer whose thickness is not δ but l, the current density and the field in the interior of the metal are considerably weaker than they are in the skin layer and decay to zero at a depth of the order of the mean free path. Thus, in the anomalous skin effect, the damping of the field is characterized by at least two complex quantities of different orders of magnitude [3] and the depths of penetration of the field and the current may differ quite considerably [7].

We shall now consider the case when a magnetic field is exactly parallel to the surface of the metal because the damping of a high-frequency field is then of a very special nature. To understand this case, we must go back to the motion of electrons along one of the orbits (orbit 1 in Fig. 87) passing through the "normal" skin layer near the surface, where the electric field is always relatively high (the static magnetic field is perpendicular to the plane of Fig. 87).

In a layer whose thickness is of the order of δ, the electrons acquire a directed velocity in an arc of length $\sqrt{r_H \delta}$ ($r_H \gg \delta$) and give rise to a current whose density is $j \sim I/\delta$.

Moving down along the orbit, the component of the electron velocity parallel to the surface changes (following the flow of the current) and the current varies proportionally to $\cos \varphi$. Moreover, the electrons spread out (for $\varphi > \sqrt{\delta/r_H}$) in a layer of thickness $\sqrt{r_H \delta} \cos \varphi$.

Thus, the density of the current generated by the electrons belonging to a given orbit is of the order of $(I/\sqrt{\delta r_H}) \cot \varphi$, i.e., it decreases rapidly with depth and at $\varphi \sim 1$ it is $\sqrt{r_H/\delta}$ times lower than in the layer δ.

At a depth $y > r_H$, the current density changes its sign but its absolute value remains small compared with I/δ and this is true as long as the angle φ does not approach φ_0 corresponding to the "bottom" of the orbit, where $|\varphi_0 - \varphi| \lesssim \sqrt{\delta/r_H}$. At a depth d, where electrons "congregate" again in a layer whose thickness is δ, the current density again rises rapidly but this time the sign of the current is different from that on the surface.

This obviously applies to all trajectories of a given radius passing through a thin skin layer parallel to the surface. Therefore, if all the electrons were to move along orbits of the same radius (i.e., if they all were to have the same velocity in the xy plane), the "glancing" electrons in the skin layer would give rise to peaks of the current and electromagnetic field at a depth y = d. Such peaks would accelerate new electrons which have "glanced" in the layer at a depth 2d, and this would be repeated at 2d, 3d, etc.

Such a situation would require the solution of the problem of a self-consistent system of currents and fields but it would explain physically the pattern shown in Fig. 88.

The phenomena observed are affected drastically by the presence of orbits of different radii (orbits 2 and 3 in Fig. 87), corresponding to different sections of the Fermi surface (we recall that, for example, in the case of free electrons $r_H = p_\perp c/eH = c/eH \sqrt{2m\varepsilon - p_z^2}$ and r_H varies from 0 to $c/eH \sqrt{2m\varepsilon}$). The scatter of the orbit radii means that only a small fraction of electrons (of the order of r_H) "congregates" at any given depth in a layer whose thickness

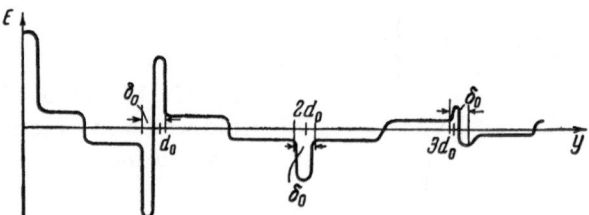

Fig. 88. Attenuation of the electric field with
depth in the presence of field peaks.

is of the order of δ and the field which is "relayed" into the interior of the metal naturally decays rapidly at each "stage."

When the cyclotron frequency Ω is independent of the section of the Fermi surface (this is true of an ellipsoidal surface) the scatter of the orbit radii in a field H parallel to the surface cannot be ignored.

However, if the cyclotron frequency depends on the section of the Fermi surface, i.e., on p_z, we can reduce considerably the scatter of the radii by means of cyclotron resonance because this resonance affects only those electrons which are moving at frequencies close to the extremal cyclotron frequencies and have a scatter of p_z of the order of $\Delta p_z \sim p_F / \sqrt{\omega\tau}$ (because $|\omega - \Omega| \sim 1/\tau$ and p_F is of the order of the Fermi momentum). The scatter of the radii is governed by $\omega\tau$ and it is quite different for the central section, which is associated (as indicated by the symmetry considerations) with an extremum of Ω and d, from the scatter for the limiting points (§§35, 36), where $d \sim \sqrt{\Delta p_z}$, and from the scatter for the other noncentral sections. At the limiting points, the scatter of the orbit diameters is $\Delta d \sim d \sim r_H / \sqrt{\omega\tau}$, so that $\Delta d \gg \delta$ and weakly attenuated current peaks cannot appear at relatively great depths. For other noncentral sections $\Delta d \sim d / \sqrt{\omega\tau}$; the field and current density peaks may appear if $\Delta d \sim \delta$, i.e., $\omega\tau \sim (r/\delta)^2$. For the central section, where $d'(0) = 0$, the scatter is $\Delta d = \frac{1}{2} d''(0) p_z^2$ and the structure shown in Fig. 88 appears if $\omega\tau > r/\delta$, $\omega \gtrsim v/\sqrt{l\delta}$, which corresponds to a wavelength of the order of 1 cm and $H \approx 10^4$ Oe for $l \approx 0.1$ cm, $\delta \approx 10^{-5}$ cm, $v = 10^8$ cm/sec, i.e., this structure should be realized under conditions frequently encountered in cyclotron resonance experiments.

It is easy to see that a departure from the inequality $\omega\tau < r_H/\delta$ reduces the amplitudes of the successive peaks in a geometrical progression in proportion to the degree of "broadening" of the current in the downward direction $\delta/\Delta d \sim (\delta/r)\omega\tau$, which naturally causes strong attenuation of the peaks. A more accurate estimate shows that the peaks corresponding to $\omega\tau < r/\delta$, $y = ad(0)$ are of the order of $E(0) a^{-1/3} (\delta/r_H)^{a/3} (\Omega\tau)^{5a/12}$. The relative rise of the current near $y = ad(0) = ad_0$ is of the order of $(\omega\tau)^{1/2}$. The relative correction to the impedance due to the field peaks is of the order of $(\delta/r_H)^{2/3} (\omega\tau)^{5/6}$ and that due to the "nonsingular" term of the expansion is of the order of $(\delta/r_H)^{1/3} (\omega\tau)^{-1/12}$. Thus, under certain conditions, a sequence of field peaks may appear in the interior of a metal.

We shall now obtain a formula for the attenuation of the field in a metal. This formula can be derived by solving Eqs. (36.5)-(36.8). For the sake of simplicity, we shall consider the structure of field peaks in the case when the resonance conditions for the central section are satisfied exactly: $\omega = q\Omega_0$, where q is an integer and $\Omega_0 = \hat{\Omega}(p_z)|_{p_z=0}$.

Bearing in mind that $\omega\tau \gg 1$ and $\omega = q\Omega_0$, we may substitute $\tau = \infty$ in the inner integrals of Eq. (36.8). Moreover, since the largest parameter is now the resonance parameter $\omega\tau$

$(\omega\tau \gg r_H/\delta)$ and the resonance occurs when $p_z \approx 0$, we may assume that, with the exception of $\exp(-2\pi i\omega/\Omega)$, the condition $p_z = 0$ is always satisfied. This allows us to modify the last of the integrals in Eq. (36.8) by replacing $\int\limits_{t-T}^{t}$ with $\int\limits_{0}^{T}$.

The extreme anomaly of the skin effect, $r_H/\delta \gg 1$, means that only two centrally symmetrical points on an orbit (the "upper" and the "lower") are important. At these points, $v_y = 0$ (it is shown in §36 that this is the consequence of the method of steepest descent). Consequently, the values of v_α and v_β are taken at the point where $\varepsilon = \varepsilon_F$, $p_z = 0$, $v_y = 0$, and, for example, $v_x > 0$; these values are taken outside the integral (the sign has to be determined separately). Selecting the α axis as the direction of velocity at this point (denoted by the subscript zero), we obtain the expression for $\varepsilon_\alpha(k)$ [$\varepsilon_\beta(k)$ is nonresonant and will not be considered]:

$$- k^2 \varepsilon_\alpha(k) - 2E'_\alpha(0) = iaK_{\alpha\alpha}(k)\,\varepsilon_\alpha(k),\qquad(38.1)$$

where

$$a = \frac{8\pi i\omega\,|\,e^3\,|\,H}{(2\pi c\hbar)^3},$$

$$K_{\alpha\alpha}(k) = v_0^2 \int\limits_{p_z^{\min}}^{p_z^{\max}} \left[1 - \exp\left(-2\pi i q\,\frac{\Omega_0}{\Omega} - \frac{2\pi}{\Omega_0\tau}\right)\right]^{-1} dp_z \left|\int\limits_{0}^{T} \operatorname{sign} v_\alpha(t)\, e^{-iky(t)-iq\Omega_0 t}\, dt\right|^2_{p_z=0}.\qquad(38.2)$$

Expanding $\Omega(p_z)$ in the form of $\Omega(p_z) = \Omega(0) + \frac{1}{2}\Omega''(0)\,p_z^2$ near the only important point $p_z = 0$, and calculating $\int dp_z\ldots$ and $\int dt\ldots$ (by the method of steepest descent), we finally obtain

$$E_\alpha(\zeta) = -\frac{2d_0}{\pi}\,E'_\alpha(0)\,I(\zeta),\qquad(38.3)$$

where

$$I(\zeta) = \int\limits_{0}^{\infty} [x^3 + ie^{-i\pi\sigma/4}\,M^4\,(1 + \sin x)]\,x\cos\zeta x\,dx,\qquad(38.4)$$

and

$$\zeta = \frac{y}{d_0},\quad \sigma = \operatorname{sign}\frac{\Omega_0}{\Omega_0''},\quad M^3 \sim 10q^3\,\sqrt{\omega\tau_0}\left(\frac{v}{\omega\delta}\right)^2.\qquad(38.5)$$

The exact expression for M^3 is quite cumbersome and will not be given; it can be found in [5].

If ζ is close to an integer a ($\zeta = a + \zeta'$, $\zeta' \ll 1$), it is convenient to transform $I(\zeta)$ in the following way:

$$I(\zeta) = \sum_{n=0}^{\infty} \int\limits_{2\pi n}^{2\pi(n+1)} dx \ldots \approx \sum_{n=0}^{\infty} \int\limits_{0}^{2\pi} \frac{2\pi n \cos(ax + 2\pi n\zeta')\,dx}{(2\pi n)^3 + i\exp(-i\pi\sigma/4)\,M^3\,(1 + \sin x)} \approx$$

$$\approx \frac{1}{2\pi M} \int\limits_{0}^{\infty} y\,dy \int\limits_{0}^{2\pi} \frac{\cos(ax + \zeta'My)\,dx}{y^3 + i\exp(-i\pi\sigma/4)\,(1 + \sin x)} = \frac{(-1)^a}{M} \int\limits_{0}^{\infty} \frac{y\cos(\zeta'My + \pi a/2)\,dy}{\sqrt{g^2-1}\,(g - \sqrt{g^2-1})^a},\qquad(38.6)$$

where $g = 1 + y^3 \exp[-i\pi(2-\sigma)/4]$. We select that square root which has the positive arithmetic value.

An analysis of the formula given above leads to the following results.

1. Near $y = ad_0$ the field is $\sqrt{M}$ times higher than the field between the maxima.

2. The distance between maximum values of the field is of the order of $d_0 M^{-1}$.

3. When we consider the maxima in ascending numerical order, we find that their amplitudes decrease proportionally to $a^{-1/3}$ (for $a \gg 1$) and their "widths" increase. At depths $y < r_H^2/\delta$ the field between the maxima is practically unattenuated. The maxima gradually decrease in amplitude and at great depths ($y \gtrsim r_H^2/\delta$) the field oscillates as $\cos(Py/r_H)$, where $P \approx 1$, and is completely attenuated at a depth of the order of $r_H M^{3/2} \approx r_H^2/\delta$.

4. At depths $y \approx 2bd_0$ ($b = 0, 1, 2, \ldots$), there are isolated extrema of the field whose signs alternate as $(-1)^b$.

5. At depths $y \approx y_b = (2b+1)d_0$ ($b = 0, 1, 2, \ldots$), there are two extrema which differ only in sign; the field near these points is antisymmetric: $E_\alpha(y_b + y') = -E_\alpha(y_b - y')$. The signs of the first pair of such neighboring extrema alternate as $(-1)^{b+1}$ and the sign of the first extremum at $y \approx d_0$ is opposite to the sign of the field at $y = +0$.

The appearance of such field peaks naturally gives rise to several new physical effects.

A. The easiest to observe is the change in the impedance of a bulk metal or in its derivative with respect to the magnetic field ($d\zeta'/dH$, $d\zeta''/dH$, $d\ln\zeta''/dH$). We must bear in mind that the impedance itself resonates only for certain polarizations of the incident electromagnetic wave.

However, reliable observations can be made only of a nonmonotonic variation in the impedance accompanied by peaks. This is because the approach of the magnetic field to its resonance value results in a resonant decrease in ζ' and ζ'' by a factor of about 1.84 due to the appearance of the peaks which reduce the total current for a given field intensity. However, this nonmonotonic dependence is not observed in all cases. It is more convenient to measure $f(H) = (H - H_{res})/d\ln\zeta'/dH$ (H_{res} is the resonance value of H); the appearance of a peak is manifested by a "hump" in the dependence on H near H_{res}.

B. The most reliable and unambiguous proof of the existence of peaks are the discontinuities in the resonance values of the impedance and its derivative in plates 10^{-3}-10^{-1} cm thick, which may be observed when the frequency is increased. These discontinuities should appear when the next peak is "cut off," i.e., in the case of resonance they are observed for $q = q\omega_{H_0}$ and $D = ad_0$.

Such a discontinuity should be accompanied by a sudden change (by unity) in the number of observed harmonics (from a to $a + 1$; Fig. 89). This number can also be used to determine

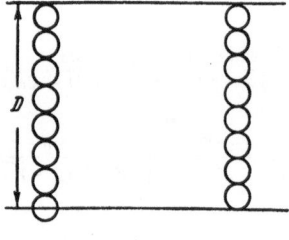

Fig. 89. a) Integral number of orbits does not fit within a plate of thickness D; b) integral number of orbits fits within the plate.

a   b

a and $d_0 = D/a = 2cp_x^{extr}/eH$, i.e., to find directly the diameter of the Fermi surface. The use of thin plates should make it possible to obtain directly the value of $(\partial S/\partial \varepsilon)_{extr}$ and, from quantum oscillations, the value of S_{extr} for each section.

This effect should be observable even for $a < M^{3/2} \sim 100$.

A similar but much more easily observed phenomenon should occur also if the frequency ω is kept fixed and a static magnetic field is rotated in the plane of a single-crystal film. The values of p_x^{extr} can be determined for those directions along which discontinuities or the appearance (disappearance) of harmonics are observed.

The Fermi surface can be constructed only if we carry out experiments on plates of various thicknesses and with different orientations of the surface relative to the crystallographic axes.

C. The third effect which results from the occurrence of field peaks is the selective transparency under resonance conditions when $D = ad_0 \sim H$ (this can be observed by varying the frequency ω or by rotating H in the plane of a film), and the appearance of a field which is "relayed" to a depth $(r_H^2/\delta)(\omega\tau)^{1/4}$. However, this effect would be very difficult to observe because of the almost perfect "specularity" of metals: just two reflections from the opposite faces of a film can reduce the intensity of the field by a factor of $(10\delta/\lambda)^4$, where λ is the wavelength of the incident wave. The field "relayed" into the metal suffers an additional attenuation by a factor of $r_H/\delta \sim v_F\lambda/2\pi c\delta$.

D. Another effect is the spatial electron echo, which is somewhat similar to the well-known spin echo. When a metal is subjected to a static magnetic field such that $\Omega_0\tau \gg 1$ and a pulse of duration $\Delta t \ll 2\pi/\Omega_0$, successive "echoes" in the form of field peaks should be observed after time intervals $2\pi/\Omega_0, 4\pi/\Omega_0, \ldots$, when the largest number of accelerated electrons congregates at the surface of a metal (in the case of a quadratic dispersion law this applies to all the electrons).

E. The fluctuations observed in a metal for $\Omega\tau \gg 1$ are also of a special nature but we shall not consider this effect in detail.

This physical analysis of the field peaks shows that they should be observed if there is a mechanism which selects — from all the electron orbits — only a small fraction of electrons whose orbit-diameter scatter is of the order of or less than the skin depth. Such a mechanism arises when a magnetic field is applied at an angle with respect to the surface of a metal. This mechanism operates also in the absence of cyclotron resonance. The field peaks can appear at high and low frequencies and the condition for the existence of the field structure shown in Fig. 88 is much less stringent than the resonance criterion [53].

We shall now consider a magnetic field inclined at an angle $\varphi \ll 1$ with respect to the surface of a metal. Then, electrons with a low drift velocity[1] $\bar{v}_z$ are subjected to conditions in a high-frequency field which are quite different from those applicable to all the other electrons. The electrons with $\bar{v}_z \sim v$ reach the skin layer only once and their contribution to the effective conductivity (of the order of $\sigma\delta/l$) is small compared with the contribution of the electrons characterized by $\bar{v}_z = 0$ because the latter return many times to the surface of the metal and if $\omega \ll \Omega$, their contribution to the effective conductivity is $\sigma_{eff} \approx \sigma(\delta/l\varphi)(\delta/r_H)$ (this can be demonstrated in a manner similar to that used in §34). If $\delta/l \ll \varphi \ll \delta/r_H$, the conductivity is dominated by the electrons whose relative velocity and, consequently, the relative distance $\Delta p_z/p_F$

---

[1] These are the electrons close to the extremal-area sections of the Fermi surface [for $S = S_{extr}$, we have $v_z = -(\partial S/\partial p_z)_\varepsilon(\partial S/\partial \varepsilon)_{p_z} = 0$] and particularly those close to the central section, where $S = S_{extr}$ because of the central symmetry of the Fermi surface.

from the extremal-area cross section, is of the order of $\delta/l\varphi$. Near the central section, where the orbit diameters are extremal and the conditions for the field peaks most favorable, this corresponds to the scatter of the diameters $\Delta d \approx r_H(\Delta p_z/p_F)^2 \approx r_H(\delta/l\varphi)^2$, which should be of the order of or smaller than δ. Consequently, the conditions for the appearance of the field peaks assume the form:

$$\frac{\delta}{r_H} \gg \varphi \gg \frac{\delta}{l}, \quad \varphi \gtrsim \left(\frac{r_H\delta}{l^2}\right)^{1/2} \tag{38.7}$$

and, necessarily,

$$\Omega\tau \gg \left(\frac{r}{\delta}\right)^{1/2}. \tag{38.8}$$

Clearly, in the case of an anomalously weak dependence of the orbit diameter on p_z (such, for example, as that obtained when the Fermi surface is close in shape to a cylinder of constant diameter), the condition for the appearance of the field peaks is much less stringent.

The presence of even weak but slowly decaying peaks, which occur every time some special orbits appear, gives rise to several important physical effects. In a parallel magnetic field, the special orbits having extremal diameters are responsible for cyclotron resonance at the relevant frequency. In a magnetic field inclined to the surface of a metal, such peaks give rise to cyclotron resonance at the frequencies indicated in §36.

The presence of field peaks in a plate results in several effects which oscillate with the magnetic field and are described in detail in review papers [54].

When the conditions (38.7)–(38.8) are not satisfied, the field peaks appear at the points where electrons congregate near some special sections but such peaks decay rapidly with depth [the weaker the dependence $d(p_z)$, the slower is the decay of the peaks]. Nevertheless, the use of a plate which "cuts off" some peaks makes it possible to observe them experimentally in the form of impedance anomalies which are periodic in H [55].

The theory of the field peaks associated with a magnetic field inclined to the surface of a metal and of the "damped" peaks can be derived from Eqs. (36.5)–(36.9) in a manner similar to that employed in §36 for the peaks associated with cyclotron resonance (for details see [53]).

§39. Free Oscillations and Weakly Damped Waves in Metals

The appearance of an anomalously slowly decaying field (i.e., a field decaying at depths large compared with δ) and of field peaks in the bulk of a metal suggests the existence of weakly damped waves, i.e., of free oscillations ($\tau = \infty$) [5]. The presence of a strong static magnetic field parallel or almost parallel to the surface of a metal means that the only characteristic dimension for the motion of electrons in a direction normal to the surface is r_H and, consequently, the kernel K in Eq. (33.7) decreases at depths of the order of r_H. The extreme anomaly of the skin effect implies that $|K_{\alpha\beta}| \approx \sigma_{eff}/r_H$, where $(4\pi\omega\sigma_{eff}/c^2)^{-1/2} \ll r_H$.

Consequently, Eq. (33.7) expressed in dimensionless variables becomes

$$\frac{1}{N} E_\alpha''(\zeta) = i \int K_{\alpha\beta}(\zeta, \zeta') E_\beta(\zeta') d\zeta', \tag{39.1}$$

where $|K_{\alpha\beta}| \sim 1$, $K_{\alpha\beta}$ decays in a depth of the order of 1, and $N \gg 1$.

If **E** decays along y at depths large compared with r_H, this means that when $N \to \infty$, Eq. (39.1) has a zero eigenvalue. Thus, for example, Eq. (39.1) for $\tau = \infty$ has an infinitely degenerate zero eigenvalue corresponding to $E_\alpha(y) = E_0 \exp(-ik_n y)$, where $K_{\alpha\alpha}(k_n) = 0$.

Under these conditions, the appearance of field peaks is a natural feature of the solution of the equation near its eigenvalue.

The existence of an eigenvalue solution of Eq. (39.1) implies the existence of undamped free oscillations in the metal. This can be demonstrated by considering the conditions under which undamped waves appear in general.

Undamped oscillations can clearly exist far from the surface of a metal only to the extent to which we can ignore the unavoidable dissipation of the momentum in the collisions of electrons with the surface. This means that Maxwell's equations (33.7) should be considered as being applicable to an infinite sample, where the plane $\xi = 0$ is not of special significance so that there are no discontinuities in $E'_\alpha(0)$. Consequently, the term with $E'_\alpha(0)$ is absent in the case of the one-dimensional oscillations represented by Eq. (39.1), which we shall regard as a special case. A nontrivial solution of this equation clearly corresponds to discrete roots of the equation

$$k_n^2 + iaK_{\alpha\alpha}(k_n) = 0, \qquad k_n = k(\omega, n). \tag{39.2}$$

The solution $E_0 \exp(ik_n y)$ can have a physical meaning only if it does not increase to $\pm\infty$ along y, i.e., it must satisfy the condition

$$\operatorname{Im} k_n = 0. \tag{39.3}$$

This condition cannot be satisfied rigorously in real metals but should be regarded as the condition which limits the range of existence of weakly damped waves. For example, Eq. (39.3) yields the physically meaningful condition $\tau = \infty$, so that weakly damped waves exist when

$$\omega\tau \gg \frac{v}{\omega\delta} \gg 1. \tag{39.4}$$

It is clear from our analysis that a similar situation applies also to the case of field peaks if $\tau = \infty$ [in this case, Eq. (39.2) is subject to the condition $aK_{\alpha\alpha}(K_n) \gg k_n^2$ and in the lowest approximation the value of k_n is determined from $K_{\alpha\alpha}(k_n) = 0$]. Free oscillations then have the following discrete eigenfrequencies [5]

$$\omega q = q\Omega_0 \qquad (q \text{ is an integer}) \tag{39.5}$$

which correspond to the discrete wave vectors of Eq. (39.3) whose values for large n are

$$k_n = \left(2n + \frac{1}{2}\right)\pi\omega_q m_0^* (2qp_x^{\text{extr}})^{-1}. \tag{39.6}$$

The nature of these free oscillations is naturally associated with the undamped rotation of electrons ($\tau = \infty$) along discrete Larmor orbits, which are defined by the resonance conditions.

In the general case of free oscillations at arbitrary frequencies (or for arbitrary wave vectors), the high-frequency equations are not as simple: under these conditions, we can only ignore the displacement current and consider the equations for an infinite sample. We shall write Maxwell's equations in terms of the Fourier components in space and time:

$$E \to E \exp(i\boldsymbol{kr} - i\omega t), \qquad H \to H \exp(i\boldsymbol{kr} - i\omega t),$$

$$\boldsymbol{j} \to \boldsymbol{j} \exp(i\boldsymbol{kr} - i\omega t), \tag{39.7}$$

$$[\boldsymbol{kE}] = -\frac{i\omega}{c} H, \qquad [\boldsymbol{kH}] = \frac{4\pi}{c} \boldsymbol{j}. \tag{39.8}$$

The linear relationship between **j** and **E** for given values of **k** and ω is, in general, of the form

$$j_i = \sigma_{is}(\omega, \ \boldsymbol{k}) E_s. \tag{39.9}$$

This relationship can be found in exactly the same manner as in §36. In the classical case, the expression for the conductivity is

$$\sigma_{is}(\omega, \ \boldsymbol{k}) = -\frac{2 |e|^3 H}{(2\pi\hbar)^3 c} \int \frac{\partial n_F}{\partial \varepsilon} \, d\varepsilon \int dp_z \int_0^T v_i(t) \, dt \int_{-\infty}^t v_s(t') \, dt' \exp\left[\int_t^{t'} \left(i\boldsymbol{kv}(t_1) + i\omega + \frac{1}{\tau} \right) dt_1 \right]. \tag{39.10}$$

The requirement for the existence of a nontrivial solution of Eqs. (39.8) and (39.9) leads to the equation

$$\left| k^2 \delta_{\alpha\beta} + \frac{4\pi i\omega}{c^2} \left(\sigma_{\alpha\beta} - \frac{\sigma_{\alpha\xi}\sigma_{\xi\beta}}{\sigma_{\xi\xi}} \right) \right| = 0 \tag{39.11}$$

(the ξ axis is selected along **k** and the angle between **k** and **H** is φ), which gives

$$\omega = \omega(\boldsymbol{k}) = \omega_1(\boldsymbol{k}) - i\omega_2(\boldsymbol{k}). \tag{39.12}$$

The real part of $\omega(\mathbf{k})$ gives the dispersion law of free oscillations and the imaginary part gives their damping [Eq. (39.7)]. It is naturally meaningful to speak of free oscillations only if there is a large number of such oscillations in any given situation, i.e., when

$$|\omega_1(\boldsymbol{k})| \gg |\omega_2(\boldsymbol{k})|. \tag{39.13}$$

This inequality contains the static magnetic field H as the parameter and it defines the range of k and H [according to Eq. (39.12) it also defines the range of ω] in which weakly damped electromagnetic waves can exist in a metal. These waves are excited under resonance conditions and, therefore, it would be interesting to consider the behavior of the impedance in the range defined by Eq. (39.13).

Clearly, the existence of weakly damped waves must always imply a fairly weak energy dissipation, i.e., a fairly large value of $\tau = 1/\nu$ in the absence of collisions with the surface. However, it is interesting to note that ν need not be the lowest characteristic frequency. In fact, even if $\nu \gg \omega$, but k is sufficiently small (y the wavelength is large enough), so that we are dealing with the normal skin effect, the lowest approximation to σ_{ik} is represented by $\omega = 0$, $k = 0$, i.e., it is the static conductivity. In this case, ω is an eigenvalue of a matrix defined by Eq. (39.11). In the lowest approximation, the damping in σ_{ik} can be ignored if $\Omega\tau \gg 1$ and then the operator σ_{ik} is found to be anti-Hermitian. This means that, in the lowest approximation, the frequency ω is an eigenvalue of a Hermitian operator [Eq. (39.11)], i.e., it is real, and this implies the existence of weakly damped oscillations (helicon waves).

The existence of weakly damped waves clearly represents the condition of high transparency of a metal to electromagnetic waves of suitable wavelength because the solution of Eq.

(39.12) for **k** and a fixed frequency ω yields a small imaginary component of **k**, which is the one responsible for the damping. This transparency was predicted in [56] for the simplest case of the normal skin effect.

The possibility of resonance excitation of free oscillations in a metal is of great interest: it allows us to obtain additional information on the electron spectra of metals.

A detailed analysis of various types of free oscillation and of the possibility of their excitation is given in [57-59]. This analysis deals with electromagnetic waves in the classical [57] and the quantum [58] cases (see also the review in [60]), and with the appearance of coupled magnetoacoustic waves [59]. Since the propagation and the absorption of ultrasound in metals are considered elsewhere in the present monograph, coupled magnetoacoustic oscillations will not be discussed here. Equation (39.11) is valid also in the quantum case but σ_{ik} of Eq. (39.10) must be replaced with a corresponding quantum formula. This formula can be taken from §31, which deals with the relationship between **j** and **E** under quantum conditions.

The presence of a large number of parameters in the problem we are considering[1] gives rise to many different limiting cases, each of which requires fairly laborious calculations. Therefore, we shall refer the reader to [57, 58, 60] and give only the main results of these investigations.

1. Classical Case [57]

A. If the wavelength is large compared with the Larmor radius and the spatial inhomogeneity of the alternating field along **H** is strong[2]:

$$\begin{aligned}
&kr_H \ll 1 \ll kv \cos \varphi \left(\frac{1}{\tau} + \omega\right)^{-1}, \\
&\cos \varphi = \frac{kH}{kH}, \quad r_H \sim \frac{v}{\Omega}.
\end{aligned}$$
(39.14)

In this limiting case, we have

$$j = \frac{(n_1 - n_2)\, ec}{kH}\, [kE] + \sum Aw\, (w^* E) + SH\, (HE)\, H^{-2} + \hat{S}E,$$
(39.15)

where

$$A = \frac{(2\pi e)^2}{(2\pi\hbar)^3} \left| \frac{m^*}{k_z \bar{v}_z' (p_z)} \right|_{\bar{v}_z = 0},$$
(39.16)

and the complex "velocity" vector **w** is defined by

$$\begin{aligned}
&w_z = -\frac{i\omega}{k_z}, \qquad w_\alpha = \overline{kv\rho_\alpha}\big|_{\bar{v}_z = 0}, \\
&\rho_x = \frac{c\, (p_y - \bar{p}_y)}{eH}, \qquad \rho_y = -\frac{c\, (p_x - \bar{p}_x)}{eH}
\end{aligned}$$
(39.17)

(the bar indicates averaging over the period of one revolution in an orbit); **w*** is the complex

---

[1] Even in the classical case, we have the following characteristic lengths: the path traversed by an electron during one field period; the mean free path; the Larmor orbit radius; the wavelength in the metal along and at right-angles to **H**; and the distance in which a wave is damped out (attenuation depth).

[2] The quasistatic case corresponding to weak spatial inhomogeneity is considered in [61] (see also [56]).

conjugate of w; n_1 and n_2 are the numbers of electrons and holes, respectively; and

$$S = \frac{2e^2}{k_z^2} \sum \left(\frac{1}{\tau} - i\omega\right) |\nu(\varepsilon_F)|,$$

$$S_{\alpha\beta} = \frac{4\pi e^2}{(2\pi\hbar)^3} \sum \left(\frac{1}{\tau} - i\omega\right) \int dp_z |m^\cdot| \overline{\rho_\alpha \rho_\beta} \qquad (39.18)$$

(the summation is carried out over different types of quasi particle and $\nu(\varepsilon_F)$ is the density of states corresponding to $\varepsilon = \varepsilon_F$).

The first term in Eq. (39.15) is the nondissipative Hall current. In the case of strong spatial dispersion, this current is orthogonal to k and E and not to H and E, which applies in the static case. The second term in j is also related to the spatial inhomogeneity of the field. The corresponding transverse components of the conductivity $\sum A w_\alpha w_\beta$ are due to the electrons moving in phase with the wave $\omega = k_z \bar{v}_z$ (near the central section of the Fermi surface $p_z = 0$). These terms represent the Cherenkov absorption of the electromagnetic field by electrons and are analogous to the well-known Landau damping in collisionless plasma.

The value of the longitudinal (relative to H) current $SH(EH)H^{-2}$ is $|k_z v|(1/\tau + \omega)^{-2}$ times weaker than that in the absence of spatial dispersion. The tensor $S_{\alpha\beta}$ represents the transverse (relative to H) conductivity of a metal in the limit of a homogeneous high-frequency field when there is no dependence on k if $n_1 = n_2$. If the magnetic field is directed along a threefold or higher symmetry axis (or if the spectrum is isotropic), we find that $\overline{v_z \rho_\alpha} = 0$, $\overline{\rho_x \rho_y} = 0$, and the tensor $S_{\alpha\beta}$ is diagonal.

a) If $n_1 \neq n_2$, we find that $|S_{\alpha\beta}|/\sum A w_\alpha w_\beta \sim (1 + \omega\tau)(k_z l)^{-1} \ll 1$, the quantities $S_{\alpha\beta}$ in Eq. (39.15) can be neglected, and the dispersion law of a weakly damped helicon wave is

$$\omega(k) = \frac{ck |kH|}{4\pi(n_1 - n_2)e}. \qquad (39.19)$$

The damping decrement and the polarization of the wave depend strongly on the shape of the Fermi surface and on the orientations of k, H, and the crystal axes relative to one another.

If H is parallel to a high-order symmetry axis, we find that

$$\frac{\omega''}{\omega} = \frac{2\pi\omega}{k^2 c^2} \sum A w_x^2 \sim k r_H \sin^2 \varphi, \quad \omega'' = -\operatorname{Im}\omega. \qquad (39.20)$$

If $k \parallel H$ ($\varphi = 0$), the spatial dispersion is unimportant and the Landau damping disappears from Eq. (39.20). The damping of the excitations is then solely due to the scattering of electrons:

$$\frac{\omega''}{\omega} \sim \frac{1}{\Omega\tau}. \qquad (39.21)$$

In this case $EH = 0$ and the vector E rotates in a plane perpendicular to H.

If the direction of H does not coincide with a symmetry axis of a crystal and the Fermi surface is singly connected, the damping of a helicon wave is nonmonotonic. If $\omega\tau < 1$ and the magnetic field is relatively weak, the damping increases proportionally to the field

$$\frac{\omega''}{\omega} \sim (k_z l\, k r_H \sin^2 \varphi)^{-1} \propto \frac{H}{k^2} \qquad (k_y^2 r_H^2 k_z l \gg 1), \qquad (39.20a)$$

where $Ew' = 0$, and then the damping reaches its maximum value $(\omega''/\omega)_{\max} \sim (k_z l k_y^2/k^2)^{-1/2}$;

in strong fields, the damping decreases monotonically:

$$\frac{\omega''}{\omega} \sim kr_{H}, \tag{39.21a}$$

where $\mathbf{EH} = 0$.

If $\omega\tau > 1$, the damping is again nonmonotonic: in this case, the maximum lies at lower magnetic fields.

In the general case of two or more groups of carriers, the damping decrement of the excitations is of the order of kr_H and it decreases monotonically with increasing H. The polarization of a helicon wave varies in the same way as in the case of one group of carriers. The physical reason for these features of the damping and the polarization of a helicon wave is the possibility of its interaction with a longitudinal wave.

b) We shall now consider the case when there is an equal number of electrons and holes: $n_1 = n_2$. In this case, high-frequency ($\omega\tau \gg 1$) magnetohydrodynamic waves may exist in a metal. The spectrum of these waves is linear, their polarization is planar, and their phase velocities are of the order of the Alfvén wave velocity $v_A = H(4\pi nm^*)^{-1/2}$.

In strong magnetic fields ($v_A \gg v$), the spatial dispersion is unimportant and the damping is determined by electron collisions: $\omega'' \sim 1/\tau$. Two types of magnetohydrodynamic wave may exist in a metal: an Alfvén wave and a fast magnetoacoustic wave (the phase velocity of the Alfvén wave does not exceed that of the magnetoacoustic wave).

In the range of magnetic fields defined by

$$1/\tau \ll \omega \ll kv \ll \Omega$$

the weakly damped waves coexist only if H is parallel to a symmetry axis. We then have two linearly polarized magnetohydrodynamic waves with identical spectra if $k \parallel H$, and one Alfvén wave with its electric field polarized along y (at right-angles to H in the k–H plane) provided k is not parallel to H.

If the range of magnetic fields is defined by

$$\frac{1}{\tau} \ll \omega \ll v \left| \frac{\omega}{cr_D} \cot \varphi \right|^{1/3} \ll \Omega \tag{39.22}$$

($r_D = \left| 4\pi e^2 \sum \nu(\varepsilon_F) \right|^{-1/2}$ is the Debye–Hückel screening radius) and if k is not parallel to H, $w_y \neq 0$, $k_y \neq 0$, and H is not parallel to a symmetry axis, a weakly damped longitudinal wave may exist in a metal and its spectrum is

$$\omega(\boldsymbol{k}) = \frac{cr_D k^2}{2} |\sin 2\varphi|. \tag{39.23}$$

B. In this case, the wavelength is short compared with the Larmor radius:

$$kr_H \gg 1. \tag{39.24}$$

If

$$\frac{1}{\Omega\tau} < \left| \frac{\pi}{2} - \varphi \right| < \frac{1}{a_n^4}, \tag{39.25}$$

where $1 \ll \alpha_n^4 \ll (v/v_A)^2$, $kr_H = \alpha_n = \pi(n + \frac{1}{4})$, it is found that the electrons close to the central section play the dominant role and the scatter of the orbit radii of these electrons is small compared with the wavelength of the electromagnetic wave. In this case, we find that weakly damped waves have a discrete spectrum:

$$\omega_n = \frac{2}{3} \pi v_A^2 v^{-2} \alpha_n^4 \left[\Omega \varphi' \left(\frac{\pi \alpha_n}{2} \right)^{1/3} - \frac{i}{\tau} \right] \quad \left(\varphi' = \frac{\pi}{2} - \varphi \right), \tag{39.26}$$

and the transverse component of the electric field of these waves is circularly polarized.

However, if

$$\varphi \ll (\Omega \tau)^{-1}, \quad 1 < \left(\frac{v_A \Omega}{v \omega} \right)^{1/3} \ll \frac{\Omega}{\omega}, \tag{39.27}$$

we find that there should exist a weakly damped wave which is polarized in the **k**–**H** plane at right-angles to **k** and has a continuous spectrum:

$$\omega = \left(\frac{2}{3} \right)^{1/3} (kr_H)^{3/2} |k_z| v_A - \frac{i}{2\tau}. \tag{39.28}$$

2. Quantum Case. One of the main reasons for the damping of free oscillations is the existence of the Landau damping, which is largely due to electrons moving in phase with the incident wave. The quantization of the electron states ensures that these electrons are located close to the Fermi energy only for certain values of the magnetic field. Therefore, the damping of free electromagnetic oscillations should be much weaker in the quantum case and considerable oscillations should be observed [58].

We shall give the results relevant to the most interesting case:

$$\hbar \omega \ll T \ll \hbar \Omega. \tag{39.29}$$

Let us assume that

$$A = \frac{c k_y e_F}{eH} \left| \frac{k_z v}{1/\tau + \omega} \right| M \ll 1, \tag{39.30}$$

where

$$M = \frac{\hbar \Omega}{8T} \sum_{\sigma = \pm 1} ch^{-2} \left(\frac{\varepsilon_{0\sigma} - \zeta}{2T} \right), \tag{39.31}$$

and $\varepsilon_{0\sigma} = \varepsilon_\sigma |_{n=n_0, \, p_z=p_{z0}}$. Here, p_{z0} is the solution of the equation $\bar{v}_z(\varepsilon_F, p_z) = -(2\pi m^*)^{-1} \frac{\partial S}{\partial p_z} = \frac{\omega}{k_z}$, and n_0 is the magnetic quantum number for which $\left| \varepsilon_F - \varepsilon_{np_{z0}}^\sigma \right|$ has its minimum value.

In this case,

$$\frac{\omega''}{\omega} \sim \sum kr_H M(H). \tag{39.32}$$

Thus, the quantization does not affect the spectrum or the polarization of free oscillations but the damping exhibits giant quantum oscillations described by the function M(H).

In the case defined by [see Eq. (39.30)]

$$A \gg 1, \tag{39.33}$$

Equation (39.32) becomes more complex and the relative damping of the wave is always of the order of $kr_H M_{min} \sim kr_H \frac{\hbar\Omega}{T} \exp\left(-\frac{\hbar\Omega}{2T}\right)$, i.e., the relative amplitude of oscillations of the damping is of the order of unity. In this case, the transverse component of the field $\mathbf{E}_\perp$ is circularly polarized. The longitudinal component $E_\parallel$ exhibits giant oscillations. Thus, in the case defined by Eq. (39.33), the quantization alters appreciably not only the damping but also the polarization of helicon waves.

We have given here the results for the most interesting effect, which is the influence of the quantization on the Landau damping. The effect of electron scattering on the quantum oscillations of a helicon wave in the case defined by Eq. (39.29) is considered in [57].

We shall not give the formulas for the surface impedance and the penetration of the field into a metal, derived in [57] for the case when an external electromagnetic field excites weakly damped waves (the penetration of the field into a metal in the case of the excitation of free oscillations is discussed in the preceding section).

High-frequency magnetohydrodynamic waves have been observed experimentally in bismuth [62]. Low-frequency helicon waves have been found in a large number of metals (sodium, indium, copper, silver, gold, lead, tin, zinc, cadmium, etc. [63]) and in degenerate In–Sb alloys [64] (under conditions such that the spatial dispersion was of no importance). Weakly damped waves near cyclotron resonance in potassium were reported in [65].

The plasma effects associated with free oscillations in metals are discussed in several reviews [60, 66].

§40. High-Frequency Fermi-Liquid Effects in Metals

The role of Fermi-liquid effects was considered in §23. We are returning to this problem because of the special nature of the high-frequency case (space and time inhomogeneity of the distribution function), which was not examined in §23.

An allowance for the collective Fermi-liquid interaction automatically implies that the one-electron approximation cannot be used. The distribution function n can be calculated by returning to the transport equation [Eqs. (23.6), (23.25), etc.].

Since the collision integral $(\partial n/\partial t)_c$ vanishes if we use the equilibrium Fermi function of the total energy ε of a quasi particle, we can follow [4] and introduce a correlation function relating (in an approximation linear with respect to the external field) the Fermi-liquid correction to the energy with the correction to $n_F(\varepsilon)$:

$$\delta\varepsilon = \int G(\mathbf{p},\ \mathbf{p}')(n - n_F(\varepsilon'))\, dp'. \tag{40.1}$$

Assuming that

$$n = n_F(\varepsilon_0(\mathbf{p})) - en' \frac{\partial n_F}{\partial \varepsilon_0}, \tag{40.2}$$

so that

$$\delta\varepsilon = \varepsilon - \varepsilon_0(\mathbf{p}) = -e \int \Phi(\mathbf{p},\ \mathbf{p}') \frac{\partial n_F}{\partial \varepsilon_0'} n'(\mathbf{p}')\, dp' \equiv -e\hat{\Phi}n' \tag{40.3}$$

and

$$n' = (1 + \hat{\Phi})^{-1} n_1, \tag{40.4}$$

we obtain the following expressions which are valid in the linear approximation:

$$\hat{G} = \hat{\Phi} (1 + \hat{\Phi})^{-1}, \tag{40.5}$$

and [$\hat{L}n_1$ is the operator obtained from $(\partial n / \partial t)_c$ in the approximation which is linear with respect to n_1]:

$$i\omega (1 - \hat{G}) n_1 + \frac{\partial n_1}{\partial r} \frac{\partial \varepsilon_0}{\partial p} - \frac{\partial n_1}{\partial p} \frac{\partial \varepsilon_0}{\partial r} + \hat{L}n_1 = \boldsymbol{v} \boldsymbol{E}, \tag{40.6}$$

$$\boldsymbol{j} = - \frac{2e^2}{(2\pi\hbar)^3} \int \boldsymbol{v} n_1 \frac{\partial n_0}{\partial \varepsilon_0} d\boldsymbol{p} = \frac{2e^2}{(2\pi\hbar)^3} \oint_{\varepsilon_0 (\boldsymbol{p}) = \varepsilon_F} \frac{\boldsymbol{v} n_1 dS}{v}. \tag{40.7}$$

The role of the Fermi-liquid interaction is demonstrated clearly by Eqs. (40.6) and (40.7). We shall now introduce the "mean free time" τ ($\hat{L}n_1 \sim n_1 / \tau$).

If $\omega\tau \ll 1$, the term proportional to ω should be dropped in the lowest approximation with respect to $\omega\tau$ and, consequently, the Fermi-liquid effects do not affect the results given in §23. However, if $\omega\tau \gtrsim 1$, we find that – as mentioned in §32 – the extreme anomalous skin effect is observed in good metals (it is understood that $\omega \ll v/\delta$; the opposite limiting case will be discussed later) and the term $\hat{G}n_1$ can be ignored in the lowest approximation with respect to the anomaly (§32). In the case of poor metals such as bismuth, in which the normal skin effect may occur, the term $\hat{G}n_1$ is small because the Fermi-liquid interaction is weak [67].

Under the anomalous skin-effect conditions, the Fermi-liquid term may be important only in the case of a very sharp cyclotron resonance (§32). In this case, the problem can be solved as follows [4]. For the sake of simplicity, we shall consider only the case when a static magnetic field is rigorously parallel to the surface of a metal whose dispersion law is quadratic. For the same reason, we shall restrict our treatment to the resonance associated with the central section. Going over to the variables ε, p_z, and t (we recall that t is the time of revolution of an electron along its orbit) in Eqs. (34.8) and (34.9), we obtain

$$i\omega (1 - \hat{G}) n_1 + v_y \frac{\partial n_1}{\partial y} + \frac{\partial n_1}{\partial t} + \hat{L}n_1 = v_\beta E_\beta + v_y E_y, \tag{40.8}$$

where

$$\beta = x, \ z,$$

and

$$\boldsymbol{j} = \frac{2 |e|^3 H}{(2\pi\hbar)^3 c} \oint_{\varepsilon = \varepsilon_F} \boldsymbol{v} n_1 \, dp_z \, dt \equiv \frac{2 |e|^3 H}{(2\pi\hbar)^3 c} \langle \boldsymbol{v} n_1 \rangle. \tag{40.9}$$

We shall assume that

$$n_1 = \nu_1 - \varphi = \nu_1 - \int_y^\infty E_y (y') \, dy', \tag{40.10}$$

which corresponds to the selection of $n_0(\varepsilon^t - \zeta_0)$ as the zeroth approximation, where ε^t is the total (kinetic and potential) energy of an electron in a field: $\varepsilon^t = \varepsilon + e\varphi$, and ζ_0 is the chemical potential in the absence of the field. This approximation is quite reasonable because the chemical potential cannot follow the rapidly varying field ($\omega\tau \gg 1$).

The function φ and, therefore, $E_y(y)$ can be found easily from the condition for the absence of uncompensated charge (§32): $\rho' = 0$, which is equivalent, according to the equation of continuity, to the condition for zero value of the normal component of the current. Since

$$\rho' = \frac{2e}{(2\pi\hbar)^3} \int (n - n_0(\varepsilon_0))\, d\mathbf{p} = \frac{2\,|e\,|^3 H}{(2\pi\hbar)^3\, c} \langle (1 - \hat{G})\, n_1 \rangle, \tag{40.11}$$

it follows that

$$\varphi = \frac{\langle (1 - \hat{G})\, \mathbf{v}_1 \rangle}{\langle (1 - \hat{G})\, 1 \rangle}. \tag{40.12}$$

Substituting Eq. (40.10) into Eqs. (40.8) and (40.9) and using Eq. (40.12), we obtain an equation which no longer includes the field E_y:

$$i\omega(1 - \hat{G})\,\mathbf{v}_1 + \left(v_y \frac{\partial}{\partial y} + \frac{\partial}{\partial t} + \hat{L} \right) \mathbf{v}_1 = v_\beta E_\beta + i\omega \frac{\langle (1 - \hat{G})\, \mathbf{v}_1 \rangle}{\langle (1 - \hat{G})\, 1 \rangle} (1 - \hat{G})\,1. \tag{40.13}$$

It is clear from Eq. (40.13) and from its derivation that the Fermi-liquid interaction plays, in a sense, a role analogous to that of the field E_y.

The function ν_1 has a sharp maximum at $v_y = 0$. We note that n_1, which contains the function φ which is independent of the angle, does not have this property. In the earlier treatment, this point was unimportant because it was permissible to ignore completely the quantity φ (§32).

Moreover, at resonance, the function ν_1 has a sharp maximum near $p_z = 0$. This circumstance, taken in conjunction with $\varepsilon = \zeta_0$, defines two centrally symmetric points on the Fermi surface at which $G(\mathbf{p}, \mathbf{p}')$ can be taken outside the integral sign. Since ν_1 has a sharp maximum, we can consistently replace $\hat{L}\nu_1$ with the relaxation time (§32). The subsequent steps of the solution are completely analogous to those given in §36 (for details see [4]).

It is thus found that the Fermi-liquid interaction affects significantly the amplitude and the width of the resonance line only if

$$\omega\tau > \left(\frac{r_H}{\delta} \right)^2, \qquad \omega > v\,(\delta^2 l)^{-1/3}. \tag{40.14}$$

The above expressions represent the conditions for a logarithmic resonance at limiting points at which the skin effect is normal if Eq. (40.14) is obeyed. An experimental investigation of the resonance under these conditions should make it possible to determine the "diagonal elements" of $G(\mathbf{p}, \mathbf{p}')$ on the Fermi surface.

The results presented in the present section and in §23 show that the Fermi-liquid interaction between the electrons does not alter the basic nature of practically any of the phenomena considered in the present monograph. This explains the success of the one-electron approximation, which is based on the representation of conduction electrons as a gas of charged fermions with a complex dispersion law $\varepsilon = \varepsilon(\mathbf{p})$. We must, however, point out that the numerical values of most of the quantities and the functions describing conduction electrons (in particular, the dispersion law) include the Fermi-liquid interaction.

Our treatment of the Fermi-liquid interaction ignores the spin correlation between electrons (this correlation is unimportant in the description of the "orbital" motion of electrons).

An allowance for the spin correlation in paramagnetic resonance (§44) gives rise to a considerable complication of the spectrum of the waves which may exist in a metal. It is found that the Fermi-liquid interaction in metals may be responsible for the propagation of special spin waves [67a], whose existence has been confirmed experimentally [67b].[1]

§41. Surface Skin Effect and Resonance at Very Low Frequencies

It is shown in §29 (and particularly in Appendix I) that even a weak inhomogeneity may alter considerably the resistance of a conductor in a strong magnetic field ($r_H/l \ll 1$). Since the term representing the inhomogeneity is proportional to r_H/l and that representing the homogeneous contribution is proportional to $(r_H/l)^2$, we find that considerable changes can be expected when the characteristic length associated with the inhomogeneity is of the order of $r_H l/r_H = l$.

The inhomogeneity may be due to various causes. One of the simplest inhomogeneous situations is that encountered in the skin effect. In this case, we can expect a different (compared with the usual skin effect, determined by the conductivity in a homogeneous field) dependence of the resistivity on the magnetic field even at very low frequencies when the usual skin depth becomes of the order of the mean free path:

$$\delta \sim \sqrt{\frac{c^2}{2\pi\omega_2\sigma(H)}} \sim l. \tag{41.1}$$

At low temperatures, the conditions necessary for the observation of this dependence can be achieved in extremely pure good metals characterized by $l \sim 1\text{-}3$ mm, corresponding to $\omega_1 \sim 1\text{-}100$ sec^{-1}.

The damping of the field and the current at frequencies $\omega \gtrsim \omega_1$ is then of a very special nature because of the influence of the static skin effect. An "almost homogeneous" electric field generates a current which practically vanishes at a depth of the order of the Larmor radius r_H (§38). Since the field and the current density have different depths of penetration, the problem can be solved by the method of successive approximations provided the frequency is not too high [68]. We shall first determine the intensity of the electric field. Since this intensity varies over distances large compared with r_H (this defines the range of "not too high" frequencies), the nature of the scattering of electrons by the surface is unimportant and the presence of the surface does not affect the conductivity tensor. This makes it possible to extend the field as an even function to the half-space outside the conductor being considered, and to write

$$j(z) = \int_{-\infty}^{\infty} \hat{\sigma}(|z - z'|) E(z') dz'. \tag{41.2}$$

Maxwell's equations for the one-dimensional case (normal incidence):

$$E_\alpha'' = \frac{4\pi i\omega}{c^2} j_\alpha \qquad (\alpha = x,\ y), \tag{41.3}$$

$$j_z = 0$$

are solved by going over to the Fourier components along the coordinate (an allowance is made for the discontinuity of E_α' at $z = 0$ due to the even extension of E_α):

---

[1] A detailed theory of spin waves in nonmagnetic metals is given in [67c].

$$- k^2 \varepsilon_\alpha(k) - 2E'_\alpha(0) = \frac{4\pi i \omega}{c^2} \, \tilde\sigma_{\alpha j}(k) \, \varepsilon_j(k), \left.\begin{array}{c} \\ \\ \end{array}\right\} \tag{41.4}$$
$$\tilde\sigma_{zj}(k) \, \varepsilon_j(k) = 0,$$

where

$$\tilde\sigma_{ij}(k) = 2 \int\limits_0^\infty \cos(kz) \, \sigma_{ij}(z) \, dz \qquad (i, \, j = x, \, y, \, z), \left.\begin{array}{c} \\ \\ \\ \\ \\ \end{array}\right\} \tag{41.4a}$$
$$\varepsilon(k) = 2 \int\limits_0^\infty \cos(kz) \, E(z) \, dz.$$

Having determined $E(z)$ from Eq. (41.4), we can obtain the next approximation in the form of a rapidly decreasing term in the expression for the current density (which naturally depends on the nature of the scattering of electrons by the surface); this can be done in the same manner as in Appendix I. In the case of interest when E varies significantly over distances large compared with r_H, the correction for the nature of the scattering is the same as in the static case (since the characteristic length in which E varies significantly is equal to δ, we may assume that E is homogeneous). It follows that the total current I_α is given by (see Appendix I):

$$I_\alpha = I_\alpha^{(0)} + I_\alpha^{\text{stat}} = (\hat Z_0^{-1})_{\alpha\beta} E_\beta(0) + (1 - q) \frac{n e^{''} r_H^2}{p} \, a_{\alpha\beta} E_\beta(0), \tag{41.5}$$

where $\hat Z_0$ is defined by Eq. (41.4); $a_{\alpha\beta} \sim 1$ and is independent of ω and l; q is the coefficient of reflection of electrons from the surface. In strong magnetic fields, the parameter q is close to zero (diffuse reflection) for good metals with deep skin layers because the angles of reflection of electrons by the surface are not small.

Using the definition of impedance $E_\alpha(0) = Z_{\alpha\beta}I_\beta$, we find that [68]

$$\hat Z^{-1} = \hat Z_0^{-1} + \frac{n e^2 r_H^2}{p} \hat a (1 - q). \tag{41.6}$$

The order of $\sigma_{\alpha\beta}(k)$ can be found easily in the same manner as before, bearing in mind that $\sigma_{\alpha\beta}(k) = \sigma_{\alpha\beta}(-k)$ (§39).

We shall not give the results of the calculations for all the frequencies and magnetic fields but we shall restrict ourselves to the characteristic resonance observed at very low frequencies (this resonance was first described in [68a] for the case of specular reflection from the surface), which occurs if $n_1 = n_2$ in a magnetic field normal to a metal whose surface coincides with one of the crystallographic planes. In this case, the electric field in the xy plane generates a current only in that plane because of the symmetry considerations and, therefore, $\sigma_{\alpha z} = \sigma_{z\alpha} = 0$, $E_z = 0$ in Eq. (41.4), and $\sigma_{\alpha\beta}$ is of the form

$$\hat\sigma = \sigma_0 \begin{pmatrix} \gamma^2 a_{11} & \gamma^2 a_{12} + \gamma(kr)^2 b_{12} \\ \gamma^2 a_{12} - \gamma(kr)^2 b_{12} & \gamma^2 a_{22} \end{pmatrix} \quad \left(\gamma = \frac{r_H}{l}\right). \tag{41.7}$$

If

$$4\pi\omega\sigma_0\gamma r_H^2/c^2 = 1, \tag{41.8}$$

the determinant of Eqs. (41.4) and (41.7) vanishes and these equations cease to have a solution. This is because of a resonance which causes one of the principal values of the impedance to become infinite. The value of $\hat{Z}_0$ can be calculated near the resonance by extending the expansion $\sigma_{xy}(\gamma, \text{k}) = \sigma_{yx}(-\gamma; \text{k}) = \sigma_{xy}(\gamma, -\text{k})$:

$$\sigma_{xy} = \sigma_0 \left(\gamma^2 a_{12} + \gamma (kr_H)^2 b_{12} + \gamma (kr_H)^4 d_{12}\right). \tag{41.9}$$

It is worth noting the meaning of this expansion: it implies, for example, that

$$j_x = \sigma_0 \left\{\gamma^2 a_{11} E_x + \gamma^2 a_{12} E_y + \gamma r_H^2 b_{12} E_y'' + \gamma r_H^4 d_{12} E_y''''\right\}, \tag{41.10}$$

i.e., that at $l = \infty$ the density of the current depends strongly on the fourth derivative of the electric field with respect to one of the coordinates.

The half-width of the resonance is of the order of $\sqrt{r_H/l}$ and the effective depth of penetration of the field under resonance conditions is $r_H(l/r_H)^{1/4}$. The real and the imaginary components of the impedance $\hat{Z}_0$ (and, therefore, of $\hat{Z}$ in the case of the specular reflection of electrons from the surface) increase at resonance by a factor $(l/r_H)^{1/4}$. In the case of nonspecular resonance [68], the weak spatial dispersion of E gives rise – as in the static case – to a strong spatial dispersion of the current density, which vanishes at a depth of the order of r_H. Consequently, the impedance calculated in the lowest approximation with respect to r_H/l is independent of the frequency and is equal to the static resistance of a plate with $l = \infty$ corresponding to $d \rightarrow \infty$ and a given value of q. In the next approximation, the impedance becomes dependent on ω and a resonance correction to the impedance is necessary: this correction is of the order of $(r_H/l)^{3/4}$, whereas the corresponding correction to Z'(H) is of the order of $(r_H/l)^{1/4}$.

The resonance is very sensitive to the geometry of the problem (this can be demonstrated by considering the general case).

The characteristic feature of the resonance is the dependence of the resonance frequency on the magnetic field. It follows from Eq. (41.8) that

$$\omega = \omega_H \left(\frac{c\omega_H}{v\omega_0}\right)^2 \sim H^3 \tag{41.11}$$

(ω_0 is the plasma frequency). The condition $\omega_H \tau \gg 1$ implies that Eq. (41.1) is applicable at frequencies exceeding $(1/\tau)(v\omega_0\tau/c)^{-2}$.

The resonance relationships are particularly clear in the case of a cubic lattice because, in this case, $a_{12} = 0$ in Eq. (41.7) and σ_{xy} is determined completely by the inhomogeneity of the alternating field. In this case, the impedance is diagonal when the incident wave is circularly polarized,

$$E_\pm = E_x \pm iE_y, \tag{41.12}$$

and the expressions in Eq. (41.4) give

$$E_\pm'' = \frac{4\pi i\omega}{c^2} j_\pm = \frac{4\pi i\omega}{c^2} \left\{\gamma^2\sigma_0 a_{11} E_\pm \pm i\sigma_0\gamma r_H^2 b_{12} E_\pm'\right\}, \tag{41.13}$$

where $E_\pm''$ drops out of Eq. (41.13) under resonance conditions so that $E_\pm'''$ has to be retained.

Outside the resonance region there is an interesting new skin effect which differs from the normal and the anomalous effects: the alternating electric field vanishes at a depth which is an order of magnitude different from that at which the magnetic field and the current decay to zero. This skin effect is observed in a wide range of frequencies or static magnetic fields corresponding to $\delta_{an} \ll r \ll \delta_n$ (δ_n and δ_{an} are the skin depths in the normal and the anomalous effects, respectively). Since the new effect is of the same nature as the static skin effect (§29), in which the electric field is homogeneous and the current decreases with increasing depth in the metal, it is natural to call it the "surface" skin effect. The theory of the surface skin effect is given in [68b] (for a semi-infinite sample) and [68c] (for a thin plate).

§42. Quantum Theory of High-Frequency Phenomena

The foregoing discussion of the high-frequency properties of metals is based on the classical ideas of the motion of conduction electrons and only the statistics of these electrons is quantum-mechanical. This approximation is fully justified because a significant contribution is made only by those electrons whose energies are close to the Fermi value and the separation between the quantum levels[1] (diamagnetic and paramagnetic spin levels) is very small compared with the Fermi energy (§1) so that the electron spectrum can, quite accurately, be regarded as continuous. However, in some cases, the discrete nature of the levels may be the dominant characteristic of a phenomenon. This applies particularly to the basically quantum phenomena such as the paramagnetic resonance, associated with resonance transitions between discrete paramagnetic spin levels; the quantum cyclotron resonance [69, 69a], due to transitions between diamagnetic levels if they are not completely equidistant (this is a result of the nonquadratic nature of the dispersion law) and the separation between neighboring resonance frequencies, is large compared with the collision-induced frequency broadening; and the combined resonance, which is due to transitions between diamagnetic levels corresponding to different values of the projection of the spin.

The discrete nature of the levels may be important also in the quantum effects which are weak but give rise to dependences significantly different from the classical case and can be observed experimentally. This applies to quantum oscillations of the surface impedance which are analogous to oscillations of the magnetization in the de Haas–van Alphen effect or of the resistance in the Shubnikov–de Haas effect. The quantum phenomena at high frequencies are discussed in §§42-43.

Before dealing with the quantum theory of the surface impedance, we shall consider one specific quantum feature of this problem [70].

As in the classical case, the problem of the determination of the surface impedance can be divided into two parts: the determination of the relationship between the current density and electric field intensity, and the subsequent determination of the relationship of Maxwell's equations subject to the first relationship. In the classical case, the magnetic induction **B** is assumed to be identical with the magnetic field **H**.

In the quantum case, the separation of the magnetic moment **M** and the current density **j** as two independent physical quantities has meaning – strictly speaking – only in the static case when we can distinguish magnetostatics and its associated thermodynamic quantity **M**, and electrostatics and its associated transport property **j**. In general, we are dealing with a typical problem in field theory, i.e., with a system of free charges in external fields **E** and **H** (the magnetic moment of atomic localized electrons does not give rise to the kind of oscillations which are of interest to us and, since this moment is small, it can be ignored). In this case, we have to determine the current density **j** which, generally speaking, cannot be divided unambiguously into the "conduction current density" and rot **M**. (This does not give rise to any misunderstand-

---

[1] The quantization may be due to the finite dimensions of a sample (§22). However, we shall restrict our treatment to the most important case, which is the quantization arising from the application of a static magnetic field.

ing in Maxwell's equations because they are of the same form as for a system of charges in vacuum with suitable boundary conditions being imposed on **E** and **B**.)

We can easily follow the effects of going over from the static case to low and high frequencies.

At low frequencies $\omega\tau \ll 1$, $\delta \gg l \gg r_H$ and in the lowest approximation with respect to $\omega\tau$, it is meaningful to write Maxwell's equations for the conduction current and the magnetic moment. We can use the formulas obtained in the static case because the system follows the variations of the fields and the skin effect is normal. The quantum oscillations of the conductivity are determined solely by the "quasiclassical parameter" and their relative amplitude is of the order of $\varkappa_1 = (eHh/cS)^{1/2}$. The amplitude of oscillations of the magnetic susceptibility $\chi_{ik} = \partial M_i/dB_k$ is of the order of $\chi \approx \varkappa_1^{-3} e^2 S^{1/2}/(m*c^2) \sim \varkappa_1^{-3}\rho^2$, where $\rho = v_F/c$. At low frequencies, the contribution of the conductivity oscillations is always small and it decreases additionally because these oscillations are usually determined not by the principal but by the very weakly populated electron groups. Therefore, the oscillations of the surface impedance at low frequencies are determined by the de Haas–van Alphen effect and the relative amplitude of such oscillations[2] is of the order of $\Delta\zeta/\zeta \sim (4\pi\chi)^{1/2}$.

These conclusions are independent of the frequency and are valid when $\omega \to 0$ because of the implicit assumption that the metal is "infinite" compared with the skin depth δ ($\delta \sim \omega^{1/2}$ at low frequencies). In a very thin plate whose thickness is of the order of $D \ll \delta\varkappa_1^4\rho^{-2} \ll \delta$, the oscillations which determine the impedance are the conductivity oscillations, i.e., the Shubnikov–de Haas effect.

The relative importance of oscillations of the magnetic moment decreases with increasing frequency. This is primarily due to the nonlocal (in the case of the anomalous skin effect) relationship between the high-frequency magnetic moment and the intensity of the high-frequency magnetic field, when only a thin skin layer is "effective." This gives rise to a small factor δ/r_H if the static magnetic field is exactly parallel to the surface of a metal and δ/l if this field is inclined. The role played by the oscillations of the magnetic moment decreases also because the system does not follow the variations of the alternating field if $\omega\tau \gg 1$ and this gives rise to an additional factor $(\omega\tau)^{-1}$ in the expression for the magnetic moment (these two factors have the same influence on the oscillating and the monotonic or classical parts of the conductivity). Therefore, at sufficiently high frequencies (which can be estimated on the basis of the foregoing discussion), the surface impedance oscillations are determined by the conduction current oscillations.[3] The quantum formula for the conduction current density can now be found quite easily. The classical formula for j has been given before; we are now interested in the quasiclassical case when $\varkappa_1 \ll 1$. All the quantities which vary over distances large compared with the "quantum" splitting (on the energy scale this corresponds to separations which are large compared with $\hbar\Omega$) should be associated with an operator whose matrix elements are equal to the Fourier components with respect to time [72]. All the quantities in the formula for j vary slowly (over distances of the order of ε_F) with the exception of $\partial n_F/\partial\varepsilon$, which varies over distances of the order of T. If $\hbar\Omega \ll$ T, the quantum oscillations are exponentially small and unimportant. If $\varepsilon_F \gg \hbar\Omega \gg$ T, we may replace the derivative $\partial n_F/\partial\varepsilon$ with the ratio

---

[2] The periodic dependence of the quantum terms on cS/ehH gives rise to characteristic effects (associated with these terms) which are nonlinear with respect to the alternating magnetic field in relatively weak alternating fields H_1 of the order of the oscillation period $H\varkappa_1^2$ [71].

[3] For a plate of given thickness, an increase in the frequency results in a change of the conductivity oscillations from the Shubnikov–de Haas effect to the de Haas–van Alphen effect, which is followed by a change back to the Shubnikov–de Haas effect.

of the finite differences

$$\frac{\Delta n_F}{\Delta \varepsilon} = \frac{n_F\left(\varepsilon^{\sigma}_{n+l,\, p_z}\right) - n_F\left(\varepsilon^{\sigma}_{n,\, p_z}\right)}{\varepsilon^{\sigma}_{n+l,\, p_z} - \varepsilon^{\sigma}_{n,\, p_z}}.$$

The corresponding rule can be obtained by noting that integration in the classical phase space corresponds to a trace (spur) in quantum mechanics. The classical formulas for the current include only integration over the momentum; integration over the whole phase space, i.e., over the momentum and the coordinates, can be carried out by introducing a delta function of the coordinates. This gives rise to the following expression (the trace is calculated also for the spin projections and, therefore, the "spin" factor of 2 disappears from the expression):

$$\boldsymbol{j}\left(\xi\right) = -\frac{2e}{(2\pi\hbar)^3}\int\frac{\partial n_F}{\partial \varepsilon}\,\boldsymbol{v}\,\Delta\varepsilon\,d\boldsymbol{p} = -\frac{2e}{(2\pi\hbar)^3}\int\frac{\partial n_F}{\partial \varepsilon}\,\boldsymbol{v}\,\Delta\varepsilon\left(\boldsymbol{r}'\right)\delta\left(\boldsymbol{r}-\boldsymbol{r}'\right)d\boldsymbol{p}\,d\boldsymbol{r}' \rightarrow$$

$$\rightarrow -e\int d p_z \sum_{\substack{n,\, l \\ \sigma=\pm 1}} \frac{n_F\left(\varepsilon^{\sigma}_{n+l,\, p_z}\right) - n_F\left(\varepsilon^{\sigma}_{n,\, p_z}\right)}{\varepsilon^{\sigma}_{n+l,\, p_z} - \varepsilon^{\sigma}_{n,\, p_z}}\left[\int_0^T \boldsymbol{v}\,\Delta\varepsilon\delta\left(\xi - \xi\left(t\right)\right)e^{-il\hbar\Omega t}\,dt\right]_{\varepsilon=\varepsilon^{\sigma}_{n,\, p_z}}. \qquad (42.1)$$

This formula will be derived later in a rigorous manner. At this point, we shall consider the influence of electron collisions with the surface on the quantum oscillations. The degeneracy in respect of P_x is lifted for those electrons which are reflected by the surface because the position of the center of the orbit relative to the surface is found to be important. Thus, additional integration with respect to P_x appears in the formula for the current density. Integration with respect to P_x and with respect to p_z involves the selection of only those electrons which have extremal periods. This reduces additionally the amplitude of the quantum oscillations corresponding to these periods. Therefore, one can simply ignore such electrons even in rigorously derived formulas.

Quantum oscillations of the surface impedance have been observed experimentally and reported in [73] (low frequencies) and [74].

We shall start developing a theory of impedance oscillations by writing down the expression for the current density $\boldsymbol{j}$. It follows from the definition of $\boldsymbol{j}$ that

$$\boldsymbol{j} = \mathrm{Sp}\,\hat{n}\hat{\boldsymbol{j}}_1, \qquad (42.2)$$

where Sp denotes a trace (spur), $\hat{n}$ is the density matrix, and $\hat{\boldsymbol{j}}_1$ is the current density operator associated with one electron.

In classical mechanics, the density of the current generated at a point $\boldsymbol{R}$ by an electron moving at a velocity $\boldsymbol{v}$ and located at a point $\boldsymbol{r}$ is

$$\boldsymbol{j}_1 = e\boldsymbol{v}\delta\left(\boldsymbol{R}-\boldsymbol{r}\right) \qquad (42.3)$$

(if $\boldsymbol{R} \neq \boldsymbol{r}$, $\boldsymbol{j}_1 = 0$, and $\displaystyle\int_{\Delta r\,\in\,R} e\boldsymbol{v}\left(\boldsymbol{R}-\boldsymbol{r}\right)d\boldsymbol{r} = e\boldsymbol{v}$).

We are only interested in the quasiclassical case. In this case, any Hermitian operator is determined unambiguously, to within $\hbar^2$, by its classical analog because all the symmetrization methods give the same result. This can be shown easily if we bear in mind [72] that, to within $\hbar^2$,

$$[\hat{a}\hat{b}] = -i\hbar\,[a,\,b] \tag{42.4}$$

($[a,\,b]$ is the classical Poisson bracket) and that various symmetrizations give rise to expressions which, after reduction to the same order of the operators by means of Eq. (42.4), cannot contain $\hbar$ because it occurs only in conjunction with i.

Therefore, to within $\hbar^2$, it follows from Eq. (42.3) that

$$j_1 = \tfrac{1}{2}\,e\,\{\hat{v}\delta\,(\boldsymbol{R}-\hat{\boldsymbol{r}}) + \delta\,(\boldsymbol{R}-\hat{\boldsymbol{r}})\,\hat{\boldsymbol{v}}\} \tag{42.5}$$

and Eq. (42.2) yields[4]

$$\boldsymbol{j}\,(\boldsymbol{R}) = \tfrac{1}{2}\,e\,\mathrm{Sp}\,\hat{n}\,\{\hat{v}\delta\,(\boldsymbol{R}-\hat{\boldsymbol{r}}) + \delta\,(\boldsymbol{R}-\hat{\boldsymbol{r}})\,\hat{\boldsymbol{v}}\}. \tag{42.6}$$

The quantity $\hat{n}$ is found using the transport equation

$$0 = \frac{d\hat{n}}{dt} = \frac{\partial\hat{n}}{\partial t} + \frac{i}{\hbar}\,[\hat{\varepsilon},\,\hat{n}] + \left(\frac{\partial\hat{n}}{\partial t}\right)_{\mathrm{c}}. \tag{42.7}$$

We shall consider only the most interesting case of high frequencies: $\omega\tau \gg 1$. In the lowest approximation, the system considered cannot follow the variations of the high-frequency field, the chemical potential remains constant, and it is natural to use $n_F(\hat{\varepsilon})$ (where n_F is the equilibrium Fermi function) as the zeroth approximation in which the collision integral vanishes. We shall linearize Eq. (42.7) by assuming that

$$\hat{n} = n_F\,(\hat{\varepsilon}) + \hat{n}', \qquad \hat{\varepsilon} = \hat{\varepsilon}_0 + \hat{\varepsilon}', \tag{42.8}$$

where $\hat{\varepsilon}_0$ is the Hamiltonian in the absence of the high-frequency field. We note that the substitution of $n_F(\hat{\varepsilon})$ in place of $\hat{n}$ in Eq. (42.6) does not give rise to zero current density. The term corresponding to the substitution represents the instantaneous-equilibrium state, and in the high-frequency case it gives rise to a magnetic moment which is unimportant if the frequency is sufficiently high and, therefore, we shall ignore it.

It follows from Eqs. (42.7) and (42.8) that

$$\frac{\partial\hat{n}'}{\partial t} + \frac{i}{\hbar}\,[\hat{\varepsilon}_0,\,\hat{n}'] + \left(\frac{\partial\hat{n}'}{\partial t}\right)_{\mathrm{c}} = -\,\frac{d}{dt}\,n_F\,(\hat{\varepsilon}). \tag{42.9}$$

The "mean free time" can be introduced for the same reasons as in the classical case, i.e., because of the extreme anomaly of the skin effect. For the sake of simplicity, we shall assume that this time constant is independent of n and p_z so that

---

[4] Equation (42.6) can be obtained also from the definition of the current density:

$$\overline{\delta\hat{\varepsilon}} = -\,\frac{1}{c}\int\boldsymbol{j}\,\delta A\,dV\,,$$

where $\hat{\varepsilon}$ is the Hamiltonian; δA is the variation of the vector potential $A(\boldsymbol{R})$; the integration is carried out over the whole space; the average $\overline{\delta\hat{\varepsilon}}$ is transformed bearing in mind the correspondence between the Hermitian operator and its classical analog mentioned earlier. Equation (42.6) can also be obtained from the quantum-mechanical formula $\boldsymbol{j} = (\psi^{*}\hat{p}\psi + \psi\hat{p}^{*}\psi^{*})/m$ by replacing $\psi^{*}(\boldsymbol{R})\psi(\boldsymbol{R}')$ with $n(\boldsymbol{R},\,\boldsymbol{R}')$ and $\hat{p}/m$ with $\hat{\boldsymbol{v}}$.

$$\left(\frac{\partial \hat{n}}{\partial t}\right)_c = \frac{\hat{n}'}{\tau}. \tag{42.10}$$

We shall write Eqs. (42.9) and (42.10) in terms of matrix elements corresponding to a system of time-independent wave functions Ψ_k:

$$\hat{\varepsilon}_0 \Psi_k = \varepsilon_k \Psi_k \tag{42.11}$$

(k is the complete set of the quantum numbers). Bearing in mind that Ψ_k is independent of time, we can rewrite the right-hand side of Eq. (42.9) retaining only the linear terms:

$$\left(\frac{d n_F(\hat{e})}{dt}\right)_{kk'} = \frac{n_F(\varepsilon_k) - n_F(\varepsilon_{k'})}{\varepsilon_k - \varepsilon_{k'}} e (\widehat{\boldsymbol{vE}})_{kk'}, \tag{42.12}$$

where

$$\widehat{\boldsymbol{vE}} = \frac{1}{2} (\hat{\boldsymbol{v}} \hat{\boldsymbol{E}} + \hat{\boldsymbol{E}} \hat{\boldsymbol{v}}).$$

Equations (42.9)-(42.12) now give (when $\hat{n}' \to \hat{n}' e^{i\omega t}$, $\boldsymbol{E} \to \boldsymbol{E} e^{i\omega t}$):

$$\left\{ i\omega + \frac{i}{\hbar} (\varepsilon_k - \varepsilon_{k'}) + \frac{1}{\tau} \right\} n'_{kk'} = - \frac{n_F(\varepsilon_k) - n_F(\varepsilon_{k'})}{\varepsilon_k - \varepsilon_{k'}} e (\widehat{\boldsymbol{vE}})_{kk'}, \tag{42.13}$$

and

$$n'_{kk'} = - \frac{n_F(\varepsilon_k) - n_F(\varepsilon'_k)}{\varepsilon_k - \varepsilon'_k} \frac{e (\widehat{\boldsymbol{vE}})_{kk'}}{i\omega + \frac{i}{\hbar} (\varepsilon_k - \varepsilon_{k'}) + \frac{1}{\tau}}. \tag{42.14}$$

Since the quasiclassical matrix elements $e\hat{\boldsymbol{v}}\hat{\boldsymbol{E}}$ are the Fourier components with respect to time and, in general, with respect to angular variables W related to action variables I (this can be shown as in [72] using $W = \partial \varepsilon_0 / \partial I$), we find that

$$n'_{kk'} = \frac{n_F(\varepsilon_k) - n_F(\varepsilon_{k'})}{\varepsilon_k - \varepsilon_{k'}} \psi_{kk'}, \tag{42.15}$$

where ψ satisfies the classical equation and $(\partial n_F / \partial \varepsilon) \psi$ is the nonequilibrium correction to the classical distribution function determined in Chap. III. If the magnetic field is parallel to the surface of a metal, Eqs. (42.15) and (42.6) yield the correspondence rule between the classical and the quantum expressions for j. The set of quantum numbers k is represented by n, p_z, σ. Equation (42.7) is evidently diagonal in respect of the spin; the spin-dependent correction to ε is always small and need only be considered in paramagnetic resonance (§§42, 45) and $\varepsilon_{n, p_z}^\sigma$ is defined by Eq. (7.3).

In an inclined magnetic field, the dependence on ξ gives rise to a dependence on y (in particular, because $A_x = -Hy$) as well as on z. This means that Eq. (42.15) contains terms which are not diagonal in respect of n and of p_z. We shall estimate the relative importance of these nondiagonal terms. Since the inhomogeneity with respect to the coordinates is represented by the skin depth δ, the corresponding indeterminacy of the momentum is of the order of $\Delta p_z \sim \hbar / \delta$ and the indeterminacy of the energy is

$$\Delta \varepsilon = \varepsilon_0 (n, p_z + \Delta p_z) - \varepsilon_0 (n, p_z) = \bar{v}_z (n, p_z) \Delta p_z.$$

Bearing in mind that

$$\left(\frac{\partial \varepsilon}{\partial p_z}\right)_n = \frac{\partial(\varepsilon, n)}{\partial(p_z, n)} = \frac{\partial(\varepsilon, S)}{\partial(p_z, S)} = \frac{\partial(\varepsilon, S)}{\partial(\varepsilon, p_z)}\bigg/\frac{\partial(p_z, S)}{\partial(\varepsilon, p_z)} = -\left(\frac{\partial S}{\partial p_z}\right)_\varepsilon\bigg/\left(\frac{\partial S}{\partial \varepsilon}\right)_{p_z},$$

and

$$\frac{\Delta p_z}{p_F} \sim \left(\frac{e\hbar H}{cS}\right)^{1/2},$$

and that the quantum oscillations correspond to $\bar{v}_z = 0$, we obtain the following expression which applies near extremal sections:

$$\Delta\varepsilon \sim v\left(\frac{e\hbar H}{cS_{\text{extr}}}\right)^{1/2}\frac{\hbar \sin \varphi}{\delta} \sim \hbar\Omega\left(\frac{e\hbar H}{cS_{\text{extr}}}\right)^{1/2}\frac{r_H \sin \varphi}{\delta},$$

where φ is the angle between H and the surface of the metal. Consequently, if

$$\left(\frac{e\hbar H}{cS_{\text{extr}}}\right)^{1/2}\frac{r_H \sin \varphi}{\delta} \ll \frac{T}{\hbar\Omega}, \tag{42.16}$$

the nondiagonality with respect to p_z can be ignored in all the terms of $\Delta n_F/\Delta\varepsilon$. However, if

$$\left(\frac{e\hbar H}{cS_{\text{extr}}}\right)^{1/2}\frac{r_H \sin \varphi}{\delta} \ll 1, \tag{42.16a}$$

the nondiagonality with respect to p_z can be ignored in all the terms which are nondiagonal with respect to n.

The physical meaning of the condition (42.16a) is very simple: it represents a large number of revolutions in the skin layer of those electrons which are responsible for the oscillations, and it corresponds to the case of very strong magnetic fields or magnetic fields almost parallel to the surface of a metal.

Far from the quantum cyclotron resonance (this resonance is discussed in the next section) when $\omega \gtrsim \omega_H$, the condition (42.16a) enables us to ignore in $n'_{kk'}$ the nondiagonality with respect to p_z and to obtain Eq. (42.1), in which the coefficient of proportionality (representing the density of states for a given value of n in the range dp_z) can be determined most conveniently by going over to the classical case.

Further calculations can be carried out using the Poisson summation formula. In this way, we obtain the following expression for the oscillatory component of the current density:

$$\Delta j_s(\xi) = \frac{(2\pi\hbar)^3}{2\pi} H^2 \sum_i \left[\chi_s\left(\frac{\partial \ln S}{\partial \varepsilon}\right)\frac{\partial \Delta M_z}{\partial H}\right]_{\varepsilon=\varepsilon_F, p_z=p_0^i}, \tag{42.17}$$

where

$$\chi_i = \frac{4\pi e^2}{(2\pi\hbar)^3 (i\omega + \tau^{-1})} \int_0^T dt v_i(t) \int_0^T v_l(t') E_l\left\{\xi - \int_{t'}^t v_\xi(t_1)\,dt_1\right\} dt',$$
$$S(p_0^i) = S_{\text{extr}},$$

and ΔM_z is the oscillatory component of the magnetic moment (§15).

The general theory of quantum oscillations of the surface impedance was developed first in [75, 76]. Somewhat later, the formulas for the simpler case of a quadratic dispersion law and a magnetic field parallel to the surface of a metal were derived independently in [77]. Near the cyclotron resonance frequency (in a magnetic field parallel to the surface of a metal), the oscillation amplitude passes through resonance when the condition (42.16a) is satisfied; this case is described by a formula which is more complex than Eq. (42.17) and which will not be given here.

Equation (42.17) can be used to derive the quantum correction to the surface impedance using the methods employed in the preceding sections. The main results given at the beginning of the present section can be confirmed in this way.

§43. Quantum Cyclotron Resonance

The cyclotron resonance described in §§35–37 is essentially of purely classical origin. In the case of an arbitrary dispersion law, the frequency Ω has a continuous spectrum $\Omega = \Omega(\varepsilon_F, p_z)$ and the resonance corresponds to the limits of the spectrum (i.e., to the limiting points of the Fermi surface) and to the critical points in the spectrum, where the number of electrons participating in resonance is relatively large because the density of states becomes infinite at a given frequency, i.e., the resonance occurs at extremal frequencies where $\Omega'(p_z) = 0$.

The quantization of the energy levels in a magnetic field [Eq. (7.4)] gives rise to a discrete spectrum of frequencies at a given energy (close to ε_F):

$$\hbar\Omega_n = \mu_n H, \tag{43.1}$$

where

$$\mu_n = \mu(\varepsilon, p_{zn}) = \frac{e\hbar}{m^* c}, \quad m^* = \frac{1}{2\pi}\frac{\partial S}{\partial \varepsilon}, \quad S(\varepsilon, p_{zn}) = n\frac{2\pi e\hbar H}{c}. \tag{43.2}$$

This means that, in principle, the resonance can occur at the discrete frequencies which satisfy the condition [69, 69a]

$$\hbar\omega = l_0 \mu_n H = l_0 \hbar\Omega_n, \tag{43.2a}$$

where l_0 is an integer. Equations (43.2) and (43.2a) define the "resonance" values of p_z and H. It is convenient to write S in Terms of ε and μ. Then, μ is defined using Eq. (43.2a): $\mu = \hbar\omega/l_0 H$, and Eq. (43.2) gives the resonance values of H. Bearing in mind that, in the lowest approximation $\varepsilon = \varepsilon_F$, we find that

$$cS(\varepsilon_F, \hbar\omega/l_0 H)/l_0 \hbar H = n. \tag{43.2b}$$

Clearly, since $n \gg 1$, the separations between the resonance levels are small compared with H; moreover, they are quasiequidistant, correspond to $\Delta n = 1$, and are defined in accordance with Eq. (43.2b) by the formula

$$\Delta\left(\frac{1}{H}\right) = \frac{2\pi e\hbar}{c}\left\{S(\varepsilon_F, \mu) + \mu\frac{\partial S(\varepsilon_F, \mu)}{\partial \mu}\right\}^{-1}_{\mu = \hbar\omega/l_0 H}. \tag{43.2c}$$

If the separation between the neighboring frequencies of the spectrum is of the order of or greater than the mean free time, i.e.,

$$\Delta \Omega_n = \Omega_{n+1} - \Omega_n \gtrsim 1/\tau, \qquad \gamma = \Delta \Omega_n \tau \gg 1, \tag{43.3}$$

the resonance is observed at each of the discrete frequencies. For the "normal" sections, $\Delta \Omega_n \approx \Omega'(n) \sim \Omega/n_0$, $n_0 \sim cS/2\pi e\hbar H$ and the resonance condition assumes the following form (because $\Omega_n \sim \omega$):

$$\omega \tau \gtrsim \frac{S}{\omega m^* \hbar} \approx \frac{\varepsilon_F}{\hbar \omega}. \tag{43.4}$$

The condition specified by Eq. (43.4) is very stringent and the resonance occurs more easily in the case of those sections for which $(\partial \Omega/\partial n)_\varepsilon \sim (\partial \Omega/\partial S)_\varepsilon = \infty$. Since $(\partial S/\partial p_z)/(\partial \Omega/\partial p_z)_\varepsilon = (\partial S/\partial \Omega)_\varepsilon$, these sections correspond to the Fermi surface sections which are extremal in area and are characterized by $(\partial S/\partial p_z)_\varepsilon = 0$, or to self-intersecting sections for which $(\partial \Omega/\partial p_z)_\varepsilon = \infty$. In the first case, when $n(\varepsilon, \Omega)$ is extremal with respect to Ω, $1 = \Delta n = \frac{1}{2}n''(\Omega)(\Delta \Omega)^2 \approx n_0(\Delta \Omega/\Omega)^2$, and the condition (43.3) becomes

$$\omega \tau \gtrsim (\varepsilon_F/\hbar \omega)^{1/2}. \tag{43.5}$$

In the case of self-intersecting sections, we have $\omega \sim \Omega_0/\ln(\varepsilon_F/\hbar \omega)$ (Ω_0 is the characteristic cyclotron frequency) and the condition for the quantization of the cyclotron resonance becomes:

$$\omega \tau \gtrsim \frac{\varepsilon_F}{\hbar \omega} e^{-\Omega_0/\omega} \ln^3 \frac{\varepsilon_F}{\hbar \omega}. \tag{43.6}$$

The sections for which the separation between the frequencies $\Delta \Omega$ is anomalously small are the least "convenient" for the observation of the cyclotron resonance. It follows from $(\partial \Omega/\partial n)_\varepsilon \sim (\partial \Omega/\partial p_z)_\varepsilon (\partial S/\partial p_z)_\varepsilon^{-1}$ that these are the noncentral sections corresponding to the frequency Ω which is extremal with respect to p_z, i.e., the sections which are close to those at which the classical cyclotron resonance is observed. In this case, the condition $\Delta \Omega \gtrsim 1/\tau$ implies $\omega \tau \gtrsim (\varepsilon_F/\hbar \omega)^2$, which is practically impossible to achieve experimentally and, therefore, the quantum cyclotron resonance is not observed near the classical resonance frequencies. The exceptions are those frequencies which correspond to the near-central sections, where $(\partial S/\partial p_z)_\varepsilon$ also vanishes and $(\partial \Omega/\partial n)_\varepsilon$ is of the same order as for the "normal" sections. Near these frequencies the quantum resonance oscillations are superimposed on the classical resonance pattern and are observed near the sections for which the density of states becomes infinite. For this reason, the amplitude of these oscillations is $(\varepsilon_F/\hbar \omega)^{1/2}$ times greater than the amplitude corresponding to the "normal" sections. The strong dependence of the classical amplitude on the magnetic field, observed near the classical resonance, may give rise, in the case $|\omega/q - \Omega| \lesssim 1/\tau$, to an aperiodicity of the quantum resonance oscillations in respect of $1/H$ and this invalidates Eq. (43.2c). If $\gamma \ll 1$ [see Eq. (43.3)], we observe the classical cyclotron resonance at selected frequencies and the amplitude of the resonance oscillations tends to zero as $\exp(-\pi \gamma^{-1})$.

For the same reasons as in the classical cyclotron resonance, the quantum resonance predicted in [69] is observed only if a static magnetic field is exactly parallel to the surface of a metal.

As in the case of quantum oscillations (§42), the amplitude of the quantum cyclotron resonance is nonexponentially small if the following inequality is satisfied:

$$2\pi^2 T \lesssim \hbar \Omega_1. \tag{43.7}$$

where

$$\Omega_l = \frac{eH}{m_l^* c}, \qquad m_l^* = \frac{1}{2\pi}\left(\frac{\partial S}{\partial \varepsilon}\right)_\Omega . \tag{43.8}$$

Since $m_l^* = m^*(\partial m^*/\partial p_z)_S/(\partial m^*/\partial p_z)_\varepsilon$, the condition of Eq. (43.7) is satisfied by the sections for which $(\partial m^*/\partial p_z)_S = 0$ and $(\partial m^*/\partial p_z)_\varepsilon \neq 0$ at practically any temperature, provided the value of τ at the temperature selected satisfies the conditions given in Eq. (43.3).

Thus, at sufficiently low temperatures, the quantum cyclotron resonance in sufficiently pure metals should be observed at discrete frequencies.

Experimental investigations of the quantum cyclotron resonance can yield very detailed information on the electron spectrum because they can be used to find directly the dependences $m^*(p_z)$, which can be deduced from the resonance frequency given by Eq. (43.2a), and $S(\mu)$, which can be found from the separation between the frequencies given by Eq. (43.2c).

The "cutoff" of the quantum resonance in a plate should make it possible to obtain an additional dependence of all the Fermi surface diameters on p_z.

In developing a theory of the quantum cyclotron resonance [69a], we shall consider only the most important case of a static magnetic field which is exactly parallel to the surface of a metal. In this case, we can use Eq. (42.1), which has been derived for the case of a parallel field (§42) without assuming that there is no resonance. Substituting the expression for $\Delta\varepsilon$, we obtain (the coefficient of proportionality can be found, as mentioned in §42, by going over to the classical case):

$$j = \frac{e^3 H}{(2\pi\hbar)^3 c} \sum_{n, l, \sigma} \int_{-\infty}^{\infty} dp_z \frac{n_F\left(\varepsilon_{n+l, p_z}^\sigma\right) - n_F\left(\varepsilon_{n, p_z}^\sigma\right)}{\varepsilon_{n+l, p_z}^\sigma - \varepsilon_{n, p_z}^\sigma} \frac{A_l(y, p_z)}{-i\omega + il\Omega_n(p_z) + 1/\tau}, \tag{43.9}$$

where

$$A_l(y, p_z) = \frac{1}{T^2} \int_0^T dt \int_0^T dt' v(t) e^{-il\Omega(t - t')} v(t') E\left(y - \int_{t'}^t v_y(t_1)\, dt_1\right).$$

The calculations are simplest at absolute zero ($T = 0$) because, in this case, the integration corresponding to Δn_F in Eq. (43.9) is carried out over the interval

$$\Delta p(n, l) = p_z(n + l) - p_z(n) \tag{43.10}$$

and Eq. (43.9) reduces to a sum of the form

$$l = \sum \varphi(n)\, \Delta p(n) = \sum g(n). \tag{43.11}$$

When the conditions of Eq. (43.3) are satisfied, it is most convenient to use the cotangent formula in the summation. The integral

$$\frac{1}{2i} \int \cot \pi z \cdot g(z)\, dz \tag{43.12}$$

is taken over a contour consisting of a semicircle, which is located in the upper half-plane and does not pass through points representing real integers, and a closing straight line located below the real axis. This integral gives residues at the points representing integers – the sum of these residues gives Eq. (43.11) – and residues (if any) at the poles of the function g(z). The integral along the straight line can be calculated by supplementing it with a semicircle in the lower half-plane (the integral along this semicircle can be found as easily as the integral along the upper semicircle). Thus, the whole integral is determined primarily by the residue of g(z) in the lower half-plane.

We thus obtain, omitting the classical term [69a]:

$$j(y) = \frac{e^3 H A_{l_s}(y, \omega/H)}{2i\hbar^3 c\pi^2 (\partial\Omega/\partial p_z)_S} \left\{ \ln \frac{\sin \pi n_2}{\sin \pi n_1} - \pi i \varkappa_1 \, \text{sign} \left(\frac{\partial S}{\partial \Omega} \right)_{\varepsilon_F} \right\}_{\substack{\varepsilon = \varepsilon_F \\ \mu\,(\varepsilon_F,\, p_z)\ H = \omega\hbar}}, \tag{43.13}$$

where

$$n_2 = n_1 - \varkappa_2, \quad n_1 = \frac{cS\,(\omega/H)}{2\pi e\hbar H} - \frac{ic\,(\partial S/\partial\Omega)}{2\pi e\hbar H\tau},$$

$$\varkappa_2 = \left(\frac{\partial\Omega}{\partial p_z} \right)_S \bigg/ \left(\frac{\partial\Omega}{\partial p_z} \right)_\varepsilon.$$

In the general case of an arbitrary temperature, it is convenient to replace the variable p_z with ε. Since we are interested only in the region close to the resonance point, we have to calculate [bearing in mind Eqs. (43.11) and (43.12)] an integral of the type

$$J = \frac{1}{i\hbar\omega} \int_{-\infty}^{\infty} \{n_F(\varepsilon + \hbar\omega) - n_F(\varepsilon)\} \left\{ \cot \pi \left(\frac{cS}{2\pi e\hbar H} + i\gamma \right) + i \, \text{sign} \, \gamma \right\} d\varepsilon, \tag{43.14}$$

$$\gamma = \frac{\pi c}{2\pi e\hbar H\tau} \left(\frac{\partial S}{\partial\Omega} \right)_\varepsilon.$$

Reducing this integral to an integral over a contour analogous to that used in the preceding paragraphs, we obtain a sum of residues and it is found (as expected) that the important value is $\varepsilon = \varepsilon_F$, so that $S = S\left(\varepsilon_F, \frac{\omega}{H}\right) + (\partial S/\partial\varepsilon)_\Omega (\varepsilon - \varepsilon_F)$. Finally, we obtain

$$J = -\frac{2\pi T}{\hbar\omega} \sum_{m=1}^{\infty} \left\{ \cot \pi \left(\frac{cS_0}{2\pi e\hbar H} + 2\pi i T \left(\frac{\partial S}{\partial\varepsilon} \right)_\Omega \frac{m^{\bullet}c}{2\pi e\hbar H} + i\gamma \right) - \cot \pi \left(\frac{cS_0}{2\pi e\hbar H} + 2\pi i T \left(\frac{\partial S}{\partial\varepsilon} \right)_\Omega \frac{m^{\bullet}c}{2\pi e\hbar H} - \left(\frac{\partial S}{\partial\varepsilon} \right)_\Omega \frac{i\omega c}{eH} + i\gamma \right) \right\},$$

$$S_0 = S(\varepsilon_F, \omega/H). \tag{43.15}$$

Having obtained the formula for the current density, we find that the remaining steps in the calculation of the impedance are exactly the same as in the classical case.

We shall not give the fairly cumbersome final formulas derived in [69a] because they confirm all the formulas, conclusions, and estimates obtained earlier on the basis of qualitative considerations (only the qualitative formulas, etc. can be checked experimentally).

§44. Paramagnetic Resonance in Metals

The application of a static magnetic field aligns the electron spins either parallel or antiparallel to the field. The energy of a spin in a magnetic field is $-\mu_0 H$ in the first case and $+\mu_0 H$ in the second (μ_0 is the magnetic moment of a free electron).

In an alternating electromagnetic field, the spin of an electron may be reversed. According to quantum mechanics, the probability of spin reversal (spin flip) is of the resonant nature and it passes through its maximum value when the energy of a quantum of the electromagnetic field is equal to the energy of spin reversal, i.e., to the separation between the energy levels corresponding to the parallel and the antiparallel spin orientations:

$$\hbar\omega = 2\mu_0 H, \qquad \omega = \Omega_0 = 2\mu_0 H/\hbar.$$

The probabilities of spin reversal are the same for both directions. However, the numbers of electrons oriented parallel (N_+) and antiparallel (N_-) to a static magnetic field are not equal because the parallel orientation is more "convenient." Under equilibrium conditions, the difference $N_+ - N_-$ is determined by the difference between the chemical potentials (Fig. 90). Since the number of spins reversed in an alternating field is proportional to the number of spins in the initial state, the application of such a field increases the number of spins oriented antiparallel to the field and reduces the difference $\Delta N = N_+ - N_-$, i.e., it results in some "depolarization" of the electron gas. Since the collisions accompanied by spin reversal tend to reestablish the state corresponding to thermal equilibrium, i.e., the difference $2\Delta N_0 = N_0(\varepsilon - \mu_0 H) - N_0(\varepsilon + \mu_0 H) = -2\mu_0 H(\partial N_0/\partial\varepsilon)$, it follows that the application of a homogeneous field establishes a difference ΔN corresponding to $\Delta N \cdot \alpha = (\Delta N_0 - \Delta N)/T_s$, i.e., $\Delta N = \Delta N_0(1 + \alpha T_s)^{-1}$, where T_s is the time between two collisions accompanied by spin reversal (the relaxation time) and α is the probability of spin reversal per unit time. At the resonance point $\alpha = 4\mu_0^2 H_1^2 T_s/\hbar^2$, where $2H_1$ is the amplitude of the alternating field.

At the resonance point, the change in the magnetic moment of the electron gas has its maximum value and, consequently, the corresponding correction to the impedance is also of the resonant nature.

In the nonlinear approximation, there is a change not only in the components of the magnetic moment which alternate in time but also in the constant component which is directed parallel to the magnetic field. Moreover, because the value of T_s is very large (reaching values of the order of 10^{-6}-10^{-7} sec, as shown in [78]), the nonlinear effects occur at alternating magnetic field intensities which are quite easy to reach in the laboratory. This can be used for the very important purpose of polarizing nuclei because the reversal of the spin of an electron as a result of its collision with a nucleus is accompanied by the reversal of the nuclear spin (this is why, in particular, the value of T_s is large) and, as shown earlier, thermal collisions under resonance conditions align more spins parallel to the field than against the field [79].

However, in the case of metals, all the formulas which employ the quantum-mechanical probability α of the spin reversal per unit time are incorrect.

In fact, the probability of spin reversal per unit time can be introduced only if an electron is subjected to an almost homogeneous field for a time which exceeds the field period. Since

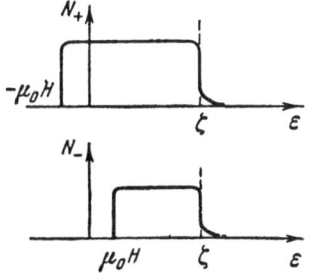

Fig. 90. Energy distribution of the "spin-up" and "spin-down" states at low temperatures.

$\delta \sim c\,[(\omega^{-1} + \tau)^{-1} 2\pi\sigma]^{-1/2}$ (σ is the conductivity of the metal) and a freely moving electron travels a distance $vT \sim v/\omega$ during the field period, we must satisfy the condition

$$v/\omega \ll c\,(2\pi\sigma)^{-1/2}(\omega^{-1} + \tau)^{1/2}, \text{ i.e., } \omega \gg v/c\,(2\pi\sigma/\tau)^{1/2} \gtrsim 10^{13} \text{ sec}^{-1}$$

and the magnetic field must be $H = \hbar\omega/2\mu_0 \gg 10^6$ Oe. This means that, in the case of realistic fields and frequencies, an electron would remain in the skin layer for a time much shorter than the period of the electromagnetic field.

In some cases, this would make it impossible to observe paramagnetic resonance in metals since an electron would not "sense" the field frequency. However, the actual situation is different for the following reason. We have mentioned earlier that the mean free time (relaxation time) of an electron corresponding to the spin reversal is very large and normally much larger than the usual mean free time τ corresponding to the momentum. Therefore, electrons suffering many collisions diffuse slowly into the metal, traveling, as a result of such diffusion, a path $\delta_{eff} = v\sqrt{\tau T_s/3}$ in a time T_s. This means that an electron re-enters the skin layer many times and spends a total time $T_s(\delta/\delta_{eff}) \sim (\delta/v)\sqrt{T_s/\tau}$ in this layer. Therefore, the resonance condition becomes $(\delta/v)\sqrt{T_s/\tau} \gtrsim 2\pi/\omega$, or $\omega \gtrsim (2\pi\sigma v^2/c^2)(\tau/T_s)$, which is quite easy to satisfy in the laboratory.

Thus, paramagnetic resonance in metals can be observed and the theory of this resonance must take account of the diffusion of spins into the metal. A comparison of the theory with the experimental results should make it possible to determine the principal parameters of the theory: the spin relaxation time T_s and the g factor (more exactly, the difference between this factor and the g factor of free electrons which is 2).

This discussion shows that the high-frequency field "carried" by the spins decays very slowly with depth and the corresponding skin depth is $\delta_{eff} = v\sqrt{\tau T_s/3}$. This means that at this depth the nuclear polarization resulting from paramagnetic resonance should also become ineffective.

It is clear from our discussion that the development of a theory of paramagnetic resonance requires the solution of two independent problems: the determination of the spin relaxation time T_s and the calculation, for a given value of T_s, of the surface impedance of a metal.

The relaxation time T_s is calculated in [79] but the main cause of the relaxation (apart from the collisions with paramagnetic impurities), which is the spin–orbit coupling of electrons with the lattice, is allowed for only in [80] and the calculated value of T_s is found to be in agreement with the values determined experimentally.

A consistent proof of the validity of the introduction of the spin relaxation time is given in [81], where the density matrix method is applied to quantum systems [82]. It is also shown in [81] that the "transverse" and the "longitudinal" spin relaxation times are identical.

Since the cited treatments have demonstrated rigorously that the relaxation time T_s is simply a parameter in the theory of paramagnetic resonance, we shall not deal with the method of calculation of T_s but give only the results. It is found that at a temperature T, which is low compared with Debye temperature Θ, the relaxation time is $T_s \approx \alpha T^{-1} \ln(Tv/\mu_0 H_s)$, and at temperatures $T \gg \Theta$, the corresponding expression is $T_s \approx T^{-1} \ln(\Theta v/\mu_0 H_s)$, where s is the velocity of sound, $\alpha \approx \rho s^2 [\omega_0 N(g-2)^2]^{-1}$, $\omega_0 = \varepsilon_F/\hbar$ is the degeneracy frequency, ρ is the density of the metal, and N is the number of electrons per unit volume.

The theory of paramagnetic absorption in a static magnetic field normal to the surface of a metal is presented in [83] for a plate of finite thickness and for a bulk metal in the case when T_s is given. The magnetic moment is determined in the same paper by solving the equation of motion for the electron spin operator [85], making allowance for the diffusion of spins in an in-homogeneous alternating field. The diffusion is allowed by considering an alternating magnetic field at the point at which a diffusing electron is located at a given moment and by subsequent averaging over all the trajectories representing random motion of an electron. This method of calculating the magnetic moment is very complex. Because of this, we shall give only the principal results and the formulas taken from [83] for the case when the surface relaxation is unimportant.

1. In the case of films which are thin compared with the skin depth δ, so that $D < 4\delta$ (here, D is the film thickness), the line representing the absorption P per unit volume has its usual symmetrical profile:

$$P = \frac{1}{4} \omega^2 H_1^2 T_s \chi \left(1 + \alpha_1^2\right)^{-1},$$

where χ is the magnetic susceptibility, $2H_1$ is the amplitude of the alternating magnetic field, $\alpha_1 = (\omega - \Omega_0) T_s$, and $\Omega_0 = 2\mu_0 H/\hbar$.

2. In the case of thick samples, the absorption line has a central structure and a width equal to the natural width $1/T_s$ with "wings" extending to $T_s^{-1} (\delta_{eff}/\delta)^2$ if $\delta < \delta_{eff}$.

3. The central structure is always much more pronounced than the "wings" and, therefore, under normal experimental conditions, the visible line width is of the order of $1/T_s$.

4. A characteristic feature of the paramagnetic resonance is that the diffusion of electrons does not broaden the absorption line but alters considerably its profile.

5. In the case of thick samples whose absorption line has a very small natural width, $D \gg \delta_{eff} \gg \delta$, the line (representing the absorption P per unit area) is asymmetric and described by the formula:

$$P = -\left(4 \sqrt{3}\right)^{-1} \omega^2 H_1^2 \chi T_s \frac{\delta^2}{\delta_{eff}^2} \operatorname{sign}(\alpha_1) \left(\sqrt{1 + \alpha_1^2} - 1\right)^{1/2} \left(1 + \alpha_1^2\right)^{-1/2}.$$

6. In the case of thick samples whose absorption line has a very large natural width, $D \gg \delta \gg \delta_{eff}$, the line (representing the absorption P per unit area) is given by the formula:

$$P \approx \frac{1}{8} \omega^2 H_1^2 \chi T_s \delta \left(1 - \alpha_1\right)\left(1 + \alpha_1^2\right)^{-1}.$$

7. The intensity of the absorption line of thick samples decreases at the center by a factor $\sim \delta/\delta_{eff}$ because of diffusion. If $\delta \ll \delta_{eff}$, the integrated line intensity is determined principally by the diffusion "wings" and not by the central part of the line.

This theory is in good agreement with the experimental results [78]. Paramagnetic resonance has been observed in lithium, sodium, and beryllium in the temperature range 4-296°K, and in potassium at 4°K. Electron paramagnetic resonance in metals was observed for the first time in 1952 [86]. Among the subsequent experimental investigations it is worth mentioning the work of Levy [87], who observed resonance in solutions of Na, Li, K, Cs, Rb, and Ca in ammonia. Investigations of paramagnetic resonance in tin, antimony, and bismuth were reported in [88, 40, 89], respectively. Lewis and Carver [90] investigated the transparency of lithium films under paramagnetic resonance conditions (this effect will be explained later). Of the recent theoretical papers, it is worth mentioning the work of Hebel, Blount, and Smith [91].

The agreement between the theory and experimental data can be seen clearly in Fig. 91. Similar agreement can be observed by comparing the theoretical curves of Fig. 92 with the experimental data of Fig. 93 (all these figures were taken from [86]).

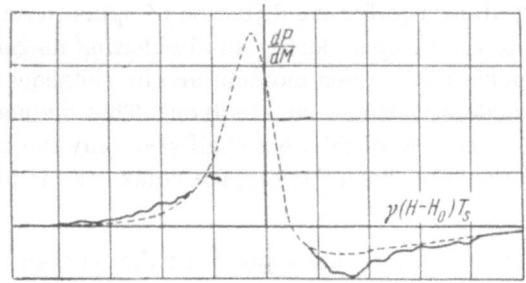

Fig. 91. Comparison of the theory (dashed curve) with the experimental results (continuous curve) on the electron paramagnetic resonance in the extreme anomalous skin effect (T = 4°K, 314 MHz).

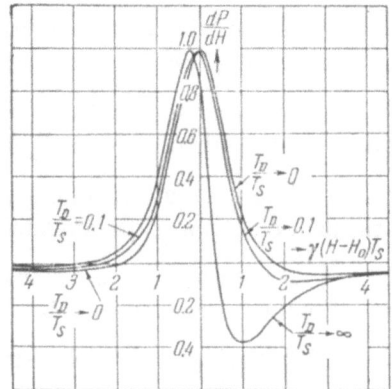

Fig. 92. Theoretical absorption curves for paramagnetic resonance in a thick lithium plate, plotted for different values of the ratio T_D/T_s (T_D is the time taken by a spin to diffuse a distance D and T_s is the spin relaxation time).

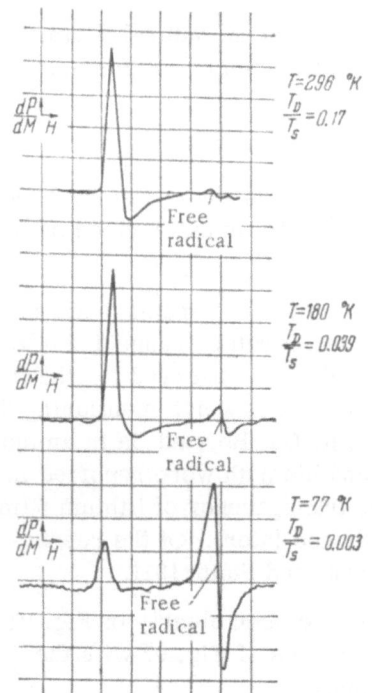

Fig. 93. Experimental absorption curves for paramagnetic resonance in a thick plate of lithium obtained for different values of the ratio T_D/T_s (320 MHz).

A comparison of the experimental results with the theory has made it possible to determine the principal parameters which characterize the paramagnetism of an electron gas: the spin relaxation time and its temperature dependence (in the case of metals with few impurities, this dependence agrees with that calculated in [80]) and the g factor (more exactly, the difference between this factor and 2) for those metals in which resonance has been observed.

The theory is further developed in [92]. This development has been necessary because of the many restrictive assumptions adopted in [83]. It is necessary to replace free electrons characterized by a dispersion law $\varepsilon = p^2/2m$ with particles having an arbitrary dispersion law, and to consider effects which are nonlinear with respect to the magnetic field H_1 (in particular, the saturation of the resonance, which is important in the polarization of nuclei but which is not considered in [83]). Moreover, the nature of the penetration of the field into a metal is not discussed in [83]. It is shown in [92] that the penetration causes selective absorption in metal films under resonance conditions. Moreover, only the case of a static magnetic field perpendicular to the surface of a metal is considered in [83].

All these limitations are avoided in [92]. The use of the transport equation makes it possible to impose consistently the boundary conditions and to analyze the nature of the approximation used in the diffusion theory.

The magnetic moment M is calculated in [92] on the basis of its relationship with the density matrix $\hat{n}$: $M = \mu_0(2\pi\hbar)^3 \int \mathrm{Sp}\,(\hat{\sigma}\hat{n})\,dp$, where $\hat{\sigma}$ is the spin operator. The density matrix $\hat{n}$ is found using the transport equation, which simplifies considerably because near the resonance point (in the lowest approximation) the quantity $\hat{n}$ can be regarded simply as the operator applicable to spins but not to the coordinates or the momentum. In this case, the Hamiltonian consists of a classical part, which gives the usual total derivative with respect to time, and the quantum term $\mu\sigma_0\mathbf{B}$, so that

$$\frac{\partial \hat{n}}{\partial t} + \boldsymbol{v}\,\frac{\partial \hat{n}}{\partial \boldsymbol{r}} + \frac{\partial \hat{n}}{\partial \boldsymbol{p}}\left\{e\boldsymbol{E} + \frac{e}{c}\,[\boldsymbol{v}\boldsymbol{B}]\right\} + \frac{i}{\hbar}\,[\mu_0\hat{\sigma}\boldsymbol{B},\,\hat{n}] + \left(\frac{\partial \hat{n}}{\partial t}\right)_c = 0$$

(B is the magnetic induction). We must bear in mind that, strictly speaking, the commutant $[\sigma\hat{\mathbf{B}}, \hat{n}] = [\boldsymbol{\sigma}, \hat{n}]\,\hat{\mathbf{B}} + \sigma\,[\hat{\mathbf{B}}, \hat{n}]$ consists of the quantum (first) term responsible for the equalization (by the alternating field) of the numbers of spins oriented parallel and antiparallel to the static magnetic field, and the classical (second) term, which reduces to the product of $\hbar/i$ and the Poisson brackets for B and n, responsible for the force acting on a spin in an inhomogeneous magnetic field. The processes represented by the quantum term result in resonant spin reversals and determine the degree of depolarization of electrons under resonance conditions. The processes represented by the classical term simply result in the alignment of the spins along the direction of B, and can be ignored in the determination of the depolarization of nuclei.

The collision integral $(\partial \hat{n}/\partial t)_c$ also consists of two terms representing the fast relaxation of the energy and the momentum (relaxation time τ) and the slow relaxation of the spins (relaxation time T_s). Since $T_s \gg \tau$, we can consider two types of collision separately. The collisions involved in the fast relaxation process cannot alter the total spin operator of the system and they are responsible for partial equilibrium corresponding to the density matrix $\hat{n}^{\text{eqm}}$, which depends only on the energy if the number of particles is fixed and the spin moment is given by

$$\mathrm{Sp}\int \hat{n}^{\text{eqm}}\,dp = \mathrm{Sp}\int \hat{n}\,dp, \quad \mathrm{Sp}\int \hat{n}^{\text{eqm}}\,\hat{\sigma}\,dp = \mathrm{Sp}\int \hat{n}\hat{\sigma}\,dp.$$

At sufficiently low temperatures and in sufficiently weak magnetic fields ($\mu_0 H \ll T$), we have

$$\hat{n}^{\text{eqm}} = \langle \hat{n} \rangle = \oint_{\varepsilon = \varepsilon_F} \hat{n}v^{-1}\,dS \Big/ \oint_{\varepsilon = \varepsilon_F} v^{-1}\,dS,$$

$$\left(\frac{\partial \hat{n}}{\partial t}\right)^{\varepsilon,\, p}_{c} = \frac{\hat{n} - \langle \hat{n} \rangle}{\tau}.$$

The collisions involved in the slow relaxation process give rise to the density matrix which represents complete equilibrium in respect of the momentum and spin:

$$\hat{n}_0 = \begin{pmatrix} n_F\,(\varepsilon - \mu_0 B) & 0 \\ 0 & n_F\,(\varepsilon + \mu_0 B) \end{pmatrix}.$$

In this case, $(\partial \hat{n}/\partial t)^{\sigma}_{c} = (\hat{n} - \hat{n}_0)/T$. The nature of the transport equation is such that we can separate the parts of the distribution function which determine the current and the magnetic moment by assuming that $\hat{n} = n_1 \hat{I} + n \hat{\sigma}$ ($\hat{I}$ is a unit matrix).

The solution of the equation obtained for **n** is quite complex and, therefore, we shall formulate only the results obtained, in the same manner as we have done earlier for the results derived in [83].

1. The total magnetic moment is $\mathbf{M} = \chi\,(\mathbf{B} - \mathbf{b})$ and if the z axis is parallel to the static magnetic field, we find that

$$b_z(\xi) = \mathrm{Re}\,\{u\,(0)\,u_0^*(\xi)\}\,\{1 + \mathrm{Re}\,[u\,(0)\,u_0^*(0)]\}^{-1},$$
$$iH_0 b = b_x + ib_y = u\,(\xi)\,\{1 + \mathrm{Re}\,[u\,(0)\,u_0^*(0)]\}^{-1}, \tag{44.1}$$

where $\mathbf{y} \perp [\mathbf{z}\boldsymbol{\xi}]$. The direction ξ coincides with the inward normal to the surface of a metal; the zero subscript of u* denotes that the resonance value of u*, corresponding to $\omega = \Omega_0$, is implied.

Near the resonance point, the function $u(\xi)$ is equal, to within numerical factors of the order of unity (in the case $\delta \ll \delta_{\mathrm{eff}}$), to

$$u\,(\xi) \approx c\,[E_x(0) + iE_y(0)]\,\tilde{\delta}_{\mathrm{eff}}\,l\,(vH_0 l r_H^2 a)^{-1}\exp\,(-\,\xi/\tilde{\delta}_{\mathrm{eff}}), \tag{44.2}$$

where

$$\tilde{\delta}_{\mathrm{eff}} = avT_s^{1/2}\{3\,[1 + i\,(\omega - \Omega_0)\,T_s]\,\tau^{-1}\}^{-1/2}.$$

In the case of strong static magnetic fields, the quantity a, which determines the magnetic moment and the depth of penetration, is quite different for fields inclined and parallel to the surface of a metal. Physically, this is due to the fact that, in a parallel field, electrons escape into the metal only as a result of collisions. Between collisions, all the electrons following closed orbits have zero average velocity along the direction ξ and if $r_H \ll l$ they diffuse in a time T_s not to a distance $l/\sqrt{T_s/3\tau}$ but to a distance l/r_H times shorter, i.e., a distance given by $r_H \sqrt{T_s/3\tau}$.

In strong fields ($r_H \ll l$) inclined at an angle $\varphi \sim 1$ with respect to the surface of the metal, the quantity a is of the order of unity, whereas if $\varphi \leq r_H/l$ this quantity is $a \sim r_H/l$.

In weak magnetic fields ($r_H \gtrsim l$), we always have $a \sim 1$. The field $E_\alpha(0)$ is determined by the amplitude of the field incident on the surface of the metal and it can be found by employing the standard formulas for the surface impedance (they have been deduced earlier for various cases).

Formulas (44.1) and (44.2) can be used in investigations of the dependence of a given parameter (for example, the absorption line profile) on the angle φ, and to study nonlinear effects.

Moreover, the same formulas can be used to calculate quite easily the surface impedance and the transparency of films [92].

To calculate the impedance, we must consider the linear approximation. In the approximation which is linear with respect to the alternating field, the formulas (44.1) and (44.2) can be used to relate the projection of the magnetic moment onto the surface of the metal to the projection $E_{\parallel}$ (where we can use the relationship between E_{ξ} and $E_{\parallel}$ found for $\mathbf{B} = \mathbf{H}$ from the equation for $j_{\xi} = 0$): $M_{\alpha}(0) = g_{\alpha\beta} E_{\beta}(0)$. Since

$$E_{\alpha}(0) = \frac{ic}{\omega} \zeta_{\alpha\beta} E'_{\beta}(0) = \xi_{\alpha\beta} B_{\beta}(0),$$

where $\zeta_{\alpha\beta}$ and $\xi_{\alpha\beta}$ are related by a self-evident expression and

$$B_{\alpha}(0) = H_{\alpha}(0) + 4\pi M_{\alpha}(0),$$

it follows that

$$E_{\alpha}(0) = \xi_{\alpha\beta} H_{\beta}(0) + 4\pi \xi^{(0)}_{\alpha\gamma} g_{\gamma\beta} E_{\beta}(0)$$

(in the second term we employ the "usual" classical form of $\xi^{(0)}_{\alpha\beta}$). On the other hand, it follows from the definition of $\xi^{(0)}_{\alpha\beta}$, that in the absence of resonance and when $\mathbf{B} = \mathbf{H}$

$$E_{\alpha}(0) = \xi^{(0)}_{\alpha\beta} H_{\beta}(0).$$

Therefore,

$$\xi_{\alpha\beta} = \xi^{(0)}_{\alpha\beta} - 4\pi \xi^{(0)}_{\alpha\gamma} g_{\gamma\gamma'} \xi^{(0)}_{\gamma'\beta},$$

which is the solution of the problem.

It is evident from Eq. (44.2) that the attenuation depth of $\mathbf{M}$ increases strongly on approach to resonance. Consequently, films become selectively transparent under paramagnetic resonance conditions because a film of thickness D ($\delta \ll D \lesssim \delta_{\text{eff}}$) transmits practically without attenuation a small fraction of the field associated with $\mathbf{M}$. Unfortunately, this effect is much weakened by the reflection from the boundaries of a film.

In the simplest case of a static magnetic field perpendicular to the surface of a metal, the resonance value of the transmission coefficient k of such a film (defined as the ratio of the intensities of the transmitted and the incident waves) is given by the following expression which is accurate to within a numerical factor of the order of unity [92]:

$$k \approx \left| 16\pi^2 \chi T_s c \zeta^{(0)'} \right|^2 \left[2\pi d \left(1 + \left| 4\pi c \zeta^{(0)} T_s H_1^{\text{inc}} / 2\pi d H_0 \right|^2 \right) \right]^{-2}.$$

The transmitted power W^{tr} has its maximum value when the intensity of the magnetic field H_1^{inc} in the incident wave is

$$H_1^{\text{inc}} = 2\pi d H_0 \left(4\pi c T_s \zeta^{(0)} \right)^{-1}.$$

In this case,

$$W^{\text{tr}}_{\text{max}} = \pi \left(2\chi \hbar c \left| \zeta^{(0)} \right| / 4\mu_0 \lambda \right)^2$$

is independent of the film thickness (λ is the wavelength of the alternating field).

A rigorous but very complex solution of the problem of paramagnetic resonance in metals is given in [78, 83, 92].

A simple approach, capable of yielding the main qualitative results of the cited investigations, is presented in [93]. In the latter paper, use is made of the equation given in [94] (this is the Bloch equation supplemented by the relaxation and diffusion terms). In a magnetic field perpendicular to the surface of a metal whose dispersion law is $\varepsilon = p^2/2m$ the equation for the alternating part of the magnetic moment is of the form

$$\frac{\partial M}{\partial t} = \gamma [M, \ H] - \frac{M}{T_s} + D \, \Delta M,$$

where $\gamma = e/mc$, $D = v^2\tau/3$. However in the approximation which is linear with respect to the alternating field H_1, this equation becomes

$$\frac{\partial M}{\partial t} = \gamma [M_0, \ H_1] + \gamma [M, \ H_0] - \frac{M}{T_s} + D \, \Delta M. \qquad (44.3)$$

Here, $M_0 = \chi H_0$, $\gamma H_0 = \Omega_0$.

The boundary condition used in [93] (which has little effect on the nature of the solution) can be replaced by extending, in an even manner, H_1 and M in Eq. (44.3) to the region where $z < 0$ (the magnetic field is, as usual, directed along the z axis).

Going over to the Fourier components, we obtain:

$$\alpha M_k - \gamma [M_k, \ H_0] = \gamma\chi [H_0, \ H_{1k}], \quad \alpha = i\omega + Dk^2 + \frac{1}{T_s},$$

and hence,

$$M_k = \chi \left(\alpha^2 + \Omega_0^2\right)^{-1} \left\{\Omega_0^2 H_{1k} + \alpha [\Omega_0, \ H_{1k}]\right\}.$$

Bearing in mind that $M_k = M_{-k}$, we can obtain an expression corresponding (in the case of H_0 perpendicular to the surface of a metal) to the principal formulas (44.1) and (44.2) used in the calculation of the impedance; this can be done bearing in mind that $\delta \ll \delta_{\text{eff}}$ and

$$и \int_0^\infty H_{1k} \, dk = \pi H_1 (0).$$

The theory presented in this section applies to the normal state of metals because paramagnetic resonance is impossible in superconductors.

The results given apply to practically all cases of paramagnetic resonance and can be used in conjunction with the experimental data in determinations of the spin relaxation time and g factor of conduction electrons.

Allowance for the diamagnetic quantization of the energy levels can give rise only to a superposition of diamagnetic oscillations on the resonance curve, provided the period of these oscillations is short compared with the resonance line width.

§45. Combined Resonance

The foregoing discussion has been concerned with resonances involving transitions between two types of level: between different diamagnetic levels corresponding to the same spin projection, as represented by the cyclotron resonance at frequencies $\omega = q\Omega_{\text{extr}}$ (these transi-

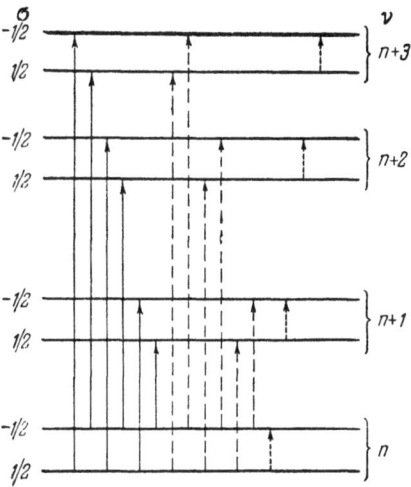

Fig. 94. Combined resonance (continuous lines), cyclotron resonance
(dashed lines), and paramagnetic
resonance (dotted lines).

tions are represented by the dashed lines in Fig. 94); and between different paramagnetic levels
with different spin projections but corresponding to the same diamagnetic level, as represented
by the paramagnetic resonance at frequencies $\omega = 2\mu_0 H/\hbar = \Omega_0$ (these transitions are represented by the dotted lines in Fig. 94).

If we ignore the spin–orbit coupling and regard the spin and the crystal momentum as two
independent quantities (in the sense that their operators commute rigorously), we find that, in
a homogeneous alternating magnetic field, the matrix elements of the Hamiltonian which are nondiagonal in respect of the spin are diagonal in respect of the number of the diamagnetic level.
Consequently, the resonance associated with those transitions which are accompanied by simultaneous changes in the spin projection and in the diamagnetic level (the combined resonance at
frequencies $\omega = q\Omega_{\text{extr}} \pm \Omega_0$ represented by the continuous lines in Fig. 94) is impossible.

In good metals, the spin–orbit coupling is usually negligible and, therefore, the combined
resonance resulting from the spin–orbit interaction is observed mainly in semiconductors and
in semimetals.

This resonance was predicted and investigated in detail in [96] (see also [97]; for experimental results refer to, for example, [98]). Since we are interested in metals, we shall not discuss this type of resonance in detail. We shall simply point out that the quasiclassical quantization represented by Eq. (7.4) cannot be used directly in the determination of the resonance
frequencies because of the spin–orbit interaction. It is necessary to find the nature of the
Hamiltonian *ab initio* using some suitable method such as the effective mass approximation
[96]. Knowing the spectrum in a magnetic field makes it possible to solve the combined resonance problem in the same way as has been done earlier for the cyclotron and the paramagnetic
resonances.

However, if the magnetic field is strongly inhomogeneous, the combined resonance may
be observed in good metals even if the spin–orbit coupling can be neglected [99]. In this case,
the spin term $\mu_0\hat{\sigma}H$ in the Hamiltonian [$\hat{\sigma}$ is the spin operator, H is the total magnetic field which
includes the time-dependent and strongly inhomogeneous component $H_1(\mathbf{r}, t)$] is found to depend
on the spin operator and on the coordinates. This means that the matrix elements of this term
are nondiagonal in respect of the spin projection σ and in respect of the diamagnetic level num-

ber ν, i.e., these terms ensure that the combined resonance is observed. This is a self-evident physical observation: the orbit of an electron in an inhomogeneous magnetic field is affected, in particular, by the forces acting on the electron spin.

The combined resonance clearly requires that the cyclotron resonance conditions (the magnetic field must be exactly parallel to the surface of a metal and $\delta \ll r_H$) as well as the paramagnetic resonance conditions ($\mathbf{H}$ and $\mathbf{H}_1$ must not be parallel) be satisfied simultaneously.

The amplitude of the combined resonance line can be determined in the same way as in the case of the paramagnetic resonance line (§44). The resonance is determined by the term $[1 - \exp(-T/\tau^*)]^{-1}$, where T is the period of revolution in a Larmor orbit and $1/\tau^* = 1/\tau + 1/T_s + i(\omega \pm \Omega_0)$, and the resonance frequency is $\omega = q\Omega_{extr} \pm \Omega_0$.

If the dispersion law is quadratic, the intensity of the combined resonance is $(1 + T_s/\tau)^{1/2}$ times weaker than the intensity of the paramagnetic resonance; in the case of an arbitrary dispersion law, the combined resonance frequency is additionally $\sqrt{\omega(\tau^{-1} + T_s^{-1})^{-1}}$ times lower. The width of the combined resonance line is $\Delta\Omega \sim \tau^{-1} + T_s^{-1}$. Since the combined resonance is related to the quasistationary nature of a state with a specified value of σ as well as a specified value of ν (i.e., with the quasistationary nature of a given orbit), the field is not "carried" into the metal, which is in contrast to the case of paramagnetic resonance.

§46. Impedance Oscillations in Weak Magnetic Fields

Only the bulk properties of metals, in particular, the electron-energy levels in an infinite sample, have been considered in the preceding sections. This approach has been used even in the treatment of the anomalous skin effect. However, we have shown in §§7 and 8 that the motion of electrons incident on the surface of a metal at small angles (and, therefore, reflected almost specularly from the surface) is quantized in a magnetic field. A characteristic system of levels, described in detail in §§7 and 8 [see Eqs. (7.25)-(7.32) and (8.22)-(8.26)], forms near the surface.

Surface levels affect the high-frequency properties of a metal if the "glancing" electrons (the electrons reflected at small angles) make a significant contribution. This does occur in the anomalous skin effect when the effective skin depth $\delta_{eff} \approx (\delta^2 l)^{1/3}$ is small compared with the mean free path l and, in the absence of a magnetic field, an important contribution is made by those electrons which reach the surface at an angle $\varphi \approx (\delta/l)^{2/3}$ (§33). At low temperatures and in high-quality samples, the angle φ may be of the order of 10^{-3}. In a weak magnetic field $(r_H \gg l)$, the important angles are $\varphi \approx (\delta_{eff}/r_H)^{1/2}$, which can be established quite easily from Fig. 95. Once again $\varphi \ll 1$.

These estimates show that the surface levels should affect the high-frequency properties of metals under the anomalous skin-effect conditions.

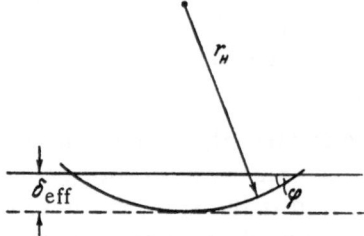

Fig. 95. Angle of incidence of an electron on the surface of a metal in a weak magnetic field.

It is evident from the results given in §§7 and 8 that the surface levels are manifested particularly strongly in the case of a cylindrical Fermi surface, i.e., when the surface has a cylindrical or nearly cylindrical region. In this case, we should observe resonance absorption of the energy of the high-frequency field, and Eq. (7.31), or a more rigorous formula which follows from Eq. (7.25), defines the resonance frequencies:

$$\omega = k\omega_n, \qquad k = 1, \ 2, \ 3, \ \ldots \tag{46.1}$$

According to Eq. (7.30), $\omega_n \propto H^{2/3}$ and, therefore, the resonance values of the magnetic field measured at a fixed frequency are proportional to the frequency raised to the power $3/2$: $H_{res} \propto \omega^{3/2}$. The resonance Larmor frequencies $\omega_H^{res} = eH_{res}/m^*c$ are very low and independent of the skin depth. The range of magnetic fields in which the energy is absorbed resonantly by the surface levels is bounded from below [see the second of the inequalities in Eq. (7.32)] and from above because the angle increases with increasing H as $\varphi_1 \propto H^{1/3}$, and if φ is sufficiently large the reflection by the surface becomes basically diffuse. This does not apply to the electrons in bismuth-type metals or the electrons in anomalously weakly populated bands of ordinary metals, for which the de Broglie wavelength is considerably larger than the lattice constant.

In the general case of an arbitrary Fermi surface, when ε_n and $\hbar\omega_{nm} = \varepsilon_n - \varepsilon_m$ are independent of p_z, resonance is observed (as in the cyclotron case) for that value of $p_z = p_0$ which corresponds to the slowest shift of the frequencies ω_{nm} away from the resonance value, i.e., when the density of states has its maximum value in relation to the frequency $\omega_H^!(p_0) = 0$. The specular nature of the reflection in the case of small angles may, in a sufficiently weak magnetic field, affect also the classical effects such as the cyclotron resonance, which disappears if q = 1 [10]. This happens when $q(\delta/l)^{1/2} \ll 1$.

The resonance line half-width is determined by the natural width of the levels and, therefore (§7),

$$\Delta\omega/\omega_n \approx q^{-1}(\varphi_n) + (\omega\tau)^{-1}. \tag{46.2}$$

The relative amplitude of the resonance line is quite different for cylindrical and noncylindrical Fermi surfaces. In the latter case, only the electrons defined by $\Delta p_z/p_0 \sim (\Delta\omega/\omega)^{1/2}$ participate in the resonance [because $\omega_n \propto (p_z - p_0)^2$] and, therefore, the relative amplitude of the resonance line is correspondingly lower.

Another important point is whether the resonance orbit extends outside the skin layer because the value of $(r_H/\delta)(\hbar\omega_H/\varepsilon_F)^{2/3}$ governs the effective path in the field as well as the relative number of field-accelerated electrons. The exact formula for the impedance can be calculated in the quasiclassical case (n ≫ 1). If $\hbar\omega \ll T$, this formula is particularly simple because we can then use the correspondence principle, replacing $1/\tau$ with $1/\tau + i\omega$ and α and ∞ in the formulas (9)-(10) in [84] for an arbitrary value of q. Moreover, the "classical" integration with respect to φ can be replaced with the "quantum-mechanical" summation with respect to φ_n and $E(\mu)$ can be taken inside the integral. If $\hbar\omega \gtrsim T$, we can replace $\partial n_F/\partial\varepsilon$ with $[n_F(\varepsilon_{k+k'}) - n_F(\varepsilon_k)]/(\varepsilon_{k+k'} - \varepsilon_k)$, where k is the set of quantum numbers n, p_z, and P_x (see also §42).

It is of considerable interest to derive the resonance curve for an arbitrary dependence $q(\varphi)$ (this is quite easy in the quasiclassical case if the points mentioned above are taken into account). A comparison of the experimental data with the theory should make it possible to obtain information on the nature of the dependence $q(\varphi)$.

The transition from the resonance orbits that do not extend outside the skin layer to the nonresonance orbits, which affects the resonance line profile, makes it possible to determine R from the relationship $y_{max} = R\varphi_n^2 \sim \delta_{eff}$, where δ_{eff} is expressed in terms of the impedance.

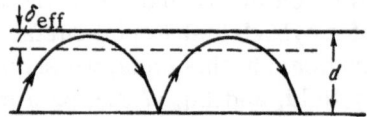

Fig. 96. Classical cyclotron resonance in a thin
(d ≪ l) plate.

This can be done because $R\varphi_n \propto R^{1/3}$ [see Eq. (7.28)]. In accordance with Eq. (7.30), the resonance frequencies should yield the value of v_x^0. Knowing R and v_x^0, we can use the dependence of the resonance line width on n in the case $\omega\tau \gg q^{-1}$ to find $q(\varphi)$ [see Eq. (46.2)].

If the magnetic field is inclined, the resonance corresponding to $p_z \neq 0$ becomes weaker (electrons escape into the interior of a metal and suffer nonperiodic collisions with the surface). However, the presence of an additional parameter in the form of the angles of inclination of the field extends the information which can be obtained, particularly in the case of the resonance corresponding to $p_z = 0$, when there is no drift into the interior.

The classical resonance should be observed in weak fields applied to a thin (d ≪ l) exactly plane-parallel plate at frequencies corresponding to $\sqrt{Rd}/v_0 = n2\pi/\omega$ when $H \propto (\omega/n)^2$ [100a]. Figure 96 explains the nature of the resonance. The same effect can be used to determine $q(\varphi)$ because the resonance line has the half-width $[1/\omega\tau + q(\frac{1}{2}\sqrt{R/d})]^{-1}$. We can find $q(\varphi)$ also from other effects which are sensitive to the angle of incidence of electrons on the surface. The search for and the investigation of such effects have only just begun.

The resonance oscillations described in the present section were discovered in 1961 [101] and have been observed several times since [102-103]. The first classical explanation relating these oscillations to the electrons which do not collide with the surface of a metal is suggested in [102]. The dominant role of the electrons reflected specularly from the surface of a metal is demonstrated in [105]; the purely quantum nature of the oscillations is shown in [106] and the correct frequency dependence of their period is obtained. The resonance nature of the oscillations is demonstrated in [107] and a suitable theory developed (a more detailed theoretical description of these oscillations is given in [113]). A quasiclassical treatment of the surface levels will be found in [104], where an experimental investigation of the oscillations in bismuth is compared with the theory. A detailed review of the theoretical and experimental investigations of the diamagnetic surface levels is presented in [114].

§47. Infrared Optics

In the preceding sections dealing with the high-frequency properties of metals, we have given special attention to the skin effect, which is extremely anomalous in the sense defined in §32 [see the inequality (32.9)]. Since the mean free path in the case $\omega\tau \gg 1$ is the path traversed by an electron in one field period v/ω and the skin depth δ tends to a constant value c/ω_0 (ω_0 is the plasma frequency of the electrons in a metal), the frequency range of the extreme anomalous skin effect is limited from above by the condition

$$\omega \ll \frac{v}{\delta} \approx \omega_0 \frac{v}{c}. \tag{47.1}$$

We shall be interested in the behavior of a metal at high frequencies under the conditions such that the smallest parameter of the dimensions of length is v/ω, i.e.,

$$\omega \gg \omega_0 \frac{v}{c}, \frac{1}{\tau}. \tag{47.2}$$

We shall first assume that the frequency is lower than the threshold of the internal photoelectric effect (this point will be considered later). The threshold of the internal photoelectric effect (photoconductivity or photo-emf), the width of the electron band (expressed in terms of the frequency), and the plasma frequency are of the same order of magnitude. Therefore, we shall assume simply that the following condition is satisfied:

$$\omega \ll \omega_0. \tag{47.3}$$

The conditions defined by Eqs. (47.2) and (47.3) are satisfied in the infrared range of wavelengths and, in the case of good metals and low temperatures, are obeyed right up to the millimeter range of wavelengths. That part of the physics of metals which deals with the range of frequencies defined by the inequalities (47.2) and (47.3) is known as the infrared optics of metals.

The largest parameter of the dimensions of reciprocal time is now the frequency. Therefore, in the lowest approximation, correction to the equilibrium distribution function is determined by the term with a partial derivative with respect to time in the transport equation; the collision integral can be ignored. When the Fermi-liquid interaction is taken into account, the correction to the distribution function (this correction is linear with respect to the electric field) has the form

$$f_1 = (i\omega)^{-1} e\boldsymbol{E} (1 + \hat{\Phi}) \, \boldsymbol{v} \, \frac{\partial f_F}{\partial \varepsilon}. \tag{47.4}$$

The operator $\hat{\Phi}$ represents the Fermi-liquid correlation in the motion of electrons (§40). In the gas approximation, we have $\hat{\Phi} \equiv 0$.

It follows from Eq. (47.4) that the current density is

$$j_i = - \frac{2e^2}{(2\pi\hbar)^3 (i\omega)} \oint_{\varepsilon(p)=\varepsilon_F} v_i (1 + \hat{\Phi}) \, v_k \frac{ds}{v} \, E_k. \tag{47.5}$$

It is evident from this formula that the current density and the electric field intensity differ in phase by $\pi/2$, i.e., the conductivity is a purely imaginary quantity. This means that, in the infrared range of frequencies, the electromagnetic properties of metals are described more conveniently by the permittivity tensor rather than by the conductivity tensor. The permittivity tensor is deduced from Eq. (47.5):

$$\varepsilon_{ik} = - \frac{8\pi e^2}{(2\pi\hbar)^3 \, \omega^2} \oint_{\varepsilon(p)=\varepsilon_F} v_i (1 + \hat{\Phi}) \, v_k \frac{dS}{v}. \tag{47.6}$$

If we restrict our treatment to this approximation, we find that the principal values of the surface impedance are purely imaginary quantities

$$\zeta_{1,2} = - i (\varepsilon_{1,2})^{-1/2}, \tag{47.7}$$

where $\varepsilon_{1,2}$ are the principal values of the tensor ε_{ik}. The principal values of ε_{ik} are of the order of magnitude of ω_0^2/ω^2, where ω_0 is the plasma frequency; $\omega_0^2 = 4\pi ne^2/m$, if we use a quadratic isotropic dispersion law and the gas approximation ($\hat{\Phi} \equiv 0$).

The purely imaginary impedance describes the total reflection of the electromagnetic wave from the surface of a metal: the reflection coefficient is unity. In the range of frequencies defined by the inequalities (47.2) and (47.3), a metal resembles very closely an electron plasma.

Unfortunately, very little information about the correlation function, which is the main characteristic of the Fermi-liquid description of the conduction electrons, can be derived from the symmetric tensor ε_{ik} whose components can be deduced from the phase shift resulting from reflection.

The absorption in the infrared range can be described using the next approximation in respect of the reciprocal frequency, and it depends strongly on the actual mechanism of the interaction of electrons with impurities, with phonons, and with other electrons. In other words, the absorption of the electromagnetic energy depends strongly on the form of the collision integral. Since the main targets of our treatment are the effects which are determined by the energy spectrum of electrons and are insensitive to the nature of collisions, we shall not consider in detail the calculations of the absorption coefficient in the infrared range (the reader is referred to [108, 109]). We shall simply estimate the order of magnitude of the absorption in the infrared range.

The principal values of the permittivity tensor in the next term of the expansion in powers of ν/ω have an imaginary correction which describes the absorption:

$$\delta\varepsilon \approx -i\frac{\nu}{\omega}\varepsilon \approx \frac{\omega_0^2\nu}{\omega^2}i; \quad \nu = \frac{1}{\tau}. \tag{47.8}$$

The real component of the surface impedance is proportional to ν/ω_0. Thus, the dependence of the absorption coefficient on the frequency, temperature, and other parameters is determined by the dependences of the collision frequency ν on these parameters (this topic is discussed in detail in [110]).

Before we estimate the collision frequency, which is naturally the sum of the frequencies of the collisions due to different mechanisms, we must make two comments concerning the estimates which will be given later:

1) Since the momentum of an electron p_F is large compared with the momentum of a photon in a metal ($p_F \gg \hbar\omega_0/c$), it follows that the frequencies of the inelastic collisions are affected by the transition to the high-frequency range, whereas the frequencies of the elastic collisions are the same as in the static case.

2) Since the Fermi energy is approximately equal to the average energy of the Coulomb interaction between electrons $\varepsilon_F \approx p_F v \approx e^2/a$, where a is the average distance between the electrons), it follows that

$$\omega_0 \approx \frac{v}{a} \approx \frac{\varepsilon_F}{\hbar}, \tag{47.9}$$

and $\varepsilon_F v/c$ is usually numerically equal to the Debye temperature, i.e., in our case,

$$\omega_0 \gg \omega \gg \nu, \frac{\theta}{\hbar}. \tag{47.3'}$$

Thus,

$$\nu = \nu_{e-i} + \nu_{e-e} + \nu_{e-ph} \tag{47.10}$$

where

$$\nu_{e-i} \approx \frac{v}{a}c_i \sim \omega_0 c_i$$

is the frequency of the collisions of electrons with impurities (c_i are the impurity concentrations);

$$\nu_{e-e} \approx \frac{\omega_0}{(\hbar\omega_0)^2}\left[T^2 + \left(\frac{\hbar\omega}{2\pi}\right)^2\right] \tag{47.11}$$

is the frequency of the collisions of electrons with one another; and ν_{e-ph} is the frequency of the collisions of electrons with phonons, which, in the case $\hbar\omega \gg \Theta$, is equal to the frequency of the electron–phonon collisions at the Debye temperature ($\nu_{e-ph} \approx \Theta/\hbar$). These formulas for the collision frequencies are suitable only for rough estimates.

Equations (47.8)–(47.11) demonstrate that the frequency dependence of the absorption in the infrared range is due to the electron–electron collisions. The constant term in the absorption coefficient is larger than the bulk (volume) absorption. The collisions of electrons with the surface of a metal (particularly those resulting in diffuse scattering) give rise to a surface component of the absorption which can be related to the spatial inhomogeneity of the field at a depth $\delta \approx c/\omega_0$. The surface absorption can be described by an effective collision frequency, which is of the order of

$$\nu_{eff} \approx \frac{v}{\delta} \approx \omega_0 \frac{v}{c}. \tag{47.12}$$

It is quite difficult to calculate the surface absorption in the case of an arbitrary dispersion law for electrons [109]. We must bear in mind that, even in the lowest approximation, the electric field cannot be regarded as varying slowly over a distance of the order of δ because, in the case of electron motion into the interior of a metal, even a homogeneous alternating field generates a current (and an associated electric field normal to the surface of the metal), which varies over distances of the order of v/ω. However, the order of magnitude of the surface losses is given correctly by the effective frequency of Eq. (47.12). It is worth noting that the real and the imaginary components of the surface impedance tensor of an anisotropic metal cannot be reduced simultaneously to the principal axes [109, 110a].

The condition specified by Eq. (47.3) is violated when the frequency is increased. This alone (apart from the photoconductivity) should alter considerably the high-frequency properties of a metal. At frequencies $\omega > \omega_0$, the permittivity of a metal is positive [we have omitted the permittivity of the ionic cores from Eq. (47.6) and this permittivity is of the order of unity far from the natural frequencies of ions] and an electromagnetic wave may travel in a metal. If we ignore the attenuation, we find that in this case the real component of the impedance is finite and that the imaginary component vanishes (Re $\zeta \neq 0$, Im $\zeta = 0$).

However, before the permittivity can change its sign, we exceed the threshold of the internal photoelectric effect ω_{th}. The direct absorption of photons by conduction electrons is due to band–band transitions (the direct intraband absorption is in conflict with the loss of conservation of energy and momentum because $v < c$). Since the momentum of a photon is negligibly small, the law of conservation of energy is

$$\varepsilon_s(p) + \hbar\omega = \varepsilon_{s'}(p) \qquad (s \neq s'). \tag{47.13}$$

Usually, min $[\varepsilon_{s'}(p) - \varepsilon_s(p)] \neq 0$ and, therefore, the absorption begins from a threshold frequency $\omega_{th} = \min[\varepsilon_{s'}(p) - \varepsilon_s(p)]$. The frequency dependence of the absorption coefficient of light Γ near the threshold ($\omega \gtrsim \omega_{th}$) has several features common to all metals and these features can be related to the energy structure of the electron spectrum [111]. An important role in an analysis of these features is played by the position of the point p_0 in the momentum space [p_0 is the point at which the function $\varepsilon_{s'}(p) - \varepsilon_s(p)$ has its minimum value]; if $\varepsilon_F - \hbar\omega < \varepsilon_s(p_0) < \varepsilon_F$, it follows that, in most cases, $\Gamma \propto \sqrt{\omega - \omega_{th}}$ (allowed transitions) but, in some cases, $\Gamma \propto (\omega - \omega_{th})^{3/2}$ (forbidden transitions). We may find that min $[\varepsilon_{s'}(p) - \varepsilon_s(p)]$ is not reached in the region of the p space bounded by the inequalities $\varepsilon_F - \hbar\omega < \varepsilon_s(p) < \varepsilon_F$. In this case, the threshold frequency is determined by the condition of contact of the surface $\varepsilon_{s'}(p) - \varepsilon_s(p) = \hbar\omega$ with one of the surfaces bounding the region mentioned above, i.e., by the surface $\varepsilon_s(p) = \varepsilon_F$ or by the surface $\varepsilon_s(p) = \varepsilon_F - \hbar\omega$. Under these conditions, the absorption coefficient is a linear function of frequency near the photoconductivity threshold: $\Gamma \propto (\omega - \omega_{th})$.

If the Fermi surface is a self-intersecting constant-energy surface, the internal photoelectric effect (absorption of photons) has no threshold. In this case, the absorption coefficient is proportional to the cube of the frequency ($\Gamma \propto \omega^3$; $\omega \gg \nu$), which is related directly to the existence of a conical point [111] (see §2).

In graphite, whose Fermi surface is self-intersecting, the cubic dependence is relatively quickly replaced by the quadratic dependence $\Gamma \propto \omega^2$. This is due to the fact that, in a rough approximation (we have ignored small interaction energies), the four conical points of the self-intersection merge to give self-intersection with tangency [112].

Beyond the photoelectric threshold, the frequency dependence of the absorption coefficient Γ is complex. Every metal has a different dependence with its own characteristic features which cannot be reduced to sufficiently general terms: the absorption lines representing transitions between different bands are superimposed and this gives rise to a complex dependence $\Gamma(\omega)$. Moreover, we must bear in mind that the intraband absorption coefficient has a complex frequency dependence when the frequency is increased beyond $\omega \ll \omega_0$. It is shown in [113] that, even beyond the photoelectric threshold (at high frequencies), the intraband transitions make a considerable contribution to the absorption because of the electron–electron interaction.

§48. Determination of Electron Energy Spectra of Metals

Most of the properties of metals described in the preceding sections have already been discovered. Some have been investigated in detail, whereas the study of others is still in its infancy. Investigations of various properties of metals can yield extensive information on conduction electrons, including their dispersion law, correlation function, mean free path, etc.

One of the foremost problems is the determination of the dispersion law, or, more generally, of the energy spectrum of a metal. A great variety of methods is now available for the determination of the energy spectra of condensed matter.

In the case of boson branches of the spectrum, the most promising methods (which also produce the most extensive information) are those which are based on the interaction between penetrating particles (neutrons, electrons of sufficiently high energy, and photons) with known energy spectra and elementary excitations. For example, when conditions analogous to those for the appearance of Cherenkov radiation are satisfied, a penetrating particle creates one elementary excitation (a boson of energy ε and a crystal or intrinsic momentum p) and, therefore, a study of inelastically scattered particles (such as neutrons, etc.) should make it possible to determine completely the dispersion law $\varepsilon(p)$ of elementary excitations. This method has been used successfully in the experimental determination of the phonon–roton spectrum of He II, the phonon spectra of some solids (deduced from the inelastic scattering of neutrons), the plasma oscillations of electrons in a metal (determined from the characteristic loss spectra of electrons transmitted by the thin films), etc.

Thermodynamic properties can be calculated if we know the density of levels $\nu(\varepsilon)$, which – in the case of boson branches of the spectrum – can be determined, in principle, from the temperature dependence of the specific heat and other thermodynamic quantities [115]. However, this method is not very reliable and a wealth of experimental data is required [116].

Let us now return to conduction electrons.

The high density of electrons in metals, which leads to the extreme degeneracy of the electron gas, allows us to restrict the problem to finding the correlation function $\Phi(\mathbf{p}, \mathbf{p}')$ for $\varepsilon(\mathbf{p}) = \varepsilon(\mathbf{p}') = \varepsilon_F$, and the function $\varepsilon = \varepsilon(\mathbf{p})$ near $\varepsilon(\mathbf{p}) = \varepsilon_F$. Assuming that $\mathbf{p} = \mathbf{p}_F + \delta\mathbf{p}$, where $\varepsilon(\mathbf{p}_F) = \varepsilon_F$, so that $\varepsilon(\mathbf{p}) = \varepsilon_F + \mathbf{v}\delta\mathbf{p}$, we conclude that it is sufficient to determine the shape of the Fermi surface $\varepsilon(\mathbf{p}) = \varepsilon_F$ and the absolute values of the electron velocities on this surface $|\partial\varepsilon/\partial\mathbf{p}|_{\varepsilon=\varepsilon_F}$ (the direction of the electron velocity is the normal to the constant-energy surface, and it is determined by the shape of the Fermi surface).

Little is known about the nature of the correlation function $\Phi(\mathbf{p}, \mathbf{p}')$. This is because, as shown earlier, the Fermi-liquid interaction is important only in the infrared part of the frequency spectrum, where the dominant factor is the permittivity tensor (very little can be said about the nature of the correlation function with four variables when only a few constants are known), and in cyclotron resonance at frequencies so high that they are still beyond the reach of experimenters (it would be very interesting to study such cyclotron resonance because it would give the values of the diagonal elements $\Phi(\mathbf{p}, \mathbf{p})|_{\varepsilon(\mathbf{p})=\varepsilon_F}$ of the correlation function).

Observations of spin waves in normal metals have yielded information on the exchange part of the correlation function of some metals [67c].

In contrast to the correlation function, the information on the dispersion law is very extensive.

Appendix III contains the principal data obtained in determinations of the dispersion law of conduction electrons, and includes an extensive bibliography.

We must bear in mind that the Fermi surfaces of the majority of metals are very complex and that they consist (even when the periodicity is ignored) of several sheets, or they may have fantastic shapes extending across the whole reciprocal lattice. Even such a basic property as the central symmetry is not satisfied by each sheet separately but by the whole multiply connected Fermi surface.

Naturally, the shapes of such complex surfaces cannot be determined without the aid of models. The most widely used model is that of almost-free electrons, according to which the principal features of the Fermi surface topology can be found simply from the spatial symmetry of the lattice. Complex Fermi surfaces can be obtained using this model by suitable "cutting" of the Fermi sphere of the free electron gas along the lines of degeneracy. A more refined theory can be used to correct the dispersion law near the points of degeneracy and, in most cases, this approach gives satisfactory agreement with the experimental data.

The various phenomena which we have described in the preceding sections are sensitive to different properties of the dispersion law. Thus, galvanomagnetic effects in strong magnetic fields can be used to determine the directions of the open sections in the case of a Fermi surface which stretches across the whole of the reciprocal lattice.

Quantum oscillations, particularly the thoroughly investigated de Haas−van Alphen effect, provide very accurate values of the extremal areas of the Fermi surface sections. A knowledge of these areas is sufficient to determine completely the topology of a convex centrally symmetric Fermi surface. In the case of surfaces not singly connected or not convex, the periods of the quantum oscillations are governed by all the extremal and singular sections. In this case, harmonic analysis, which would be necessary for the identification of individual periods, is fairly difficult. However, this analysis can be carried out "physically" using a plate in which sections of different diameters can be "cut off" in succession.

The de Haas−van Alphen effect can be used also to determine the effective mass of conduction electrons but the value obtained is very inaccurate because it is found from the oscillation amplitude, which is very sensitive to departures of the crystal lattice from periodicity and such departures are difficult to control.

The extremal effective mass can be found more conveniently and much more accurately from cyclotron resonance because this mass is deduced from the resonance values of the magnetic field. In the determination of the shape of the Fermi surface, it is sufficient to know the effective mass for all the directions if we wish to find the velocities on the Fermi surface under the conditions which correspond to the resonance values of the magnetic field. However, one must bear in mind that the extremal effective mass and the extremal cross-sectional area may correspond to different sections.

At elliptical limiting points the effective mass (also determined in cyclotron resonance) governs the value of $v\sqrt{\varkappa}$, where v is the velocity and $\varkappa$ is the Gaussian curvature at the point in question.

The average value of the Gaussian curvature can be found from the anomalous skin effect in the absence of a static magnetic field. The same curvature at the elliptical points determines the shape of the Fermi surface. The velocity v can be found from the Doppler splitting of the cyclotron resonance frequency in a static magnetic field applied at a small angle to the surface of a metal.

Ultrasonic measurements and the "cutoff" of resonance and nonresonance orbits in plates enable us to find directly the extremal diameters of the Fermi surface. In the case of a central section, such a diameter gives the shape of the surface without additional calculations.

Experimental investigations of the quantum variant of cyclotron resonance should make it possible to determine the areas of all (and not only the extremal) sections of the Fermi surface and all the diameters of the surface (this can be done by means of plates in which orbits are "cut off").

The use of different experimental methods and of theoretical models which represent correctly the symmetry of the problem should make it possible to determine accurately the shape of the Fermi surface.

The limited precision of the experimental data and the theoretical models poses the question whether, in principle, it is sufficient to use the experimental data which can be obtained by the known methods to determine reliably the Fermi surface and the velocities on the surface. The importance of this question can be illustrated by the following very rough approximation.

In two dimensions, a surface corresponds to a curve and an extremal area to an extremal diameter. Let us assume that all the extremal diameters are equal. Then, in the case of a convex centrally symmetric curve, we obtain a definite circumference and, in the case of a curve without central symmetry (even a convex one), we obtain some constant-width figure (for example, that shown in Fig. 97). If the curve in question is centrally symmetrical but not convex, the constancy of the extremal diameters does not give sufficient information to determine the shape

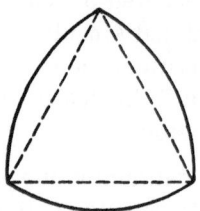

Fig. 97. Constant-width figure.

of the curve. This example shows that reliable results can be obtained only if we combine data obtained in different experiments.

Determinations of some of the Fermi surfaces of metals have demonstrated that the electron energy spectra can be very complex: in addition to the principal groups of electrons (with about one electron per atom), most of the metals have small groups ($\sim 10^{-3}$-10^{-5} electrons per atom) and the energy barriers which separate the various parts of the surface can be small. In view of this, every method for investigation of the electron energy spectrum should be used bearing in mind that an external agency can alter the structure of the spectrum quite easily. A good example of this is the phenomenon known as magnetic breakthrough (breakdown) which may induce open directions in a closed Fermi surface and closed trajectories in an open surface.

The role of small electron groups is not yet fully understood. It is quite likely that these groups determine the properties of metals which are quire different from those considered in the present monograph: for example, they may determine the plasticity and the elasticity, the changes in the properties caused by doping, etc. This is because electrons belonging to small groups have anomalously small masses. Moreover, their de Broglie wavelengths $\hbar/p_F$ are anomalously long and they are affected more than the other types of electron by external agencies.

Determination of the energy spectrum represents only the first step in the study of a metal. We would still need to determine the correlation function of a given metal, study the processes of the interaction of electrons with other quasi particles, and, particularly, investigate all the new properties of metals which can be predicted and calculated only if the electron energy spectrum is known.

QUANTUM OSCILLATIONS OF THE RESISTANCE OF METALS AT LOW FREQUENCIES

In the static case, when the electrostatics and the magnetostatics can be considered separately, we can deal independently with the quantum oscillations of the magnetic moment and those of the electrical resistance. In an alternating electromagnetic field, we determine only one quantity, which is the surface impedance tensor relating the field intensity on the surface of a conductor to the total current flowing through it. In the case of good conductors with a high charge density, it is usually permissible to ignore the displacement current in Maxwell's equations. Therefore, in such cases, the total current is determined by the intensity of the alternating magnetic field on the surface. In the one-dimensional case of a semi-infinite conductor,[1] the impedance $\hat{Z}$ is defined by the following expression (t denotes the tangential component, s the surface, and n the normal to the surface; $\mathbf{E}$ and $\mathbf{H}$ are the intensities of the alternating electric and magnetic fields):

$$E_t^{(s)} = \frac{c}{4\pi}\,\hat{Z}\,[\mathbf{H}\mathbf{n}]^{(s)}. \tag{I.1}$$

Knowledge of the tensor $\hat{Z}$ allows us to reduce the solution to a purely external problem of the field in vacuum with the boundary conditions set by Eq. (I.1) and to express the complex reflection coefficient of an electromagnetic wave in terms of $\hat{Z}$.

The quantum oscillations of $\hat{Z}$ are determined by the quantum oscillations of the magnetic moment $\mathbf{M}$ and of the total conduction current $\mathbf{j}$. Strictly speaking, these two quantities are inseparable under nonstationary conditions. We are dealing here with a typical problem in field theory, which is the behavior of free charges in the fields $\mathbf{E}$ and $\mathbf{B}_0 + \mathbf{B}$ (§19); $\mathbf{B}_0$ is determined by the static magnetic field and $\mathbf{B}$ by the alternating field. In this case, Maxwell's equations can be written in the same form as for a vacuum:

$$\operatorname{rot}\mathbf{E} = -\frac{i\omega}{c}\,\mathbf{B}, \qquad \operatorname{rot}\mathbf{B} = \frac{4\pi}{c}\,\mathbf{j}_{\text{tot}}\,, \tag{I.2}$$

where $\mathbf{j}_{\text{tot}}$ is related to the density matrix $\hat{n}$ in the external fields $\mathbf{E}$ and $\mathbf{B}$ by the expression:

$$j(\mathbf{r}) = \frac{1}{2}\,e\,\operatorname{Sp}\{\hat{v}\delta(\mathbf{r} - \hat{\mathbf{r}}) + \delta(\mathbf{r} - \hat{\mathbf{r}})\,\hat{v}\}\,\hat{n}. \tag{I.3}$$

The physical meaning of this formula is self-evident: in the classical case, $ev\delta[\mathbf{r} - \mathbf{r}(t)]$ is the current density established at a point $\mathbf{r}$ by a charge located at a point $\mathbf{r}(t)$. In the quasiclassical

---

[1] This case is the most important in actual applications. A high electrical conductivity implies a short wavelength in a conductor. Therefore, all the characteristic dimensions can be regarded as infinite compared with the wavelength.

approximation, the formula can be deduced directly from the definition of j: $j = -c\delta\bar{\bar{\mathcal{H}}}/\delta A$, where $\bar{\bar{\mathcal{H}}} = \mathrm{Sp}\,(\mathcal{H}\hat{n})$ is the average value of the Hamiltonian operator.

At low frequencies ($\omega \ll \nu$, where ν is the characteristic collision frequency) in the lowest approximation, we may assume that $\hat{n} = \hat{n}_0 + \hat{n}_1$, where $\hat{n}_0$ is an equilibrium function corresponding to the "instantaneous" values of the fields. In this case, the conduction current density j is determined by $\hat{n}_1$ and $c\,\mathrm{rot}\,M = j_0$, $\delta\bar{\bar{\mathcal{H}}} = \int M\delta B\,dV$; the system is almost capable of following the instantaneous values of the alternating fields and the quantities M, j can be represented, in the lowest approximation with respect to ω/ν, by the formulas applicable to stationary inhomogeneous fields when the values of the alternating fields at a given moment are substituted into these formulas.

In the case of frequencies sufficiently low for the normal skin effect to occur (§§32, 33), the skin depth δ in an isotropic conductor is of the form

$$\sqrt{\frac{c^2}{2\pi\omega\sigma\,(1 + 4\pi\varkappa)}}\;. \tag{I.4}$$

The order of magnitude of the magnetic susceptibility $\varkappa$ has been estimated earlier (§§15, 20). Similar considerations can be used to obtain the relative amplitude of the quantum oscillations of the conductivity $\Delta\sigma/\sigma$ (§31). The dependence of this amplitude on T and τ is the same as for $\varkappa$; the principal contribution to $\Delta\sigma$, as in the case of $\varkappa$, is made by the fraction $(\hbar\Omega/\varepsilon_0)^{1/2}$ of electrons near the extremal sections but the factor v/c^2 is missing because $\Delta\sigma$ (in contrast to $\varkappa$) is not of relativistic origin. Consequently, we obtain

$$\frac{\Delta\sigma}{\sigma} \sim \left(\frac{\hbar\Omega}{\varepsilon_0}\right)^{1/2} \exp\left(-\frac{2\pi^2 T}{\hbar\Omega} - \frac{2\pi^2}{\Omega\tau}\right) \ll 1. \tag{I.5}$$

In fact, the value of $\Delta\sigma/\sigma$ may be even smaller because σ is determined by all the electron groups and the oscillations $\Delta\sigma$ are usually associated with the anomalously weakly populated bands.

Comparing $\Delta\sigma/\sigma$ and $\varkappa$, we can show that

$$\frac{4\pi\varkappa}{\Delta\sigma/\sigma} \sim \left(\frac{v}{c}\frac{\varepsilon_0}{\hbar\Omega}\right)^2 = S. \tag{I.6}$$

This estimate of $\varkappa$ is invalid if $4\pi\varkappa \lesssim 1$ but, in this case, the contribution $\Delta\sigma/\sigma \ll 1$ is known to be small compared with the contribution of $4\pi\varkappa$.

In the usually employed magnetic fields $\hbar\Omega \ll \varepsilon_0 v/c$ and oscillations of the surface impedance are determined by oscillations of the magnetic moment.

The situation changes only at sufficiently high frequencies and at extremely low frequencies. At low frequencies ω for which $\delta \gg Sd \gg d$ (d is the thickness of the conductor in question), Eq. (I.1) includes the difference between the values of H_t at the surfaces of a sample (this difference determines the total current). Estimates show that oscillations of σ are the dominant effect (the Shubnikov–de Haas effect).

At sufficiently high frequencies, corresponding to the anomalous skin effect and $\delta < r$, the magnetic moment is determined by the alternating field only over a small part of the orbit and $\varkappa$ decreases by a factor of δ_{eff}/r (δ_{eff} is the effective depth at which B vanishes). If $\omega\tau \gg 1$, the value of $\varkappa$ decreases additionally by a factor $\omega\tau$ because the magnetic moment cannot follow the changes in **B**.[2] Consequently, at sufficiently high frequencies, the oscillations of $\hat{Z}$ are determined by the oscillations of j.

A characteristic feature of the quantum oscillations in alternating fields is the existence of nonlinear effects even if the incident wave is relatively weak. If $\omega\tau \ll 1$, the expression for $\varkappa$ includes the "total" magnetic induction $\mathbf{B} = \mathbf{B}_0 + \mathbf{B}$. Therefore, the magnetic induction in the incident wave need be only of the order of the period of the quantum oscillations, i.e.,

$$B \sim B_0 \frac{\hbar\Omega}{\varepsilon_0} \sim \Delta B, \tag{I.7}$$

for the nonlinear effects to become significant. The case $B \gg \Delta B$ is of particular interest because of a characteristic "pseudoresonance": a sharp maximum of the reflection coefficient of the n-th harmonic ($n \gg 1$) is observed when $B/\Delta B \sim n$. As $n \to \infty$, the width of this maximum tends to zero and its relative amplitude to infinity.

---

[2] These two factors reduce also the quantum oscillations of Δj but the reduction is the same as in the classical value of j and, therefore, the relative quantity $\Delta j/j$ remains unchanged.

ROLE OF ELECTRONS IN THE PROPAGATION AND THE ABSORPTION OF SOUND IN METALS

At first sight, the propagation of sound in metals, as in other solids, would seem to be entirely a lattice property. Any role played by the conduction electrons, which are in a sense "free" from interaction with ions in an ideal crystal lattice, would seem unimportant. However, direct experiments carried out at low temperatures ($T \ll \Theta$) demonstrate that this is not true. The most striking observation is the large decrease in the absorption coefficient of sound when a metal becomes superconducting. It is known that only the state of the conduction electrons changes at the superconducting transition temperature $T = T_c$ and that no changes occur in the crystal lattice. The influence of the conduction electrons on the propagation of sound is due to the role played by these electrons in the forces acting between ions in a crystal lattice. The elastic properties of a metal are determined, to a considerable degree, by electrons. Without electrons a stable system of identically charged ions could not exist at all.

One of the possible ways of approaching studies of the dynamic properties of the crystal lattice — stimulated recently by the use of computers — is to calculate the phonon spectra, making direct allowance for the interaction of ions with one another, of electrons with ions, and of electrons with one another [1]. Naturally, one has to use a reasonably realistic model such as one of the modifications of the almost-free electron approximation.

A different approach is basically semiphenomenological. The motion in a metal is described by the introduction of quasi particles which obey the Bose–Einstein statistics (phonons), and of quasi particles which obey the Fermi–Dirac statistics (conduction electrons). The dispersion laws of the phonons and the conduction electrons are assumed to be given. Therefore, it is natural to assume that the velocity of sound is given because a macroscopic acoustic wave represents a coherently excited flux of long-wavelength phonons. In this semiphenomenological approach, which we shall employ, the problem is not to calculate the velocity of sound s but the absorption coefficient and the dispersion of sound resulting from the interaction of an acoustic wave with electrons. In good metals (in which the number of electrons is approximately one per atom), the electronic mechanism of the absorption and the dispersion of sound is dominant. The interaction of an acoustic wave with thermal phonons can be ignored.

We have just said that a macroscopic acoustic wave represents a coherently excited flux of long-wavelength phonons. However, under normal experimental conditions, we excite so many phonons that the classical approach is quite justified. Therefore, an acoustic wave can be represented by a classical field in an analysis of the interaction between this wave and the conduction electrons. We must bear in mind that the velocity of an acoustic wave is considerably lower than the velocity of the Fermi electrons ($s \ll v_F$, where $s \approx 10^5$ cm/sec and $v_F \approx 10^8$ cm/sec). Therefore, "from the point of view of electrons," an acoustic wave of frequency ω in a metal generates an alternating and inhomogeneous (but practically "immobile") field of forces with a wavelength $\lambda_{ac} = 2\pi s/\omega$.

A consistent determination of the role of electrons in the propagation of sound is closely related to the derivation of the equations of elasticity for metals or, which is equivalent, to the determination of the temporal and the spatial dispersions of the elastic moduli governed by the dynamic properties of electrons. Since the dynamics of the conduction electrons depends strongly on the applied magnetic field, the dynamic elastic moduli should also depend on the magnetic field. We shall not give the derivation of the equations of elasticity of a metal (the reader is referred to, for example, [2]) and the consequences that these equations have on different relationships between physical parameters. This would require as much space as our analysis of the high-frequency properties of metals (Chap. IV). We shall give only the principal results applicable to the propagation of sound in metals, and we shall explain their physical meaning.

We shall begin by considering the absorption of low-frequency sound. In dividing the acoustic vibrations into low- and high-frequency branches, we must bear in mind the following: the natural parameters of the electron gas, with which we must compare the frequency ω and the acoustic wavelength λ_{ac}, are the relaxation frequency $1/\tau$ and the mean free path l. Since $v_F \gg s$, it follows that

$$kl = v_F \omega \tau / s \gg \omega \tau \qquad (k = 2\pi/\lambda_{ac} = \omega/s). \qquad (II.1)$$

Therefore, the spatial dispersion appears only at very low frequencies. The relationship between the mean free path and the acoustic wavelength is important at lower frequencies than those affected by the relationship between the frequency of sound and the quantity $1/\tau$. Moreover, it is known [3] that when $kl \gg 1$ the absorption coefficient of sound depends weakly on the product $\omega\tau$ and this allows us to extend classical calculations right up to frequencies which are high compared with the relaxation frequency (we shall return to this point later).

Thus, we shall regard an acoustic wave as having a low frequency if it obeys the inequality $kl \ll 1$. This does not mean that the dispersion of the absorption coefficient Γ is absent at all frequencies satisfying $kl \ll 1$ and that this coefficient (as indicated by macroscopic considerations) is proportional to the square of the frequency [4]. The macroscopic approach to the calculation of the absorption coefficient Γ is permissible at low frequencies defined as above. This means that Γ can be expressed in terms of the quasistatic ("hydrodynamic") characteristics of a metal such as its viscosity, thermal conductivity, the coefficient representing heat transfer from the electrons to the lattice, etc.[1] However, the large number of relaxation mechanisms, each proceeding at a different rate, gives rise to a complex frequency dependence of the absorption coefficient at low frequencies [5]. In some limiting cases, when the dissipation mechanisms differ considerably in their relaxation time, the absorption coefficient $\Gamma(\omega)$ can be written as follows:

$$\Gamma(\omega) = \sum_j A_j \frac{\omega^2 \tau_j}{1 + \omega^2 \tau_j}, \qquad (II.2)$$

where A_j is a coefficient representing a change in the j-th macroscopic parameter under the action of an acoustic wave; τ_j is the relaxation time of this parameter. For example, if the absorption of sound is governed by heat conduction, the relaxation time is $\varkappa/C_V s^2$, where $\varkappa$ is the thermal conductivity and C_V is the specific heat per unit volume. An investigation of the frequency dependence of the absorption coefficient and, in particular, the determination of the

---

[1] The temperature and the magnetic-field dependences of these macroscopic parameters can be established by employing a kinetic approach.

relaxation resonance maxima in the dependence $\Gamma = \Gamma(\omega)$ can be used to find information on a great variety of interactions (including weak ones) which dissipate acoustic energy.

The absorption coefficient of high-frequency sound ($kl \gtrsim 1$) can be found only by the kinetic approach. The passage of an acoustic wave through a metal increases the energy of an electron ε by an amount $\delta\varepsilon$ which is proportional to the space and time derivatives of the displacement vector [2, 3, 6]:

$$\delta\varepsilon = \lambda_{ik}(p) \frac{\partial u_i}{\partial x_k} + (p - mv)\, \boldsymbol{u}. \tag{II.3}$$

Here, $\lambda_{ik}(p) = \left[\dfrac{\partial \varepsilon(p)}{\partial u_{ik}} \right]_{u_{ik}=0} + p_i v_k$ is a second-rank tensor representing the change in the dispersion law of a conduction electron in the case of an inhomogeneous deformation of the crystal lattice and the transfer of momentum from one point in space to another (it is assumed that the local periodicity of the lattice is preserved over distances of the order of the de Broglie wavelength). The tensor $\lambda_{ik}(\boldsymbol{p})$ acts effectively as the "phonon charge": it describes the interaction of electrons with phonons. The antisymmetric part of the tensor $\partial u_i / \partial x_k$ is present in the above equation because, during the passage of an acoustic wave characterized by an arbitrary wave vector $\boldsymbol{k}$, an element of volume of the crystal lattice undergoes not only translational motion but also rotates at an angular velocity $\frac{1}{2}$ rot $\dot{\boldsymbol{u}}$. The antisymmetric part of the tensor vanishes in the case of an isotropic dispersion law. A comparison of the terms containing the time derivative with the terms containing the space derivatives shows that the latter are v_F/s times larger. Therefore, we can simply omit the term $(\boldsymbol{p} - m\boldsymbol{v})\dot{\boldsymbol{u}}$ which describes the inertial effects.

The time dependence of the electron energy (resulting from the time dependence contained in $\boldsymbol{u}(\boldsymbol{r}, t) = \boldsymbol{u}_0 e^{-i(\omega t - \boldsymbol{kr})}$) disturbs the equilibrium of the electron system. Various dissipative processes (collisions of electrons with impurities, with thermal phonons, and with one another, discussed in §24) are the cause of the attenuation of sound and the absorption coefficient Γ is determined by the dissipative function ($T\dot{S}$) calculated per unit volume:

$$\Gamma = T\dot{S}/2E_{\mathrm{ac}}, \tag{II.4}$$

where E_{ac} is the energy of an acoustic wave per unit volume and $\dot{S}$ is the rate of change of entropy per unit volume. The rate of change of entropy can be expressed, by means of statistical mechanics formulas [7], in terms of the time derivative of the distribution function. This derivative can be found from the transport equation. The external "force" is represented by the derivatives of $\delta\varepsilon$ [see Eq. (II.3)].

In writing the transport equation, we must bear in mind that oscillations of the electron gas give rise to an electromagnetic field which can be found using Maxwell's equations containing the current density and the charge density, and derived by suitable averaging of the electron distribution function.[2]

The dissipative function is the sum of two terms. The first term represents the "viscous" losses and is directly due to the time dependence of the strain tensor. This term has the structure $\frac{1}{2}\overline{\Lambda_{iklm}\dot{u}_{ik}^*\dot{u}_{lm}}$ and can be regarded as the definition of the dynamic viscosity tensor of the metal $\Lambda_{iklm}(\omega, \boldsymbol{k}/k)$, where the bar denotes averaging over the volume and period $2\pi/\omega$. The second term describes the Joule losses, i.e., it depends on the strain tensor only through the

---

[2] It must be mentioned that Maxwell's equations should be used to find the transverse components of the fields. The longitudinal electric field is found, as usual (§24), from the condition $\rho' = 0$.

electric field (a suitable substitution makes it possible to transform this term so that it acquires the structure of the viscous term and to "renormalize" the viscosity tensor). The relationship between the viscous and the Joule losses depends strongly on the relationship between the acoustic wavelength $\lambda_{ac} = 2\pi s/\omega$ and the electromagnetic wavelength in the metal $\lambda_{em} = c/\sqrt{2\pi\sigma(k)\,\omega}$, which should be calculated taking into account the spatial dispersion of the electrical conductivity. If $kl \gg 1$, it follows from §33 that $\sigma(k) \approx \sigma_0/kl$, and $\lambda_{em} \approx c/\omega_0 \cdot (v_F/s)^{1/2}$, where $\omega_0 = 4\pi ne^2/m$ is the plasma frequency of the electron gas (in rough estimates, we can use the quadratic and isotropic dispersion law). If $\lambda_{ac} \ll \lambda_{em}$, the viscous losses predominate ($\Gamma \approx \Gamma_{visc}$; in the opposite limiting case, the Joule losses are comparable with the viscous term ($\Gamma_J \lesssim \Gamma_{visc}$) [2].

If $kl \gg 1$, an acoustic wave does not interact with all the electrons whose energy is of the order of the Fermi value but only with those which move together with the acoustic wave, i.e., the electrons which satisfy $v_n = s$, where v_n is the projection of the electron velocity along the wave vector $\mathbf{k}$ of the acoustic wave. Since $v_F \gg s$, we may assume that only those electrons which satisfy $v_n = 0$ interact actively with the acoustic wave. A natural consequence of this conclusion is the expression for the absorption coefficient $\Gamma(\omega)$ in the form of an integral over a strip $\mathbf{v}_F\mathbf{k} = 0$ on the Fermi surface [3, 8]:

$$\Gamma(\omega) \approx \frac{\pi\omega}{(2\pi\hbar)^3\,\rho s} \int\limits_0^{2\pi} \frac{|\Lambda|^2\,d\varphi}{v^2\varkappa(\varphi)}. \qquad (II.5)$$

Here, ρ is the density of the metal, $\varkappa(\varphi)$ is the Gaussian curvature of the Fermi surface over the strip $\mathbf{v}_F\mathbf{k} = 0$, and $|\Lambda|^2$ represents, symbolically, a basically positive quantity depending quadratically on the components of the tensor λ_{ik}. The special role of the electrons moving at right-angles to the wave vector of the acoustic wave resembles the situation in the anomalous skin effect (§33), in which only those electrons are effective which move parallel to the surface of the metal. As in the skin effect, at high frequencies the absorption increases more slowly, because of the acoustic field inhomogeneity, than that at low frequencies. The correct order of magnitude of $\Gamma(\omega)$ in the case $kl \gg 1$ can be obtained if the expression for $\Gamma(\omega)$, obtained in the kinetic theory for the case $kl \ll 1$ [6],

$$\Gamma(\omega) \approx \frac{N\varepsilon_F\omega^2\tau}{\rho s^2}, \qquad (II.6)$$

is divided by kl ($1/kl$ is the fraction of electrons participating in the dissipation of the acoustic energy in the case $kl \gg 1$).

Let us again consider the Joule losses. Since the viscous losses in the case $kl \gg 1$ are given by Eq. (II.5), which is independent of the relationship between λ_{ac} and λ_{em}, and since $\Gamma_J(\omega) \ll \Gamma_{visc}$ if $\lambda_{ac} \ll \lambda_{em}$, where $\Gamma_J(\omega) \sim \Gamma_{visc}$ if $\lambda_{ac} \gg \lambda_{em}$, the total absorption coefficient is an approximately linear function of the frequency in both limiting cases. The factor in front of the frequency changes when $\lambda_{ac} \approx \lambda_{em}$, i.e., when $\omega \approx \omega_0(s^3/c^2 v_F)^{1/2}$. The transition in the dependence $\Gamma(\omega)$ from a given slope to a less steep one should be accompanied by an increase in the dispersion of the velocity of sound. It is indeed found that when $\omega \approx \omega_0(s^3/v_F c^2)^{1/2}$ the change in the velocity of sound $\Delta s/s$ reaches a value $\Gamma(\omega)/\omega$ [2].

At very high frequencies ($\omega\tau \gg 1$), the absorption coefficient can be calculated using a perturbation theory and considering $\delta\varepsilon$ of Eq. (II.3) as the cause of transitions in a system of equilibrium electrons. In this way, we can easily obtain [3] the following expression[3]:

---

[3] We shall omit the factors in which frequency does not occur. The frequency in front of the integral sign appears because, in the approach adopted, the absorption coefficient is proportional to the energy absorbed in transitions and $\varepsilon' - \varepsilon = \hbar\omega$.

$$\Gamma(\omega) \sim \omega \int |\Lambda|^2 (n_F - n_F') \delta(\varepsilon + \hbar\omega - \varepsilon') \, d\tau_p, \tag{II.7}$$

where $\varepsilon' \equiv \varepsilon(\mathbf{p} + \hbar\mathbf{k})$ and $n_F^! = n_F(\varepsilon')$. Since, in accordance with the law of conservation of energy, $\varepsilon' = \varepsilon + \hbar\omega$ and $\hbar\omega$ is known to be considerably smaller than ε_F (even in the case of Debye phonons because $\Theta \ll \varepsilon_F$), it follows that

$$\Gamma(\omega) \sim -\omega^2 \int |\Lambda|^2 \frac{\partial n_F}{\partial \varepsilon} \delta(\varepsilon + \hbar\omega - \varepsilon') \, d\tau_p. \tag{II.8}$$

The argument of the δ function in the case $\hbar k \ll p_F$ can be written in the form $\hbar k v - \omega = \hbar k v (\cos\theta - s/v)$. Integration, in which account is taken of two δ functions $[-\partial n/\partial\varepsilon = \delta(\varepsilon - \varepsilon_F)]$, yields the formula (II.5), which is therefore valid in a very wide range of frequencies [3].

The high-frequency limit of our analysis is associated with the assumed form of interaction between electrons and the lattice. It is evident from the formula itself that the expression (II.3) should be valid as long as $ka \ll 1$ (a is the lattice constant). Thus, Eq. (II.5), which is supported by the experimental results (the linear frequency dependence of the absorption coefficient has been observed for various metals at low temperatures[4]), makes it possible to estimate the lifetime of the Debye phonons ($\hbar\omega \sim T \ll \Theta$) and the experimental investigations of the absorption of high-frequency sound ($kl \gg 1$) can be used to determine directly the constant of the interaction between electrons and long-wavelength phonons.

If the condition $ka \ll 1$ is not satisfied (this condition is known to be obeyed in the case of the external excitation of sound), the structure of the absorption coefficient of phonons becomes very similar to Eq. (II.7):

$$\Gamma(k) = \int |M|^2 (n_F - n_F') \delta(\varepsilon + \hbar\omega - \varepsilon') \, d\tau_p. \tag{II.9}$$

All the factors (such as $2\pi/\hbar$, etc.) are included in the square of the matrix element $|M|^2$ whose dependence on the momenta of the electrons and the phonons cannot be obtained in the general form. In the case of small phonon momenta, we can use the macroscopic approach and, in accordance with Eq. (II.3), $|M|^2 \sim k^2/\omega(\mathbf{k})$.

Equation (II.9) can be used to establish the presence of singularities of $\Gamma(\mathbf{k})$ at the values of $\mathbf{k}$ equal to the diameter of the Fermi surface [9].

Since $\hbar\omega(\mathbf{k}) \ll \varepsilon_F$ for all the values of $\mathbf{k}$, it follows from Eq. (II.9) that

$$\Gamma(k) \approx \hbar\omega(k) \int |M|^2 \delta(\varepsilon_p - \varepsilon_F) \delta(\varepsilon_{p+\hbar k} - \varepsilon_F) \, d\tau_p \tag{II.10}$$

and, consequently, the absorption coefficient does not vanish only if the Fermi surface $\varepsilon_p = \varepsilon_F$ intersects its analog shifted by $-\mathbf{k}$. In general, the geometrical locus of the singularities of $\Gamma(\mathbf{k})$ is the set of those values of the vector $\mathbf{k}$ for which the surfaces $\varepsilon_p = \varepsilon_F$ and $\varepsilon_{p+\hbar k} = \varepsilon_F$ are in contact [10]. The nature of the contact determines the type of singularity [11, 12]. In the case of ordinary point contact, the absorption coefficient suddenly vanishes [9]. In the case of a "closer" contact (along a line or along a flattened area), the absorption coefficient $\Gamma(\mathbf{k})$ becomes infinite. A more rigorous analysis which does not involve expansion of $n_F = n_F(\varepsilon + \hbar\omega)$ in powers of $\hbar\omega$ smooths out the singularities somewhat and, in particular, $\Gamma(\mathbf{k})$ does not become infinite but has a sharp maximum [12]. The singularities of the absorption coefficient of phonons

---

[4] Low temperatures are necessary to satisfy the condition $kl \gg 1$.

appear because of an inhomogeneous distribution of electrons in the **p** space. The temperature and the various electron scattering mechanisms smooth out these singularities. In the range $T \gg \hbar/\tau$, the dominant factor is the temperature, whereas in the opposite limiting case the scattering is more important ($\hbar/\tau \approx 1°K$ for $l \approx 10^{-3}$ cm).

An inhomogeneous distribution of momenta in the **p** space gives rise to singularities not only in the absorption coefficient but also in the phonon dispersion law $\omega = \omega(\mathbf{k})$. This can be deduced directly [9, 11] or we can use relationships between the phonon energy $\hbar\omega(\mathbf{k})$ and the absorption coefficient $\Gamma(\mathbf{k})$ (these relationships are similar to the well-known Kramers–Kronig relations, as shown, for example, in [13]). The singularities of $\omega(\mathbf{k})$ are known as the Migdal–Kohn singularities and are located at the same values of the phonon momentum $\hbar\omega$ as the singularities of the absorption coefficient $\Gamma(\mathbf{k})$ so that they can be classified in accordance with the nature of the contact between the surfaces $\varepsilon_p = \varepsilon_F$ and $\varepsilon_{p+\hbar k} = \varepsilon_F$ [12]. In the case of ordinary contact, the group velocity of the phonons $\partial\omega(\mathbf{k})/\partial\mathbf{k}$ has a logarithmic singularity [9]. In the case of closer contact, the singularity becomes more pronounced.

It follows from our analysis that the geometrical locus of the singularities $\omega(\mathbf{k})$ and $\Gamma(\mathbf{k})$ is a form of image of the Fermi surface of a metal but we must bear in mind that the singularities should be observed at all values of the phonon momentum equal to the values of the inner and the outer limiting diameters of the Fermi surface. In the case of a periodic Fermi surface, this circumstance greatly complicates the situation. So far, only the cases of closer contact (along a line or over a part of a plane) have been observed experimentally.

We have not yet considered the influence of a static magnetic field on the propagation and the absorption of sound in metals. Although the most interesting results (such as the applications of electroacoustic effects in the interpretation of the electron energy spectra of metals) are obtained in strong magnetic fields, we shall consider this problem only very briefly because a detailed discussion would be outside the scope of this appendix.

First, the application of a magnetic field requires the re-examination of the "boundary" separating the low-frequency from the high-frequency sound because – in addition to the relaxation time τ and the mean free path l – we now have new characteristic quantities ω_H and r_H. However, we are interested primarily in the strong magnetic fields defined by

$$r_H \ll l, \quad \omega_H \gg \tau^{-1}, \tag{II.11}$$

and, therefore, the condition $kl \ll 1$ formulated earlier automatically leads to the following two inequalities:

$$kr_H \ll kl \ll 1 \text{ and } \omega \ll \omega_H \ll \tau^{-1}. \tag{II.12}$$

In the case of propagation of low-frequency sound, a magnetic field simply alters those macroscopic ("hydrodynamic") quantities which determine the absorption coefficient and the dispersion of sound (they include the thermal conductivity, the electrical conductivity, etc.). These changes need not be small. It follows from the results given in §25 that, for example, the thermal conductivity of a metal subjected to a strong magnetic field may change by a large factor $\varkappa_H \sim \varkappa_0 (r_H/l)^2$ and this results in a corresponding increase in the relaxation time of the acoustic energy.

At low energies, when the quantization of the electron energy in the magnetic field is significant ($\hbar\omega_H \lesssim T$), we observe quantum oscillations whose nature is the same as that of oscillations of the magnetic moment (§15) or the conductivity (§31): the absorption coefficient and the velocity of sound vary periodically with the reciprocal of the magnetic field.

The greatest interest lies, naturally, in the high-frequency case. All the features of the dynamics of electrons with a complex dispersion law in a magnetic field are manifested also in the interaction of electrons with an acoustic wave. Since an acoustic wave in a metal produces a quasistatic field of wavelength λ_{ac}, the situation is similar to that in the size quantization effect. We mentioned earlier that the absorption of the acoustic energy is dominated by those Fermi electrons whose velocity is perpendicular to the wave vector. However, if electrons move in a static magnetic field, the direction of their motion is always varying. Let us assume that the magnetic field is perpendicular to the acoustic wave vector. Then, the effectiveness of the interaction between the electrons and the acoustic wave depends strongly on the relationship between the extremal orbit diameter $(\Delta p_x^F)_{extr}$ and the acoustic wavelength λ_{ac} (as usual, the z axis is directed along the magnetic field and the y axis along the wave vector).

If

$$\frac{c}{eH}\left(\Delta p_{x}^{F}\right)_{extr} = \left(n + {}^{1}/_{2}\right)\lambda_{ac} \text{ for transverse sound,}$$

$$\frac{c}{eH}\left(\Delta p_{x}^{F}\right)_{extr} = n\lambda_{ac} \text{ for longitudinal sound} \qquad (II.13)$$

$$(n = 1,\ 2,\ 3,\ \ldots),$$

the interaction between the electrons and the acoustic wave becomes stronger. However, if the conditions of Eq. (II.13) are not satisfied, the interaction is not enhanced. Since the electrons interact effectively with the sound only during a small fraction of their period (when their velocity is approximately perpendicular to the wave vector), the conditions of Eq. (II.13) do not give rise to resonance peaks but to a periodic dependence of the absorption coefficient on the reciprocal of the magnetic field. The period of this dependence is [14, 15]

$$\Delta \frac{1}{H} = \frac{e\lambda_{ac}}{c\left(\Delta p^F\right)_{extr}}. \qquad (II.14)$$

The phenomenon described here is known as the geometrical or Pippard resonance, and it provides a convenient method for the determination of the extremal diameters of the Fermi surface. We must draw attention to the fact that there is a diameter which is perpendicular to the wave vector. This is a manifestation of the correspondence between the trajectories in the coordinate and the momentum spaces described in §4 (the trajectories are similar to within a rotation by $\pi/2$ and a conversion factor c/eH). This correspondence is manifested particularly clearly in the case of an open Fermi surface. Let us assume that $\mathbf{k}$ ($k_y = k$) and $\mathbf{H}$ ($H_z = H$) are still perpendicular to each other and oriented in such a way that the open direction can be made to coincide with the p_x axis. Then, the electrons in the momentum space move mainly along the p_x axis, whereas in the coordinate space they move mainly along the y axis. If the period of the open Fermi surface is Δp_x^F (this period is usually equal to the corresponding reciprocal-lattice dimension and is independent of p_z), a particularly favorable situation obtains in the case when the trajectory period in the coordinate space is equal to the acoustic wavelength or is a multiple of this wavelength [16]:

$$\frac{c\,(\Delta p_x)}{eH} = n\lambda_{ac}, \quad n = 1,\ 2,\ \ldots. \qquad (II.15)$$

This condition leads to the previous formula (II.14) for the period but the intensity of the interaction is now much stronger. In the absence of dissipative processes, the absorption coefficient would become infinite [16] but allowance for these processes makes the absorption coefficient a finite quantity ($\Gamma \sim \tau$). (This characteristic size effect may be called the "giant" geometrical resonance.)

As expected, the spatial and not the temporal periodicity of an acoustic wave is important when electrons interact with sound in a magnetic field. However, if

$$\omega - \bar{k}\bar{v} = \frac{1}{n}\,\omega_H, \qquad n = 1,\,2,\,\ldots, \tag{II.16}$$

we may observe the acoustic cyclotron resonance [17] (the term $\bar{k}\bar{v}$ describes the Doppler frequency shift). The acoustic cyclotron resonance, like other resonances, is observed only if fairly rigorous conditions are satisfied ($\omega\tau \gg 1$ etc.).

This cursory description of the interaction between electrons and sound in $H \neq 0$ is not exhaustive in the sense of elucidating all the features of the electroacoustic phenomena in a magnetic field. In particular, we have practically ignored all those changes in the absorption of high-frequency sound which result from the quantization of the electron energy in a magnetic field. In this case, sound must be described in a quantum manner. In other words, we must formulate the conditions for the absorption of a phonon in a quantized electron gas. The laws of conservation of energy and of the z-th projection of momentum require that the following equality be satisfied:

$$\varepsilon_n(p_z) + \hbar\omega = \varepsilon_m(p_z + \hbar k_z). \tag{II.17}$$

Since an allowance for the quantization is necessary only in strong fields ($\hbar\omega_H > T$), the energy of a phonon under these conditions is always considerably less than the separation between electron levels ($\hbar\omega \ll \Delta\varepsilon = \hbar\omega_H$). Therefore, the condition of Eq. (II.17) need be satisfied only for m = n. The energy of an electron before and after the absorption of a phonon [compare with Eq. (II.10)] should be equal to the Fermi energy. This, in combination with the law of conservation, gives rise to two equations for the same quantity p_z. In the case of an arbitrary value of the wave vector and of the magnetic field, these equations have no solution. However, if the wave vector is fixed, there is a discrete set of values of the magnetic field at which these equations have solutions and a phonon may be absorbed [18]. Consequently, the dependence of the absorption coefficient Γ on the magnetic field is very special: the value of Γ increases at fixed values of the magnetic field and the peaks are separated by deep troughs. The lower the temperature and the longer the relaxation time, the more pronounced are these "giant oscillations" of the absorption coefficient.[5]

It is evident from Eq. (II.17) that the giant oscillations are basically due to the existence (in a quantizing magnetic field) of discrete electron energy levels corresponding to a fixed value of p_z. The quantization of the electron energy is a consequence of the finite nature of the motion of an electron in a plane perpendicular to the magnetic field. However, if the Fermi surface is open, we find that, for certain directions of the magnetic field, the electron motion in the xy plane is infinite and this corresponds to a characteristic energy band structure (obtained for p_z = const): wide bands of allowed energies $\delta\varepsilon \sim \hbar\omega_H$ are separated by narrow forbidden gaps (§8). Under these conditions, the absorption pattern is the reverse of that obtained in the case of giant oscillations: the absorption coefficient Γ decreases strongly near certain fixed values of the magnetic field and at intermediate values it depends weakly on the field [19].

---

[5] This is the name given to the effect just described. A theory of this effect [18] is in good agreement with the experimental results.

APPENDIX III

TOPOLOGY OF FERMI SURFACES OF METALS (SUMMARY TABLE)[1]

The Fermi surfaces of most metals have been investigated reasonably thoroughly. In particular, it is known whether the Fermi surface of a given metal is closed or open. The topological types of open Fermi surface have been determined.

This appendix has been compiled to give a general idea of the topology of the Fermi surfaces of metals.

The appendix consists of three parts: a table, figures showing the principal types of open Fermi surface discovered experimentally, and a bibliography. For a given metal the table presents information on the topological type of its Fermi surface and on the relationship between the numbers of electrons n_1 and holes n_2. It also includes references to the principal experimental and theoretical investigations of the band structure of metals.

In view of the existence of the magnetic breakthrough (breakdown) effect, the concepts of the "open" and "closed" surfaces are arbitrary: in a magnetic field, a closed Fermi surface may become open and an open surface may become closed. Therefore, Table 1 lists the approximate value of the magnetic breakthrough field and mentions changes in the Fermi surface topology induced by such breakthrough.

The appendix gives no references to papers which simply describe the experimental methods or which do not give any information in addition to that obtainable from simpler methods. There are also no references to the methods which have not found wide application or which have yielded relatively little information.

The few investigations of the anomalous skin effect (abbreviated to ASE in Table 1) and of the positron annihilation (PA) are cited at the end of the bibliography and the references to them are given in Table 1 under the column headed "Remarks."

The metals are arranged alphabetically in Table 1. At the end of the table, data are presented on several relatively thoroughly investigated intermetallic compounds.

The bibliography in this appendix is not comprehensive but the intention is to refer to all the principal investigations of the Fermi surfaces of metals.

---

[1] Compiled by Yu. P. Gaidukov.

TABLE 1

Metal	Crystal lattice	Topological type and other information on Fermi surface	Method of investigating Fermi surface						Remarks
			galvanomagnetic effects	de Haas–van Alphen and Shubnikov–de Haas effects	cyclotron resonance	magnetoacoustic effect	rf size effect	band structure calculations and comparison with experiment	
Aluminum	fcc	$n_1 \neq n_2$; H < 30 kOe, closed; H > 30 kOe, magnetic breakthrough	1-4	5-8, 332	9-13	14-17, 338	18	19-21, 324	314 (ASE)
Antimony	rhombohedr.	$n_1 = n_2$; closed	254	255-260	261	262-266	367	46,267	
Arsenic	rhombohedr.	$n_1 = n_2$; closed	168,330	169-171, 336	172,354	173,174		46, 175-177	
Beryllium	hcp	$n_1 = n_2$; H < 50 kOe, closed; H > 50 kOe, magnetic breakthrough in basal plane, open; planar network of corrugated cylinders along $\langle 1\bar{2}10 \rangle$ and $\langle 1\bar{0}10 \rangle$	23,339	23,339				24-26	318,319 (PA) (Fig. 1)
Bismuth	rhombohedr.	$n_1 = n_2$, closed	30	31-36, 340	37,38	39-42, 341	43,44	45-47	315 (ASE)
Cadmium	hcp	$n_1 = n_2$; open; corrugated cylinder along [0001] axis	102,103, 346	104-106	107,108	109-112, 347	113-117	118,119	(Fig. 4)
Calcium	fcc	open (calculation), similar to Fermi surface of lead		134				135,329, 349	
Cesium	bcc	$n_1 = 1$ electron/atom, $n_2 = 0$; closed sphere		123,290 291		370		131,132, 291	
Chromium	bcc	$n_1 = n_2$; H > 60 kOe, magnetic breakthrough, open	279, 283	284-286		287,288		167,289	(Fig. 12)
Cobalt	fcc							77	
Copper	fcc	$n_1 = 1$ electron/atom, $n_2 = 0$; open; three-dimensional network of corrugated cylinders along $\langle 111 \rangle$ axes, similar to Fermi surface of gold	148-151	80,82, 152-154	155,156	85,157, 158,338 351	352	86,159 326,327, 343,353	316 (ASE) (Fig. 3)
Gadolinium	hcp							55-57	
Gallium	body-centered orthorhomb.	$n_1 = n_2$, open; corrugated cylinder along c axis	58-60	61	62,63	64	65	66,325	320 (PA)

Metal	Structure	Description							
Gold	fcc	$n_1 = 1$ electron/atom, $n_2 = 0$; open; three-dimensional network of corrugated cylinders along $\langle111\rangle$ axes (principal open direction) as well as along $\langle110\rangle$ and $\langle100\rangle$ axes (secondary open directions)	78,79	80-82	83	84,85		86,326, 327,343	(Fig. 3)
Graphite		closed self-intersecting							
Indium	tetrag.	$n_1 = n_2$; closed	4,87-89	67,68 90-92, 344	69,70 93	94,95	96	71,342 93,97	321 (PA) (Fig. 2)
Iron	bcc	$n_1 = n_2$; open; three-dimensional network of corrugated cylinders along $\langle001\rangle$ axes	72-74	74				75-77	
Lead	fcc	$n_1 = n_2$; open; three-dimensional network of corrugated cylinders along $\langle111\rangle$ axes	234,235, 332	236-238	239,240	95,241		242,243	(Fig. 10)
Lithium	bcc	$n_1 \neq n_2$; closed	120					131,132, 350	
Magnesium	hcp	$n_1 = n_2$; H > 5 kOe, magnetic breakthrough in (0001) plane, open; planar network of corrugated cylinders along $\langle1\bar{2}10\rangle$ and $\langle10\bar{1}0\rangle$; H > 70 kOe, magnetic breakthrough, open directions along [0001] axis	136,137	138-141	142	143-144		145-147	314 (ASE) (Fig. 1)
Mercury	rhombohedr.	$n_1 = n_2$; open; open directions parallel to $\langle100\rangle$ and $\langle011\rangle$ axes	226-228	229	230,361	362		231	322 (PA) (Fig. 9)
Molybdenum	bcc	$n_1 = n_2$; closed	47,48, 160	49,161	162	163,164	165	166,167	
Nickel	fcc	$n_1 = n_2$; open; open three-dimensional network of corrugated cylinders along $\langle111\rangle$ axes, similar to Fermi surface of gold	180	181-184				185,186, 355	(Fig. 3)
Niobium	bcc	$n_1 \neq n_2$; magnetic breakthrough, open; network of corrugated cylinders along $\langle001\rangle$, $\langle110\rangle$, and $\langle111\rangle$ axes	187-189	190,356				29,187, 357	(Fig. 5)
Osmium	hcp	$n_1 = n_2$; open; planar network of corrugated cylinders parallel to (0001) plane and [0001] axis, magnetic breakthrough	206	358					
Palladium	fcc	$n_1 = n_2$; open; three-dimensional network of corrugated cylinders along $\langle001\rangle$ axes	207	208-210				211,331	(Fig. 7)
Platinum	fcc	$n_1 = n_2$; open; three-dimensional network of corrugated cylinders along $\langle001\rangle$ axes; similar to Fermi surface of palladium	136,212	213-215, 359		216		217,331, 360	(Fig. 7)

TABLE 1 (Cont'd)

Metal	Crystal lattice	Topological type and other information on Fermi surface	Method of investigating Fermi surface						Remarks
			galvanomagnetic effects	de Haas–van Alphen and Shubnikov–de Haas effects	cyclotron resonance	magnetoacoustic effect	rf size effect	band structure calculations and comparison with experiment	
Potassium	bcc	$n_1 = 1$ electron/atom, $n_2 = 0$; $H < 50$ kOe, closed sphere; $H > 50$ kOe, magnetic breakthrough, open	120–122	123	124	125–128	129,130, 348	131,133, 328	
Rhenium	hcp	$n_1 = n_2$; $H < 30$ kOe, open, corrugated cylinder along [0001] axis; $H > 30$ kOe, magnetic breakthrough, additional open directions along [0001] and [10$\bar{1}$0] axes	218,219	190,220		221,222		222,223	(Fig. 8)
Rhodium	fcc	closed (calculation)						224	
Rubidium	bcc	$n_1 = 1$ electron/atom, $n_2 = 0$; closed sphere		224,225 123,232				131–133, 328	
Ruthenium	hcp							233	
Scandium	polymorphic (fcc, hcp)	$n_1 \neq n_2$; closed	251	233,363				101,252	according to calculations, Fermi surface should be open
Silver	fcc	$n_1 = 1$ electron/atom, $n_2 = 0$; open; three-dimensional network of corrugated cylinders along [111] axes, similar to Fermi surface of gold	244,245	80,82, 246	247	85,248		86,87, 249,250, 327,333, 343	(Fig. 3)
Sodium	bcc	$n_1 = 1$ electron/atom, $n_2 = 0$; closed sphere	120,178	179	124			131,132, 328	
Strontium	polymorphic	open (calculation)						253	
Tantalum	bcc	$n_1 \neq n_2$; magnetic breakthrough, open; three-dimensional network of corrugated cylinders along $\langle 001 \rangle$ axes, similar to Fermi surface of niobium	187	190,276, 277	278			29,357	(Fig. 5)
Thallium	hcp	$n_1 \neq n_2$; open: two corrugated (0001) planes joined by narrow necks along [0001] axis; $H > 30$ kOe, magnetic breakthrough	102,268, 269,270, 364	271,365, 366		272–274	368	275	(Fig. 11)

Element	Structure	Description							Notes
Thorium	polymorphic								
Tin	tetrag.	n₁ = n₂; open planar network of correlated cylinders along ⟨010⟩ and ⟨110⟩ axes; H > 50 kOe, magnetic breakthrough	191–193	280,281 194–197	198–200	201	202,203	282 196,204,205	317 (ASE) (Fig. 6) (see also figures in 196,197)
Titanium	hcp	n₁ = n₂; closed	279					101,369	according to calculations, Fermi surface should be open
Tungsten	bcc	n₁ = n₂; closed	27,48	49,50	51,52			29,54	according to 27, Fermi surface is closed
Vanadium	bcc	n₁ ≠ n₂, open (?)	27,28				53	29	
Ytterbium	fcc	closed	98	99,345				100,101 118,119, 305	according to free-electron model, Fermi surface should be open
Yttrium	hcp	open (calculation)	102,	31,					(Fig. 13)
Zinc	hcp	n₁ = n₂; H < 2.5 kOe, open; corrugated cylinder along [0001] axis; H > 2.5 kOe, magnetic breakthrough, open directions along ⟨12̄10⟩ and ⟨101̄0⟩ axes	292–294	295–297	298–301	111,112, 302–304			
Zirconium	hcp	open (calculation)		306				307	323 (PA)
AuSn	hexag.	open	308						
AuAl₂	fcc	open, similar to Fermi surface of gold	309						(Fig. 3)
AgZn, CuZn, PdIn	β-brass	open (calculation)		310				311,371	
AuGa₂, AuIn₂	fcc	open, similar to Fermi surface of gold	309,334	312				335	(Fig. 3)
MgZn₂	hexag.			313					

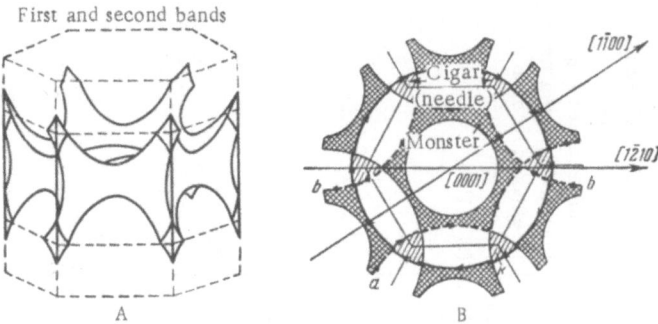

Fig. 1. A) Open Fermi surface ("monster") for beryllium and magnesium (obtained by ignoring the spin-
orbit interaction of electrons) [118]; B) formation of
the ⟨1Ī00⟩ and the ⟨1Ī10⟩ directions as a result of
magnetic breakthrough between the two parts of the
Fermi surface known as the "monster" and the "cigar"
("needle") observed in beryllium, magnesium, and
zinc. Figure 1B shows a section by the (0001) plane;
aa and bb are open trajectories along the [1Ī00] and
[1Ī10] directions.

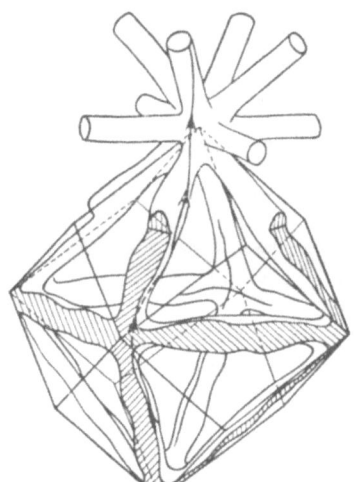

Fig. 2. One of the variants of
the open Fermi surface of iron
[75]. The arrowed curve is an
open trajectory along the [001]
axis.

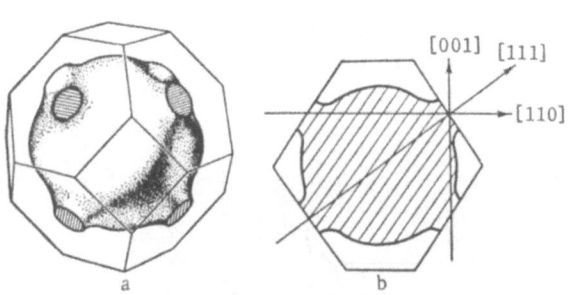

Fig. 3. a) Open Fermi surface of gold, copper, and
silver [316]; b) section by the (110) plane showing
open directions along the [111], [110], and [001] axes
[244].

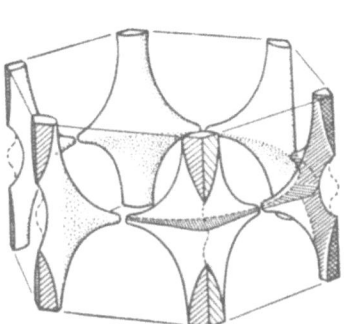

Fig. 4. Open surface of cadmium. Discontinuities in the basal plane prevent the appearance of open directions along the ⟨12̄10⟩ and ⟨11̄00⟩ axes (Fig. 1B) [107].

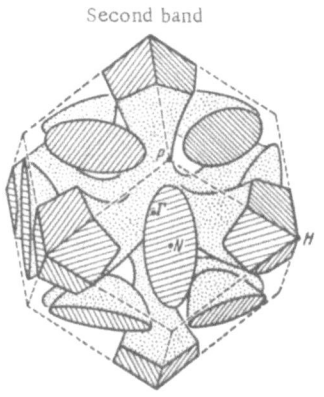

Fig. 5. Open Fermi surface of group VA metals (V, Nb, Ta) according to calculations reported in [187].

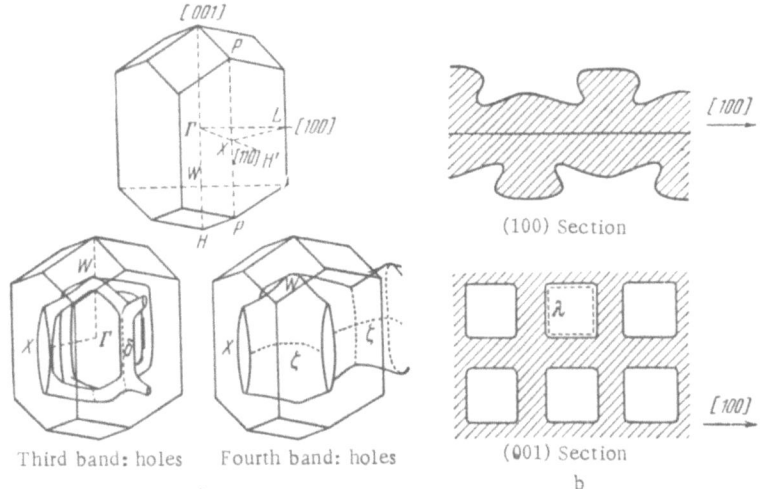

Fig. 6. Open Fermi surfaces of tin (calculations confirmed by experiments reported in [191, 194, 196, 198, 202, 203]): a) open hole surfaces in the third and the fourth bands [194] (see also figures in [196, 197]); b) sections of an open electron surface in the fifth band.

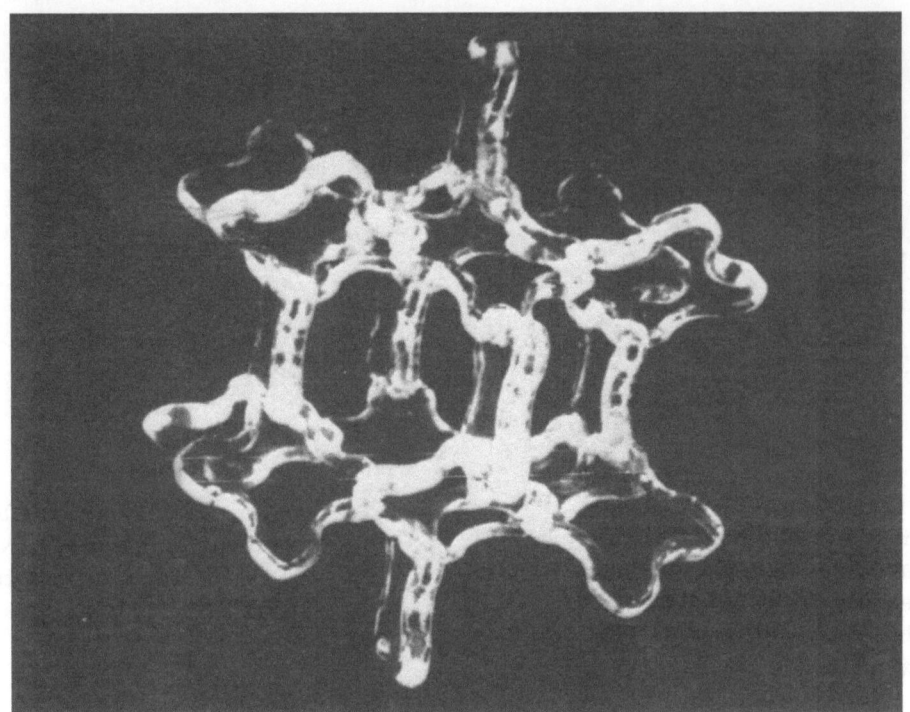

Fig. 7. Model of the open Fermi surface of palladium and platinum [212].

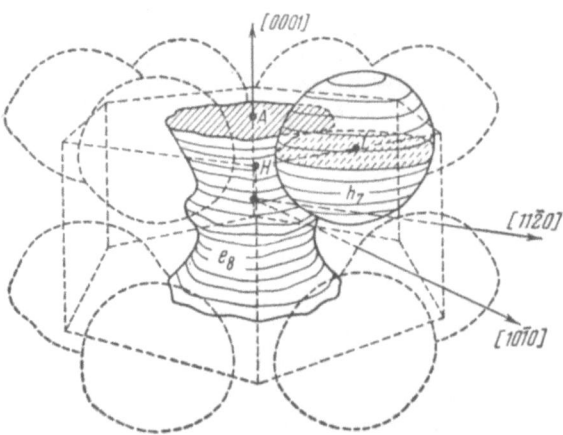

Fig. 8. Fermi surface of rhenium according to calculations reported in [222]. Here, e_8 is an electron surface in the eighth band open along the [0001] direction and h_7 is a closed hole surface in the seventh band. Magnetic breakthrough between e_8 and h_7 gives rise to open directions along the $\langle 1\bar{1}00 \rangle$ axes.

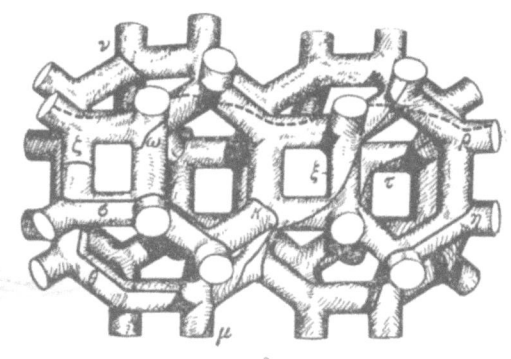

a

b

Fig. 9. Open electron Fermi surface of lead (third band); a) lines μ and ρ represent open trajectories [236, 237]; b) the same surface in one reciprocal-lattice cell constructed to scale, using all experimentally obtained dimensions (M. S. Khaikin and R. T. Mina).

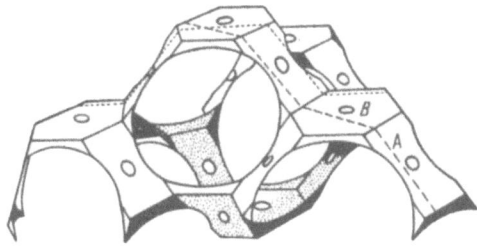

Fig. 10. Open hole Fermi surface of mercury in the first band, according to calculations reported in [231]. The dotted and the dashed lines represent open trajectories (see also figures in [230]).

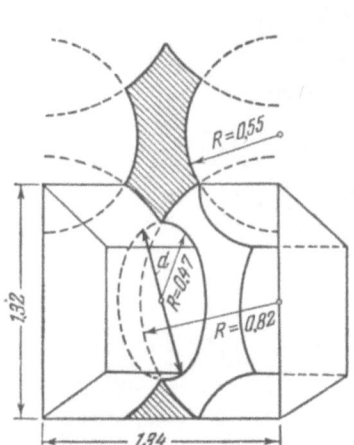

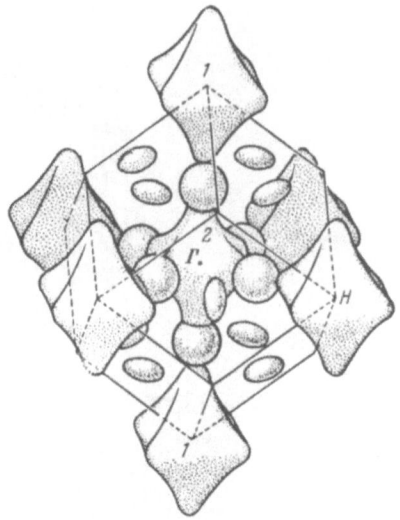

Fig. 11. Open hole Fermi surface of thallium according to
Harrison's model [102, 118],
showing a section by the (10$\bar{1}$0)
plane. According to measurements, the diameter d is considerably smaller than the
calculated value: d_{exp} = 0.1
[all dimensions are given in
units of b = 1.16 (2π/a), where
a = 3.45 Å].

Fig. 12. Fermi surface of chro-
mium-type metals [29]. Open
directions appear along the ⟨001⟩
axes because of magnetic breakthrough between the closed surfaces 1 and 2.

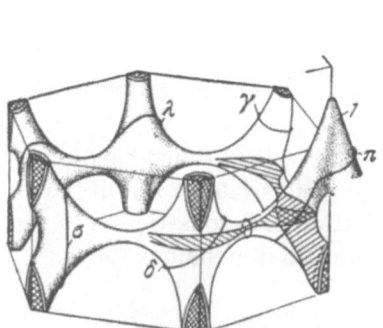

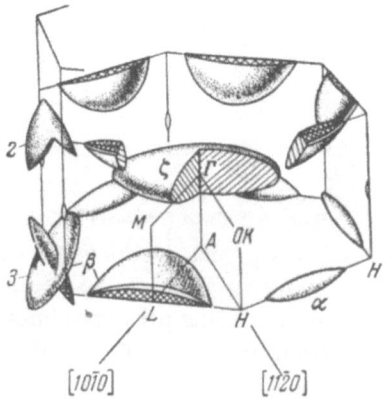

Fig. 13. Schematic representation of the completely determined
Fermi surface of zinc [118, 297]. 1) Open hole surface in the second band; in contrast to Cd, there are no discontinuities in the basal
plane and, therefore, magnetic breakthrough gives rise to open directions along the [1$\bar{2}$10] and [1$\bar{1}$00] axes (Fig. 1B); 2) closed parts
of the surface located in different bands, constructed making allowance for magnetic breakthrough between the surfaces α and β ("cigar" and "butterfly"); 3) same as 2, but constructed by ignoring magnetic breakthrough between the "cigar" and the "butterfly" surfaces.

LITERATURE CITED

Chapter I

1. J. M. Ziman, Principles of the Theory of Solids, Cambridge University Press (1964).
2. L. D. Landau, Zh. Éksp. Teor. Fiz., 30:1058 (1956).
3. V. M. Galitskii and A. B. Migdal, Zh. Éksp. Teor. Fiz., 34:139 (1958).
4. W. A. Harrison, Pseudopotentials in the Theory of Metals, Benjamin, New York (1966).
5. G. E. Zil'berman and Ya. E. Aizenberg, Fiz. Metal. Metalloved., 4:216 (1957).
6. F. G. Moliner, Phil. Mag., 3:207 (1958).
7. H. Jones, The Theory of Brillouin Zones and Electronic States in Crystals, North-Holland, Amsterdam (1960).
8. É. I. Rashba and V. I. Sheka, Fiz. Tverd. Tela, Sbornik (Suppl.) 2:162 (1959).
9. A. M. Ermolaev, Fiz. Tverd. Tela, 8:560 (1966).
10. L. Van Hove, Phys. Rev., 89:1189 (1953).
11. A. A. Slutskii, Zh. Éksp. Teor. Fiz. Pis. Red., 5:90 (1967).
12. A. A. Slutskii, Zh. Éksp. Teor. Fiz., 45:978 (1963).
13. L. D. Landau and E. M. Lifshitz, The Classical Theory of Fields, 2nd ed., Pergamon Press, Oxford (1962).
14. V. M. Nabutovskii and A. A. Slutskii, Fiz. Metal. Metalloved., 12:170 (1961).
15. I. M. Lifshits, A. A. Slutskii, and V. M. Nabutovskii, Zh. Éksp. Teor. Fiz., 41:939 (1961).
16. I. M. Lifshits, A. A. Slutskii, and V. M. Nabutovskii, Dokl. Akad. Nauk SSSR, 137:553 (1961).
17. M. I. Kaganov, I. M. Lifshits, and V. B. Fiks, Fiz. Tverd. Tela, 6:2723 (1964).
18. L. A. Fal'kovskii, Zh. Éksp. Teor. Fiz., 49:609 (1965).
19. G. E. Zil'berman, Zh. Éksp. Teor. Fiz., 32:296 (1957).
20. M. Ya. Azbel', Zh. Éksp. Teor. Fiz., 39:1276 (1960); 46:929 (1964); Dokl. Akad. Nauk SSSR, 159:703 (1964).
21. G. E. Zil'berman, Zh. Éksp. Teor. Fiz., 32:296 (1957); 33:387 (1957); 34:515 (1958).
22. See, for example, Yu. A. Bogod and V. V. Eremenko, Paper presented at Fifteenth Conf. on Low-Temperature Physics [in Russian].
23. É. I. Rashba, Zh. Éksp. Teor. Fiz., 48:1427 (1965); Z. S. Gribnikov, V. A. Kochelap, and É. I. Rashba, Zh. Éksp. Teor. Fiz., 51:266 (1966).
24. I. M. Lifshits and M. I. Kaganov, Usp. Fiz. Nauk, 69:419 (1959).
25. I. M. Lifshits, Dokl. Akad. Nauk SSSR, 48:83 (1945); Zh. Éksp. Teor. Fiz., 17:1017 (1947).
26. A. A. Slutskii, Zh. Éksp. Teor. Fiz., 53:767 (1967)
27. E. I. Blount, Phys. Rev., 126:1636 (1962).
28. L. M. Falicov and R. W. Stark, Progr. Low Temp. Phys., 5:235 (1967).

Chapter II

1. A. A. Abrikosov, L. P. Gor'kov, and I. E. Dzyaloshinskii, Methods of Quantum Field Theory in Statistical Physics, Prentice-Hall, Englewood Cliffs, N. J. (1963).

2. See, for example, R. E. Peierls, Quantum Theory of Solids, Clarendon Press, Oxford (1955).

3. W. A. Harrison, Pseudopotentials in the Theory of Metals, Benjamin, New York (1966).

4. F. G. Moliner, Phil. Mag., 3:207 (1958).

5. G. E. Zil'berman and Ya. E. Aizenberg, Fiz. Metal. Metalloved., 4:216 (1957).

6. Yu. P. Gaidukov, Appendix III, this volume.

7. A. A. Abrikosov and L. A. Fal'kovskii, Zh. Éksp. Teor. Fiz., 43:1089 (1962).

7a. R. E. Peierls, Quantum Theory of Solids, Clarendon Press, Oxford (1955), Chap. V, §§3, 4.

7b. L. A. Fal'kovskii, Usp. Fiz. Nauk, 94:3 (1968).

8. R. G. Arkhipov, Zh. Éksp. Teor. Fiz., 43:349 (1962).

9. J. W. McClure, Phys. Rev., 108:612 (1957); P. Nozières, Phys. Rev., 109:1510 (1958).

10. L. D. Landau and E. M. Lifshitz, Statistical Physics, 2nd ed., Pergamon Press, Oxford (1969).

11. L. D. Landau, Zh. Éksp. Teor. Fiz., 30:1058 (1956).

12. S. S. Nedorezov, Zh. Éksp. Teor. Fiz., 51:868, 1575 (1966).

13. E. W. Elcock, Proc. Roy. Soc., London, A222:239 (1954); G. E. Zil'berman and F. I. Itskovich, Zh. Éksp. Teor. Fiz., 32:158 (1957).

14. I. M. Lifshits, Zh. Éksp. Teor. Fiz., 38:1569 (1960).

15. N. B. Brandt and L. G. Lyubutina, Zh. Éksp. Teor. Fiz., 52:686 (1967).

16. V. G. Bar'yakhtar, B. G. Lazarev, and V. I. Makarov, Fiz. Metal. Metalloved., 24:829 (1967).

17. See, for example, F. Blokh (Bloch), Molecular Theory of Magnetism, Gosnauchtekhizdat Ukrainy, Kiev (1964).

18. E. N. Adams II, Phys. Rev., 89:633 (1953); M. V. Nitsovich, Fiz. Metal. Metalloved., 7:641 (1959).

19. L. D. Landau and E. M. Lifshitz, Quantum Mechanics: Nonrelativistic Theory, 2nd ed., Pergamon Press, Oxford (1965).

20. See, for example, R. E. Peierls, Electron Theory of Metals [Russian translation], IL, Moscow (1947).

21. M. H. Cohen and E. I. Blount, Phil. Mag., 5:115 (1960); M. H. Cohen, Phys. Rev., 121:387 (1961); N. B. Brandt, T. F. Dolgolenko, and N. N. Stupochenko, Zh. Éksp. Teor. Fiz., 45:1319 (1963); L. A. Fal'kovskii, Zh. Éksp. Teor. Fiz., 49:609 (1965).

22. B. I. Verkin and I. V. Svechkarev, Zh. Éksp. Teor. Fiz., 47:404 (1964); B. I. Verkin, I. V. Svechkarev, L. B. Kuz'micheva, and I. B. Svechkarev, Zh. Éksp. Teor. Fiz., 54:74 (1968).

23. R. Courant and D. Hilbert, Methods of Mathematical Physics, Vol. 1, Interscience, New York (1953).

24. I. M. Lifshits and A. M. Kosevich, Zh. Éksp. Teor. Fiz., 29:730 (1955).

25. Yu. A. Bychkov, Zh. Éksp. Teor. Fiz., 39:1401 (1960).

26. R. B. Dingle, Proc. Roy. Soc., London, A211:517 (1952).

27. M. Ya. Azbel', Zh. Éksp. Teor. Fiz., 39:878 (1960).

28. L. M. Falicov and R. W. Stark, Progr. Low-Temp. Phys., 5:235 (1967).

29. M. G. Priestley, Proc. Roy. Soc., London, A276:258 (1963).

29a. A. P. Kochkin, Zh. Éksp. Teor. Fiz., 54:603 (1968).

30. L. D. Landau, Zh. Éksp. Teor. Fiz., 32:59 (1957).

31. Yu. A. Bychkov and L. P. Gor'kov, Zh. Éksp. Teor. Fiz., 41:1592 (1961).

32. S. S. Nedorezov, Zh. Éksp. Teor. Fiz., 51:1575 (1966).

33. I. I. Kulik, Zh. Éksp. Teor. Fiz. Pis. Red., 6:652 (1967).

34. I. M. Lifshits and A. M. Kosevich, Dokl. Akad. Nauk SSSR, 91:795 (1953).

35. M. Ya. Azbel', Zh. Éksp. Teor. Fiz., 34:754 (1958).

36. I. M. Lifshits and A. V. Pogorelov, Dokl. Akad. Nauk SSSR, 96:1143 (1954).

37. A. V. Gold, Phil. Trans. Roy. Soc., London, A251:85 (1958); A. V. Gold and M. G. Priestley, Phil. Mag., 5:1089 (1960); E. M. Gunnersen, Phil. Trans. Roy. Soc., London, A249:299 (1957).

38. I. M. Lifshits and A. M. Kosevich, Izv. Akad. Nauk SSSR, Ser. Fiz., 19:395 (1955).

39. D. Shoenberg, Phil. Trans. Roy. Soc., London, A255:85 (1962); D. Shoenberg and J. J. Vuillemin, Proc. Tenth Intern. Conf. on Low-Temperature Physics, Moscow, 1966, Vol. 3, publ. by VINITI, Moscow (1967), p. 67.

40. A. B. Pippard, Proc. Roy. Soc., London, A272:192 (1963).

41. M. Ya. Azbel' and E. G. Skrotskaya, Zh. Éksp. Teor. Fiz., 47:1958 (1964).

42. A. S. Davydov and I. Ya. Pomeranchuk, Zh. Éksp. Teor. Fiz., 9:1294 (1939).

43. I. M. Tsidil'kovskii, V. I. Sokolov, and M. M. Aksel'rod, Fiz. Metal. Metalloved., 16:318 (1963).

44. C. G. Grenier, J. M. Reynolds, and J. R. Sybert, Phys. Rev., 132:58 (1963); E. P. Vol'skii, Zh. Éksp. Teor. Fiz., 46:2035 (1964); N. E. Alekseevskii and T. I. Kostina, Zh. Éksp. Teor. Fiz., 48:1209 (1965); N. B. Brandt and L. G. Lyubutina, Zh. Éksp. Teor. Fiz., 47:1711 (1964); G. E. Smith, G. A. Baraff, and J. M. Rowell, Phys. Rev., 135:A1118 (1964).

45. M. Ya. Azbel' and N. B. Brandt, Zh. Éksp. Teor. Fiz., 48:1206 (1965).

46. N. B. Brandt, L. G. Lyubutina, and N. A. Kryukova, Zh. Éksp. Teor. Fiz., 53:134 (1967); N. B. Brandt, E. A. Svistova, and G. Kh. Tabieva, Zh. Éksp. Teor. Fiz. Pis. Red., 4:27 (1966); N. B. Brandt, E. A. Svistova, and T. V. Gorskaya, Zh. Éksp. Teor. Fiz., 53:1274 (1967).

46a. N. B. Brandt, E. A. Svistova, Yu. G. Kashirskii, and L. V. Lin'ko, Zh. Éksp. Teor. Fiz., 56:65 (1969).

46b. N. B. Brandt, E. A. Svistova, and R. G. Valeev, Zh. Éksp. Teor. Fiz. Pis. Red., 6:724 (1967).

47. L. D. Landau and E. M. Lifshitz, Electrodynamics of Continuous Media, Pergamon Press, Oxford (1960), §30.

48. J. H. Condon, Phys. Rev., 145:526 (1966).

49. I. A. Privorotskii, Zh. Éksp. Teor. Fiz. Pis. Red., 5:280 (1967); Zh. Éksp. Teor. Fiz., 52:1755 (1967).

49a. J. H. Condon and R. E. Walstedt, Phys. Rev. Lett., 21:612 (1968).

49b. I. A. Privorotskii, Usp. Fiz. Nauk, 97:547 (1969).

50. M. Ya. Azbel', Zh. Éksp. Teor. Fiz. Pis. Red., 5:282 (1967); Zh. Éksp. Teor. Fiz., 53:1751 (1967).

51. J. J. Quinn, J. Phys. Chem. Solids, 24:933 (1963).

52. M. Ya. Azbel' and V. G. Peschanskii, Zh. Éksp. Teor. Fiz., 49:572 (1965).

53. A. M. Kosevich, Zh. Éksp. Teor. Fiz., 35:738 (1958).

54. M. Ya. Azbel', Zh. Éksp. Teor. Fiz., 53:2131 (1967).

55. I. M. Lifshits and A. M. Kosevich, Zh. Éksp. Teor. Fiz., 29:730 (1955); 33:88 (1957); J. Phys. Chem. Solids, 4:11 (1958).

56. I. A. Privorotskii and M. Ya. Azbel', Zh. Éksp. Teor. Fiz., 56:388 (1969).

57. M. Ya. Azbel', Zh. Éksp. Teor. Fiz., 46:929 (1964); Dokl. Akad. Nauk SSSR, 159:703 (1964).

58. F. I. Itskovich, Zh. Éksp. Teor. Fiz., 50:1425 (1966).

59. F. I. Itskovich, Zh. Éksp. Teor. Fiz., 51:301 (1966).

60. F. I. Itskovich, Zh. Éksp. Teor. Fiz., 52:1720 (1967).

61. F. I. Itskovich, Zh. Éksp. Teor. Fiz., 53:705 (1967).

62. M. I. Kaganov, I. M. Lifshits, and K. D. Sinel'nikov, Zh. Éksp. Teor. Fiz., 32:605 (1957).

63. I. O. Kulik and G. A. Gogadze, Zh. Éksp. Teor. Fiz., 44:530 (1963).

64. A. R. Calawa, R. H. Rediker, B. Lax, and A. L. McWhorter, Phys. Rev. Lett., 5:55 (1960).

65. V. G. Bar'yakhtar and V. I. Makarov, Dokl. Akad. Nauk SSSR, 146:63 (1962).

Chapter III

1. L. D. Landau and E. M. Lifshitz, Electrodynamics of Continuous Media, Pergamon Press, Oxford (1958).
2. V. I. Smirnov, Course of Higher Mathematics, Vol. 4, Gostekhizdat, Moscow (1951).
3. I. M. Lifshits and M. I. Kaganov, Usp. Fiz. Nauk, 87:389 (1965).
4. L. D. Landau, Zh. Éksp. Teor. Fiz., 30:1058 (1956); 32:59 (1957).
5. V. P. Silin, Zh. Éksp. Teor. Fiz., 33:495 (1957).
6. L. D. Landau, Zh. Éksp. Teor. Fiz., 35:97 (1958).
7. J. M. Ziman, Electrons and Phonons, Clarendon Press, Oxford (1960).
8. E. G. Brovman and Yu. Kagan, Zh. Éksp. Teor. Fiz., 52:557 (1967).
9. M. I. Kaganov, I. M. Lifshits, and L. V. Tanatarov, Zh. Éksp. Teor. Fiz., 31:232 (1956).
10. E. S. Borovik, Dokl. Akad. Nauk SSSR, 91:771 (1953); V. L. Ginzburg and V. P. Shabanskii, Dokl. Akad. Nauk SSSR, 100:445 (1955); M. I. Kaganov and V. G. Peschanskii, Zh. Éksp. Teor. Fiz., 33:1261 (1957).
11. R. E. Peierls, Quantum Theory of Solids, Clarendon Press, Oxford (1955).
12. R. N. Gurzhi, Zh. Éksp. Teor. Fiz., 47:1415 (1964).
13. L. D. Landau and I. Ya. Pomeranchuk, Zh. Éksp. Teor. Fiz., 7:379 (1937).
14. I. Ya. Pomeranchuk, Zh. Éksp. Teor. Fiz., 35:992 (1958).
15. Yu. Kagan and A. P. Zhernov, Zh. Éksp. Teor. Fiz., 53:1744 (1967).
16. Yu. Kagan and A. P. Zhernov, Zh. Éksp. Teor. Fiz., 50:1107 (1966).
17. J. Kondo, Progr. Theor. Phys., 32:37 (1964); A. A. Abrikosov, Usp. Fiz. Nauk, 97:403 (1969).
17a. A. A. Abrikosov, Physics, 2:5, 61 (1965); S. V. Maleev, Zh. Éksp. Teor. Fiz., 53:1038 (1967); S. L. Ginzburg and S. V. Maleev, Zh. Éksp. Teor. Fiz., 55:1483 (1968).
17b. M. I. Kaganov and M. Ya. Azbel', Zh. Éksp. Teor. Fiz., 27:762 (1954); K. Fuchs, Proc. Cambridge Phil. Soc., 34:100 (1938); R. B. Dingle, Proc. Roy. Soc., London, A201:546 (1950); B. N. Aleksandrov and M. I. Kaganov, Zh. Éksp. Teor. Fiz., 41:1333 (1961).
17c. R. N. Gurzhi, Usp. Fiz. Nauk, 94:689 (1968).
18. V. L. Ginzburg, in: Commemorative Collection in Honour of A. A. Andronov, Izd. AN SSSR, Moscow (1955), p. 622.
19. S. R. de Groot and P. Mazur, Non-Equilibrium Thermodynamics, North-Holland, Amsterdam (1962).
20. M. Köhler, Ann. Phys. (Leipzig), 32:211 (1938).
21. See, for example, J. L. Olsen, Helv. Phys. Acta, 31:713 (1958).
22. M. C. Jones and E. H. Sondheimer, Phys. Lett., 11:122 (1964).
23. E. S. Borovik, V. G. Volotskaya, and N. Ya. Fogel', Zh. Éksp. Teor. Fiz., 45:46 (1963).
24. S. V. Vonsovskii and Ya. S. Shur, Ferromagnetism, OGIZ, Moscow (1948); M. I. Kaganov, in: Magnetic Structure of Ferromagnets, Izd. AN SSSR, Moscow (1960), p. 79.
25. M. I. Kaganov and V. G. Peschanskii, in: Investigations of Energy Spectra of Electrons in Metals, Naukova Dumka, Kiev (1965), p. 4.
26. L. S. Dubinskaya, Dissertation, Institute of Semiconductors, Academy of Sciences of the USSR, Leningrad (1969).
27. M. Ya. Azbel', Zh. Éksp. Teor. Fiz., 44:983 (1963); M. Ya. Azbel' and V. G. Peschanskii, Zh. Éksp. Teor. Fiz., 49:572 (1965).
28. I. M. Lifshits, M. Ya. Azbel', and M. I. Kaganov, Zh. Éksp. Teor. Fiz., 31:63 (1956).
29. R. V. Coleman, A. J. Funes, J. S. Plaskett, and C. M. Tapp, Phys. Rev., 133:A521 (1964).
30. W. G. Chambers, Proc. Phys. Soc., London, 81:877 (1963).
31. E. S. Borovik, Zh. Éksp. Teor. Fiz., 30:262 (1956).
32. V. G. Volotskaya, Zh. Éksp. Teor. Fiz., 45:49 (1963); C. C. Grimes and A. F. Kip, Phys. Rev., 132:1991 (1963); A. C. Thorsen and T. G. Berlincourt, Phys. Rev. Lett., 6:617

(1961); K. Okumura and I. M. Templeton, Phil. Mag., 7:1239 (1962); D. Shoenberg and P. J. Stiles, Proc. Roy. Soc., London, A281:62 (1964).

33. P. B. Alers and R. T. Webber, Phys. Rev., 91:1060 (1953); G. N. Rao, N. H. Zebouni, C. G. Grenier, and J. M. Reynolds, Phys. Rev., 133:A141 (1964).

34. F. G. Bass, M. I. Kaganov, and V. V. Slezov, Fiz. Metal. Metalloved., 5:406 (1957).

35. E. S. Borovik, Zh. Éksp. Teor. Fiz., 23:91 (1952).

36. L. Esaki, Phys. Rev. Lett., 8:4 (1962).

37. V. P. Kalashnikov, Fiz. Tverd. Tela, 6:2435 (1964).

38. I. M. Lifshits and V. G. Peschanskii, Zh. Éksp. Teor. Fiz., 35:1251 (1958).

39. I. M. Lifshits and V. G. Peschanskii, Zh. Éksp. Teor. Fiz., 38:188 (1960).

40. V. G. Peschanskii, Dissertation, Khar'kov State University (1959).

41. N. E. Alekseevskii, Yu. P. Gaidukov, I. M. Lifshits, and V. G. Peschanskii, Zh. Éksp. Teor. Fiz., 39:1201 (1960).

42. E. S. Borovik, Zh. Éksp. Teor. Fiz., 23:83, 91 (1952).

43. Yu. P. Gaidukov, Zh. Éksp. Teor. Fiz., 37:1281 (1959).

44. W. A. Harrison, Pseudopotentials in the Theory of Metals, Benjamin, New York (1966).

45. M. I. Kaganov and V. G. Peschanskii, Zh. Éksp. Teor. Fiz., 35:1052 (1958).

46. N. E. Alekseevskii, N. B. Brandt, and T. I. Kostina, Dokl. Akad. Nauk SSSR, 105:46 (1955).

47. M. I. Kaganov, A. M. Kadigrobov, and A. A. Slutskii, Zh. Éksp. Teor. Fiz., 53:1135 (1967).

48. E. S. Borovik and V. G. Volotskaya, Zh. Éksp. Teor. Fiz., 48:1554 (1965).

49. M. Ya. Azbel', Zh. Éksp. Teor. Fiz., 44:983 (1963).

50. G. A. Zaitsev, Zh. Éksp. Teor. Fiz., 45:1266 (1963); G. A. Zaitsev, S. V. Stepanova, and V. I. Khotkevich, Zh. Éksp. Teor. Fiz., 48:760 (1965).

51. M. Ya. Azbel' and V. G. Peschanskii, Zh. Éksp. Teor. Fiz., 49:572 (1965).

52. M. I. Kaganov, A. M. Kadigrobov, I. M. Lifshits, and A. A. Slutskii, Zh. Éksp. Teor. Fiz. Pis. Red., 5:218 (1967).

53. L. M. Falicov and P. R. Sievert, Phys. Rev., 138:A88 (1965).

54. V. G. Peschanskii, Zh. Éksp. Teor. Fiz., 52:1312 (1967).

55. E. H. Sondheimer, Phys. Rev., 80:401 (1950).

56. E. N. Adams and T. D. Holstein, J. Phys. Chem. Solids, 10:254 (1959).

57. M. Ya. Azbel' and V. G. Peschanskii, Zh. Éksp. Teor. Fiz., 52:1003 (1967).

58. V. G. Peschanskii, Zh. Éksp. Teor. Fiz., 54:137 (1969).

59. V. G. Peschanskii and M. Ya. Azbel', Zh. Éksp. Teor. Fiz., 52:1003 (1967); 55:1980 (1968).

60. G. A. Zaitsev, S. V. Stepanova, and V. I. Khotkevich, Abstracts of Conf. Papers, Moscow 1966.

61. G. A. Zaitsev, S. V. Stepanova, and V. I. Khotkevich, Proc. Intern. Conf., Zurich, 1968.

62. Yu. A. Bogod, V. V. Eremenko, and L. K. Chubova, Phys. Status Solidi, 28:K155 (1968); Zh. Éksp. Teor. Fiz., 56:32 (1969).

63. W. S. Boyle, F. S. L. Hsu, and J. E. Kunzler, Phys. Rev. Lett., 4:278 (1960); J. Lepage, M. Garber, and F. J. Blatt, Phys. Lett., 11:102 (1964).

64. M. Ya. Azbel', M. I. Kaganov, and I. M. Lifshits, Zh. Éksp. Teor. Fiz., 32:1188 (1957).

65. Yu. A. Bychkov, L. É. Gurevich, and G. M. Nedlin, Zh. Éksp. Teor. Fiz., 37:534 (1959).

66. L. É. Gurevich and G. M. Nedlin, Zh. Éksp. Teor. Fiz., 37:765 (1959).

67. M. S. Bresler, R. V. Parfen'ev, and S. S. Shalyt, Fiz. Tverd. Tela, 8:1776 (1966).

68. I. M. Lifshits, Zh. Éksp. Teor. Fiz., 32:1509 (1957).

69. A. A. Abrikosov, Zh. Éksp. Teor. Fiz., 56:1391 (1969).

70. A. M. Kosevich and V. V. Andreev, Zh. Éksp. Teor. Fiz., 38:882 (1960); V. V. Andreev and A. M. Kosevich, 43:1060 (1962).

71. N. N. Bogolyubov, "Problems of a dynamical theory in statistical physics," in: Studies in Statistical Mechanics (ed. by J. H. De Boer and G. E. Uhlenbeck), Vol. 1, Wiley, New York (1961), pp. 5-515; K. P. Gurov, Zh. Éksp. Teor. Fiz., 17:614 (1947).

72. S. Titeica, Ann. Phys. (Leipzig), 22:129 (1935).

73. N. W. Ashcroft, Phil. Mag., 8:2055 (1963); W. Harrison, Phys. Rev., 118:1182 (1960).

74. Yu. N. Obraztsov, Fiz. Tverd. Tela, 6:414 (1964); 7:573 (1965).

Chapter IV

1. H. London, Proc. Roy. Soc. London, A176:522 (1940).

2. A. B. Pippard, Proc. Roy. Soc. London, A191:385 (1947).

3. G. E. Reuter and E. H. Sondheimer, Proc. Roy. Soc. London, A195:336 (1948).

4. M. Ya. Azbel', Zh. Éksp. Teor. Fiz., 39:1138 (1960).

5. M. Ya. Azbel', Zh. Éksp. Teor. Fiz., 39:400 (1960).

6. É. A. Kaner, Zh. Éksp. Teor. Fiz., 34:658 (1958).

7. M. Ya. Azbel' and É. A. Kaner, Zh. Éksp. Teor. Fiz., 32:896 (1957); J. Phys. Chem. Solids, 6:113 (1958).

8. R. G. Chambers, Proc. Phys. Soc., London, A65:458 (1952).

9. M. I. Kaganov and M. Ya. Azbel', Dokl. Akad. Nauk SSSR, 102:49 (1956).

10. A. B. Pippard, Phil. Trans. Roy. Soc. London, A250:325 (1957).

11. M. S. Khaikin, Zh. Éksp. Teor. Fiz., 39:212 (1960).

12. Proc. Seventh Intern. Conf. on Low Temperature Physics, University of Toronto, 1960, publ. by University of Toronto Press (1961); A. F. Kip, D. N. Langenberg, B. Rosenblum, and G. Wagoner, Phys. Rev., 108:494 (1957); E. Fawcett and W. M. Walsh, Jr., Phys. Rev. Lett., 8:476 (1962); J. F. Koch and A. F. Kip, Proc. Ninth Intern. Conf. on Low Temperature Physics, Columbus, Ohio, 1964, Part B, publ. by Plenum Press, New York (1965), p. 818.

13. M. Ya. Azbel', Dokl. Akad. Nauk SSSR, 100:437 (1955).

14. É. A. Kaner, Dokl. Akad. Nauk SSSR, 119:471 (1958); V. F. Gantmakher and É. A. Kaner, Zh. Éksp. Teor. Fiz., 45:1430 (1963); D. M. Sparlin and D. S. Schreiber, Proc. Ninth Intern. Conf. on Low Temperature Physics, Columbus, Ohio, 1964, Part B, publ. by Plenum Press, New York (1965), p. 823.

15. V. F. Gantmakher, Zh. Éksp. Teor. Fiz., 42:1416 (1962); 44:811 (1963).

16. Ya. G. Dorfman, Dokl. Akad. Nauk SSSR, 81:765 (1951).

17. R. B. Dingle, Proc. Roy. Soc., London, A212:38 (1952).

18. G. Dresselhaus, A. F. Kip, and C. Kittel, Phys. Rev., 92:827 (1953); B. Lax, H. J. Zeiger, R. N. Dexter, and E. S. Rosenblum, Phys. Rev., 93:1418 (1954).

19. M. Ya. Azbel' and M. I. Kaganov, Dokl. Akad. Nauk SSSR, 95:41 (1953).

20. R. G. Chambers, Phil. Mag., 1:459 (1956).

21. É. A. Kaner, Zh. Éksp. Teor. Fiz., 33:1472 (1957).

22. V. G. Peschanskii and V. S. Lekhtsier, Zh. Éksp. Teor. Fiz., 46:764 (1964).

23. M. Ya. Azbel' and V. G. Peschanskii, Zh. Éksp. Teor. Fiz. Pis. Red., 5:26 (1967); Zh. Éksp. Teor. Fiz., 54:477 (1968).

24. B. Lax, Rev. Mod. Phys., 30:122 (1958).

25. E. Fawcett, Phys. Rev., 103:1582 (1956).

26. P. A. Bezuglyi and A. A. Galkin, Zh. Éksp. Teor. Fiz., 33:1076 (1957); 34:237 (1958); A. F. Kip, D. N. Langenberg, B. Rosenblum, and G. Wagoner, Phys. Rev., 108:494 (1957); M. S. Khaikin, Zh. Éksp. Teor. Fiz., 37:1473 (1959); 39:513 (1960); 42:27 (1962); 43:59 (1962).

27. D. N. Langenberg, A. F. Kip, and B. Rosenblum, Bull. Amer. Phys. Soc., 3:416 (1958); A. F. Kip, D. N. Langenberg, and T. W. Moore, Phys. Rev., 124:359 (1961); D. N. Langenberg and T. W. Moore, Phys. Rev. Lett., 3:328 (1959); J. F. Koch, R. A. Stradling, and A. F. Kip, Phys. Rev., 133:A240 (1964).

28. P. A. Bezuglyi and A. A. Galkin, Zh. Éksp. Teor. Fiz., 34:236 (1959); R. C. Young, Phil. Mag., 7:2065 (1962); J. E. Aubrey, Phil. Mag., 5:1001 (1960); M. S. Khaikin and R. T. Mina, Zh. Éksp. Teor. Fiz., 42:35 (1962); R. T. Mina and M. S. Khaikin, Zh. Éksp. Teor. Fiz., 45:1304 (1963); 51:62 (1966).

29. P. A. Bezuglyi and A. A. Galkin, Zh. Éksp. Teor. Fiz., 37:1480 (1959); J. G. Castle, Jr., B. S. Chandrasekhar, and J. A. Rayne, Phys. Rev. Lett., 6:409 (1961); R. T. Mina and M. S. Khaikin, Zh. Éksp. Teor. Fiz., 48:111 (1965); Zh. Éksp. Teor. Fiz. Pis. Red., 1(2):34 (1965).

30. J. K. Galt, F. R. Merritt, W. A. Yager, and K. W. Dail, Jr., Phys. Rev. Lett., 2:292 (1959); J. K. Galt and F. R. Merritt, Fermi Surface (Proc. Intern. Conf., Cooperstown, N. Y., 1960), publ. by Wiley, New York (1960), pp. 159, 174 (discussion).

31. D. N. Langenberg and T. W. Moore, Phys. Rev. Lett., 3:137 (1959); E. Fawcett, Phys. Rev. Lett., 3:139 (1959); F. W. Spong and A. F. Kip, Phys. Rev., 137:A431 (1965); T. W. Moore and F. W. Spong, Phys. Rev., 125:846 (1962); C. C. Grimes, A. F. Kip. F. W. Spong, R. A. Stradling, and P. Pincus, Phys. Rev. Lett., 11:455 (1963); A. A. Galkin, V. P. Naberezh-nykh, and V. L. Mel'nik, Fiz. Tverd. Tela, 5:201 (1963).

32. S. Foner, H. J. Zeiger, R. L. Powell, W. M. Walsh, Jr., and B. Lax, Bull. Amer. Phys. Soc., 1:117 (1956); M. S. Khaikin and V. S. Edel'man, Zh. Éksp. Teor. Fiz., 47:878 (1964); V. S. Edel'man and M. S. Khaikin, Zh. Éksp. Teor. Fiz., 49:107 (1965); L. C. Hebel, Phys. Rev., 138:A1641 (1965); Yi-Han Kao, Phys. Rev., 129:1122 (1963); B. Lax, K. J. Button, H. J. Zeiger, and L. M. Roth, Phys. Rev., 102:715 (1956); J. E. Aubrey, J. Phys. Chem. Solids, 19:321 (1961); J. E. Aubrey and R. G. Chambers, J. Phys. Chem. Solids, 3:128 (1957); J. K. Galt, W. A. Yager, F. R. Merritt, B. B. Cetlin, and H. W. Dail, Jr., Phys. Rev., 100:748 (1955); J. K. Galt, W. A. Yager, F. R. Merritt, B. B. Cetlin, and A. D. Brailsford, Phys. Rev., 114: 1396 (1959).

33. A. E. Dixon and W. R. Datars, Solid State Commun., 3:377 (1965); A. Ron and M. Revzen, Phys. Lett., 20:106 (1966).

34. T. W. Moore, Phys. Rev. Lett., 16:581 (1966).

35. D. G. Howard, Phys. Rev., 190:A1705 (1965).

36. D. N. Langenberg and S. M. Marcus, Phys. Rev., 136:A1383 (1964).

37. C. C. Grimes and A. F. Kip, Phys. Rev., 132:1991 (1963).

38. E. Fawcett and W. M. Walsh, Jr., Phys. Rev. Lett., 8:476 (1962); W. M. Walsh, Jr., Phys. Rev. Lett., 12:161 (1964).

39. J. K. Galt, F. R. Merritt, and P. H. Schmidt, Phys. Rev. Lett., 6:458 (1961); C. C. Grimes, A. F. Kip, F. Spong, R. A. Stradling, and P. Pincus, Phys. Rev. Lett., 11:455 (1963); M. P. Shaw and T. G. Eck, Proc. Ninth Intern. Conf. on Low Temperature Physics, Columbus, Ohio, 1964, Part B, publ. by Plenum Press, New York (1965), p. 761.

40. W. R. Datars, Phys. Rev., 126:975 (1962); Can. J. Phys., 39:1922 (1961); 40:1784 (1962); W. R. Datars and R. N. Dexter, Phys. Rev., 124:75 (1961); W. R. Datars and J. Vanderkooy, Bull. Amer. Phys. Soc., 9:598 (1964).

41. T. G. Eck and M. P. Shaw, Proc. Ninth Intern. Conf. on Low Temperature Physics, Colum-bus, Ohio, 1964, Part B, publ. by Plenum Press, New York (1965), p. 759.

42. R. G. Chambers, Can. J. Phys., 34:1395 (1956).

43. E. L. Blount, Phys. Rev. Lett., 4:114 (1960).

44. G. E. Everett, Phys. Rev., 128:2564 (1962); J. Kirsch and A. G. Redfield, Bull. Amer. Phys. Soc., 7:477 (1962); J. F. Koch and A. F. Kip, Phys. Rev. Lett., 8:473 (1962).

45. R. G. Chambers, Proc. Phys. Soc. London, 86:305 (1965); 88:695, 701 (1966).

46. C. C. Grimes, A. F. Kip, F. W. Spong, R. A. Stradling, and P. Pincus, Phys. Rev. Lett., 11:455 (1963); F. W. Spong and A. F. Kip, Phys. Rev., 137:A431 (1965).

47. P. E. Bloomfield, Proc. Tenth Intern. Conf. on Low Temperature Physics, Moscow, 1966, Vol. 3, publ. by VINITI, Moscow (1967), p. 111; É. A. Kaner and A. Ya. Blank, ibid., Vol. 3, p. 116; J. Phys. Chem. Solids, 28:1735 (1957).

48. M. S. Khaikin, Zh. Éksp. Teor. Fiz., 42:27 (1962); R. T. Mina and M. S. Khaikin, Zh. Éksp. Teor. Fiz. Pis. Red., 1:34 (1965); D. G. Howard, Phys. Rev., 140:A1705 (1965).

49. M. Ya. Azbel' and V. G. Peschanskii, Zh. Éksp. Teor. Fiz. Pis. Red., 5:26 (1967); Zh. Éksp. Teor. Fiz., 54:477 (1968).

50. M. S. Khaikin, Zh. Éksp. Teor. Fiz., 41:1773 (1961); 42:27 (1962); 43:59 (1962).

51. H. Suzuki and K. Ariyama, J. Phys. Soc. Jap., 17:704 (1962); H. Suzuki, J. Phys. Soc. Jap., 17:1077 (1962).

52. W. M. Walsh, Jr., C. C. Grimes, G. Adams, and L. W. Rupp, Jr., Proc. Ninth Conf. on Low Temperature Physics, Columbus, Ohio, 1964, Part B, publ. by Plenum Press, New York (1965), p. 765.

53. É. A. Kaner, Zh. Éksp. Teor. Fiz., 44:1036 (1963).

54. V. F. Gantmakher, Progr. Low Temp. Phys., 5:181 (1967); É. A. Kaner and F. V. Gantmakher, Usp. Fiz. Nauk, 94:193 (1968).

55. V. F. Gantmakher, Zh. Éksp. Teor. Fiz., 43:345 (1962).

56. O. V. Konstantinov and V. I. Perel', Zh. Éksp. Teor. Fiz., 38:161 (1960).

57. É. A. Kaner and V. G. Skobov, Zh. Éksp. Teor. Fiz., 45:610 (1963); 46:1106 (1964).

58. V. G. Skobov and É. A. Kaner, Zh. Éksp. Teor. Fiz., 46:1809 (1964).

59. V. G. Skobov and É. A. Kaner, Zh. Éksp. Teor. Fiz., 46:273 (1964).

60. É. A. Kaner and V. G. Skobov, Usp. Fiz. Nauk, 89:367 (1966).

61. F. G. Bass, A. Ya. Blank, and M. I. Kaganov, Zh. Éksp. Teor. Fiz., 45:1081 (1963).

62. J. E. Aubrey and R. G. Chambers, J. Phys. Chem. Solids, 3:128 (1957); S. J. Buchsbaum and J. K. Galt, Phys. Fluids, 4:1514 (1961); J. Kirsh, R. R. Haering, and P. B. Miller, Phys. Rev. Lett., 9:421 (1962); P. B. Miller and R. R. Haering, Phys. Rev., 128:126 (1962); M. S. Khaikin, V. S. Édel'man, and R. T. Mina, Zh. Éksp. Teor. Fiz., 44:2190 (1963); M. S. Khaikin and V. S. Édel'man, Zh. Éksp. Teor. Fiz., 49:1695 (1965); V. S. Édel'man and M. S. Khaikin, Zh. Éksp. Teor. Fiz., 45:826 (1963); M. S. Khaikin, L. A. Fal'kovskii, V. S. Édel'man, and R. T. Mina, Zh. Éksp. Teor. Fiz., 45:1704 (1963).

63. F. E. Rose, M. T. Taylor, and R. Bowers, Phys. Rev., 127:1122 (1962); M. T. Taylor, J. R. Merritt, and R. Bowers, Phys. Rev., 129:2525 (1963); J. J. Quinn and S. Rodriguez, Phys. Rev., 128:2487 (1962); A. B. Pippard, Rep. Progr. Phys., 23:176 (1960); R. Bowers, C. Lengendy, and F. E. Rose, Phys. Rev. Lett., 7:339 (1961); P. Cotti, P. Wyder, and A. Quattropani, Phys. Lett., 1:50 (1962).

64. A. Libchaber and R. Veilex, Phys. Rev., 127:774 (1962).

65. W. M. Walsh, Jr. and P. M. Platzman, Phys. Rev. Lett., 15:784 (1965).

66. A. K. Jonscher, Brit. J. Appl. Phys., 15:365 (1964).

67. A. A. Abrikosov and L. A. Fal'kovskii, Zh. Éksp. Teor. Fiz., 43:1089 (1962).

67a. V. P. Silin, Zh. Éksp. Teor. Fiz., 35:1243 (1958).

67b. S. Schultz and G. Dunifer, Phys. Rev. Lett., 18:283 (1967).

67c. V. P. Silin, Spin Waves in Nonferromagnetic Metals, Suppl. to A. I. Akhiezer, V. G. Bar'yakhtar, and S. V. Peletminskii, Spin Waves, Wiley, New York (1968).

68. M. Ya. Azbel' and S. Ya. Rakhmanov, Zh. Éksp. Teor. Fiz. Pis. Red., 9:252 (1969).

68a. É. A. Kaner and V. G. Skobov, Zh. Éksp. Teor. Fiz., 46:1106 (1964).

68b. M. Ya. Azbel' and S. Ya. Rakhmanov, Zh. Éksp. Teor. Fiz., 57:983 (1969).

68c. M. Ya. Azbel' and S. Ya. Rakhmanov, Fiz. Tverd. Tela, 11:3183 (1969).

69. I. M. Lifshits, Zh. Éksp. Teor. Fiz., 40:1235 (1961).

69a. I. M. Lifshits, M. Ya. Azbel', and A. A. Slutskin, Zh. Éksp. Teor. Fiz., 43:1464 (1962).

70. M. Ya. Azbel' and G. A. Begiashvili, Zh. Éksp. Teor. Fiz. Pis. Red., 3:201 (1966).

71. M. Ya. Azbel' and L. B. Dubovskii, Zh. Éksp. Teor. Fiz. Pis. Red., 5:414 (1967).

72. L. D. Landau and E. M. Lifshitz, Quantum Mechanics: Non-Relativistic Theory, 2nd ed., Pergamon Press, Oxford (1965).

73. E. P. Vol'skii, Zh. Éksp. Teor. Fiz., 43:1120 (1962); 46:2035 (1964).

74. M. S. Khaikin, R. T. Mina, and V. S. Édel'man, Zh. Éksp. Teor. Fiz., 43:2063 (1962); N. W. A. Marsh and J. E. Aubrey, Nature, 205:894 (1965).
75. M. Ya. Azbel', Zh. Éksp. Teor. Fiz., 34:969 (1958).
76. M. Ya. Azbel', Zh. Éksp. Teor. Fiz., 34:1158 (1958).
77. D. C. Mattis and G. Dresselhaus, Phys. Rev., 111: 403 (1958).
78. G. Feher and A. F. Kip, Phys. Rev., 98:337 (1955).
79. A. W. Overhauser, Phys. Rev., 89:689 (1953); 92:411 (1953).
80. R. J. Elliott, Phys. Rev., 96: 266 (1954).
81. V. V. Andreev and V. I. Gerasimenko, Zh. Éksp. Teor. Fiz., 35:1209 (1958).
82. N. N. Bogolyubov and K. P. Gurov, Zh. Éksp. Teor. Fiz., 17:614 (1947).
83. F. J. Dyson, Phys. Rev., 98:349 (1955).
84. M. Ya. Azbel', Dokl. Akad. Nauk SSSR, 99:519 (1954).
85. D. I. Blokhintsev, Principles of Quantum Mechanics, Allyn and Bacon, Rockleigh, N. J. (1964), §62.
86. T. W. Griswold, A. F. Kip, and C. Kittel, Phys. Rev., 88:951 (1952).
87. R. A. Levy, Phys. Rev., 102:31 (1956).
88. M. S. Khaikin, Zh. Éksp. Teor. Fiz., 39:899 (1960).
89. G. E. Smith, J. K. Galt, and F. R. Merritt, Phys. Rev. Lett., 4:276 (1960).
90. R. B. Lewis and T. R. Carver, Phys. Rev. Lett., 12:693 (1964); N. S. VanderVen and R. T. Schumacher, Phys. Rev. Lett., 12:695 (1964).
91. L. C. Hebel, E. I. Blount, and G. E. Smith, Phys. Rev., 138:A1636 (1965).
92. M. Ya. Azbel', V. I. Gerasimenko, and I. M. Lifshits, Zh. Éksp. Teor. Fiz., 31:357 (1956); 32:1212 (1957); 35:691 (1958); I. M. Lifshits, M. Ya. Azbel', and V. I. Gerasimenko, J. Phys. Chem. Solids, 1:164 (1956).
93. J. I. Kaplan, Phys. Rev., 115:575 (1959).
94. H. C. Torrey, Phys. Rev., 104:563 (1956).
95. M. Ya. Azbel' and I. M. Lifshits, Zh. Éksp. Teor. Fiz., 33:792 (1957).
96. É. I. Rashba, Usp. Fiz. Nauk, 84:557 (1964).
97. P. Bloomfield, Bull. Amer. Phys. Soc., 8:206 (1963).
98. G. E. Smith, L. C. Hebel, and S. J. Buchsbaum, Phys. Rev., 129:154 (1963).
99. M. Ya. Azbel', Fiz. Tverd. Tela, 4:569 (1962).
100. É. A. Kaner, Zh. Éksp. Teor. Fiz., 33:1472 (1957).
100a. V. G. Peschanskii, Zh. Éksp. Teor. Fiz. Pis. Red., 7:489 (1968).
101. M. S. Khaikin, Zh. Éksp. Teor. Fiz., 39:212 (1960).
102. J. F. Koch and A. F. Kip, Proc. Ninth Intern. Conf. on Low Temperature Physics, Columbus, Ohio, 1964, Part B, publ. by Plenum Press, New York (1965), p. 818.
103. J. F. Koch and C. C. Kuo, Phys. Rev., 143:470 (1966).
104. K. S. Khaikin, Zh. Éksp. Teor. Fiz., 55:1696 (1968).
105. M. S. Khaikin, Zh. Éksp. Teor. Fiz. Pis. Red., 4:164 (1966).
106. A. P. Van Gelder, Phys. Lett., 22:7 (1966).
107. Tsu-Wei Nee and R. E. Prange, Phys. Lett., 25A:582 (1967); R. E. Prange, Phys. Rev., 171:737 (1968).
108. L. P. Pitaevskii, Zh. Éksp. Teor. Fiz., 34:942 (1958).
109. R. N. Gurzhi, M. Ya. Azbel', and Hao Pai Lin, Fiz. Tverd. Tela, 5:759 (1963).
110. R. N. Gurzhi, Zh. Éksp. Teor. Fiz., 35:965 (1958).
110a. M. I. Kaganov and V. V. Slezov, Zh. Éksp. Teor. Fiz., 32:1496 (1957).
111. M. I. Kaganov and I. M. Lifshits, Zh. Éksp. Teor. Fiz., 45:948 (1963).
112. J. C. Slonczewski and P. R. Weiss, Phys. Rev., 109:272 (1958).
113. R. N. Gurzhi and M. I. Kaganov, Zh. Éksp. Teor. Fiz., 49:941 (1965).
114. M. S. Khaikin, Usp. Fiz. Nauk, 96:409 (1968).
115. I. M. Lifshits, Zh. Éksp. Teor. Fiz., 26:551 (1954).
116. R. G. Chambers, Proc. Phys. Soc., London, 78:941 (1961).

APPENDIX II

1. E. G. Brovman and Yu. Kagan, Zh. Éksp. Teor. Fiz., 52:557 (1967).
2. V. M. Kontorovich, Zh. Éksp. Teor. Fiz., 45:1638 (1963).
3. A. I. Akhiezer, M. I. Kaganov, and G. Ya. Lyubarskii, Zh. Éksp. Teor. Fiz., 32:837 (1957).
4. L. D. Landau and E. M. Lifshitz, Theory of Elasticity, 2nd ed., Pergamon Press, Oxford (1970).
5. M. I. Kaganov and V. G. Peschanskii, Fiz. Tverd. Tela, 9:3570 (1967); 5:3215 (1963).
6. A. I. Akhiezer, Zh. Éksp. Teor. Fiz., 8:1330 (1938).
7. L. D. Landau and E. M. Lifshitz, Statistical Physics, 2nd ed., Pergamon Press, Oxford (1969), §54.
8. A. B. Pippard, Phil. Mag., 46:1104 (1955).
9. A. B. Migdal, Zh. Éksp. Teor. Fiz., 34:1438 (1958); W. Kohn, Phys. Rev. Lett., 2:393 (1959).
10. P. L. Taylor, Phys. Rev., 131:1995 (1963).
11. A. M. Afanas'ev and Yu. Kagan, Zh. Éksp. Teor. Fiz., 43:1456 (1962).
12. M. I. Kaganov and A. I. Semenenko, Zh. Éksp. Teor. Fiz., 50:630 (1966).
13. L. D. Landau and E. M. Lifshitz, Electrodynamics of Continuous Media, Pergamon Press, Oxford (1960), §62.
14. A. B. Pippard, Phil. Mag., 2:1147 (1957).
15. V. L. Gurevich, Zh. Éksp. Teor. Fiz., 37:71 (1959).
16. É. A. Kaner, V. G. Peschanskii, and I. A. Privorotskii, Zh. Éksp. Teor. Fiz., 40:214 (1961).
17. É. A. Kaner, Zh. Éksp. Teor. Fiz., 43:216 (1962).
18. V. L. Gurevich, V. G. Skobov, and Yu. A. Firsov, Zh. Éksp. Teor. Fiz., 40:786 (1961).
19. Yu. M. Gal'perin, S. V. Gantsevich, and V. L. Gurevich, Zh. Éksp. Teor. Fiz., 56:1728 (1969).

APPENDIX III

1. R. J. Balcombe, Proc. Roy. Soc., London, A275:113 (1963).
2. E. S. Borovik and V. G. Volotskaya, Zh. Éksp. Teor. Fiz., 48:1554 (1965).
3. R. A. Parker and R. J. Balcombe, Phys. Lett., 27A:197 (1968).
4. R. Lück, Phys. Status Solidi, 18:49 (1966).
5. E. M. Gunnersen, Phil. Trans. Roy. Soc. London, A249:299 (1957).
6. M. G. Priestley, Phil. Mag., 7:1205 (1962).
7. E. P. Vol'skii, Zh. Éksp. Teor. Fiz., 46:123 (1964).
8. C. O. Larson and W. L. Gordon, Phys. Rev., 156:703 (1967).
9. T. W. Moore and F. W. Spong, Phys. Rev., 125:846 (1962).
10. A. A. Galkin, V. P. Naberezhnykh, and V. A. Mel'nik, Zh. Éksp. Teor. Fiz., 44:127 (1963).
11. V. P. Naberezhnykh and V. P. Tolstoluzhskii, Zh. Éksp. Teor. Fiz., 46:18 (1964).
12. F. W. Spong and A. F. Kip, Phys. Rev., 137:A431 (1965).
13. R. T. Mina, V. S. Édel'man, and M. S. Khaikin, Zh. Éksp. Teor. Fiz., 51:1363 (1966).
14. P. A. Bezuglyi, A. A. Galkin, and A. I. Pushkin, Zh. Éksp. Teor. Fiz., 44:71 (1963).
15. G. N. Kamm and H. V. Bohm, Phys. Rev., 131:111 (1963).
16. K. Fossheim and T. Olsen, Phys. Status Solidi, 6:867 (1964).
17. H. P. Aubauer, Phys. Rev., 155:673 (1967).
18. J. F. Koch and T. K. Wagner, Bull. Amer. Phys. Soc., 11:170 (1966).
19. V. Heine, Proc. Roy. Soc. London, A240:361 (1957).
20. B. Segall, Phys. Rev., 131:121 (1963).
21. N. W. Ashcroft, Phil. Mag., 8:2055 (1963).
22. N. E. Alekseevskii and V. S. Egorov, Zh. Éksp. Teor. Fiz., 55:1153 (1968).

23. B. R. Watts, Phys. Lett., 3:284 (1963); Proc. Roy. Soc. London, A282:521 (1964).
24. T. L. Loucks and P. H. Cutler, Phys. Rev., 133:A819 (1964).
25. T. L. Loucks, Phys. Rev., 134:A1618 (1964).
26. J. H. Terrell, Phys. Rev., 149:526 (1966).
27. N. E. Alekseevskii and V. S. Egorov, Zh. Éksp. Teor. Fiz., 46:1205 (1964).
28. K. S. Nelson, J. L. Stanford, and F. A. Schmidt, Phys. Lett., 28A:402 (1968).
29. L. F. Mattheiss, Phys. Rev., 139:A1893 (1965).
30. R. N. Zitter, Phys. Rev., 127:1471 (1962).
31. J. S. Dhillon and D. Schoenberg, Phil. Trans. Roy. Soc. London, A248:1 (1955).
32. L. S. Lerner, Phys. Rev., 130:605 (1963).
33. N. B. Brandt, T. F. Dolgolenko, and N. N. Stupochenko, Zh. Éksp. Teor. Fiz., 45:1319 (1963).
34. Y. Eckstein and J. B. Ketterson, Phys. Rev., 137:A1777 (1965).
35. C. G. Grenier, J. M. Reynolds, and J. R. Sybert, Phys. Rev., 132:58 (1963).
36. R. J. Balcombe and A. M. Forrest, Phys. Rev., 151:550 (1966).
37. Yi-Han Kao, Phys. Rev., 129:1122 (1963).
38. V. S. Édel'man and M. S. Khaikin, Zh. Éksp. Teor. Fiz., 49:107 (1965).
39. D. H. Reneker, Phys. Rev., 115:303 (1959).
40. A. P. Korolyuk, Zh. Éksp. Teor. Fiz., 49:1009 (1965); 51:697 (1966).
41. Y. Sawada, E. Burstein, and L. Testardi, J. Phys. Soc. Jap., Vol. 21, Suppl., p. 760 (1966).
42. S. Mase, Y. Fujimori, and H. Mori, J. Phys. Soc. Jap., 21:1744 (1966).
43. M. S. Khaikin and V. S. Édel'man, Zh. Éksp. Teor. Fiz., 47:878 (1964).
44. V. F. Gantmakher, Progr. Low Temp. Phys., 5:181 (1967).
45. L. S. Lerner, Phys. Rev., 127:1480 (1962).
46. A. A. Abrikosov and L. A. Fal'kovskii, Zh. Éksp. Teor. Fiz., 43:1089 (1962).
47. L. G. Ferreira, J. Phys. Chem. Solids, 29:357 (1968).
48. E. Fawcett, Phys. Rev., 128:154 (1962).
49. D. M. Sparlin and J. A. Marcus, Phys. Rev., 144:484 (1966).
50. R. F. Girvan, A. V. Gold, and R. A. Phillips, J. Phys. Chem. Solids, 29:1485 (1968).
51. E. Fawcett and W. M. Walsh, Jr., Phys. Rev. Lett., 8:476 (1962).
52. P. Herrmann (Gerrmann) and V. S. Édel'man, Zh. Éksp. Teor. Fiz., 53:1563 (1967).
53. W. M. Walsh, Jr., C. C. Grimes, G. Adams, and L. W. Rupp, Jr., Proc. Ninth Intern. Conf. on Low Temperature Physics, Columbus, Ohio, 1964, Part B, publ. by Plenum Press, New York (1965), p. 765.
54. T. L. Loucks, Phys. Rev., 143:506 (1966).
55. J. O. Dimmock, A. J. Freeman, and R. E. Watson, J. Appl. Phys., 36:1142 (1965).
56. A. J. Freeman, J. O. Dimmock, and R. E. Watson, Phys. Rev. Lett., 16:94 (1966).
57. S. C. Keeton and T. L. Loucks, Phys. Rev., 146:429 (1966); 168:672 (1968).
58. N. E. Alekseevskii and Yu. P. Gaidukov, Zh. Éksp. Teor. Fiz., 37:672 (1959).
59. W. A. Reed and J. A. Marcus, Phys. Rev., 126:1298 (1962).
60. J. A. Munarin, J. A. Marcus, and P. E. Bloomfield, Phys. Rev., 172:718 (1968).
61. A. Goldstein and S. Foner, Phys. Rev., 146:442 (1966).
62. T. W. Moore, Phys. Rev. Lett., 18:310 (1967); Phys. Rev., 165:864 (1968).
63. M. Surma, Acta Phys. Pol., 32:677 (1967).
64. P. A. Bezuglyi, A. A. Galkin, and S. E. Zhevago, Zh. Éksp. Teor. Fiz., 47:825 (1964); Fiz. Tverd. Tela, 7:480 (1965).
65. A. Fukumoto and M. W. P. Strandberg, Phys. Rev., 155:685 (1967).
66. J. C. Slater, G. F. Koster, and J. H. Wood, Phys. Rev., 126:1307 (1962).
67. D. E. Soule, J. W. McClure, and L. B. Smith, Phys. Rev., 134:A453 (1964).
68. S. J. Williamson, S. Foner, and M. S. Dresselhaus, Phys. Rev., 140:A1429 (1965).
69. S. J. Williamson, M. Surma, H. C. Praddaude, R. A. Patten, and J. K. Furdyna, Solid State Commun., 4:37 (1966).

70. R. F. O'Brien and S. Foner, Phys. Lett., 25A:310 (1967).

71. M. S. Dresselhaus and J. G. Mavroides, IBM J. Res. Develop., 8:262 (1964).

72. E. Fawcett and W. A. Reed, Phys. Rev. Lett., 9:336 (1962); Phys. Rev., 131:2463 (1963); W. A. Reed and E. Fawcett, Phys. Rev., 136:A422 (1964).

73. A. Isin and R. V. Coleman, Phys. Rev., 137:A1609 (1965).

74. A. V. Gold, J. Appl. Phys., 39:768 (1968).

75. S. Wakoh and J. Yamashita, J. Phys. Soc. Jap., 21:1712 (1966).

76. J. H. Wood, Phys. Rev., 126:517 (1962).

77. L. F. Mattheiss, Phys. Rev., 134:A970 (1964).

78. E. Justi, Phys. Z., 41:563 (1940).

79. Yu. P. Gaidukov, Zh. Éksp. Teor. Fiz., 37:1281 (1959).

80. D. Shoenberg, Phil. Trans. Roy. Soc. London, A255:85 (1962).

81. A. S. Joseph, A. C. Thorsen, and F. A. Blum, Phys. Rev., 140:A2046 (1965).

82. J. P. Jan and I. M. Templeton, Phys. Rev., 161:556 (1967).

83. D. N. Langenberg and S. M. Marcus, Phys. Rev., 136:A1383 (1964).

84. R. W. Morse, A. Myers, and C. T. Walker, Phys. Rev. Lett., 4:605 (1960).

85. H. V. Bohm and V. J. Easterling, Phys. Rev., 128:1021 (1962).

86. D. J. Roaf, Phil. Trans. Roy. Soc. London, A255:135 (1962).

87. E. S. Borovik, Dokl. Akad. Nauk SSSR, 69:767 (1949).

88. E. S. Borovik and V. G. Volotskaya, Zh. Éksp. Teor. Fiz., 38:261 (1960).

89. Yu. P. Gaidukov, Zh. Éksp. Teor. Fiz., 49:1049 (1965).

90. G. B. Brandt and J. A. Rayne, Phys. Lett., 12:87 (1964); Phys. Rev., 132:1512 (1963).

91. W. J. O'Sullivan, J. E. Shirber, and J. R. Anderson, Phys. Lett., 27A:144 (1968).

92. A. J. Hughes and A. H. Lettington, Phys. Lett., 27A:241 (1968).

93. R. T. Mina and M. S. Khaikin, Zh. Éksp. Teor. Fiz., 48:111 (1965); 51:62 (1966).

94. J. A. Rayne and B. S. Chandrasekhar, Phys. Rev., 125:1952 (1962); J. A. Rayne, Phys. Rev., 129:652 (1963).

95. R. J. Balcombe, E. W. Guptill, and M. H. Jericho, Phys. Lett., 13:287 (1964).

96. V. F. Gantmakher and I. P. Krylov, Zh. Éksp. Teor. Fiz., 49:1054 (1965).

97. N. W. Ashcroft and W. E. Lawrence, Phys. Rev., 175:938 (1968).

98. W. R. Datars and S. Tanuma, Phys. Lett., 27A:182 (1968).

99. S. Tanuma, Y. Ishizawa, H. Nagasawa, and T. Sugawara, Phys. Lett., 25A:669 (1967).

100. T. L. Loucks, Phys. Rev., 144:504 (1966).

101. S. L. Altman and C. J. Bradley, Proc. Phys. Soc. London, 92:764 (1967).

102. N. E. Alekseevskii and Yu. P. Gaidukov, Zh. Éksp. Teor. Fiz., 43:2094 (1962).

103. D. C. Tsui and R. W. Stark, Phys. Rev. Lett., 19:1317 (1967).

104. J. G. Anderson and W. F. Love, Bull. Amer. Phys. Soc., 8:258 (1963).

105. A. D. C. Grassie, Phil. Mag., 9:847 (1964).

106. D. C. Tsui and R. W. Stark, Phys. Rev. Lett., 16:19 (1966).

107. J. K. Galt, F. R. Merritt, and J. R. Klauder, Phys. Rev., 139:A823 (1965).

108. M. P. Shaw, T. G. Eck, and D. A. Zych, Phys. Rev., 142:406 (1966).

109. J. D. Gavenda and B. C. Deaton, Phys. Rev. Lett., 8:208 (1962).

110. M. R. Daniel and L. Mackinnon, Phil. Mag., 8:537 (1963).

111. D. F. Gibbons and L. M. Falicov, Phil. Mag., 8:177 (1963).

112. B. C. Deaton and J. D. Gavenda, Phys. Rev., 136:A1096 (1964).

113. N. H. Zebouni, R. E. Hamburg, and H. J. Mackey, Phys. Rev. Lett., 11:260 (1963).

114. A. A. Mar'yakhin and V. P. Naberezhnykh, Zh. Éksp. Teor. Fiz. Pis. Red., 3:205 (1966).

115. V. P. Naberezhnykh, A. A. Mar'yakhin, and V. L. Mel'nik, Zh. Éksp. Teor. Fiz., 52:617 (1967).

116. R. G. Goodrich and R. C. Jones, Phys. Rev., 156:745 (1967).

117. R. C. Jones, R. G. Goodrich, and L. M. Falicov, Phys. Rev., 174:672 (1968).

118. W. A. Harrison, Phys. Rev., 118:1190 (1960).
119. R. W. Stark and L. M. Falicov, Phys. Rev. Lett., 19:795 (1967).
120. R. G. Chambers and B. K. Jones, Proc. Roy. Soc. London, A270:417 (1962).
121. P. A. Penz, Phys. Rev. Lett., 20:725 (1968).
122. J. R. Reitz and A. W. Overhauser, Phys. Rev., 171:749 (1968).
123. D. Shoenberg and P. J. Stiles, Proc. Roy. Soc. London, A281:62 (1964).
124. C. C. Grimes and A. F. Kip, Phys. Rev., 132:1991 (1963).
125. H. J. Foster, P. H. Meijer, and E. V. Mielczarek, Phys. Rev., 139:A1849 (1965).
126. J. Trivisonno, M. S. Said, and L. A. Pauer, Phys. Rev., 147:518 (1966).
127. T. G. Blaney, Phil. Mag., 17:405 (1968); 20:23 (1969).
128. J. R. Peverley, Phys. Rev., 173:689 (1968).
129. J. F. Koch and T. K. Wagner, Phys. Rev., 151:467 (1966).
130. P. S. Peercy, W. M. Walsh, Jr., L. W. Rupp, Jr., and P. H. Schmidt, Phys. Rev., 171:713 (1968).
131. F. S. Ham, Phys. Rev., 128:2524 (1962).
132. V. Heine and I. V. Abarenkov, Phil. Mag., 9:451 (1964).
133. N. W. Ashcroft, Phys. Rev., 140:A935 (1965).
134. J. H. Condon and J. A. Marcus, Phys. Rev., 134:A446 (1964).
135. S. L. Altmann and A. P. Cracknell, Proc. Phys. Soc. London, 84:761 (1964).
136. N. E. Alekseevskii and Yu. P. Gaidukov, Zh. Éksp. Teor. Fiz., 38:1720 (1960).
137. R. W. Stark, T. G. Eck, W. L. Gordon, and F. Moazed, Phys. Rev. Lett., 8:360 (1962); R. W. Stark, T. G. Eck, and W. L. Gordon, Phys. Rev., 133:A443 (1964).
138. M. G. Priestley, Proc. Roy. Soc. London, A276:258 (1963).
139. M. G. Priestley, L. M. Falicov, and G. Weisz, Phys. Rev., 131:617 (1963).
140. R. W. Stark, Bull. Amer. Phys. Soc., 11:169 (1966).
141. R. W. Stark, Phys. Rev., 162:589 (1967).
142. T. G. Eck and M. P. Shaw, Proc. Ninth Intern. Conf. on Low Temperature Physics, Columbus, Ohio, 1964, Part B, publ. by Plenum Press, New York (1965), p. 759.
143. L. E. Hartmann and J. M. Luttinger, Phys. Rev., 151:430 (1966).
144. J. B. Ketterson and R. W. Stark, 156:748 (1967).
145. L. M. Falicov, Phil. Trans. Roy. Soc. London, A255:55 (1962).
146. L. M. Falicov and M. H. Cohen, Phys. Rev., 130:92 (1963).
147. L. M. Falicov, A. B. Pippard, and P. R. Sievert, Phys. Rev., 151:498 (1966).
148. N. E. Alekseevskii and Yu. P. Gaidukov, Zh. Éksp. Teor. Fiz., 37:672 (1959).
149. J. E. Kunzler and J. R. Klauder, Phil. Mag., 6:1045 (1961).
150. A. J. Funes and R. V. Coleman, Phys. Rev., 131:2084 (1963).
151. J. R. Klauder, W. A. Reed, G. F. Brennert, and J. E. Kunzler, Phys. Rev., 141:592 (1966).
152. D. Shoenberg, Nature, 183:171 (1959).
153. A. S. Joseph and A. C. Thorsen, Phys. Rev., 134:A979 (1964).
154. A. S. Joseph, A. C. Thorsen, E. Gertner, and L. E. Valby, Phys. Rev., 148:569 (1966).
155. D. N. Langenberg and T. W. Moore, Phys. Rev. Lett., 3:328 (1959).
156. A. F. Kip, D. N. Langenberg, and T. W. Moore, Phys. Rev., 124:359 (1961).
157. J. R. Boyd and J. D. Gavenda, Phys. Rev., 152:645 (1966).
158. R. E. McFarlane, J. A. Rayne, and C. K. Jones, Phys. Lett., 24:197 (1967).
159. E. I. Zornberg and F. M. Mueller, Phys. Rev., 151:557 (1966).
160. V. E. Startsev, N. V. Vol'kenshtein, and N. A. Novoselov, Zh. Éksp. Teor. Fiz., 51:1311 (1966).
161. G. Leaver and A. Myers, Phil. Mag., 19:465 (1969).
162. R. Herrmann, Phys. Status Solidi, 25:661 (1968).
163. P. A. Bezuglyi, S. E. Zhevago, and B. I. Denisenko, Zh. Éksp. Teor. Fiz., 49:1457 (1965).
164. A. A. Galkin, S. E. Zhevago, T. F. Butenko, and E. P. Degtyar, Ukr. Fiz. Zh., 13:1104 (1968).

165. V. V. Boiko, V. A. Gasparov, and I. G. Gverdtsiteli, Zh. Éksp. Teor. Fiz. Pis. Red., 6:737 (1967); Zh. Éksp. Teor. Fiz., 56:489 (1969).
166. W. M. Lomer, Proc. Phys. Soc. London, 84:327 (1964).
167. T. L. Loucks, Phys. Rev., 139:A1181 (1965).
168. A. P. Jeavons and G. A. Saunders, Phys. Lett., 27A:19 (1968).
169. M. G. Priestley, L. R. Windmiller, J. B. Ketterson, and Y. Eckstein, Phys. Rev., 154:671 (1967).
170. J. Vanderkooy and W. R. Datars, Phys. Rev., 156:671 (1967).
171. Y. Ishizawa, J. Phys. Soc. Jap., 25:160 (1968).
172. W. R. Datars and J. Vanderkooy, J. Phys. Soc. Jap., Vol. 21, Suppl., p. 657 (1966).
173. Y. Shapira and S. J. Williamson, Phys. Lett., 14:73 (1965).
174. J. B. Ketterson and Y. Eckstein, Phys. Rev., 140:A1355 (1965).
175. L. M. Falicov and S. Golin, Phys. Rev., 137:A871 (1965).
176. S. Golin, Phys. Rev., 140:A993 (1965).
177. P. J. Lin and L. M. Falicov, Phys. Rev., 142:441 (1966).
178. N. E. Alekseevskii and Yu. P. Gaidukov, Zh. Éksp. Teor. Fiz., 36:447 (1959).
179. M. J. G. Lee, Proc. Roy. Soc. London, A295:440 (1966).
180. E. Fawcett and W. A. Reed, Phys. Rev. Lett., 9:336 (1962); W. A. Reed and E. Fawcett, J. Appl. Phys., 35:754 (1964).
181. A. S. Joseph and A. C. Thorsen, Phys. Rev. Lett., 11:554 (1963).
182. D. C. Tsui and R. W. Stark, Phys. Rev. Lett., 17:871 (1966).
183. D. C. Tsui, Phys. Rev., 164:669 (1967).
184. J. Ruvalds and L. M. Falicov, Phys. Rev., 172:508 (1968).
185. J. Yamashita, M. Fukuchi, and S. Wakoh, J. Phys. Soc. Jap., 18:999 (1963).
186. J. C. Phillips, Phys. Rev., 133:A1020 (1964).
187. E. Fawcett, W. A. Reed, and R. R. Soden, Phys. Rev., 159:533 (1967).
188. N. E. Alekseevskii, K. H. Bertel, A. V. Dubrovin, and G. É. Karstens, Zh. Éksp. Teor. Fiz. Pis. Red., 6:637 (1967).
189. W. A. Reed and R. R. Soden, Phys. Rev., 173:677 (1968).
190. A. C. Thorsen and T. G. Berlincourt, Phys. Rev. Lett., 7:244 (1961).
191. N. E. Alekseevskii, Yu. P. Gaidukov, I. M. Lifshits, and V. G. Peschanskii, Zh. Éksp. Teor. Fiz., 39:1201 (1960).
192. R. C. Young, Phys. Rev., 152:659 (1966).
193. J. G. Anderson and R. C. Young, Phys. Rev., 168:696 (1968).
194. A. V. Gold and M. G. Priestley, Phil. Mag., 5:1089 (1960).
195. J. A. Woollam, Phys. Lett., 27A:246 (1968).
196. M. D. Stafleu and A. R. de Vroomen, Phys. Status Solidi, 23:675, 683 (1967).
197. J. E. Craven and R. W. Stark, Phys. Rev., 168:849 (1968).
198. J. F. Koch and A. F. Kip, Phys. Rev. Lett., 8:473 (1962).
199. M. S. Khaikin, Zh. Éksp. Teor. Fiz., 42:27 (1962).
200. M. S. Khaikin and S. M. Cheremisin, Zh. Éksp. Teor. Fiz., 54:69 (1968).
201. T. Olsen, J. Phys. Chem. Solids, 24:649 (1963).
202. M. S. Khaikin, Zh. Éksp. Teor. Fiz., 41:1773 (1961).
203. V. F. Gantmakher, Zh. Éksp. Teor. Fiz., 44:811 (1963); 46:2028 (1964); V. F. Gantmakher and É. A. Kaner, Zh. Éksp. Teor. Fiz., 48:1572 (1965).
204. M. Miasek, Phys. Rev., 130:11 (1963).
205. G. Weisz, 149:504 (1966).
206. N. E. Alekseevskii and N. N. Mikhailov, Zh. Éksp. Teor. Fiz., 46:1979 (1964); N. E. Alekseevskii, A. V. Dubrovin, G. É. Karstens, and N. N. Mikhailov, Zh. Éksp. Teor. Fiz., 54:350 (1968).
207. N. E. Alekseevskii, G. É. Karstens, and V. V. Mozhaev, Zh. Éksp. Teor. Fiz., 46:1979 (1964).

208. J. J. Vuillemin and M. G. Priestley, Phys. Rev. Lett., 14:307 (1965).
209. J. J. Vuillemin, Phys. Rev., 144:396 (1966).
210. F. M. Mueller and M. G. Priestley, Phys. Rev., 148:638 (1966).
211. A. J. Freeman, A. M. Furdyna, and J. O. Dimmock, J. Appl. Phys., 37:1256 (1966).
212. N. E. Alekseevskii, G. É. Karstens, and V. V. Mozhaev, Proc. Tenth Intern. Conf. on Low Temperature Physics, Moscow, 1966, Vol. 3, publ. by VINITI, Moscow (1967), p. 169.
213. M. D. Stafleu and A. R. de Vroomen, Phys. Lett., 19:81 (1965).
214. J. B. Ketterson, M. G. Priestley, and J. J. Vuillemin, Phys. Lett., 20:452 (1966).
215. L. R. Windmiller and J. B. Ketterson, Phys. Rev. Lett., 20:324 (1968).
216. R. Fletcher, L. Mackinnon, and W. D. Wallace, Phys. Lett., 25A:395 (1967).
217. A. R. Mackintosh, Bull. Amer. Phys. Soc., 11:215 (1966).
218. N. E. Alekseevskii, V. S. Egorov, and B. N. Kazak, Zh. Éksp. Teor. Fiz., 44:1116 (1963).
219. W. A. Reed, E. Fawcett, and R. R. Soden, Phys. Rev., 139:A1557 (1965).
220. A. S. Joseph and A. C. Thorsen, Phys. Rev. Lett., 11:67 (1963); Phys. Rev., 133:A1546 (1964); A. C. Thorsen, A. S. Joseph, and L. E. Valby, Phys. Rev., 150:523 (1966).
221. C. K. Jones and J. A. Rayne, Phys. Rev., 139:A1876 (1965).
222. L. R. Testardi and R. R. Soden, Phys. Rev., 158:581 (1967).
223. L. F. Mattheiss, Phys. Rev., 151:450 (1966).
224. P. T. Coleridge, Phys. Lett., 15:223 (1965); Proc. Roy. Soc. London, A295:458 (1966).
225. J. B. Ketterson, L. R. Windmiller, and S. Hörnfeldt, Phys. Lett., 26A:115 (1968).
226. J. M. Dishman and J. A. Rayne, Phys. Lett., 20:348 (1966).
227. W. R. Datars and A. E. Dixon, Phys. Rev., 154:576 (1967).
228. J. M. Dishman and J. A. Rayne, Phys. Rev., 166:728 (1968).
229. G. B. Brandt and J. A. Rayne, Phys. Rev., 148:644 (1966).
230. A. E. Dixon and W. R. Datars, Solid State Commun., 3:377 (1965); Phys. Rev., 175:928 (1968).
231. S. C. Keeton and T. L. Loucks, Phys. Rev., 152:548 (1966).
232. K. Okumura and I. M. Templeton, Phil. Mag., 7:1239 (1962).
233. P. T. Coleridge, Phys. Lett., 22:367 (1966).
234. H. E. Alekseevskii and Yu. P. Gaidukov, Zh. Éksp. Teor. Fiz., 41:354 (1961).
235. J. E. Schirber, Phys. Rev., 131:2459 (1963).
236. A. V. Gold, Phil. Trans. Roy. Soc. London, A251:85 (1958).
237. J. R. Anderson and A. V. Gold, Phys. Rev., 139:A1459 (1965).
238. P. J. Tobin, D. J. Sellmyer, and B. L. Averbach, Phys. Lett., 28A:723 (1969).
239. M. S. Khaikin and R. T. Mina, Zh. Éksp. Teor. Fiz., 42:35 (1962).
240. R. C. Young, Phil. Mag., 7:2065 (1962).
241. A. R. Mackintosh, Proc. Roy. Soc. London, A271:88 (1963).
242. T. L. Loucks, Phys. Rev. Lett., 14:1072 (1965).
243. K. Sh. Agababyan, R. T. Mina, and V. S. Pogosyan, Zh. Éksp. Teor. Fiz., 54:721 (1968).
244. N. E. Alekseevskii and Yu. P. Gaidukov, Zh. Éksp. Teor. Fiz., 42:69 (1962).
245. H. J. Fink, Phys. Lett., 13:105 (1964).
246. A. S. Joseph and A. C. Thorsen, Phys. Rev., 138:A1159 (1965).
247. D. G. Howard, Phys. Rev., 140:A1705 (1965).
248. V. J. Easterling and H. V. Bohm, Phys. Rev., 125:812 (1962).
249. S. Chatterjee and S. K. Sen, Proc. Phys. Soc. London, 87:779 (1966); 91:749 (1967).
250. N. E. Christensen, Phys. Status Solidi, 31:635 (1968).
251. Yu. A. Bogod and V. V. Eremenko, Phys. Status Solidi, 11:K51 (1965).
252. G. S. Fleming and T. L. Loucks, Phys. Rev., 173:685 (1968).
253. A. P. Cracknell, Phys. Lett., 24A:263 (1967).
254. S. Epstein and H. J. Juretschke, Phys. Rev., 129:1148 (1963).
255. L. S. Lerner and P. C. Eastman, Can. J. Phys., 41:1523 (1963).
256. J. B. Ketterson and Y. Eckstein, Phys. Rev., 132:1885 (1963).

257. G. N. Rao, N. H. Zebuoni, C. G. Grenier, and J. M. Reynolds, Phys. Rev., 133:A141 (1964).
258. L. R. Windmiller, Phys. Rev., 149:472 (1966).
259. S. Tanuma, Y. Ishizawa, and S. Ishiguro, J. Phys. Soc. Jap., Vol. 21, Suppl., p. 662 (1966).
260. N. B. Brandt, N. Ya. Minina, and Chên Kang-Ch'on, Zh. Éksp. Teor. Fiz., 51:108 (1966).
261. W. R. Datars and J. Vanderkooy, IBM J. Res. Develop., 8:247 (1964).
262. Y. Eckstein, Phys. Rev., 129:12 (1963).
263. O. Beckman, L. Eriksson, and S. Hörnfeldt, Solid State Commun., 2:7 (1964).
264. L. Eriksson, O. Beckman, and S. Hörnfeldt, J. Phys. Chem. Solids, 25:1339 (1964).
265. T. Fukase and T. Fukuroi, J. Phys. Soc. Jap., Vol. 21, Suppl., p. 751 (1966).
266. A. R. Korolyuk and L. Ya. Matsakov, Zh. Éksp. Teor. Fiz., 52:415 (1967).
267. L. M. Falicov and P. J. Lin, Phys. Rev., 141:562 (1966).
268. A. R. Mackintosh, L. E. Spanel, and R. C. Young, Phys. Rev. Lett., 10:434 (1963).
269. J. C. Milliken and R. C. Young, Phys. Rev., 148:558 (1966).
270. R. C. Young, Phys. Rev., 163:676 (1967).
271. M. G. Priestley, Phys. Rev., 148:580 (1966).
272. J. A. Rayne, Phys. Rev., 131:653 (1963).
273. Y. Eckstein, J. B. Ketterson, and M. G. Priestley, Phys. Rev., 148:586 (1966).
274. J. B. Coon, C. G. Grenier, and J. M. Reynolds, J. Phys. Chem. Solids, 28:301 (1967).
275. P. Soven, Phys. Rev., 137:A1706, A1717 (1965).
276. A. C. Thorsen and T. G. Berlincourt, Phys. Rev. Lett., 7:244 (1961).
277. J. H. Condon, Bull. Amer. Phys. Soc., 11:170 (1966).
278. W. L. Dahlquist and R. G. Goodrich, Phys. Rev., 164:944 (1967).
279. N. E. Alekseevskii and V. S. Egorov, Zh. Éksp. Teor. Fiz. Pis. Red., 1:31 (1965).
280. A. C. Thorsen, A. S. Joseph, and L. E. Valby, Phys. Rev., 162:574 (1967).
281. D. J. Boyle and A. V. Gold, Phys. Rev. Lett., 22:461 (1969).
282. R. P. Gupta and T. L. Loucks, Phys. Rev. Lett., 22:458 (1969).
283. A. J. Arko, J. A. Marcus, and W. A. Reed, Phys. Lett., 23:617 (1966); Phys. Rev., 176:671 (1968).
284. B. R. Watts, Phys. Lett., 10:275 (1964).
285. G. B. Brandt and J. A. Rayne, Phys. Rev., 132:1945 (1963).
286. J. E. Graebner and J. A. Marcus, Phys. Rev., 175:659 (1968).
287. W. D. Wallace, N. Tepley, H. V. Bohm, and Y. Shapira, Phys. Lett., 17:184 (1965).
288. W. D. Wallace and H. V. Bohm, J. Phys. Chem. Solids, 29:721 (1968).
289. L. M. Falicov and M. J. Zuckermann, Phys. Rev., 160:372 (1967).
290. K. Okumura and I. M. Templeton, Phil. Mag., 8:889 (1963).
291. K. Okumura and I. M. Templeton, Proc. Roy. Soc. London, A287:89 (1965).
292. W. A. Reed and G. F. Brennert, Phys. Rev., 130:565 (1963).
293. R. W. Stark, Phys. Rev., 135:A1698 (1964).
294. A. C. Thorsen, L. E. Valby, and A. S. Joseph, Proc. Ninth Intern. Conf. on Low Temperature Physics, Columbus, Ohio, 1964, Part B, publ. by Plenum Press, New York (1965), p. 867.
295. A. S. Joseph and W. L. Gordon, Phys. Rev., 126:489 (1962).
296. R. J. Higgins, J. A. Marcus, and D. H. Whitmore, Phys. Rev., 137:A1172 (1965).
297. V. A. Venttsel', A. I. Likhter, and A. V. Rudnev, Zh. Éksp. Teor. Fiz., 53:108 (1967).
298. J. K. Galt, F. R. Merritt, W. A. Yager, and H. W. Dail, Jr., Phys. Rev. Lett., 2:292 (1959).
299. V. P. Naberezhnykh and V. A. Mel'nik, Zh. Éksp. Teor. Fiz., 47:873 (1964).
300. M. P. Shaw, P. I. Sampath, and T. G. Eck, Phys. Rev., 142:399 (1966).
301. J. O. Henningsen, Phys. Status Solidi, 22:441 (1967).
302. A. A. Galkin and A. P. Korolyuk, Zh. Éksp. Teor. Fiz., 38:1688 (1960).
303. D. F. Gibbons and L. M. Falicov, Phil. Mag., 8:177 (1963).
304. A. Myers and J. R. Bosnell, Phil. Mag., 13:1273 (1966).

305. W. A. Harrison, Phys. Rev., 126:497 (1962); 129:2512 (1963).

306. A. C. Thorsen and A. S. Joseph, Phys. Rev., 131:2078 (1963).

307. T. L. Loucks, Phys. Rev., 159:544 (1967).

308. D. J. Sellmyer and P. A. Schroeder, Phys. Lett., 16:100 (1965).

309. J. T. Longo, P. A. Schroeder, and D. J. Sellmyer, Phys. Lett., 25A:747 (1967).

310. J. P. Jan, W. B. Pearson, and Y. Saito, Proc. Roy. Soc. London, A297:275 (1967).

311. H. Amar, K. H. Johnson, and K. P. Wang, Phys. Rev., 148:672 (1966).

312. J. P. Jan, W. B. Pearson, Y. Saito, M. Springford, and I. M. Templeton, Phil. Mag., 12:1271 (1965).

313. B. W. Veal and J. A. Rayne, Phys. Lett., 6:12 (1963).

314. E. Fawcett, J. Phys. Chem. Solids, 18:320 (1961).

315. G. E. Smith, Phys. Rev., 115:1561 (1959).

316. A. B. Pippard, Phil. Trans. Roy. Soc. London, A250:325 (1957).

317. E. Fawcett, Proc. Roy. Soc. London, A232:519 (1955).

318. A. T. Stewart, J. B. Shand, J. J. Donaghy, and J. H. Kusmiss, Phys. Rev., 128:118 (1962).

319. S. Berko, Phys. Rev., 128:2166 (1962).

320. R. W. Williams and A. R. Mackintosh, Phys. Rev., 168:679 (1968).

321. S. Berko and J. Zuckerman, Phys. Rev. Lett., 13:339a (1964).

322. D. R. Gustafson, A. R. Mackintosh, and D. J. Zaffarano, Phys. Rev., 130:1455 (1963).

323. R. P. Gupta and T. L. Loucks, Phys. Rev., 176:848 (1968).

324. E. C. Snow, Phys. Rev., 158:683 (1967).

325. J. H. Wood, Phys. Rev., 146:432 (1966).

326. S. Chatterjee and S. K. Sen, J. Phys. C (Solid State Phys.), 1:759 (1968).

327. J. P. Jan, J. Phys. Chem. Solids, 29:561 (1968).

328. M. J. G. Lee and L. M. Falicov, Proc. Roy. Soc. London, A304:319 (1968); M. J. G. Lee, Phys. Rev., 178:953 (1969).

329. B. Vasvari, A. O. Animalu, and V. Heine, Phys. Rev., 154:535 (1967).

330. J. R. Sybert, H. J. Mackey, and K. L. Hathcox, Phys. Rev., 166:710 (1968).

331. O. K. Andersen and A. R. Mackintosh, Solid State Commun., 6:285 (1968).

332. R. Lück, Phys. Status Solidi, 18:59 (1966).

333. P. E. Lewis and P. M. Lee, Phys. Rev., 175:795 (1968).

334. J. T. Longo, P. A. Schroeder, and D. J. Sellmyer, Phys. Rev., 182:658 (1969).

335. J. P. Jan, W. B. Pearson, Y. Saito, M. Springford, and I. M. Templeton, Phil. Mag., 12:1271 (1965).

336. C. Miziumski and A. W. Lawson, Phys. Rev., 180:749 (1969).

337. J. R. Anderson and S. S. Lane, Phys. Rev. B, 2:298 (1970).

338. A. G. Beattie, Phys. Rev., 184:668 (1969).

339. W. A. Reed and J. H. Condon, Phys. Rev. B, 1:3504 (1970).

340. R. D. Brown, Phys. Rev. B, 2:928 (1970).

341. J. W. Dooley and N. Tepley, Phys. Rev., 187:781 (1969).

342. W. van Haeringen and H. G. Junginger, Solid State Commun., 7:1723 (1969).

343. W. J. O'Sullivan, A. C. Switendick, and J. E. Schirber, Phys. Rev. B, 1:1443 (1970).

344. A. J. Hughes and J. P. G. Shepherd, J. Phys. C (Solid State Phys.), 2:661 (1969).

345. S. Tanuma, W. R. Datars, H. Doi, and A. Dunsworth, Solid State Commun., 8:1107 (1970).

346. W. R. Datars and J. R. Cook, Phys. Rev., 187:769 (1969).

347. J. D. Gavenda and F. H. S. Chang, Phys. Rev., 186:630 (1969).

348. A. Libchaber, G. Adams, and C. G. Grimes, Phys. Rev. B, 1:361 (1970).

349. S. Chatterjee and D. K. Chakraborti, Metal Phys. [Suppl. to J. Phys. C (Solid State Phys.)], 3:120 (1970).

350. R. E. Borland and J. R. A. Cooper, Metal Phys. [Suppl. to J. Phys. C (Solid State Phys.)], 3:253 (1970).

351. G. N. Kamm, Phys. Rev. B, 1:554 (1970).

352. B. Perrin, G. Weisbuch, and A. Libchaber, Phys. Rev. B, 1:1501 (1970).

353. M. J. G. Lee, Phys. Rev., 187:901 (1969).

354. C. S. Ih and D. N. Langenberg, Phys. Rev. B, 1:1425 (1970).

355. E. I. Zornberg, Phys. Rev. B, 1:244 (1970).

356. M. H. Halloran, J. H. Condon, J. E. Graebner, J. E. Kunzler, and F. S. L. Hsu, Phys. Rev. B, 1:366 (1970).

357. L. F. Mattheiss, Phys. Rev. B, 1:373 (1970).

358. G. N. Kamm and J. R. Anderson, Phys. Rev. B, 2:2944 (1970).

359. J. B. Ketterson and L. R. Windmiller, Phys. Rev. B, 2:4813 (1970).

360. J. B. Ketterson, F. M. Mueller, and L. R. Windmiller, Phys. Rev., 186:656 (1969).

361. V. L. Mel'nik and I. V. Svechkarev, Phys. Status Solidi, 33:33 (1969).

362. T. E. Bogle, J. B. Coon, and C. G. Grenier, Phys. Rev., 177:1122 (1969).

363. P. T. Coleridge, J. Low Temp. Phys., 1:577 (1969).

364. R. E. Hamburg, C. G. Grenier, and J. M. Reynolds, Phys. Rev. Lett., 23:236 (1969).

365. Y. Ishizawa and W. R. Datars, Phys. Rev. B, 2:3875 (1970).

366. F. A. Capocci, P. M. Holtham, D. Parsons, and M. G. Priestley, J. Phys. C (Solid State Phys.), 3:2081 (1970).

367. W. L. Dahlquist and R. G. Goodrich, Phys. Rev., 164:944 (1967).

368. J. C. Shaw and G. E. Everett, Phys. Rev. B, 1:537 (1970).

369. E. M. Hygh and R. M. Weich, Phys. Rev. B, 1:2424 (1970).

370. J. Trivisonno and J. A. Murphy, Phys. Rev. B, 1:3341 (1970).

371. F. J. Arlinghaus, Phys. Rev., 186:609 (1969).